Proceedings of the
Twelfth International Conference on the

Physics of
Semiconductors

July 15 – 19, 1974 Stuttgart

Editor: M. H. Pilkuhn

Springer Fachmedien Wiesbaden GmbH

The Conference was substantially supported
by the following Institutions:

International Union of Pure and Applied Physics

Deutsche Physikalische Gesellschaft

Bundesministerium für Forschung und Technologie

Ministerium für Wirtschaft, Mittelstand und Verkehr
des Landes Baden-Württemberg

Stadt Stuttgart

The following Industrial Organizations made significant
financial contributions: Allgemeine Elektrizitätsgesellschaft
AEG Telefunken, Badische Anilin- & Sodafabrik AG,
E. Merck AG, IBM Deutschland GmbH, Intermetall Halb-
leiterwerk Deutsche ITT Industries GmbH, Kodak AG,
Physik Instrumente GmbH, Robert Bosch GmbH,
Siemens AG, Spectra-Physics, Spex Industries GmbH,
Standard Elektrik Lorenz AG, Valvo GmbH, Volkswagen-
werk AG, Wacker Chemitronic GmbH, Wandel & Golter-
mann KG

ISBN 978-3-519-03013-3 ISBN 978-3-322-94774-1 (eBook)
DOI 10.1007/978-3-322-94774-1

© Springer Fachmedien Wiesbaden 1974
Originally published by B.G. Teubner Stuttgart in 1974
Softcover reprint of the hardcover 1st edition 1974
Cover design: W. Koch, Sindelfingen

PREFACE

The Twelfth International Conference on the Physics of
Semiconductors was held at Stuttgart, Federal Republic of
Germany, from July 15 to 19, 1974.

The Conference was sponsored by the International Union of Pure
and Applied Physics and we wish to thank this Organization as
well as all the other Institutions and Companies listed on the
preceding page for their substantial support.

About 700 scientists from 31 countries came to Stuttgart in
order to attend the Conference.

Following the example of the previous Conference at Warsaw, a
distinction was made between plenary invited and invited papers
presented at parallel sessions. Altogether, the Proceedings
contains 27 invited and 206 contributed papers. The members of
the International Program Committee had the difficult task of
making a selection from 550 abstracts. Their work was essential
for the success of the Conference and it is gratefully acknow-
ledged.

The number of pages alloted to each paper had to be limited in
order to keep the Proceedings within reasonable size. In general,
the papers in the Proceedings have been arranged according to
the Conference Program except for the following changes: Each
plenary invited paper was associated with an appropriate
session, the sessions were put into a slightly different order
and, in a few cases, new titles were introduced. These changes
were made because the Proceedings will serve as a general text-
book describing the latest progress in the field of semiconduc-
tor physics and it will also be used by scientists who did not
attend the Conference.

The members of the Conference Committees are listed on the fol-
lowing page. I am very grateful to them as well as to all the
authors of papers for their work and cooperation. Finally, I
would like to thank all the numerous other people whose names do
not appear but who have worked hard for the success of the
Twelfth Semiconductor Conference.

Manfred H. Pilkuhn

ORGANIZATION

Twelfth International Conference on the Physics of Semiconductors

ORGANIZING COMMITTEE:

Chairman:	O. Madelung, Marburg
Secretary:	O.G. Folberth, Boeblingen
Program:	H.J. Queisser, Stuttgart
Publications:	M. Pilkuhn, Stuttgart
Satellite Conferences:	A. Seeger, Stuttgart
Local Arrangements:	H. Düker, Stuttgart
Finance:	H. Wilde, Stuttgart

INTERNATIONAL ADVISORY COMMITTEE:

M. Balkanski	(France)
F. Bassani	(Italy)
W. Brauer	(GDR)
R. Grigorovici	(Rumania)
C. Hilsum	(Great Britain)
E.O. Kane	(USA)
H. Kawamura	(Japan)
R.C.C. Leite	(Brazil)
A. Many	(Israel)
M. Matyás	(CSSR)
W.J. Merz	(Switzerland)
N.I. Meyer	(Denmark)
L. Sosnowski	(Poland)
G. Szigeti	(Hungary)
V.M. Tuchkevich	(USSR)
H.J. Vink	(Netherlands)
J.C. Woolley	(Canada)

PROGRAM COMMITTEE:

C.Benoit à la Guillaume	(Paris)
V.L. Bronch-Bruevich	(Moscow)
M. Cardona	(Stuttgart)
E. Gutsche	(Berlin/GDR)
C.R. Pidgeon	(Edinburgh)
H.J. Queisser, Chairman	(Stuttgart)
F. Stern	(Yorktown Heights/N.Y.)
W. Zawadzki	(Warsaw)

TABLE OF CONTENTS

HIGH DENSITY EFFECTS, BIEXCITONS

BANDS AND BONDS

IMPURITIES

RAMAN SCATTERING

ELECTRON-TWO PHONON INTERACTION

MAGNETO-OPTICS

PHASE TRANSITIONS

LAYER COMPOUNDS

SURFACES

SPIN-DEPENDENT PROPERTIES AND SPIN-FLIP

HOT ELECTRONS

ACOUSTO- AND PIEZOELECTRIC EFFECTS

MAGNETIC SEMICONDUCTORS

EXCITONS

DISORDERED SEMICONDUCTORS

HEAVILY DOPED MATERIALS

SMALL-GAP SEMICONDUCTORS

OPTICAL PROPERTIES, RECOMBINATION

PHOTOEMISSION

CLOSING SESSION

ELECTRON-HOLE DROPS

CONTENTS

NEGATIVE AND POSITIVE MAGNETORESISTANCE RELATED TO SURFACE STATES IN SILICON INVERSION LAYERS

G. DORDA and I. EISELE

Forschungslaboratorien der Siemens AG, München, F. R. Germany

Magnetoresistance measurements of (100), (110) and (111) p- and n-channel silicon inversion layers have been carried out at 4.2 K. The dependence on carrier concentration and magnetic field orientation suggests a positive effect which is caused by a mass anisotropy and a negative effect depending on spin scattering in connection with surface states.

According to the usual description the magnetoresistance effect is determined by a mass anisotropy factor and by the magnetoresistance coefficient which is obtained from the solution of the Boltzmann equation. At sufficiently low temperatures the Fermi level for the inversion layer lies within the conduction or valence band, respectively. In this case the magnetoresistance coefficient becomes zero and only a mass anisotropy leads to a positive magnetoresistance [1]. For Si surfaces we always deal with nonspherical energy surfaces except for the (100) n-channel. In this special case due to quantization phenomena no magnetoresistance effect should appear because at low temperatures all electrons are restricted to the lowest subband corresponding to the isotropic effective mass [2].

Measurements at 4.2 K have been carried out on (100), (110) and (111) n- as well as p-channel Si (7 - 10 Ωcm) MOSFET's. The oxide thickness was 120 nm and the geometry of the channel (channel length 400 μm, channel width 40 μm) allowed additional information about the Hall mobility.

As can be seen from Fig. 1 the p-channel magnetoresistance for all three surface orientations has the expected positive

FIG. 1: Magnetoresistance vs. surface hole concentration N_{inv} for various surface orientations. Magnetic field B = 1 Tesla; T = 4.2 K

FIG. 2: (100) magnetoresistance vs. direction of the magnetic field for two different surface hole concentrations. T = 4.2 K; B = 1 Tesla. α is the angle between magnetic field B and channel current J.

behaviour over a wide concentration range. Only for low hole concentrations a negative effect appears. In order to explain the origin of this unexpected effect the magnetic field orientation was changed with respect to the direction of the channel current. For a (100) orientation the results for $N_{inv} = 2 \times 10^{16}$ m^{-2} and $N_{inv} = 4 \times 10^{15}$ m^{-2} are presented in Fig. 2. In accordance with the model of a two-dimensional hole gas the positive effect has a maximum at $\alpha = 0^{\circ}$ and a minimum at $\alpha = 90^{\circ}$. In contrary the negative effect is completely independent of the magnetic field direction suggesting two possible mechanisms: (a) the conductivity changes from a band type transport to a hopping process in which the magnetic field increases the hopping probability of the holes [3] or (b) a localized spin model in which the scattering by localized spins is reduced by the magnetic field in aligning the spins. Because of the short spin relaxation time of the holes the model (b) seems to be less probable.

Whereas the p-channel results confirm the above discussed theories the measurements for n-type samples show anomalous behaviour in almost any case (see Fig. 3). Despite the expected absence of a magnetoresistance effect for the (100) orientation we find pronounced positive as well as negative resistance changes. Magnitude and direction depend on the sample quality, i.e. on the density of surface states. The surface state concentrations have been detected analizing the ratio between Hall- and effective mobility [4].Typical results for a high and low quality sample are presented in Fig. 3: profile (a) with $N_{ss} = 1 \times 10^{15}$ m^{-2} and profile (b) with $N_{ss} = 2 \times 10^{16}$ m^{-2}. The angle dependence of the positive effect is almost sinusoidal with a maximum for the magnetic field perpendicular to the surface and a minimum which is close to zero if field and channel direction are parallel. This behaviour can be explained best with the help of mass anisotropy and quantization. In agreement with piezoresistance measurements [5] we conclude that despite of quantization even at low gate voltages the other 4 anisotropic valleys are partially populated yielding a positive magnetoresistance effect. The sharp decrease at lowest electron concentrations is in agreement with the p-channel results and can be attributed to the existence of a hopping transport.

The obvious connection between the magnetoresistance effect and the concentration of the surface states seems also to be the key for the analysis of the negative effect. The hopping process cannot be the only origin for the observed negative effect because this kind of transport is restricted to low carrier concentrations [6]. On the other hand the influence of the magnetic field on electron spin scattering cannot be neglected as it was for the p-channel because of the larger electron spin relaxation time. The importance of the spin scattering in semiconductors at low temperatures has been pointed out recently [7]. The application of this theory to the electron inversion layer seems to be appropriate for our results because the surface state densities N_{ss} for low quality (100) samples ($N_{ss} \geq 10^{16}$ m^{-2}) are comparable to the

FIG. 3: Magnetoresistance vs. surface electron concentration N_{inv} for various surface orientations. B = 1 Tesla; T = 4.2 K; sample a: $N_{ss} = 1 \times 10^{15} \text{ m}^{-2}$, sample b: $N_{ss} = 2 \times 10^{16} \text{ m}^{-2}$.

FIG. 4: (100) magnetoresistance vs. direction of the magnetic field for two different samples: a) $N_{ss} = 1 \times 10^{15} \text{ m}^{-2}$; b) $N_{ss} = 2 \times 10^{16} \text{ m}^{-2}$; B = 1 Tesla; T = 4.2 K; $N_{inv} = 1.2 \times 10^{16} \text{ m}^{-2}$.

bulk impurity concentrations in doped semiconductors [3]. An existing phenomenological theory [8] describing a negative magnetoresistance by the interaction between localized spin centers and conduction electrons has been modified introducing Fermi statistics for the quantum limit case [9]. For the conductivity change we obtained:

$$\frac{\Delta G}{G_o} = \frac{\hbar^2}{16\, m_T a^2 (E_F - E_o)} \left(\frac{g \mu_B B}{kT} \right)^2$$

where a is the spacing between localized spins (surface states), g the Landé factor, μ_B the Bohr magneton, B the magnetic field, m_T the transverse effective mass, E_F the Fermi energy, and E_o the minimum energy of the lowest subband. Measurements of the magnetic field dependence and temperature dependence of the negative effect are in accordance with the postulated model and proof its validity. The constance of the

angle dependence in Fig. 4 over a wide range also supports
the spin scattering model. The breakdown in the vicinity of
$\alpha = 90^{\circ}$ cannot be explained sufficiently at the moment.
For (110) and (111) surface orientations only negative effects
have been observed (Fig.3) despite of the mass anisotropy of
the subbands. These results are not unexpected if one con-
siders the high surface state densities N_{ss} which are known
to increase with an increasing concentration of dangling
bonds at the surface: $N_{ss}(100) < N_{ss}(110) < N_{ss}(111)$. This
sequence holds for n-channel as well as p-channel surface
layers and elucidates in both cases the magnitude of the
negative effect for different surface orientations.

In summary it should be noted that the negative magneto-
resistance is connected with surface states which act as spin
scattering centers. The positive effect results from the
influence of the Lorentz force on the carrier motion.

REFERENCES

[1] Smith, R.A.: Semiconductors, Cambridge University Press,
 Cambridge (1968)
[2] For a review see: Dorda, G. in "Festkörperprobleme XIII",
 Pergamon Press, Vieweg (1973)
[3] Woods, J.F.; Chen, C.Y.: Phys. Rev. 135, A1462 (1964)
[4] Eisele, I.; Dorda, G.: submitted to Solid State Comm.
[5] Dorda, G; Eisele, I.: phys. stat. sol.(a) 20, 263 (1973)
[6] Stern, F.: to be published in Phys. Rev.
[7] Solomon, I.: Proc. 11th Int. Conf. Phys. Sem., Warsaw
 (1972), p. 27.
[8] Boon, M.R.: Phys. Rev. B7, 761 (1973)
[9] Eisele, I; Dorda, G.: submitted for publication

SPECTROSCOPY OF SPACE CHARGE LAYERS ON n-TYPE Si

AVID KAMGAR, P. KNESCHAUREK, W. BEINVOGL, AND J.F. KOCH

Physik-Department, Technische Universität München, Garching
Fed. Rep. Germany

We study the infrared absorption due to resonant
transitions between quantized, electric field in-
duced surface levels in space charge layers on n-
type (100) Si. Both electron levels in accumula-
tion and hole levels in inversion layers have been
observed. The separation of the groundstate and
next higher subband is determined for a number of
different gate voltages. We find that a magnetic
field applied in the plane of the sample increases
the energy splitting of the levels.

1. INTRODUCTION

In a recent publication[1] we reported the observation of reso-
nant transitions between electron levels in an accumulation
layer on n-type (100) Si. Such measurements have now been exten-
ded to higher frequencies. In addition, with some minor modifi-
cations of the apparatus, we have observed resonances due to
hole levels in a p-type inversion layer on the same samples.
The hole resonances are observed for infrared energies of 7.23
meV and higher. The calculation of the hole levels, because of
the complications of the nonparabolic band structure, are still
an outstanding theoretical challenge. Experimental studies of
magnetoconductance in p-channels on Si (100) have recently been
reported[2,3].

Stern and Howard[4] predicted that a magnetic field applied
parallel to the surface of a semiconductor would increase the
splitting of energy levels in an inversion layer. Such shifts
have been observed by Tsui in InAs[5]. We have examined the
effects of a magnetic field on electrons in an accumulation
layer on Si.

2. EXPERIMENTAL NOTES

The experimental arrangement is in essence that discussed in
ref.[1]. Infrared radiation from a molecular gas laser is chan-
nelled through a tapered section to a parallel plate trans-
mission line. The sample, a 6x9 mm slab of thickness 0.2 mm, is
placed between the two copper plates that form the transmission

line. The radiation is polarized predominantly perpendicular to the Si-SiO$_2$ interface. Transmitted power is detected by a Ge bolometer.

The samples* are of nominal 10 Ω-cm material (phosphorus doping, N_D-N_A = 5×10^{14}cm^{-3}), with thermally grown oxide layers of thickness about 2300 Å**.

For positive gate voltages on our samples we find that the space charge layer is in equilibrium with the applied voltage. This is not the case for negative voltages. No signal was found in our previous work for negative gate voltages. We have modified the light pipe arrangement in the present experiments to admit a small amount of visible light. This enables us to achieve a minority carrier concentration in the inversion layer in equilibrium with the applied voltage.

For the magnetic field runs the transmission line assembly is placed in a 50 kOe superconducting solenoid.

3. SPECTROSCOPY OF LEVELS IN INVERSION AND ACCUMULATION LAYERS

Shown in Fig.1 are resonance signals observed for two different frequencies. For accumulation of electrons we observe at 10.45 meV the signal previously published [1]. At the higher

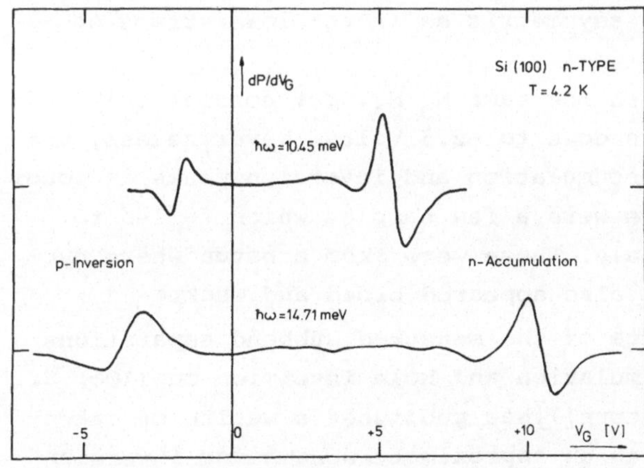

FIG. 1:
Resonance signals for subband to subband transitions in n-type (100) Si. Positive voltages V$_G$ give electrons in accumulation layers; negative voltages correspond to holes in inversion layers. dP/dV$_G$ is the power absorption derivative in arbitrary units.

* Samples have been supplied to us by the Siemens Semiconductor Research Division, München.

** The oxide thickness in the sample used in the experiment of ref. [1] was remeasured and found to be closer to 2300 Å instead of the 2100 Å stated.

frequency of $\hbar\omega = 14.71$ meV the resonance appears at increased V_G. We interpret this resonance as arising from transitions from the lowest filled subband with energy E_0 to the next higher subband E_1. Other transitions, such as $0 \to 2,3$ etc., cannot be resolved in the spectrum. Transitions $1 \to 2,3$ etc. are not expected, because the n = 1 subband is not occupied [6] until much higher values of V_G.

When a small amount of visible light is admitted to the transmission line, a signal appears also for negative gate voltages. From the C-V curve at 300°K we determine the threshold voltage for inversion to be -0.7 V. The flat band condition is close to 0. V. The inversion signal appears first at a laser energy of 7.23 meV at -1.1 V. For $\hbar\omega = 10.45$ and 14.71 meV the resonance grows stronger and moves to higher voltages as in Fig.1. This behaviour is consistent with an interpretation of the signal as due to transitions between the two lowest subbands of holes. The signal is similar to that for electron accumulation, except for the fact that it is not observed until a certain threshold energy on the order of 7 meV is reached. Electron accumulation signals are observed down to 5.64 meV and lower. A peculiarity of the hole resonance line is that it broadens with increasing gate voltage and becomes asymmetric as in the lower trace of the figure.

We have run samples with the same $N_D - N_A$, but considerably different threshold voltages down to -2,5 Volts. Nevertheless, the separation between the accumulation and inversion peaks is found to remain constant. There were a few samples which failed to show good resonance signals. These were from a batch where cyclotron resonance signals also appeared broad and weak.

In Fig.2 we present data of the measured subband separations $E_1 - E_0$, for electron accumulation and hole inversion on (100) Si. For electron inversion Stern [7] has published a wealth of calculations. He has also given an approximate scheme for inversion layer calculations. Because of the complicated, nonparabolic valence band of Si, this calculation cannot be expected to give the correct description of hole levels. We use it only as an estimate in order to compare with the experimental points of Fig.2. Taking $N_D - N_A = 5 \times 10^{14} cm^{-3}$ and a hole mass of 0.5 m_e

gives the theoretical curve of Fig.2. There is reasonable numerical agreement. The calculated level spacing rises with voltage, but it does so more slowly than the experimental points.

FIG. 2:
Energy splittings (E_1-E_0) as a function of gate voltage for a sample with oxide thickness d = 2320 Å. The solid line gives a theoretical energy estimate for holes in inversion using a doping level of 5×10^{14} cm^{-3} and a hole mass of 0.5 m_e; threshold voltage is -0.9 V. An approx. theoretical value for 1×10^{12} electrons/cm^2 due to Duke [6] is also shown.

This tendency is also found in our very recent measurements of electrons in inversion layers.

4. MAGNETIC FIELD DEPENDENCES

The binding energy of electrons in the surface of a semiconductor is expected to increase with the application of a parallel magnetic field. According to Stern and Howard [4] this increase is given by

$$\Delta E_n = \frac{e^2 H^2}{2m^* c^2} \left(\langle z_n^2 \rangle - \langle z_n \rangle^2 \right) \tag{1}$$

where $\langle z_n \rangle$ is the coordinate of the electron normal to the surface in the nth state. The spread of the electronic wave function of the first excited state is larger than that of the groundstate. It follows from Eq.(1) that the pair of levels move apart with H^2. At fixed energy $\hbar\omega$ the resonance must shift to lower V_G. Such a magnetic shift confirms our interpretation of the resonance. It would also provide a measure of the spread of the wave functions.

Fig.3 shows the experimental results for the case of electron accumulation at a frequency of $\hbar\omega$ = 10.45 meV. The line is observed to shift down from 5.6 V at H = 0 to about 4.3 V at a field of 50.7 kOe. Together with the shift we find a marked change in lineshape. The shift varies approximately as H^2.

From the curves in Fig.2 one finds that a shift of 1.3 Volts corresponds to an energy change $E_1-E_0 = 1.83$ meV. Substituting this value in Eq.(1), together with H = 50.7 kOe and $m^* = 0.2m_e$ yields for the differences of the square of the wave function spreads the number $1.6 \times 10^{-12} cm^2$. To translate this number into a $\langle z_0 \rangle$ or $\langle z_1 \rangle$ requires a knowledge of the detailed shape of the wave functions. Without these, we note that our observed shift implies something like 100 Å for $\langle z_1 \rangle$.

5. ACKNOWLEDGEMENTS

We thank Dr. G. Dorda for many stimulating discussions. We also acknowledge his collaboration in many aspects of sample preparation and characterization.

6. REFERENCES

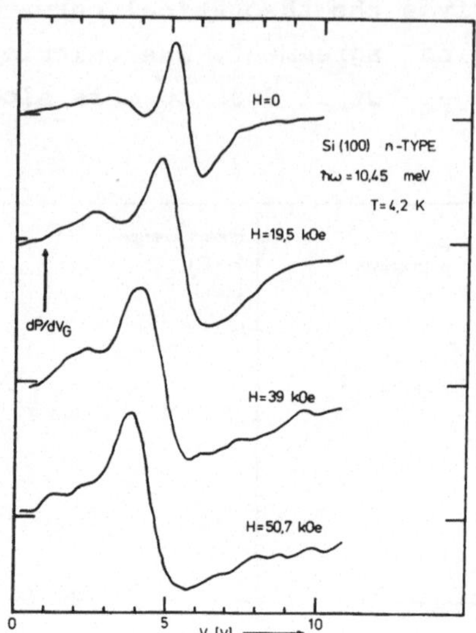

FIG. 3:
Magnetic field dependence of the resonance signal for electron accumulation at $\hbar\omega$ = 10.45 meV.

[1] Kamgar, A.; Kneschaurek, P.; Dorda, G.; and Koch, J.F.: Phys. Rev. Lett. **32**, 1251 (1974)
[2] von Klitzing, K.; Landwehr, G.; and Dorda, G.: Solid State Commun. **14**, 387 (1974)
[3] Lakhani, A.A.; Stiles, P.J.; and Cheng, Y.C.: Phys. Rev. Lett. **32**, 1003 (1974)
[4] Stern, F.; and Howard, W.E.: Phys. Rev. **163**, 816 (1967)
[5] Tsui, D.C.: Solid State Commun. **9**, 1789 (1971)
[6] Duke, C.B.: Phys. Rev. **159**, 632 (1967)
[7] Stern, F.: Phys. Rev. B **5**, 4891 (1972)

SELF CONSISTENT CALCULATIONS OF ELECTRIC SUBBANDS IN P-TYPE SILICON INVERSION LAYERS

E.BANGERT, K.v.KLITZING, G.LANDWEHR

Physikalisches Institut der Universität Würzburg

Fed.Rep.Germany

For (110)-p-type silicon inversion layers, electric subbands were calculated in a self consistent fashion, including all three valence bands. The effective cyclotron masses were computed for various surface electric fields. The agreement between theory and experiment is very good.

INTRODUCTION

Recently magneto-quantum oscillations in p-type silicon inversion layers have been observed [1] for (110) and also [2] for (111) and (100) surfaces. From the temperature dependence of Shubnikov-de Haas (SdH) oscillations the cyclotron mass m_c was deduced. In all cases m_c increased with increasing surface electric field. It became clear, that the data could not be explained on the basis of the non-parabolicity of the valence bands. Thus it was highly desirable to calculate the electric subbands and cyclotron masses. This task is considerably more complicated than the calculation of electric subbands in n-type silicon inversion layers [3], because of the complex valence band structure of silicon.

Because the data for (110) surfaces were of the highest quality and especially interesting due to the presence of a second SdH-period, we first performed calculations for this orientation. The small spin orbit splitting of 44 meV made it mandatory to include all three valence bands.

CALCULATION

Our calculation is based on the effective mass approximation in the formulation of Kohn and Luttinger [4]. The E($\underline{k}$)-relations for the three spin degenerated bands are the eigenvalues of the following 6 x 6-matrix:

$$H_0(k_x, k_y, k_z) = \begin{pmatrix} H_1 & R+i\Delta & S & 0 & 0 & \Delta \\ R-i\Delta & H_2 & T & 0 & 0 & i\Delta \\ S & T & H_3 & -\Delta & -i\Delta & 0 \\ 0 & 0 & -\Delta & H_1 & R-i\Delta & S \\ 0 & 0 & i\Delta & R+i\Delta & H_2 & T \\ \Delta & -i\Delta & 0 & S & T & H_3 \end{pmatrix} \tag{1}$$

$$H_1 = 1/2\, L(k_z-k_x)^2 + 1/2\, M(k_x^2+k_y^2+k_z^2+2k_xk_z) \qquad R = 1/2\, N(k_z^2-k_x^2)$$

$$H_2 = 1/2\, L(k_z+k_x)^2 + 1/2\, M(k_x^2+k_y^2+k_z^2-2k_xk_z) \qquad S = 1/\sqrt{2}\, N k_y(k_z-k_x)$$

$$H_3 = Lk_y^2 + M(k_x^2+k_z^2) \qquad T = 1/\sqrt{2}\, N k_y(k_z+k_x)$$

Here k_x, k_y, k_z are the components of the $\underline{k}$-vector with respect to the $[\bar{1}10]$, [001], and [110] direction, 3Δ is the spin-orbit energy of 44 meV and the parameters $L = -6.5\, \hbar^2/2m_0$, $M = -2.7\, \hbar^2/2m_0$, $N = -7.3\, \hbar^2/2m_0$ were determined from microwave cyclotron resonance experiments on bulk silicon[5]. For a (110) surface the surface potential $V(z)$ varies only along the z-direction. Therefore the wave equation for the z-dependent part of the envelope function is given by:

$$\left\{ H_0\left(k_x, k_y, \frac{1}{i}\frac{\partial}{\partial z}\right) + V(z) \right\} \Psi_{i,k_x,k_y}(z) = E_i(k_x, k_y)\, \Psi_{i,k_x,k_y}(z) \tag{2}$$

Here the k_z-component in the matrix H_0 is replaced by the operator $-i\partial/\partial z$. The energies $E_i(k_x,k_y)$ and the wavefunctions $\Psi_{i,k_x,k_y}(z)$ belong to the i-th electric subband. The electrostatic potential $V(z)$ that originates from the charges in the states $\Psi_{i,k_x,k_y}(z)$ and the unpopulated donor states of concentration N_D-N_A in the depletion layer region is determined from Poisson's equation:

$$\frac{d^2V(z)}{dz^2} = \frac{e^2}{\varepsilon\varepsilon_0}\left(\sum_{i,k_x,k_y} \left| \Psi_{i,k_x,k_y}(z)\right|^2 + \left(N_D-N_A\right)\right) \tag{3}$$

where the sum covers all populated states. ε is the dielectric constant of Si. Self consistent solutions of (2) and (3) are obtained, if the potential entering equ. (2) is reproduced by equ. (3). Because of the complicated structure of equs. (2), (3) some simplifications were made:

a) our calculations were performed for the temperature T=0 K, which is well
 justified, because the experiments were carried out at liquid helium
 temperatures,

b) although we want to calculate cyclotron masses for comparison with ex-
 periments, no magnetic field is introduced. This is a deficiency, because
 spin effects cannot be explained from this theory. But incorporating high
 magnetic fields would enlarge the calculation substantially, because of
 the structure of H_0,

c) the boundary conditions for the wavefunction are simplified to be
 $\Psi(z = \infty)= 0$ and $\Psi(z = 0) = 0$. The last condition neglects the possibili-
 ty, that Ψ may penetrate somewhat into the oxide.

Now the calculation is straightforward. Expanding Ψ into a series of N basis
functions $u_j(z)$, 6xN-dimensional matrices were diagonalised for about 60
different pairs of k_x and k_y. From the eigenvalues we know the $E(k_x,k_y)$-
relations within the whole two-dimensional Fermi-areas of each subband. From
the eigenvectors the charge distribution and from this the surface potential
can be calculated. The u_j were chosen as Airy functions, which give a very
rapid convergence, such that N=6 could be used. The Fermi energy was deter-
mined from the requirement, that for each induced single charge one state on
the two-dimensional Fermi-area is available.

RESULTS

 A self consistent potential for a surface carrier concentration of
$N_{inv.}$ = 7×10^{12} cm^{-2} and $N_D-N_A=10^{15}$ cm^{-3} is shown in fig. 1a. E_0, E_0' and E_1
denotes the tops of the three highest electric subbands. A comparison of
the eigenvectors of the subbands with suitable eigenvectors of the bulk
states indicates, that E_0 and E_1 belong to the heavy hole band, whereas E_0'
originates from the light hole band. One could also imagine this from
$|\Psi_{i,k_x=0,k_y=0}(z)|^2$, which are plotted in fig. 1b. Obviously, the form of
E_0 and E_0' corresponds to that of a ground state, whereas E_1 looks like a
first excited state. In this special case of $N_{inv.}=7 \times 10^{12}$ cm^{-2} two electric
subbands were populated. Fig. 1c shows the corresponding Fermi surfaces,
which are simple curves in the k_x,k_y-plane for both subbands. The twofold
spin degeneracy of the bulk states only remains at $k_x=k_y=0$ and is lifted
more and more for increasing k_x and k_y. Therefore two Fermi curves arise
from each subband. It should be noticed, that due to this lifting of the
spin degeneracy, which is already present without a magnetic field, the
energy difference between spin split Landau levels in a magnetic field can-

not be described by a simple g-factor. The dashed line of constant energy in the k_x-k_y-plane, obtained from the bulk $E(\underline{k})$-function, shows, that there is a substantial influence from the surface potential. This line is constructed

Fig.1 $N_{inv.}=7\times10^{12}$ cm^{-2} a) the selfconsistent potential with the three highest electric subbands, b) $|\Psi(z)|^2$ at $k_x=k_y=o$ for these subbands, c) Fermi contours in the E_0 and E_1 subbands (——) and the bulk analogy to the E_0 band (---).

using the conditions that the enclosed area and the energy at $k_x=k_y=o$ are the same as for the E_0 electric subband. Therefore this curve can be considered as an analogy to the Fermi contour of the E_0 electric subband.

Fig.2 Theoretical and experimental cyclotron masses at different surface carrier concentrations.

The cyclotron mass m_c^i of the i-th electric subband is approximately given by

$$m_c^i(E) = \frac{\hbar^2}{2\pi} \cdot \frac{dF_i(E)}{dE}$$

where $F_i(E)$ is the area in the k_x,k_y-plane enclosed by the constant energy contour. For $E=E_F$ a comparison of experimental and theoretical m_c values is shown in fig. 2c. The self consistent calculations are indicated by stars and the drawn line is an interpo-

lation. The open points are the experimentally determined masses. The upper line corresponds to the first electric subband which is the heavy hole band whereas the lower line is due to the second electric subband, the light hole band. Furthermore fig.2 indicates that population of the second subband starts at $N_{inv.}=3.8\times10^{12}$ cm^{-2}, which is close to the extrapolated experimental value of $N_{inv.}=3.2\times10^{12}$ cm^{-2}.

The good agreement between the measured and the calculated effective masses has encouraged us to extend the calculations to the (111) and (100) orientations in the near future.

REFERENCES

[1] v.Klitzing,K., Landwehr,G., Dorda,G.: Solid State Commun.14,387 (1974)

[2] v.Klitzing,K., Landwehr,G., Dorda,G.: Solid State Commun.15 (1974)
 Lakhani A.A., Stiles P.J., Cheng Y.C: Phys.Rev.Letters 32,1003 (1974)

[3] Stern,F.: Phys.Rev.B 5, 4891 (1972)

[4] Luttinger, J.M., Kohn,W.: Phys.Rev. 97, 869 (1955)

[5] Dexter, R.N., Lax,B.: Phys.Rev. 96, 223 (1954)

VALLEY-ORTIT SPLITTING OF THE STATES
IN SURFACE INVERSION LAYERS

K. NARITA AND E. YAMADA

Central Research Laboratory, Hitachi Ltd.
Kokubunji, tokyo, Japan

The mechanism of valley-orbit splitting for the states
in surface inversion layer is investigated. The imp-
roved effective mass equation which couples the wave-
functions associated with different valleys is derived
from the Schrodinger equation. The equation can be
solved regarding the coupling as a small perturbation.
The magnitude of the splitting is consistent with that
infered from the experiments.

1. INTRODUCTION

An oscillatory magnetoconductance has been observed by several
authors for the inversion layers of silicon (100) surfaces [1]
[2]. In their experiments, the magnetic field was applied
perpendicular to the surface in the electric quantum limit in
which all the electrons occupy only the lowest sub-band.

Therefore, each Landau level associated
with the sub-band has four-fold dege-
neracy due to the spin and the valley.
As the magnetic field is increased,
the each peak of conductivity osci-
llation becomes to split into two due
to the removal of the spin degeneracy.
At higher magnetic field a further
splitting is observed [1]. This is
considered to be associated with the
removal of the valley degeneracy. The
splitting will come from the breakdown
of the validity of simple effective

Fig. 1. The Fourier transform of the solution of the one-
valley effective mass equation. A(k) is calculated for
$F_s = 3.0 \times 10^5$ V/cm. The corresponding function for the shallow
donor state in silicon is also shown for the comparison.

mass theory. The application of one-valley effective mass
theory to the surface problem shows that the wave function in
k-space spreads over the other valley as shown in Fig. 1.
We improved the effective mass equation to include the inter-
valley effect. The improved equation results the removal of
the valley degeneracy and the magnitude of the splitting is
shown to be consistent with that infered from the experiment.

2. IMPROVED EFFECTIVE MASS EQUATION

Let us consider a hypothetical one-dimensional semiconductor,
which has two equivalent minima at the points k_0 and $-k_0$ in
the conduction band. If the potential given by

$$V = \begin{cases} \infty & \text{for } z < 0 \\[2ex] -U_0 + eF_s z \, . & \text{for } z \geqq 0 \end{cases} \qquad (1)$$

is applied to the system, the Hamiltonian for an electron can
be expressed as

$$H = H_0 + V. \qquad (2)$$

We are interested in the lowest two states, which are degene-
rate if the valley splitting is absent, of the Schrodinger
equation

$$H\Psi_{0,i} = W_{0,i}\Psi_{0,i}. \qquad i=1,2 \qquad (3)$$

The wave functions are assumed to be expanded in terms of the
Bloch functions of the band as,

$$\Psi_{0,i} = \sum_k c_i(k)\phi_k \, . \qquad (4)$$

If the $c_i(k)$ corresponding to k near different minima are only
weakly coupled, it may be a good approximation to write

$$c_i(k) = a(k) \pm b(k), \qquad (5)$$

where i = 1 and 2 correspond to plus and minus sign in the equation, respectively. The functions a(k) and b(k) are supposed to be centered at k_0 and $-k_0$, respectively. Substituting the eq.(4) into eq.(3) and taking the scalar product with ϕ_k, we obtain

$$[\varepsilon(k)-W_{0,i}]a(k) + \sum_{k'} (k|V|k')_0 a(k') + \sum_{k'} (k|V|k')_1 a(k')$$
$$\pm[\varepsilon(k)-W_{0,i}]b(k) \pm \sum_{k'} (k|V|k')_0 b(k') \pm \sum_{k'} (k|V|k')_1 b(k') = 0. \quad (6)$$

Here, we divided the matrix element $(k|V|k')$ in to two parts, that is intravalley matrix element $(k|V|k')_0$ and intervalley matrix element $(k|V|k')_1$. Keeping the main terms of eq.(6) and transforming it into the real space with usual approximations, we can obtain

$$[\Omega - W_{0,i}]f_a(z) = \mp \mu^0_{kk'}G_b(z), \quad (7)$$

where

$$\Omega = (\hbar^2/2m^*)\partial^2/\partial z^2 + V(z), \quad G_b(z) = \exp(-2ik_0 z)v(z)f_b(z),$$

$$f_a(z) = \sum_k \exp[i(k-k_0)z]a(k), \quad f_b(z) = \sum_k \exp[i(k+k_0)z]b(k).$$

$\mu^\nu_{kk'}$ are the coefficient in the expansion of the product of periodic parts of the Bloch functions for the intervalley matrix elements;

$$u_k(z)u_{k'}(z) = \sum_\nu \exp(iK\nu z). \quad (8)$$

The magnitude of the coefficient $\mu^0_{kk'}$ should be in between 0 and 1 and we treat it as a parameter. If we neglect the intervalley term, or equivalently if we put $G_b(z)$ to be zero in eq.(7), the equation is reduced to the one-valley effective mass equation;

$$[\Omega - W_{0,i}]v_0(z) = 0. \quad (9)$$

Regarding the intervalley terms as a small perturvation we can

get the approximate solution
of the equation. The mag-
nitude of the splitting is
given by

$$\Delta W = \left| W_{0,1} - W_{0,2} \right|$$

$$= 2\mu_{kk'}^{0} \left| \int v_0(0) G_b(z) \, dz \right|.$$

If $G_b(z)$ is small pertur-
bation, the function $f_b(z)$
in $G_b(z)$ may be approximated
by $v_0(z)$.
We use a tractable function

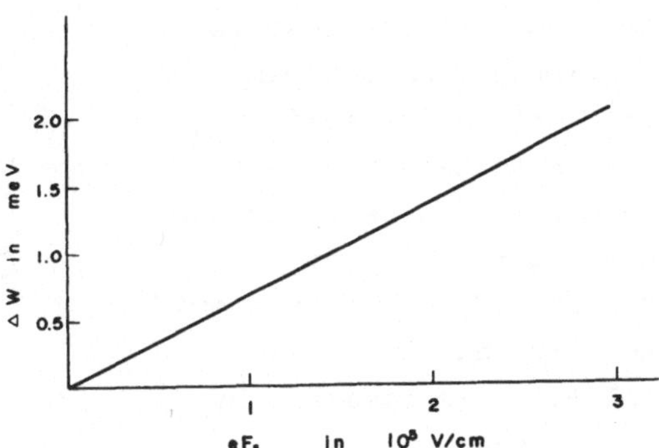

Fig. 2. The calculated value
of valley splitting for the
parameter $\mu_{kk'}^{0}=1$, as a function
of surface field eF_s.

$$g(z) = (b^{3/2}/\sqrt{2}) \, z \, \exp(-bz/2)$$

for $v_0(z)$. The parameter b in the function can be determined
by the variational procedure as,

$$b = (12m^* eF_s/\hbar^2)^{1/3}.$$

The numerical value of the splitting for $\mu_{kk'}^{0}=1$ is shown in
Fig. 2 against the eF_s, making use of the following constants:

$$U_0 = 1 \text{ eV}, \qquad k_0 = 0.5 \times 10^8, \qquad \text{and } m = 0.2m^*.$$

It should be noted that this is the maximum value for the
splitting, since we have taken as $\mu_{kk'}^{0}=1$.

3. DISCUSSIONS

The exact magnitude of the splitting has not yet been deter-
mined by the experiments. However the splitting should be in
between kT and $\hbar\omega_c$ at which the splitting in peaks of magneto-
conductivity oscillation was observed. This suggests that
the valley splitting of Landau states in surface inversion
layers may be of the order of 1 meV for the typical cases.

As seen from the results in section 2, the maximum value of valley splitting due to surface field is calculated to be 2 meV for surface field F_s of 3×10^5 V/cm. This seems to show that the valley splitting is possible by this mechanism.
The formalism used to calculate the valley splitting in this article is similar to that of Baldereschi for

Fig.3. The calculated value of valley splitting for the parameter $\mu_{\perp}^{0}=1$, as a function of 1/b which is the measure of the spreading of the wave function.

shallow donor state [3]. However, in the case of Baldereschi the wave-number-dependence of dielectric functions of silicon seems to be essential, and our mechanism gives very small splitting for such large orbit of the wave function. In order to see that this is the case, the splitting is plotted against 1/b, which is a measure of the orbital radius, in Fig. 3.
As seen from the figure, the magnitude of the splitting is decreased as the value of the 1/b is increased and it is as small as 0.2 meV for 1/b~20A. Finally it should be noted that P. J. Stiles presented an idea that the enhancement of the valley splitting may occur in a similar way as the spin splitting of Landau level in the inversion layers [4]. However, further investigation for this problem is wanted.
The authors wish to express warm thanks to Miss R. Bohta for assistance in the numerical calculation. They also indebted to Profs. P. J. Stiles and A. Morita for helpful discussions.

4. REFERENCES

[1] Fowler, A. B.; Fang, F.F.; Howard, W. E.; and Stiles,P. J.
 J. Phys. Soc. Japan 21 Supplement, 331 (1966).

[2] Komatsubara, K. F.; Narita, K.; Katayama, Y.; Kotera, N.;
 and Kobayashi, M,: to be published in J. Phys. Chem. Solid.

[3] Baldereschi, A.: Phys. Rev. B1, 4673 (1970).

[4] Stiles, P. J. : private communication.

THEORY OF CYCLOTRON RESONANCE LINE SHAPE IN MOS INVERSION LAYERS

Tsuneya ANDO and Yasutada UEMURA

Department of Physics, University of Tokyo
Bunkyo-ku, Tokyo 113, Japan

A theory is presented for the cyclotron reso-
nance line shape in a two-dimensional system.
The line shape depends on the range of scat-
terers strongly. By assuming short-ranged scat-
terers, the dynamical conductivity is calculated
explicitely as a function of the applied magnet-
ic field in n-channel inversion layers on Si
(100) surface. At low temperatures, Schubnikov-
de Haas type oscillation appears. The line
shape becomes asymmetric around the resonance
magnetic field as a result of the fact that the
level width of each Landau level depends on the
magnetic field.

1. INTRODUCTION

Recently, the cyclotron resonance (CR) has been observed ex-
perimentally in a two-dimensional system such as inversion lay-
ers on semiconductor surfaces [1][2]. Under strong magnetic
fields applied perpendicularly to the surface, the energy spec-
trum of this system becomes discrete because of the complete
quantization of the orbital motion. In such a singular system,
the assumption of the constant relaxation time τ to the level
broadening of each Landau level is not valid and therefore, the
classical formula of the CR line shape is not directly applica-
ble. Further, the singular nature of our system is expected to
play some important role. In this report, a theory of the CR
line shape in the two-dimensional system is presented and it is
applied to inversion layers.

2. CHARACTERISTICS OF THE CR LINE SHAPE

We consider the same two-dimensional system with randomly
distributed scatterers as that treated previously [3]. In a
similar manner to that in ref. [3], one starts with the Kubo
formula and expresses the dynamical conductivity in terms of the
Green's function and the current vertex part. Those are calcu-
lated in the simplest approximation without the difficulty of
divergence [3]. In this approximation, scattering from a sin-

gle scatterer is taken into account in the lowest Born approxi-
mation, while the broadening effects are included in a self-con-
sistent way (self-consistent Born approximation). It gives the
semi-elliptic form of the density of states of each Landau level
under strong magnetic fields. Characteristics of the CR line
shape deduced from the obtained expression of the conductivity

FIG. 1: The dynamical conductivity Re $\sigma_{xx}(\omega)$ for scatterers
with the Gaussian potential at T=0 K. Γ_N is the level
width of the N-th Landau level and $\alpha = d/\ell$ where $\ell^2 = c\hbar/eH$.
(a) Filled Landau level. (b) Half-filled Landau level.

are summarized as follows.

(i) Short-ranged scatterers $d \ll \ell$ (d is the range of scatterers and ℓ is the radius of the ground cyclotron orbit given by $\ell^2 = c\hbar/eH$). The broadening is essentially the life-time broadening and transitions between all states in adjacent Landau levels become allowed. The resonance width is determined by the level width Γ, which is nearly given by $\Gamma \propto (\hbar\dot{\omega}_c \cdot \hbar/\tau)^{1/2}$ in terms of the relaxation time under no magnetic field.

(ii) Long-ranged scatterers $d \gg \ell$. The broadening is essentially the so-called inhomogeneous one and only the transition between the states with nearly the same relative energy measured from the center of each Landau level. The level width is determined by the fluctuation of the local potential energy $V(r)$, *i.e.* $\Gamma \propto [<(V(r)-<V(r)>)^2>]^{1/2}$, while the width of the resonance $\Delta(\hbar\omega)$ is determined by the fluctuation of the gradient of $V(r)$, *i.e.* $\Delta(\hbar\omega) \propto <(\ell \text{grad } V(r))^2>/\Gamma$.

In order to see such dependence on the range of scatterers explicitly, we assume a model scatterer with the Gaussian potential ($\propto \exp(-r^2/d^2)$) and calculate the line shape. Examples are shown in Fig. 1. The line shape is strongly dependent on the position of the Fermi level, which is particularly remarkable in case of short-ranged scatterers.

3. APPLICATION TO MOS INVERSION LAYER

In n-channel layers on Si (100) surface, the scatterers are considered to be short-ranged. We calculate the conductivity as a function of the magnetic field, keeping the total number of electrons constant ($m=0.195m_0$). Examples are shown in Fig. 2. At low temperatures, an oscillatory behavior appears. This is caused by the fact that the conductivity depends on the position of the Fermi level. Each dip in the low field side and each peak in the high field side correspond to the case in which each Landau level is filled. This can easily be understood by comparing (a), (b) in Fig. 1 and the position of the Fermi level shown in Fig. 3. One can regard such oscillation as the Schubnikov-de Haas oscillation. The reason why the line shape is asymmetric around $\omega_c = \omega$ is that the level width depends on the magnetic field H.

FIG. 2: Several examples of Re $\sigma_{xx}(\omega)$ as a function of the applied magnetic field H in n-channel layers on Si (100) surface. Short-ranged scatterers are assumed for main scatterers. $m^*=0.195m_0$, $\hbar\omega=5.63$ meV and $g^*=2$. The total number of electrons N_{inv} and the mobility are shown.

FIG. 3: Landau levels and the Fermi level in n-channel layers on Si (100) surface at T=0 K. Solid and broken straight lines are Landau levels of spin up and down, respectively. Solid curves show the Fermi level. N_{inv} (cm^{-2}) is shown in the figure.

The width of the CR is usually larger than the classical width $\Delta(\hbar\omega_c)=\hbar/\tau$. Notice that the width of (a) in Fig. 2 is smaller than that of (b) at low temperatures even if the mobilities are about the same. In case that the total number of electrons is sufficiently small, electrons occupy only the lower part of the lowest Landau level and the width becomes effectively smaller. These facts agree with the experimental result of Allen *et al.* [2].

Our approximation is the simplest one and not sufficient at spectral edges. Therefore, the detailed form of the conductivity can not directly be compared with experimentally observed one. The overall behavior is, however, considered to be reasonable and the oscillatory effect is expected to continue to exist even if higher approximations are employed. Careful experiments at low temperatures are highly expected.

4. REFERENCES

[1] Abstreiter, G.; Kneschaurek, P.; Kotthaus, J. P.; Koch. J. F.: Phys. Rev. Letters 32, 104 (1974)

[2] Allen, J. S.; Tsui, D. C.; Dalton, J. V.: Phys. Rev. Letters 32, 107 (1974)

[3] Ando, T.; Uemura, Y.: J. Phys. Soc. Japan 36, 959 (1974)

THE INFLUENCE OF LANDAU LEVEL BROADENING ON THE SHUBNIKOV-DE HAAS EFFECT IN TWO-DIMENSIONAL CONDUCTORS

L.M. BLIEK

Physikalisch-Technische Bundesanstalt, Braunschweig

Fed. Rep. Germany

The density of states at the Fermi energy for two-dimensional conductors in a perpendicular magnetic field is calculated for Lorentzian broadening of the Landau levels. The results are compared to published experimental data on the magnetoresistance of field effect devices.

1.INTRODUCTION

Inversion layers in field effect devices behave essentially like two-dimensional conductors [1]. The Shubnikov-De Haas effect in such systems has been studied by different authors [1-6]. For zero absolute temperature and without Landau level broadening, due to the dependence of the Fermi energy on magnetic field and gate voltage, the fields and voltages for which peaks in the density of states at this energy occur, do not depend on the size of the spin splitting of the energy levels. Consequently the spin splitting observed in Shubnikov-De Haas experiments should be independent of the Landé g-factor. Broadening of the energy levels will reduce the field dependence of the Fermi energy. In order to find out if it may become sufficiently constant to justify the determination of g from the size of the spin splitting, the density of states at the Fermi energy in a magnetic field was calculated, assuming a Lorentzian line shape for the Landau levels. It has been shown previously for the three-dimensional case that this density of states closely resembles the observed transverse magnetoresistance [7]. It was found that g cannot be determined from the size of the spin splitting and also that the period of the oscillations depends on the broadening of the energy levels.

2. CALCULATION OF THE DENSITY OF STATES AT THE FERMI ENERGY

The density of states $D(E)$ is found by summing the Lorentzian

$$\frac{\hbar/\pi\tau}{(E-E_i)^2 + (\hbar/\tau)^2}$$

over all possible quantum states E_i. Here

the broadening parameter is written as $\hbar/\tau$, where τ is an
effective lifetime. If, for simplicity, a parabolic and iso-
tropic energy band is assumed, the E_i are given as usual by
$E_{n,s}$ = (n+1/2+ $sgm_c/4m_o$) $\hbar\omega$ with $\omega = eB/m_c$, n=o,1,2,... and
s=+1,-1. there are $eBA_s/2\pi\hbar$ degenerate states for each pair of
Landau and spin quantum numbers n and s, if A_s is the area of
the two-dimensional sample. Consequently one finds

$$D(E) = \frac{\hbar\omega}{2} D_o \sum_{n,s} \frac{\hbar/\pi\tau}{(E-E_{n,s})^2+(\hbar/\tau)^2} \tag{1}$$

where D_o is the density of states for B=0 and without level
broadening which is equal to $D_o = A_s m_c/\pi\hbar^2$.
The number of states below E, n(E), is obtained by integrating
(1) from E_o to E, where E_o is a cut-off energy well below the
lowest Landau level, introduced to improve the convergence of
the series. One obtains

$$n(E) = \frac{\hbar\omega}{2} D_o \sum_{n,s} \frac{1}{\pi} \left[arctg(\frac{E-E_{n,s}}{\hbar/\tau}) - arctg(\frac{E_o-E_{n,s}}{\hbar/\tau}) \right] \tag{2}$$

In the limit $T\to 0$ n(E) is equal to n_o, the number of electrons
if E equals the Fermi energy E_F. In field effect devices n_o is
determined by the applied voltage according to

$$V = \frac{en_o d}{\varepsilon_o \varepsilon_r A_s} \tag{3}$$

where d is the thickness of the insulating layer. If V is kept
constant and hence n_o remains constant we have

$$1 = \frac{\hbar\omega}{2E_{Fo}} \sum_{n,s} \frac{1}{\pi} \left[arctg(\frac{E_F-E_{n,s}}{\hbar/\tau}) - arctg(\frac{E_o-E_{n,s}}{\hbar/\tau}) \right] \tag{4}$$

where E_{Fo} is equal to n_o/D_o. For $\hbar/\tau \to 0$ this leads to

$$1 = (\hbar\omega/2E_{Fo}) \sum_{n,s} \Theta(E_F-E_{n,s}) \tag{5}$$

with $\Theta(x)=1$ for x>0 and $\Theta(x)=0$ for x<0. This means that in
this limit, on changing B, E_F jumps from one Landau Level to
the next if $E_{Fo}/\hbar\omega$ = 1/2, 1, 1 1/2, 2,... while it sticks to
one level in between.
The density of states at the Fermi energy is given by (1) with
$E=E_F$, E_F being determined by (4). This means that if kT and $\hbar/\tau$
are both small compared to $\hbar\omega$ and to the spin splitting of the
Landau levels a peak(n,s) appears in $D(E_F)$ when

FIG. 1: Density of states at the Fermi energy in units of its broadening- and magnetic field-free value as a function of the reciprocal of the magnetic induction in units of $\hbar e/m_c E_{FO}$ for different values of the level broadening parameter $A = \hbar/\tau E_{FO}$. E_{FO} is the field- and broadening-free Fermi energy

$$E_{FO}/\hbar\omega = (2n+1+ 1/2\ s)/2 \qquad (6)$$

independent of the size of the g-factor.

A finite value of $\hbar/\tau$ tends to reduce the amplitude of the oscillations of the Fermi energy. It will therefore bring the positions of the spin split peaks closer to the g-dependent ones that were to be expected for a constant Fermi energy.

Fig. 1 shows the results of a numerical calculation of $D(E_F)$ in units of D_O as a function of the reciprocal of the magnetic induction in units of $\hbar e/m_c E_{FO}$ for a number of values of the broadening parameter $A = \hbar/\tau E_{FO}$. The curves[+] show that where a spin splitting is observable, its value is far away from the value of 0.2 $\hbar\omega$ chosen arbitrarily for the splitting of the Landau levels. Therefore it appears that g-factors cannot be derived from the magnitudes of the spin splittings observed in measurements of the magnetoresistance.

Obviously the period of the oscillations in the density of

[+] The assistance of Dipl.-Phys. G. Hein and the computer staff of the PTB in producing the computer plots shown in figs. 1 and 2 is gratefully acknowledged.

states shown in fig. 1 depends on the magnitude of the
broadening parameter. It is given approximately by 1+0.5A.
In a field effect device one can produce a Shubnikov- De Haas
effect also by keeping the magnetic field constant and
changing the gate voltage. Since according to (3) the number of
carriers is proportional to the gate voltage one may compare
the results of such experiments to plots of the density of
states at the Fermi energy vs. the number of electrons.

FIG. 2:

Density of states
at the Fermi
energy in units of
its broadening-
and magnetic
field-free value
D_O, as a function
of the number of
electrons in units
of $\hbar\omega D_O$

Fig. 2 shows curves calculated using (1) and (2) with $E = E_F$
for a number of values of the broadening parameter A which is
now equal to $\hbar/\tau$ in units of $\hbar\omega$. The spin splitting of the
Landau levels was again chosen as 0.2 $\hbar\omega$. Once more it is seen
that the size of the spin splitting is almost entirely
independent of the splitting of the Landau levels. For small
values of the broadening parameter the splitting of the peaks
in the density of states amounts to one half of the oscillation
period, as in the previous case.
The oscillation pattern is not exactly periodical. Upon
developing the arctg functions in (2) into power series one

finds that the distance between two neighbouring peaks with equal s but different n is given approximately by

$$n_{O_{n+1,s}} - n_{O_{n,s}} = 1 - 1/\pi(n+1)\omega\tau \tag{7}$$

in good agreement with the numerical results.

3. COMPARISON WITH EXPERIMENTAL WORK

In studying the published results of experiments on inversion layers on Si surfaces [1-5] one finds in good agreement with the results presented here that whenever a clear spin splitting is observed, it amounts to one half of the oscillation period and that only for weakly developed splittings smaller values occur. Consequently the g-factors derived in [5] from the size of the splittings are probably in error. The authors find $gm_c/m_O \approx 1$, equivalent of a splitting of half a period. By comparing the magnetic induction for which the spin splitting disappears to that for which the oscillations disappear altogether one finds $gm_c/m_O \approx 0.6$ which leads to $g \approx 2$. The complexity of the Landau level spectrum in Si makes it difficult to compare such properties as periodicity and amplitude of the oscillations. Relation (6) was confirmed in experiments on $Hg_xCd_{1-x}Te$ [6].

4. REFERENCES

[1] Fowler, A.B.; Fang, F.F.; Howard, W.E.; Stiles, P.J.: J. Phys. Soc. Japan 21 Suppl. , 331 (1966)
[2] Fang, F.F.; Stiles, P.J.: Phys. Rev. 174, 823 (1968)
[3] Kobayashi, M.; Komatsubaru, K.F.: Solid State Comm. 13, 293 (1973)
[4] Lakhani, A.A.; Stiles, P.J.: Phys. Rev. Letters 31, 25 (1973)
[5] Klitzing, K.v.; Landwehr, G.; Dorda, G.: Solid State Comm. 14,387 (1974)
[6] Tasch Jr. A.F.; Buss, D.D.; Bate, R.T.; Breazeale, B.H.: Proc. 10th Int. Conf. Phys. Semiconductors, Cambridge (Mass.) 1970, p.458
[7] Bliek, L.M.; Landwehr, G.; Ortenberg, M.v.: Proc 10th Int. Conf. Phys. Semiconductors, Moscow, 1968, p.458

SURFACE STATES AT CLEAN AND CESIATED Ge (111) SURFACES

J. VON WIENSKOWSKI, W. MÖNCH

2. Physikalisches Institut and Sonderforschungsbereich "Festkörper-

elektronik", RWTH Aachen, Fed. Rep. Germany

Clean as well as cesiated, cleaved Ge (111)
surfaces are found to be p-type. With increasing
Cs coverage the field effect mobility decreases
and thus the density of surface states near the
Fermi-level E_F increases. - At the clean,
cleaved surface the field effect mobility and thus the
density of surface states near the Fermi-level
are mainly determined by the density of cleavage
steps.

1. INTRODUCTION

The properties of metal-semiconductor contacts are mostly discussed

in terms of surface states in the semiconductor band gap. The first

stages in the formation of a junction may be studied by following the

changes of typical surface properties, as for example band bending,

field effect of surface conductivity, work function and LEED, in depen-

dence of the metal coverage. In this paper results of such measure-

ments with clean, cleaved and subsequently cesiated Ge (111) surfaces

are reported.

2. EXPERIMENTAL RESULTS

The cleaved surfaces were obtained with samples (insert of Fig. 1) cut

from p-type single crystals with a resistivity of 14 Ω cm at room tem-

perature. After cleavage in UHV (5 to 8 x 10^{-11} Torr) cesium was depo-

sited in situ on the cleaved surfaces up to a monolayer coverage. The

monolayer $\Theta = 1$ is defined such, that there is one adsorbed Cs atom for

each substrate site in an ideal Ge (111) bulk plane, i. e. 7. 23 x 10^{14}

atoms/cm^2. Details of the sample preparation and of the cesium source

have been published previously [1] .

In Fig. 1 the measured change of surface conductivity $\Delta\sigma$ and field effect mobility $\mu_{FE} = -\dfrac{d(\Delta\sigma)}{d\,Q_{in}}$ are plotted versus the Cs coverage. Q_{in} is the charge induced in the surface by an external electrode.

FIG. 1

Change of surface conductivity and field effect mobility versus cesium coverage.

Samples: ✕ 29, △ 33,
 ○ 37, □ 41.

With increasing coverage the surfaceconductivity passes through a more or less pronounced minimum while the field effect mobility decreases monotonously but remains negative. This means, that the cleaved surface is p-type independent of Cs-coverage.

3. DISCUSSION

3.1 ELECTRONIC PROPERTIES OF CESIATED Ge SURFACES

From the change in $\Delta\sigma$ observed with increasing coverage the correspon-ding surface potential may be calculated since the surface conductivity of the clean, cleaved surface had been determined previously [2, 3] . The mobility of the holes in the surface space charge layer was evaluated

by assuming diffuse surface scattering [4] . The resulting distance
between Fermi-level and valence band edge increases up to a coverage
of 0. 1 of a monolayer but decreases for higher values (Fig. 2).

The same dependence has been observed with Ge (111) surfaces cleaned
by ion-bombardment and annealing [5] while for Ge (100) surfaces the
Fermi-level was found to coincide with the valence band edge [6] . For
Si (111) and (100) surfaces [1, 6] the surface potential changes monoto-
nously with increasing coverage while cesiated GaAs (110) surfaces exhibit
the same behaviour as the Ge (111) surfaces [7] .

FIG. 2

Surface potential of
cesiated Ge surfaces.
□ after Riach and Peria
[5]
+ after Jeanes and
Mularie [6]
x sample 29,
o sample 37.

Up to a monolayer coverage the work function is monotonously lowered
by about 3. 2 eV indicating the formation of a dipole layer at the surface
with the cesium adsorbed as positive ions. From this it may be concluded
that the compensating negative charges occupy surface states, since the
positive space charge of the clean surface only decreases by $3 \times 10^{+12}$
e_o / cm^2 up to $\Theta = 0.1$. The increase of the distance $(E_F - E_v)_s$ up to this
coverage may thus be explained by the filling of acceptor-like surface
states above the Fermi-level position at the clean surface. Since after
the deposition of a Cs monolayer Fermi-level and valence band edge
coincide the formation of new acceptor-like surface states below the
valence band edge has to be assumed with increasing coverage beyond
$\Theta = 0.1$.

This change in surface state distribution could not be correlated with a

structural conversion. The observed LEED patterns only show the extra-spots of a 2x1 cleavage structure to decrease in intensity with increasing coverage. Above $\Theta = 0.6$ only normal order spots are visible.

However, there have been raised objections to the widely accepted idea to discuss the properties of Schottky barriers in terms of surface states in the semiconductor band gap. Recent calculations by Inkson [8] have revealed that the band gap of a covalent semiconductor should disappear at the interface of an intimate metal-semiconductor junction. The increase of the surface conductivity for coverages $\Theta > 0.1$ could then be caused by a gradual shrinkage of the gap. Also the simultaneous decrease of the field effect mobility might be explained by this model since due to the reduced band gap the density of states near the Fermi-level will be enlarged.

3.2 CORRELATION BETWEEN SURFACE STEPS AND SURFACE STATES

The different values of the field effect mobility found with the clean Ge surfaces (Fig. 1) are correlated with surface steps. In Fig. 3 the reciprocal of the field effect mobility is plotted versus $\langle \tan \alpha \rangle$, where α is the angle of misorientation between the (111) plane and the cleaved surface *).
Since the steps are a double layer spacing in height [10] $\langle \tan \alpha \rangle$ is proportional to the density of atoms at the step edges. The charge Q_{in} influenced in

FIG. 3

Field effect mobility of clean, cleaved Ge (111) surfaces in dependence of cleavage steps.
Samples:
X 29, △ 33, ◇ 34a,
○ 37, ▽ 40, □ 41.

*) For the determination of $\langle \tan \alpha \rangle$ see [9].

the semiconductor surface distributes between surface states and space charge and thus the reciprocal of the field effect mobility may be written as

$$- \frac{1}{\mu_{FE}} = \frac{dQ_{in}}{d(\Delta\sigma)} = \frac{dQ_{sc}}{d(\Delta\sigma)} + e_o \cdot N_s^* (E_F)/ \frac{d(\Delta\sigma)}{dV_s}$$

where Q_{sc} is the surface space charge, $N_s^* (E_F)$ is the density of surface states near to the Fermi-level and V_s is the band bending. The slopes $d(\Delta\sigma)/dQ_{sc}$ and $d(\Delta\sigma)/dV_s$ were evaluated by solving Poissons equation assuming that both expressions are independent of the step density and that the surface scattering is diffuse. The resulting values for $N_s^*(E_F)$ are given on the right-hand side of Fig. 3. If the scattering at the surface would be specular then $N_s^*(E_F)$ would be larger by a factor of about 4 to 10 in the range covered by the experimental points. The scattering mechanism at cleaved Ge surfaces has not been studied in detail, but the hitherto existing experimental results reveal no marked correlation between surface conductivity and surface steps [11, 12]. Therefore, it can be stated that the surface states at cleaved Ge surfaces near to the Fermi-level are mainly determined by surface steps. Since at these surfaces the distance $(E_F - E_v)_s$ is small the surface states correlated with cleavage steps are close to the valence band edge. Recently, at cleaved Si surfaces an extra photoemission intensity from surface states near to the valence band edge has been found, which is correlated with the surface step density [13, 14].

4. REFERENCES

[1] Mönch, W.: phys. stat. sol. 40, 257 (1970)
[2] Henzler, M.: Surface Sci. 9, 31 (1968)
[3] Wienskowski, J. v.; Mönch, W.: phys. stat. sol. 45, 583 (1971)
[4] Many, A.; Goldstein, Y.; Grover, N. B.: Semiconductor Surfaces,
 Ch. 8, North Holland Publishing, Amsterdam 1965
[5] Riach, G. E.; Peria, W. T.: Surface Sci. 40, 479 (1973)
[6] Jeanes, M. R.; Mularie, W. M.: 31st Physical Electronics Conf., 1971
[7] Clemens, H. J.; Mönch, W.; Wienskowski, J. v.: to be published
[8] Inkson, J. C.: J. Phys. C 6, 1350 (1973)
[9] Ibach, H.; Horn, K.; Dorn, R.; Lüth, H.: Surface Sci. 38, 433 (1973)
[10] Henzler, M.: Surface Sci. 19, 159 (1970)
[11] Henzler, M.: private communication
[12] Wienskowski, J. v.: to be published
[13] Rowe, J. H.: private communication
[14] Sebenne, C.; Bolmont, D; Guichar, G.; Balkanski, M.:
 This conference, paper 19 D 05

SPIN-DEPENDENT PROPERTIES AND SPIN-FLIP

CONTENTS

OPTICAL PUMPING IN SEMICONDUCTORS

G. LAMPEL

Laboratoire P.M.C., Ecole Polytechnique, Paris, France[*]

The principles of spin orientation by optical pumping in semiconductors are recalled. Several aspects of optical pumping of free carriers are presented : determination of lifetimes and spin relaxation times, hot electron effects, spin dependent recombination, optical detection of E.S.R. and N.M.R. The extension of these methods to excitonic orientation is also summarized.

1. INTRODUCTION

The transfer of angular momentum from light to matter by absorption of circularly polarized photons, leading to a situation where the populations of the Zeeman levels are significantly different from their equilibrium values, has been called "optical pumping" [1]. This effect was first extensively used to study atomic states in gases [2] and localized states in solids [3] The first experiment where spin polarized conduction electrons were obtained by optical pumping in a semiconductor was reported in 1968 [4]. It was very soon recognized that a convenient way to study the polarization of the carriers was to measure the polarization of the recombination light [5] [6] [7] [8]. By now these methods have proven to be very fruitful in the physics of semiconductors [9]. We will here recall briefly the principles of optical pumping in semiconductors. Specific effects such as energy relaxation, spin dependent recombination, optical detection of E.S.R. and N.M.R., will be mentioned. Finally the extension of optical pumping to the study of excitonic orientation will be quoted.

2. PRINCIPLES OF OPTICAL PUMPING IN SEMICONDUCTORS

Due to the spin-orbit interaction, which splits the valence bands by Δ (at the maximum of the bands in $\vec{k}$-space) absorption of cir-

[*] Equipe de Recherche du Centre National de la Recherche Scientifique.

cularly polarized σ_+ light of energy $h\nu$ leads to the preferential orientation of the spins of the photoexcited carriers when $E_G < h\nu < E_G + \Delta$ where E_G is the band gap [4][10]. Thus an initial polarization $P_i = \left(\dfrac{n_+ - n_-}{n_+ + n_-}\right)_i$ is achieved for the photoexcited electrons in the conduction band, $n_+ (n_-)$ being the number of electrons with spin $+ (-)$ along the direction of light propagation. P_i depends only on the symmetries of the valence and conduction bands. In III-V compounds with zinc-blende symmetry like GaSb, GaAs, InP..., $P_i = -0,5$ [5][11][12]. In II-VI wurtzite-type compounds like CdSe and CdS, $P_i = -1$ [13][14]. When $h\nu > E_G + \Delta$ some electrons are excited to the conduction band from the split-off valence band, leading to an opposite contribution to the polarization and finally for $h\nu \gg E_G + \Delta$, $P_i \to 0$. Solving kinetic equations with the lifetime τ and the spin relaxation time T_1 [4][5][6], one obtains : $P = P_i T_1 / (T_1 + \tau)$. For a given electronic polarization P, the polarization of the luminescence light $\rho = \dfrac{L_{F+} - L_{F-}}{L_{F+} + L_{F-}}$ is proportional to P [5], L_{F+} (L_{F-}) being the intensity of the luminescence with σ_+ (σ_-) polarization. If the recombination occurs between the same states as the absorption, then $\rho = P_i P = P_i^2 T_1 / (T_1 + \tau)$ [5] and a measurement of ρ, knowing P_i, yields the ratio T_1 / τ. This equality holds also for band-shallow acceptor recombination, which is the main luminescence channel at low temperature for p-type doped semiconductors where excitonic effects are not present. If the recombination light is gathered at an angle θ with that of the polarization $\rho(\theta) = \rho(0) \cos\theta$ [11].

Application of a transverse magnetic field H_τ induces precession of the spins at a rate $\omega = \gamma H_\tau$, γ being the gyromagnetic ratio of the electrons. It is easily shown [5] that the motion of the spins results in a steady-state polarization $\rho(H_\tau)$:
$\rho(H_\tau) = \rho(0) / (1 + \omega^2 T_{1e}^2)$ where the total polarization lifetime T_{1e} is : $1/T_{1e} = 1/T_1 + 1/\tau$. A measurement of the depolarization curve as a function of the transverse magnetic field gives a Lorentzian profile with a half-width $\Delta H = 1/\gamma T_{1e}$. This experiment is very similar to the Hanle effect in atomic physics [2] and, together with the measurement of ρ, yields separately T_1 and τ provided P_i and γ are known [9][12][15][16][17].

3. OPTICAL PUMPING OF FREE CARRIERS

3.1. HOT ELECTRON EFFECTS

Optical pumping of hot electrons, that is of electrons created above the bottom of the conduction band has been reported in various cases. All the experiments can be interpreted in the following way [11] [18] : When $h\nu > E_G$, electrons are initially excited at an energy $\varepsilon_i = \frac{(h\nu\ -E_G)}{1+m_c^*/m_v^*}$ where m_c^*/m_v^* is the ratio of the effective masses in the conduction and valence bands. The electrons will relax their energy to reach the bottom of the conduction band, where they recombine emitting luminescence radiation. In the process of thermalization during a time $\tau(\varepsilon_i)$ the carriers may lose their polarization if their spin have time to relax during $\tau(\varepsilon_i)$: On the contrary, they keep their initial polarization when $\tau(\varepsilon_i)$ is much shorter than their average spin relaxation time during thermalization. In GaSb, oscillations of the polarization as a function of $h\nu$ have been reported [15] [19] . These oscillations are attributed to fast energy relaxation by phonon emission and acceptor ionization, at well defined excitation energies. In GaAs no such oscillations have been observed but instead P reverses sign when $h\nu > E_G+\Delta$ [20] [21] . Electrons excited from the upper valence band Γ_8 are created high in the conduction band and lose their initial polarization P_i during energy relaxation. Electrons excited from the split-off band Γ_7 are created near the bottom of the conduction band with an initial polarization P_i', opposite to P_i, which they do not lose completely before recombination.

A theory of the polarization loss during thermalization has been given in [11] and [22] , where the main spin relaxation mechanism is attributed to the k^3 splitting of the conduction band, resulting from the noncentrosymmetric character of III-V compounds.

A discussion of the spin relaxation mechanisms for GaAs has been given in [12]. It was found that $1/T_1(\varepsilon) \simeq 10^{18}\,\varepsilon^3\tau_p(\varepsilon)$, where the kinetic energy ε is expressed in meV and τ_p, momentum relaxation time, in seconds. For energy and momentum relaxations due to optical phonons ($\tau(\varepsilon) \simeq \tau_p \simeq 10^{-13}$s) one finds $\tau(\varepsilon) < T_1(\varepsilon)$ as long as $\varepsilon < 5.10^2$ meV, which indicates that no spin relaxation occurs during thermalization in this energy range, in

disagreement with experimental results. Thus, the origin of the spin relaxation for these energies is still an open problem.

3.2. SPIN DEPENDENT RECOMBINATION

Another striking feature related to optical pumping is spin dependent recombination [23] [24] . In the case where spin up electrons do not recombine at the same rate than spin down electrons the total intensity of the luminescence depends on the polarization of the excitation light. Moreover, due to the different recombination rates, the electronic spins tend to accumulate in one state, leading to an overall polarization which may be closer to unity than P_i. An electronic polarization $P = 70$ % has been measured in p-type $Ga_{0,6}Al_{04}As$ [24] where $P_i = 50$ %.

3.3. OPTICAL DETECTION OF E.S.R.

The very large sensitivity of optical detection of electronic polarization has permitted the detection of electron spin resonance in cases where conventional resonance techniques have failed [25] [26] . Figure 1 shows the resonance signal of 10^8 conduction electrons in p-type GaSb. The g*-factors of conduction electrons have been obtained in GaSb ($|g^*| = 9.3$), GaAs ($|g^*| = 0.51$) and InP ($|g^*| = 1.26$).

FIG. 1 : Resonant change in the luminescence polarization induced by CESR.R.F. power = 14 W.

3.4. OPTICAL DETECTION OF N.M.R.

The hyperfine coupling between polarized electrons and the nuclei of the lattice leads to dynamic nuclear polarization by Overhauser effect [4] [27] . Conversely, the polarized nuclei create a magnetic field $\vec{H}_N$ along the direction of the electronic polarization, acting on the electronic spins. $\vec{H}_N$ may be in some experimental cases of the order of a few kG [28] [29] . When the electronic polarization P depends on $\vec{H}_N$, any change in the

nuclear polarization induced by a dc transverse field [30] [31] or a resonant r.f. field [31] [32] at the nuclear frequency, may cause a change in the electronic polarization P and will be detected by a corresponding change in the polarization of the luminescence (or in its intensity under conditions of spin dependent recombination [24]). Optical detection of NMR has been carefully studied in $Ga_x Al_{1-x} As$ [28] where the resonance of Ga^{69}, Ga^{71}, Al^{75} has been observed. The nuclear relaxation time was $\simeq$ 20 s, T_1 was $\simeq 10^{-9}$s. Recently NMR of Ga^{69}, Ga^{71} and As^{75} has also been optically detected in GaAs [33]. The coupling between nuclear and electronic spins has been discussed in [29] [34]. If the electronic relaxation time T_1 depends on the total magnetic field $\vec{H}_t = \vec{H}_o + \vec{H}_N$ ($\vec{H}_o$ is the external longitudinal field), a change in $\vec{H}_N$ will result in a change of $P=P_i T_1(\vec{H}_t)/(T_1(\vec{H}_t)+\tau)$. Moreover, when the direction of $\vec{H}_N$ is varied, $\vec{H}_N$ acts as a Hanle field for the electrons and the luminescence is partially depolarized. Let us point out that this last process is very general and should be present in every cases when large optical dynamic nuclear polarization occurs.

4. OPTICAL PUMPING OF EXCITONS

The fundamental characteristic of excitonic orientation is that wavefunctions of free excitons have <u>integer</u> angular momentum symmetries because they are formed from the coupling of electrons and holes with half integer angular momentum [35]. Absorption of a photon in a pure polarization state (that is a J = 1 pure quantum state) creates a corresponding pure excitonic J = 1 state. On the contrary, when free pairs are created, the loss of coherence between electrons and holes due to their separate evolution results generally in a statistical superposition [36]. As a consequence, optical pumping of free excitons can be obtained with linearly as well as circularly polarized light [35] [37]. This effect does not occur for individual carriers or complexes with an odd number of particles, such as excitons bound to shallow neutral impurities where no linear polarization has ever been observed [37]. In order to deal with excitons in pure quantum states and observe their specific

orientation properties, it is necessary to create them resonantly, without initial free carrier states. The direct absorption of resonant light would render quite impossible the observation of the excitonic luminescence at the same wavelength [14] . An elegant way to overcome these experimental difficulties is to choose a quite impure sample where most of the excitations decay through non-radiative paths. Only some elementary excitations are transferred from their initial state to their final radiative state through a limited number of resonant channels.
Up to now, only in CdS [14] [37] [38] has it been possible to meet such a situation. The experiments extensively discussed in [37] have provided a great number of results about various relaxation times and lifetimes of free and bound excitons and transfer mechanisms from primary excited states to luminescent ones. A careful analysis of the signs and amplitudes of the degree of polarization has been given, taking into account band-mixing and depolarization by longitudinal-transverse splitting for circularly and linearly polarized light.
Optical pumping of excitons has also been obtained in other non cubic semiconductors such as CdSe [13] and GaSe [39]. In pure cubic semiconductors orientations of free and bound excitons have been observed in CdTe [40] , GaAs [12] [26] [40] and InP [26] [40] [41] . In all these cases free electron hole pairs are primarily excited and the analysis of the polarization of the excitonic lines yields information about the formation of free excitons and complexes. To deal with the complicated experimental situation, a phenomenological density matrix formalism was introduced [36].

5. CONCLUSION

In an optical pumping experiment, measurement of the degree of polarization ρ and of the Hanle line-width ΔH provides very simply informations about lifetimes and spin relaxation times. Moreover a great variety of results in the physics of semiconductors have already been obtained by these techniques, including hot electron effects, surface effects, optical detection of electronic and nuclear resonances. It is to be hoped that

resonant excitonic orientation will help to clarify the decay processes of elementary excitations in semiconductors. Detection of excitonic magnetic resonance may give detailed informations about the structure of free or bound excitons. Furthermore, one can optically achieve important inversion of Zeeman populations in an external magnetic field $\vec{H}_o$ and this opens the possibility of tunable lasers in the far infra-red at frequencies $\nu = \gamma/(2\pi) H_o$.

6. ACKNOWLEDGMENTS

It is a pleasure to thank the members of the Polytechnique Optical Pumping group, G. Fishman, C. Hermann, D. Paget and C. Weisbuch for their collaboration and Professor Benoit à la Guillaume's group for many discussions. The difficult task of typing the manuscript has been kindly achieved by Mrs A.M.Hernec.

7. REFERENCES

[1] Kastler,A.: J. Phys. Radium 11, 255 (1950).
[2] Cohen-Tannoudji,C.; Kastler,A.: Progress in Optics (E. Wolf ed.) 5, 1 (1966).
[3] Margerie,J.: Proc. of the XIth Colloque Ampère, Bordeaux, France (1963).
[4] Lampel,G.: Phys. Rev. Lett. 20, 491 (1968).
[5] Parsons,R.R.: Phys. Rev. Lett. 23, 1152 (1969).
[6] Weisbuch,C.: Thèse de 3ème Cycle, Paris (1970).
[7] Ekimov, A.I.; Safarov,V.I.: JETP Lett. 12, 198 (1970). (Russian ed. 12, 293).
[8] Zakharchenya,B.I.; Fleisher,V.G.; Dzhioev,R.I.;Veshchunov, Yu.P.; Rusanov,I.B.: JETP Lett. 13, 137 (1971).(Russian ed. 13, 195).
[9] Zakharchenya,B.I.; Proc. of the XIth Int. Conf. on the Physics of Semiconductors, Warsaw, Poland (1972) p. 1315.
[10] Lampel,G.: Thèse d'Etat, Orsay (1968). Lampel,G.: Proc. of the Xth Int. Conf. on the Physics of Semiconductors, Moscow, USSR (1970) p. 1139.
[11] D'yakonov,M.I.; Perel',V.I.: Sov. Phys. JETP 33, 1053 (1971). (Russian ed. 60, 1954).
[12] Fishman,G.: Thèse d'Etat, Orsay (1974).
[13] Gross,E.F.; Ekimov,A.I.; Razbirin,B.S.; Safarov,V.I.: JETP Lett. 14, 70 (1971). (Russian ed. 14, 108).
[14] Bonnot,A.: Thèse d'Etat, Paris VII (1972).
[15] Parsons,R.R.: Can. J. Phys. 49, 1850 (1971).
[16] Garbuzov,D.Z.; Ekimov,A.I.; Safarov,V.I.: JETP Lett. 13, 24 (1971). (Russian ed. 13, 36).
[17] Bichard,R.; Lavallard,P.; Benoit à la Guillaume,C.: this Volume.
[18] Fishman,G.; Hermann,C.: Phys. Stat. Sol.(b) 63, 307 (1974).

[19] Parsons,R.R.: Proc. of the IX[th] Int. Conf. on the Physics of Semiconductors, Boston, USA (1970) p. 814. Parsons,R.R.: Can. J. Phys. 51, 718 (1973).

[20] Ekimov,A.I.; Safarov,V.I.: JETP Lett. 13, 495 (1971).(Russian ed. 13, 700).

[21] Dzhioev,R.I.; Zakharchenya,B.P.; Fleisher,V.G.: JETP Lett. 14, 381 (1971).(Russian ed. 14, 553).

[22] D'yakonov,M.I.; Perel',V.I.: Sov. Phys. Solid State 13, 3023 (1972).(Russian ed. 13, 3581).

[23] Solomon,I.: Ref. 9, p. 27.

[24] Weisbuch,C.: J. Phys. (Paris) 35, C3-21 (1974). Weisbuch,C.; Lampel,G.: Solid State Commun. 14, 141 (1974).

[25] Hermann,C.; Lampel,G.: Phys. Rev. Lett. 27, 373 (1971).

[26] Weisbuch,C.; Hermann,C.; Fishman,G.: this Volume.

[27] Overhauser,A.: Phys. Rev. 92, 411 (1953).

[28] Berkowits,V.L.; Ekimov,A.I.; Safarov,V.I.: Zh. Eksp. & Teor. Fiz. 65, 346 (1973).

[29] D'yakonov,M.I.; Perel,V.I.: Zh. Eksp. & Teor. Fiz. 65, 362 (1973).

[30] Ekimov,A.I.; Safarov,V.I.: JETP Lett. 15, 179 (1972).(Russian ed. 15, 257).

[31] Ekimov,A.I.; Safarov,V.I.: ref. 9, p. 1351.

[32] Ekimov,A.I.; Safarov,V.I.: JETP Lett. 15, 319 (1972).(Russian ed. 15, 453).

[33] Paget,D.: to be published.

[34] D'yakonov,M.I.; Perel',V.I.: Sov. Phys. JETP 36, 995 (1973). (Russian ed. 63, 1883).

[35] Bir,G.L.; Pikus,G.E.: JETP Lett. 15, 516 (1972).(Russian ed. 15, 730). Bir,G.L.; Pikus,G.E.: ref. 9, p. 1341.

[36] Fishman,G.; Hermann,C.; Lampel,G.: J. Phys. (Paris) 35, C3-13 (1974).

[37] Bonnot,A.; Planel,R.; Benoit à la Guillaume,C.: Phys. Rev. B9, 690 (1973). Benoit à la Guillaume,C.: J. Phys. (Paris) 35 C3-1 (1974).

[38] Bonnot,A., Planel,R.; Benoit à la Guillaume,C.; Lampel,G.: ref. 9, p. 1334.

[39] Veshchunov, Yu.P.; Zakharchenya,B.P.; Leonov,E.M.: Sov. Phys. Solid State 14, 2312 (1973).(Russian ed. 14, 2678).

[40] Fishman,G.; Hermann,C.; Lampel,G.; Weisbuch,C.: J. Phys. (Paris) 35, C3-7 (1974).

[41] Weisbuch,C.; Lampel,G.: ref. 9, p. 1327.

INTERIMPURITY LUMINESCENCE IN SILICON WITH HIGHLY SPIN-POLARIZED CARRIERS*

A. HONIG AND P. E. VANIER

Syracuse University, Syracuse, N. Y. 13210, U. S. A.

Luminescence associated with electron transfer from shallow donors
to shallow acceptors in silicon is analyzed with respect to elec-
tron and hole spin-polarization, interimpurity separation, and de-
lay time between optical excitation of the impurity pair system
and the observation of the luminescence. Spin-polarizations of up
to 97% are achieved at the low temperatures and high magnetic
fields employed, resulting in up to 90% quenching of the lumines-
cence. The hole g-factor is obtained in the absence of applied stress.

1. INTRODUCTION

Spin-polarization of the carriers in semiconductors gives rise to very
pronounced effects. For a number of years, the dependence on polarization, $\mathcal{P}$,
of the transport properties in silicon have been investigated, with $\mathcal{P}$ values
in excess of 99% achieved at readily available magnetic fields and low temper-
atures. As an example, group V neutral donor capture cross-sections [1] for
electrons depend on $\mathcal{P}$ because the bound state of the donor negative ion, D^-,
is a singlet (S=0), and hence no capture occurs when the neutral donors and
conduction electrons are fully spin-polarized ($\mathcal{P}$=1). Although the group III
acceptors are more complex than the group V donors in silicon, similar spin-
dependent selection rules apply, and the shallow donor-acceptor interimpurity
luminescence can be almost entirely quenched [2] when the bound donor elec-
trons and the bound acceptor holes are highly spin-polarized. We demonstrate
in this report that the rate of radiative decay can be controlled through the
values of the electron and hole spin-polarizations, $\mathcal{P}_e$ and $\mathcal{P}_h$, respectively,
and that as a result, new aspects of the luminescence process can be probed.
The understanding of the luminescence spectra as a function of magnetic field
strength, H, and temperature, T, allows the dependence of $\mathcal{P}_e$ and $\mathcal{P}_h$ on H and
T to be determined, and the hole g-factor to be calculated, without recourse
to resonance measurements. For silicon, this permits a determination of the
hole g-factor in the absence of external stress.

The observation of spin-polarization dependent luminescence has been re-
ported for a single line in GaP and CdS [3] at $\mathcal{P}$ values of $\sim$60%. Optical
pumping techniques have been used to achieve $\sim$70% polarization in Ga$_{.6}$Al$_{.4}$As,
and partial quenching of the luminescence was observed [4]. In the present
study, polarizations up to 97% are utilized, and the entire spectrum, compris-
ing a large range of donor-acceptor pair separations, is investigated. A pre-
vious extensive study [5] of the zero-polarization case provides a useful

* Research supported in part by the National Science Foundation.

frame of reference.

2. EXPERIMENTAL METHOD

The apparatus is similar in conceptual features to that described in [5]. The principal point of departure in the present system is that two separate fiber optics cables, on opposite sides of the $\sim$1mm thick polished sample, are used for the excitation light and the luminescence light. This necessitates use of two chopper wheels properly phased with respect to each other to control the delay time, t_d, between the end of the intrinsic light excitation and the average time of observation of the luminescence. The intrinsic excitation source was a 1 kw Xenon arc lamp, and a short-pass filter restricted the radiation to $\lambda < 900$ nm. A superconducting Nb-Ti solenoid provided H values up to 45 kOe, and the $\vec{H}$ direction was parallel to the sample [111] orientation. The silicon sample contained $6\times10^{16}/cm^3$ and $2\times10^{16}/cm^3$ phosphorus donor and boron acceptor concentrations, respectively.

3. DONOR-ACCEPTOR PAIR LUMINESCENCE IN SILICON

At zero polarization, impurity pair spectra have been observed and analyzed for most of the group V - group III donor-acceptor combinations [5]. For a given donor-acceptor pair separation, R, luminescence spectra appear at photon energies, E(R), given by

$$E(R) = E_g - (E_D + E_A) + e^2/\epsilon R - E_{phonon} , \qquad (1)$$

where E_g is the indirect band gap energy, E_D and E_A are respectively the donor and acceptor ionization energies, and $e^2/\epsilon R$ is the Coulomb energy of the ionized donor-acceptor pair, with ϵ the dielectric constant of silicon. E_{phonon} is the nearly monochromatic energy of emitted phonons of appropriate wavevector. The results to be presented are on the highest intensity phonon component, the 0.058 eV TO phonon assisted line, although similar results have been obtained with other phonon (and no-phonon) components. The Coulomb energy term broadens the lines, and luminescence is seen for pairs ranging in R value from about 3.0 nm (E=1.057 eV) to about 25.0 nm (E=1.020 eV). Since distant pairs undergo electron transfer more slowly than the close ones, the line width decreases and the line peak position moves toward lower energies as the delay time, t_d, between excitation (neutralization of charged impurities) and luminescence observation increases.

A quantitative theory based on an approximate model in which a given acceptor is surrounded by a large number of randomly located donors has been put forth [3], and when modified [5] to take into account the anisotropy of the donor wave function, accounts satisfactorily for the luminescence spectra in silicon [5]. For the moment, we neglect the anisotropy and rewrite the

derived [3] expression for the luminescence intensity per unit energy at energy E, time t_d, magnetic field strength H, and temperature T:

$$I_E(t_d;0,T) = (4\pi\epsilon/e^2)N(D)R^4(W_o e^{-2R/a})\ (e^{-W_o t_d\ \exp(-2R/a)})\ \langle Q(t_d)\rangle\ , \qquad (2)$$

where N(D) is the donor concentration, a is the donor Bohr radius, $W_o e^{-2R/a}$ is the electron transfer rate at impurity separation R, and R is related to E through Eq. (1). $Q(t_d)$, the probability of a given acceptor being neutral at time t_d, is given by

$$Q(t_d) = \exp\left[-\sum_j W(R_j)t_d\right]\ , \qquad (3)$$

reflecting the decay of $Q(t_d)$ via many channels of neighboring donors located at R_j with respect to the acceptor. $\langle\ \rangle$ in Eq.(2) denotes an ensemble average.

In the presence of a strong magnetic field, the donor electron and acceptor hole become substantially spin polarized. In Fig. 1, we consider their separate energy levels, assuming the presence of a small local strain. To compute the donor-acceptor transition rate, the electric dipole oscillator strength is multiplied by a factor($\leqslant 1$) which depends on the occupied magnetic states of electrons and holes. This factor is calculated under thermal equilibrium conditions for the almost equivalent case of GaP in [3]. It is

Fig.1. Bound electron and hole magnetic energy levels in Si.

straightforward to derive a somewhat simpler, but equivalent, expression in terms of separate electron and hole magnetic polarizations:

$$W(R;H,T) = W(R;0,T)(1 - P_e P_h) = W_o\exp(-2R/a)(1 - P_e P_h)\ , \qquad (4)$$

where equilibrium values of P_e and P_h are respectively defined as

$$P_e^{eq} = -\sum_j m_j(e)N(m_j^e)\ /\ \tfrac{1}{2}\sum_j N(m_j^e);\quad P_h^{eq} = -\sum_j m_j(h)N(m_j^h)\ /\ \tfrac{3}{2}\sum_j N(m_j^h)\ . \qquad (5)$$

In the absence of strain, P_e and P_h depend on their respective g-factors, g_e and g_h, and H/T. With strain present, Eq. (4) is still applicable, but the value of P_h at given H and T is modified. Even though W(R;0,T) is modified when account is taken of the donor wave-function anisotropy [5], the reduction factor due to the electron and hole spin-polarizations should remain the same, which is the justification for having conducted the above discussion in terms of the simpler isotropic model.

4. EXPERIMENTAL RESULTS

The luminescence spectra were taken over a range of H, T, and t_d. A

typical set of spectra is shown in Fig. 2, where 4.4 ms delayed luminescence is given for three H/T values and two temperatures. The field values at the two temperatures were chosen to yield the same H/T, and although the small difference between the two curves is real, the closeness of the two spectra indicates that the spectral changes are primarily due to spin-polarization rather than to explicit H or T dependence. At 40 kOe and 1.3K, where $\mathscr{P}_e$ = 97% and $\mathscr{P}_h$ is at least 85%, the ratio $I_E(t_d;H,T)/I_E(t_d;0,T)$ is seen to vary with E, resulting in a shift of the intensity peak position toward lower energies as H/T increases. The reason

Fig.2. Luminescence spectra at various H and T. t_d = 4.4 ms.

for this is given below; but actually, for reasons associated with the complicated dynamics due to the large spread in W(R;H,T) values, a better approach for analyzing the luminescence is available. Let us assume for the moment that all pairs have the same magnetic level structure and that the same $(1 - \mathscr{P}_e \mathscr{P}_h)$ attenuation factor applies to all W(R;0,T). Then, referring to Eqs. (2) and (3), for a t_d' equal to $t_d(1 - \mathscr{P}_e \mathscr{P}_h)^{-1}$, W(R;0,T)$t_d$ equals W(R;H,T)t_d', Q(t_d) = Q'(t_d'), and $I_E(t_d;0,T)$ should be equal to $I_E(t_d';H,T)\dfrac{t_d'}{t_d}$.

Thus, the spectral shapes for (0,T) at t_d should be identical to those for (H,T) at t_d', but the intensity should be greater by t_d'/t_d.

In Fig. 3, the superposed spectra for t_d and t_d' are shown, with the values of H and T chosen so that $(1 - \mathscr{P}_e \mathscr{P}_h)^{-1}$ is near the t_d'/t_d ratio of 4. If the assumptions we made are applicable, the t_d' spectrum multiplied by t_d'/t_d (=4) should coincide with the t_d spectrum , if $(1 - \mathscr{P}_e \mathscr{P}_h)$ equals 1/4. We see that the high energy portion of the spectra agree fairly well, but the low energy portion, corresponding to distant pairs, does not agree at all. We attribute these results to short spin-lattice relaxation times, T_1, of donors belonging to close pairs [6],

g.3. Superposed luminescence spectra at $(t_d;0,T)$ and $(t_d';H,T)$. t_d'/t_d = 4.

leading to thermal equilibrium values of $\mathcal{P}_e$, and much longer T_1 values for the donors from distant pairs, leading to large departures from thermal equilibrium. The acceptor holes have short T_1's and are very probably in equilibrium. The intrinsic light excitation is known to produce a high spin temperature, and in the 4.4 ms prior to observation of luminescence, the donors from distant pairs, expected to have T_1 of about 0.5 sec [7], would not recover polarization, but the ones from close pairs probably would. These experiments, in fact, should yield T_1 values for donors as a function of the donor separation from a nearby neutral acceptor.

The fit at 34 kOe for the t_d and t_d' spectra at high E appears best for a multiplication of the t_d' spectrum by about 4.75, rather than 4. A 30 kOe t_d' spectrum (not shown) was found to fit best for a multiplying factor of about 3. Thus, the shape discrimination is fairly sensitive. Using a mean value of 32 kOe, the g_h calculated from Eq. (5), neglecting strain, is 0.90 ±0.05. The g_h can be obtained in another way which also tests the validity of Eq. (4). At E = 1.143 eV, R is 5.0 nm, and $W(5.0nm;H,T)^{-1}$ is greater than 10 ms [5]. Therefore, the luminescence intensity at 4.4 ms delay should be proportional to $W(5.0nm;H,T)$, or to $(1 - \mathcal{P}_e \mathcal{P}_h)$. In Fig. 4,

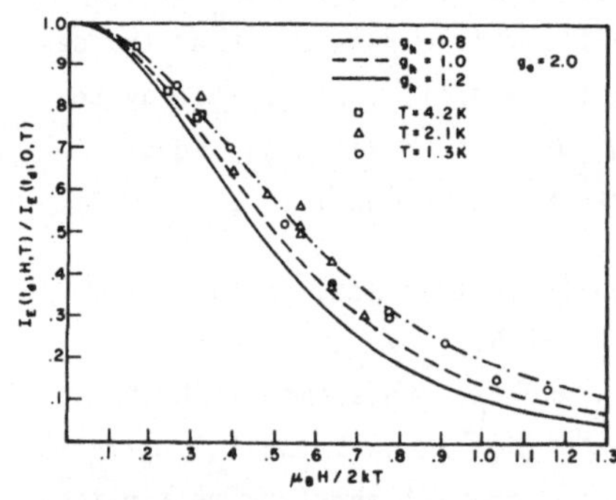

$I_{1.143}(4.4ms;H,T)$ as a function of H/T is illustrated for temperatures between 1 and 4K and H values up to 45 kOe. The theoretical curve which fits best corresponds to $g_h = 0.85\pm0.05$, in good agreement with the value derived from fitting the line shape in Fig. 3.

Fig.4. Luminescence intensity vs. H/T. t_d=4.4 ms. Curves based on Eq.(4).

5. CONCLUSIONS

The quenching of interimpurity luminescence in silicon has been demonstrated over a spin-polarization range up to 97%. The donors belonging to close pairs achieve thermal equilibrium, which results in strong quenching, and enables one to determine the hole g-factor.

6. REFERENCES

[1] D. D. Thornton and A. Honig, Phys. Rev. Lett. 30, 909 (1973).
[2] P. E. Vanier and A. Honig, Bull. Am. Phys. Soc. 19, 92 (1974).
[3] D.G.Thomas,J.J.Hopfield and W.M.Augustyniak, Phys.Rev.140,A202(1965).
[4] G. Weisbuch and G. Lampel, Sol. St. Comm. 14, 141 (1974).
[5] R. C. Enck and A. Honig, Phys. Rev. 177, 1182 (1969).
[6] G. Yang and A. Honig, Phys. Rev. 168, 271 (1968).
[7] A. Honig and E. Stupp, Phys. Rev. 117, 69 (1960).

SPIN-DEPENDENT CONDUCTIVITY AND MICROWAVE HOT ELECTRON EFFECT
NEAR THE METAL-NONMETAL TRANSITION OF HEAVILY DOPED SILICON

N. KISHIMOTO, K. MORIGAKI, and M. ONDA

Institute for Solid State Physics, University of Tokyo
Roppongi, Tokyo 106, Japan

Resistivity changes due to ESR and also due to
microwave irradiation without ESR have been obser-
ved in heavily phosphorus-doped silicon. The
observed results are discussed in connection with
the electronic states involved near the metal-
nonmetal transition.

1. INTRODUCTION

We have done the measurements of the spin-dependent conductivity
and also the change in the electrical resistivity due to micro-
wave irradiation(the microwave hot electron effect) in phospho-
rus doped silicon whose donor concentration ranges from 1.8 X
$10^{18}cm^{-3}$ to 1.8 X $10^{19}cm^{-3}$. The purpose of this work is to ob-
tain information about the mechanism underlying the metal-non
metal(M - NM) transition and also the electronic states involved
near the transition in heavily doped semiconductors.

2. EXPERIMENTAL

For samples with $N_D <$ 3.7 X $10^{18}cm^{-3}$, the resistivity change due
to ESR at 9 GHz was measured with a technique described in [1].
For samples with $N_D \geq$ 3.7 X $10^{18}cm^{-3}$, in order to detect a small
change in the resistivity due to ESR, we used the modulation
technique in which the microwave power is modulated in a square
waveform with the frequency of 100 KHz and the phase-sensitive
detection method is employed. This technique was also applied
to measure the change in the resistivity due to microwave irra-
diation at 9 GHz without ESR.

3. RESULTS AND DISCUSSION

3.1 SPIN-DEPENDENT CONDUCTIVITY MEASUREMENT

As was shown in our previous experiments[2,3], the resistivity
decreases with ESR in samples belonging to the intermediate con-
centration region(1.5 X $10^{18}cm^{-3} \lesssim N_D <$ 3.7 X $10^{18}cm^{-3}$). Figure 1
shows the magnitude of the resistivity decreases due to ESR,

FIG. 1: $(\Delta\rho/\rho)_{ESR}$ plotted as a function of N_D at 1.5 K and at the microwave power level of -12 dB for Si:P.

$(\Delta\rho/\rho)_{ESR}$, as a function of donor concentration, N_D, at 1.5 K and at the microwave power level of -12 dB (the microwave power at 0 dB is about 100mW) in P-doped silicon. These results have been interpreted in terms of the spin energy transfer model as follows; two types of electron systems, that is, the donor spin system responsible for ESR and the mobile electron system contributing to the charge transport exist in the crystal and the spin energy transfer from the former system to the latter one causes the resistivity to decrease with ESR. In a previous work [1,2], the case that the mobile electron system is non-degenerate has theoretically been treated. But, when N_D exceeds about $3 \times 10^{18} cm^{-3}$, the number of the mobile electrons rapidly increases with N_D, so that they become degenerate at 1.5 K. We have extended the previous treatment of $(\Delta\rho/\rho)_{ESR}$ to the degenerate case. On taking into account that ESR arises from a coupled of the above two electron systems in the donor concentration of the present concerns and also on assuming the relation between the mobility μ and the electron temperature T_e such as $\mu \propto T_e^a$ $(a > 0)$, we obtain

$$(\Delta\rho/\rho)_{ESR} = -\frac{3a}{\pi^2}(\chi_d + \chi_e) H^2 \frac{s}{k_B^2 T_e^2 N(E_F)} (1/T_{1eff})/(1/\tau), \tag{1}$$

where χ_d and χ_e are the paramagnetic susceptibilities of the donor spin system and the mobile electron system respectively, H, k_B, $N(E_F)$, τ, and s designate respectively the strength of resonant magnetic field, the Boltzmann constant, the density of states at the Fermi level, the relaxation time of the kinetic energy of mobile electrons, and the saturation factor given by eq.(8) of [1], and T_{1eff} is approximately given by [4]

$$(1/ T_{1eff}) = \cdot (1/\chi_r)(1/ T_{1e}), \qquad (2)$$

where χ_r and T_{1e} designate χ_d/χ_e and the spin-lattice relaxa-time of the mobile electron respectively.

As seen from Fig.1, the magnitude of $(\Delta\rho/\rho)_{ESR}$ rapidly decreases with N_D just before the M-NM transition occurs. This feature can be qualitatively understood in terms of eq.(1), considering the variation of the parameters involved in eq.(1) with N_D. In a quantitative sense, we can explain the magnitude of $(\Delta\rho/\rho)_{ESR}$ for a sample with $N_D = 3.02 \times 10^{18} cm^{-3}$ if we take as the value of $N(E_F)$ that of two order of magnitudes smaller than that for the free electron model. However, as has been pointed out in [3], the magnitude of $(\Delta\rho/\rho)_{ESR}$ will be reduced from that calculated by eq.(1) with the relation of $\mu \propto T_e^a$. Probably in samples whose N_D is close to the critical concentration of the M-NM transition, the enhancement of the variable-range hopping type conductivity due to ESR [5] should be taken into account to understand a quantitative behaviour of $(\Delta\rho/\rho)_{ESR}$.

For the metallic region, we have looked for a change in the resistivity due to ESR to obtain information about the s-d type scattering responsible for the negative magnetoresistance effect observed in this region[6]. But, we could not observe any change in the resistivity due to ESR at donor concentrations ranging from $4 \times 10^{18} cm^{-3}$ to $1.8 \times 10^{19} cm^{-3}$ within the sensitivity of our apparatus of 10^{-7}.

3.2 MICROWAVE HOT ELECTRON EFFECT

The resistivity change due to microwave irradiation without ESR at zero magnetic field has been measured in liquid helium temperature range. For samples belonging to the intermediate concentration region, the resistivity always decreases with microwave irradiation, but at $N_D = 3.7 \times 10^{18} cm^{-3}$ it increases in the whole region of static electric field used. But, in the metallic region, the resistivity increase due to microwave irradiation occurs in higher static electric fields, while the resistivity decrease again appears in lower static electric fields, as shown in Fig.2. The critical static electric field for transformation from the resistivity decrease $(\Delta\rho/\rho < 0)$ to the resistivity inc-

FIG. 2: $(\Delta\rho/\rho)$ due to micro-
wave irradiation plotted as a
function of static electric
field at the microwave power
level of -6.3 dB for sample N_D
$=5.7 \times 10^{18} cm^{-3}$. Curves with
$(+)$ and $(-)$ correspond to $(\Delta\rho/\rho)$
> 0 and $(\Delta\rho/\rho) < 0$ respectively.

FIG. 3: Concentration depen-
dence of the magnitude of the
resistivity increase taken by
its maximum value for static
electric field at 1.5 K and
at the microwave power level
of 0 dB.

rease $(\Delta\rho/\rho > 0)$ increases with N_D. Finally, at N_D=1.8 X 10^{19}
cm^{-3}, only the resistivity decrease component appears in the
whole region of static electric field used. Figure 3 shows the
concentration dependence of the magnitude of the resistivity
increase taken by its maximum value for static electric field.
As seen from this figure, the resistivity increase component
suddenly appears near the M-NM transition. We note that the
appearance of the resistivity increase component is only limited
to the metallic region where the Fermi level lies in the impuri-
ty band. In P-doped silicon, it is known that the Fermi level
enters into the conduction band of host silicon at N_D=2 X 10^{19}
cm^{-3}[7].

The above complicated behaviour of the resistivity change due to
microwave irradiation is interpreted in terms of the localized
or delocalized characters of the impurity band and the scatter-
ing mechanism relevant to the metallic region. Just after the
M-NM transition occurs with N_D, the impurity band is assumed to
be rather narrow and separated from the conduction band. The
Fermi level lies in the middle position of the impurity band.

With microwave irradiation, some of electrons are excited from
the Fermi level to the higher energy region of the impurity band
in which the mobility is lower than in the middle position be-
cause of the localized character. Thus, it seems to us that
the microwave irradiation causes the resistivity to increase.
As N_D increases, the impurity band is broadened and overlapped
into the conduction band, so that the resistivity increase com-
ponent due to microwave irradiation becomes weak with N_D, as
shown in Fig.3.

4. REFERENCES

[1] Morigaki, K.; Onda, M.: J. Phys. Soc. Japan 33, 1031
(1972)

[2] Morigaki, K.; Toyotomi, S.: J. Phys. Soc. Japan 30, 1207
(1971)

[3] Morigaki, K.; Onda, M.: J. Phys. Soc. Japan 36, 1049
(1974)

[4] Hasegawa, H.: Progr. Theoret. Phys. 21, 483 (1959);
Ue, H.; Maekawa, S.: Phys. Rev. 3, 4232 (1971)

[5] Mott, N. F.: Advance in Phys. 21, 785 (1972)

[6] Yamanouchi, C; Mizuguchi, K.; Sasaki, W.: J. Phys. Soc.
Japan 22, 859 (1967)

[7] Alexander, M. N.; Holcomb, D. F.: Rev. Mod. Phys. 40,
815 (1968)

DYNAMICS OF EXCITONIC COMPLEXES AND DETECTION OF
ELECTRON SPIN RESONANCE BY OPTICAL SPIN ORIENTATION TECHNIQUES

C. WEISBUCH, C. HERMANN

Laboratoire P.M.C., Ecole Polytechnique, Paris, France

G. FISHMAN

Groupe de Physique des Solides de l'E.N.S., Paris, France

Analysis of spin orientation by optical pumping in pure GaAs and InP shows that, although photoluminescence is dominated by excitonic lines, many features of the spin memory are due to the conduction electrons. We are thus able to study the formation process of excitonic complexes and to perform the optical detection of conduction electron spin resonance.

1. INTRODUCTION

Spin orientation by optical pumping [1] in pure GaAs and InP is of special interest because of the situation prevailing at low temperatures : although free electron-hole pairs are created by light of energy larger than the bandgap, the photoluminescence is not mainly due to free carriers, but rather to free or bound excitons. The comparison between the spin polarization of the complexes and that of free carriers gives information on the dynamics and mechanism of formation of such complexes. Another remarkable feature of our experiments is the sensitivity of the optical detection of electron spin resonance which permits the measurement of the g^* factor of conduction electrons in GaAs and InP.

2. PHOTOLUMINESCENCE AND POLARIZATION

Figure 1 shows the photoluminescence spectrum $(L_{F+} + L_{F-})$ and the polarization spectrum $(L_{F+} - L_{F-})$ of GaAs at 1.7 K excited by 10^{19} photons/cm^2.s at 1.64 eV. $L_{F+}(L_{F-})$ is the intensity of luminescence with σ_+ (σ_-) circular polarization under σ_+ excitation. The results are similar in InP. The identifi-

cation [2] of the various lines is given in Fig. 1. A complete
discussion of polarization resúlts will be presented elsewhere[3],

FIG. 1 : Photoluminescence and polarization spectra of GaAs.
Note the change in photon energy scale. The indicated lines are
the following : X : free exciton ; D°-X : exciton bound to a
neutral donor ; D°-h : free hole-neutral donor ; A°-X : exci-
ton bound to a neutral acceptor ; e-A° : free electron-neutral
acceptor ; D°-A°: donor-acceptor pairs ; X-LO : phonon replica
of X.

but one puzzling result can already be noted : the polariza-
tion of the e-A° line, which reflects the polarization of free
electrons [4], is much smaller than that of the excitonic lines
and of the D°-h line. One should normally find a reverse si-
tuation, because the complexes are formed from free electrons
and should retain less spin memory. Our results may be due to

a hot electron capture mechanism like that invoked by Ulbrich[5] in the case of donors : hot electrons are preferentially captured by impurities or by holes to form excitons. These non-thermalized electrons have a shorter lifetime than the thermalized electrons of the e-A° line and thus retain a larger spin-memory.

The difference in polarization of the various lines can have two origins : (i) the transfer of polarization from free electrons to bound complexes must be analyzed in terms of density matrix to account for the coherence between interacting electrons and holes. It can then be shown that this transfer varies from one complex to another because of the different symmetries of the wavefunctions [6]. (ii) After complex formation, there could be some depolarization. This must be ruled out in the present case because the spin relaxation time of complexes is of the order of a few nanoseconds or more (see below), while their lifetime is shorter [7]: once formed, the complexes recombine before any spin relaxation has time to occur [8].

3. TRANSVERSE MAGNETIC FIELD EFFECT

The application of a transverse magnetic field causes spin depolarization because of precession [4]. The half-height width ΔH of the depolarization curve is given by

$\gamma \Delta H = (1/T_1 + 1/\tau)$, where γ is the gyromagnetic factor, T_1 the spin relaxation time and τ the lifetime. All the luminescence lines have the same magnetic width ΔH at a given intensity. This shows clearly that the transverse magnetic field depolarization occurs during the step of formation which is common to all the lines observed, that is when the electrons are free. It is also the longest step in the process :

EXCITATION → FREE CARRIERS → COMPLEXES → RECOMBINATION because it is the only one during which significant precession has time to occur. From the analysis [1] of the degree of circular polarization $\rho = (L_{F+} - L_{F-})/(L_{F+} + L_{F-})$ and of ΔH one deduces that T_1 is of the order of nanoseconds and τ of tens of nanoseconds for free electrons.

Figure 2 shows the variation of ρ and ΔH with light intensity. ΔH is mainly determined by the shortest of the two times τ and T_1, which is here T_1, the smallness of ρ indicating that significant depolarization occurs on free electrons before they get bound. The variation of ΔH shows that T_1 decreases with increasing intensity, and the roughly similar variation of ρ shows that τ decreases even more rapidly. A possible explanation for the variation of T_1 is that the spin relaxation is determined by the exchange interaction of electrons with photoexcited holes [8]

FIG. 2 : Degree of circular polarization and half-height depolarizing transverse magnetic field as a function of excitation intensity. D°-h line was chosen as an example for all near-edge lines which have the same behaviour. Unit excitation is 5.10^{19} photons/cm^2.s.

4. ELECTRON SPIN RESONANCE RESULTS

A longitudinal magnetic field splits the spin levels. By applying a transverse r.f. field, we induce transitions between the levels, and we observe a resonant decrease of the polarization [10] as shown in Fig. 3 for InP. An important confir-

FIG. 3 : Resonant decrease of the polarization of the D°-X line by conduction electron spin resonance in InP. Resonances at the same field of 86 G are also observed on the other luminescence lines. Signal averaging time is two hours.

mation of the formation model described above is that the resonance occurs at the same field independently of the luminescence line under observation : the r.f. field is strong enough to induce a spin-flip only during the longest step of the formation process common to all lines. This implies that the resonance we perform is that of the conduction electrons. The resonant depolarization of the complexes once formed would require much larger r.f. fields because of the shorter times involved, but we would then observe different g^* values for the different lines.

We have measured $g^*(InP) = 1.26 \pm 0.05$ and $g^*(GaAs) = 0.51 \pm 0.03$. The result for InP is the first resonant determination of g^* and is in fair agreement with $\vec{k}.\vec{p}$ perturbation theory $g^* = 1.2$ [11]. For GaAs our measurement has to be compared to $g^* = 0.52$ obtained by ordinary E.P.R. techniques in doped samples where an impurity band could be present [12]. Our resonant determinations of g^* are in very good agreement with those obtained from the Zeeman splitting of the A°-X line for Sn doped GaAs [13] and Bi doped InP [14].

5. ACKNOWLEDGMENTS

It is a pleasure to thank Drs Dean and White (RRE), and Dr Cardwell (Plessey Company) for the gift of very pure samples and Dr Lampel for many discussions and suggestions.

6. REFERENCES

[1] Lampel, G.: this volume p.
[2] See, for instance : White, A.M.; Dean,P.J.; Taylor,L.L.; Clarke,R.C.; Ashen,D.J.; Mullin,J.B.: J. Phys. C5,1727(1972)
[3] Weisbuch,C.; Hermann,C.; Fishman,G.: to be published.
[4] Parsons,R.R.: Phys. Rev. Lett. 23, 1152 (1969).
[5] Ulbrich,R.: Phys. Rev. Lett. 27, 1512 (1971).
[6] Fishman,G.; Hermann,C.; Lampel G.: J. Phys.(Paris) 35, C3-13
[7] Hwang,C.J.: Phys. Rev. B8, 646 (1973).
[8] Fishman,G.; Hermann,C.; Lampel,G.; Weisbuch,C.: J. Phys. (Paris) 35, C3-7 (1974).
[9] Bir,G.L.; Pikus,G.E.: private communication.
[10] Hermann,C.; Lampel,G.: Phys. Rev. Lett. 27, 373 (1971).
[11] Lawaetz,P.: Phys. Rev. B4, 3460 (1971).
[12] Duncan,W.; Schneider,E.E.: Phys. Lett. 7, 23 (1963).
[13] White,A.M.; Hinchliffe,I.; Dean,P.J.; Greene,P.D.: Solid State Commun. 10, 497 (1972).
[14] White,A.M.; Dean,P.J.; Fairhurst,K.M.; Bardsley,W.; Day,B.: J. Phys. C7, L35 (1974).

OPTICAL PUMPING OF A DEGENERATED DISTRIBUTION OF ELECTRONS IN InSb

R. BICHARD, P. LAVALLARD, C. BENOIT A LA GUILLAUME

Groupe de Physique des Solides de l'E.N.S.[+]
Paris, France

Experiments of spin orientation by optical pumping in InSb ($p = 2 \times 10^{15}$ cm^{-3}) with a CO laser are presented. Agreement with experimental spectra is obtained when the two spin gases are described by quasi Fermi levels and slightly different temperatures. The relevant mechanism of spin relaxation is shown to be a scattering by injected holes.

1. INTRODUCTION

Spin orientation of electrons was studied recently in many semiconductors |1|. The case of InSb is of special interest since in this material a degeneracy of the photo-electrons distribution is easily obtained at low temperature. In our experiments, the source is a CW CO laser cooled with liquid nitrogen. It operates in a single line. By varying the energy of this line near the band gap of InSb, we change both the quasi Fermi level energy and the effective temperature of the electrons.

2. EXPERIMENTAL RESULTS

Figure 1a shows the photoluminescence spectrum of a sample of InSb ($p = 2 \times 10^{15}$ cm^{-3}) immersed in pumped liquid helium. The excitation line is seen at 237.4 meV. Spin orientation of electrons is observed when the sample is excited with circularly polarized light and the recombination light analyzed with a circular or linear polarizer. The effect of a sufficiently high transverse magnetic field is a complete spin depolarization due to precession (Hanle effect). We use a square wave modulated magnetic field and detect the intensity of polarized luminescence at the same frequency. Figure 1b shows the spectra we

+ Laboratoire associé au C.N.R.S..

obtained with σ_+ and σ_- circular polarization under σ_+ excitation. The difference $(L_{F+} - L_{F-})$ between intensities of σ_+ and σ_- components of the luminescence is not proportional to the intensity of the recombination line as it would be in the case of a Boltzmann distribution of electrons (Fig. 1a). The curve peaks on the high energy side of the recombination line.

FIG. 1: Positions of crosses are calculated with fitting parameters.
a : Solid curve is the luminescence spectrum. Note the CO line at 237.4 meV. The dotted curve is the difference $(L_{F+} - L_{F-})$ between σ_+ and σ_- components of the luminescence.
b : $\left(L_{F+}(H=0) - L_{F+}(H\infty)\right)$ and $\left(L_{F-}(H=0) - L_{F-}(H\infty)\right)$. These spectra are obtained through a modulation of the spin polarization with a transverse magnetic field.

3. INTERPRETATION

We suppose that the energy relaxation mechanism due to electron-electron collision is efficient enough to legitimate the use of an effective electronic temperature. In the steady state and without transverse magnetic field, the distribution of electrons is described by two quasi Fermi levels E_{F+} and E_{F-}, one for each spin gas. With a transverse magnetic field, high enough to destroy the spin polarization, the distribution is described by only one quasi Fermi level E_{FO}. The differences between the recombination light intensities at zero and "infinite" magnetic field are respectively expressed for σ_+ and σ_- polarization by :

$$L_{F+}(0) - L_F(H\infty) \simeq \exp - \frac{E}{\sigma kT_h} E^{1/2} \left[3F(E_{F-}) + F(E_{F+}) - 4F(E_{F0}) \right]$$

$$L_{F-}(0) - L_F(H\infty) \simeq \exp - \frac{E}{\sigma kT_h} E^{1/2} \left[3F(E_{F+}) + F(E_{F-}) - 4F(E_{F0}) \right]$$

where $\sigma = \dfrac{m_e}{m_h}$ is the mass ratio of electrons and holes,
$E = \dfrac{\hbar\omega - E_g}{1 + \sigma}$ the kinetic energy of electrons, and kT_h the hole temperature. F is the Fermi-Dirac function.

If the electronic temperatures are equal in the three distributions and if the differences $(E_{F-} - E_{F0})$ and $(E_{F0} - E_{F+})$ are much less than E_{F0}, then the function $(L_{F+} - L_{F-})$ is roughly the derivative of a Fermi-Dirac function. Qualitatively, our experimental results are in good agreement with this conclusion. Still, a "negative polarization" can be seen on Fig. 1a on the low energy side of the spectrum. That fact cannot be explained by this model. A slight difference between electronic temperatures could explain the observed spectra. We tried to fit experimental and theoretical curves with the use of three different temperatures, kT_+, kT_- and kT_0. The results are presented on Fig. 1 for the following values of the parameters :

$$kT_0 = 0.34 \pm 0.02 \text{ meV} ; \quad E_{F0} + E_g = 236.05 \pm 0.05 \text{ meV}$$

$$\Delta E_{F-} \simeq -\Delta E_{F+} = 0.013 \pm 0.005 \text{ meV} ; \quad \frac{\Delta T_-}{T_0} = (0.7 \pm 0.2)\times 10^{-2} ;$$

$$\frac{\Delta T_+}{T_0} = -(0.4 \pm 0.1)\times 10^{-1}.$$

The values of the spin relaxation time T_1 and lifetime τ are determined by the width of the depolarization curve

$$\gamma_e \Delta H = \frac{1}{\tau} + \frac{1}{T_1}$$

where the gyromagnetic ratio γ_e in InSb is $4.4 \ 10^8 \ G^{-1} \ sec.^{-1}$, and by the degree of circular polarization

$$\rho = \frac{L_{F+}(0) - L_{F-}(0)}{L_F(H = \infty)} = 0.25 \ \frac{T_1}{T_1 + \tau}$$

ρ is determined by the ratio of the area under the two curves of

figure 1a. With the experimental conditions of figure 1, we found :

$$T_1 = 2.7 \times 10^{-9} \text{ s and } \tau = 8 \times 10^{-8} \text{ s.}$$

We did not fit experimental and theoretical curves below a photon energy of 235 meV because there is no theory available for the low energy side of the recombination line. If this low energy tail is not taken into account, the degree of circular polarization is enhanced by a factor 2.

4. SPIN RELAXATION MECHANISM

Because of the degeneracy of the electron distribution the contribution of electron-electron collisions |2| to momentum and spin relaxation is weaker than the contribution of electron-hole collisions.

FIG. 2: Square root of the inverse of the spin relaxation time T_1 versus Fermi level. We vary the Fermi level by changing the energy of the CO line.

We evaluated the magnitude of the spin relaxation time by electron-hole scattering. The calculus is done by analogy with the scattering on ionized impurities |3| :

$$\frac{1}{T_1} \simeq \frac{1}{\tau_k} \frac{2}{3} \left(\frac{\gamma E_F}{E_g} \right)^2$$

γ is a function of the spin-orbit splitting Δ and of the band gap E_g. The inverse of momentum relaxation time τ_k is proportional to the ratio of the ionized impurities and electronic densities, $\frac{n_i}{n}$.

For electron-hole scattering, we replace n_i by $p \simeq n$ and find $\frac{1}{\tau_k}$ constant. As a consequence, $\frac{1}{T_1}$ is proportional to E_F^2. One can see on figure 2 a good agreement with this variation. Taking $\frac{1}{\tau_k} = 2 \times 10^{12}$ sec.$^{-1}$ |3|, we find $\frac{1}{\gamma_e T_1} = 0.2$ G for $E_F = 2.1$ meV, which is to be compared to the experimental

value :

$$\frac{1}{\gamma_e T_1} = 0.8 \ G$$

We may note that T_1 is the spin relaxation time at the Fermi level since there is almost no loss of energy in an electron-hole collision.

5. DISCUSSION

The use of an effective electronic temperature to describe the distribution of photo-excited electrons assumes that inter-electronic collisions control energy exchanges. The temperature of electrons is then determined by the interplay between the power supplied to them by the exciting light and the power they lose to the lattice.

The same mechanism (electron-electron collisions) which ensures thermalization of one gas of electrons would ensure the thermalization between the two gases. We are able to put into evidence a very weak difference of temperature $\frac{T_- - T_+}{T_o} \approx 5 \ \%$. That shows the limit of validity for the use of a Fermi-Dirac function to describe the electronic distribution.

6. ACKNOWLEDGEMENTS

It is a pleasure to thank Dr. Henry and his laboratory for helping us to build the CO laser. This study has benefited from many discussions and suggestions of Dr. Lampel.

7. REFERENCES

|1| Lampel, G.: this volume p.
 2 Pines, D.: <u>Solid State Physics</u>, Vol. 1 (Seitz and Turnbull, Academic Press, 1955), p. 415.
|3| Chazalviel, J.N.: to be published.

STRONG MAGNETIC FIELDS OF OPTICALLY POLARIZED
NUCLEAR SPINS IN SEMICONDUCTORS

M.I. DYAKONOV, V.I. PEREL,
V.L. BERKOVITS, A.I. EKIMOV, V.I. SAFAROV

A.F. Ioffe Physical-Technical Institute,
Academy of Sciences of the USSR, Leningrad, USSR

Pronounced optical effects due to nuclear
polarization by optically pumped spin-
oriented electrons in a semiconductor were
investigated.

1. Spin-oriented electrons created in a semiconductor by
circularly polarized light induce strong dynamic polarization
of the lattice nuclei[1-3]. Due to the hyperfine interaction
the polarized nuclei produce a strong effective magnetic field
$\vec{H}_N$ acting on the electron spins.

$$\vec{H}_N = (\mu_0 g)^{-1} \sum_{\infty} A_\alpha <\vec{I}_\alpha> \tag{1}$$

where $\mu_0 g$ is the electronic gyromagnetic ratio, A_α is the
hfs constant, $<\vec{I}_\alpha>$ is the mean spin vector of the nuclear
species α, the sum goes over the nuclei in the unit cell.
If isotops are present, their natural abundance should be
taken into account. The nuclear field $\vec{H}_N$ acts on the electro-
nic spins in the same way as the external magnetic field $\vec{H}_0$.
Thus it should manifest itself in all spin-dependent phenomena
which are sensitive to the magnetic field.

2. An example of such a phenomenon provides the electronic
spin relaxation which may be slowed down essentially in a
longitudinal magnetic field. In experiments on optical
pumping of spin-polarized electrons in semiconductors[4] this
leads to an increase of the stationary degree of orientation
resulting in a corresponding increase of the degree of the
circular polarization of the luminescence P [2,3,5]. The
nuclear field $\vec{H}_N$ weakens or intensifies the effect of the
external magnetic field $\vec{H}_0$ depending on their mutual orient-
ation[2,3]. The dependence $P(H_0)$ for weakly doped crystal
$Ga_{0.7}Al_{0.3}As$ under excitation by σ^+ and σ^- circularly

polarized light is depicted in Fig.1. The difference between curves 1 and 2 is caused by the nuclear field, which is antyparallel to the external field for curve 1, and parallel to it for curve 2. The sign of the nuclear field is determined by the direction of the electronic spin orientation which has opposite signs for σ^+ and σ^- excitation. After the excitation is switched on it takes some time (of the order of 10-100sec. depending on the light intensity and magnetic field) for the stationary value of $\mathcal{P}$ to be reached. The values of $\mathcal{P}$, obtained immediatly after the light was switched on,are present in Fig.1 (dotted line). These values correspond evidently to the absence of the nuclear field ($H_N=0$).

One can see from Fig.1 that the nuclear field reaches the value of about 5 kGs. The nuclear polarization, created by light was conserved in strong fields for about one hour after the excitation was switched out. Under NMR conditions pronounced variations of the luminescence polarization due to variations of the direction and magnitude of the nuclear field were also investigated[2,3].

3. Strong magnetic field of polarized nuclei manifests itself also in the incomplete Hanle effect (depolarization of electrons by a magnetic field directed under some angle θ to the exciting light beam, $0 < \theta < 90^\circ$). In this case the degree of circular polarization of the luminescence,observed in the direction of the excitation has the following dependence on the magnetic field H

$$P(H) = P(0)\left[\frac{\sin^2\theta}{1 + (g\mu_0 H\tau/\hbar)^2} + \cos^2\theta\right] \tag{2}$$

were τ is the decay time of the electron orientation due to recombination and spin relaxation. If nuclear polarization exists $H = H_0 + H_N$. The nuclear field $\vec{H}_N$ can be regarded as colinear to $\vec{H}_0$. In our experiments the value of the depolarizing field for nuclei was much lesser then for electrons. Thus the component of the mean nuclear spin transverse to $\vec{H}_0$ disappears alredy in such small fields , which do not affect the electron orientation. The longitudinal to $\vec{H}_0$ component of nuclear spin is conserved and contributes to the

Fig.1. The degree of circular polarization of the luminescence $\mathcal{P}$ as a function of the longitudinal magnetic field H_o. $Ga_{0,7}Al_{0,3}As$. $p \sim 10^{16} cm^{-3}$. $T=4,2^oK$

1,2 - excitation by σ^+ and σ^- circularly polarized light. Dotted line - see main text.

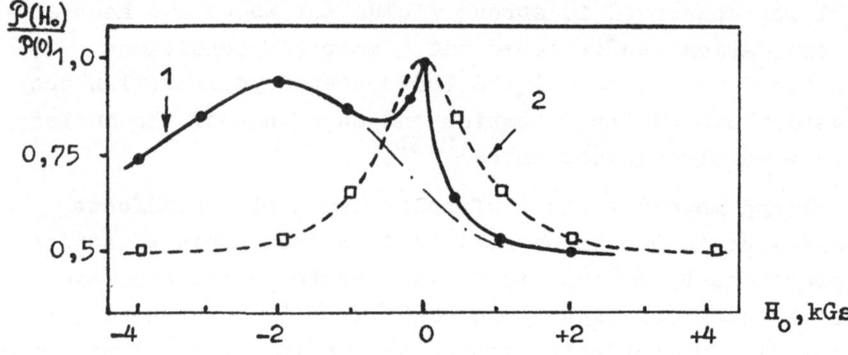

Fig.2. The degree of circular polarization of the luminescence $\mathcal{P}$ as a function of the external field H_o in case of incomplete Hanle effect. $Ga_{0,7}Al_{0,3}As$. $p \simeq 9 \cdot 10^{18} cm^{-3}$. $T=4,2^oK$. $\theta = 45^o$.

1 - stationary excitation by σ^+ polarized light, 2 - the excitation by light with alternating circular polarization.

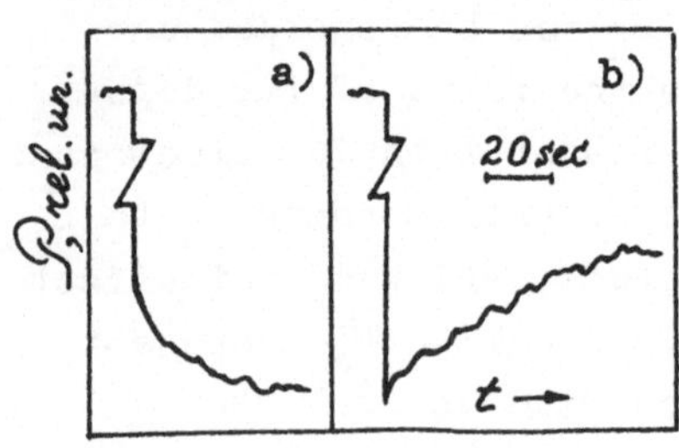

Fig.3. Transition processes observed after switching on the external field. Incomplete Hanle effect. $\theta = 45^o$.

a) $H_o = +1,0$ kGs, b) $H_o = -3,5$ kGs

nuclear field H_N. For the standart geometry of the Hanle -
-effect experiment (θ =90°) no nuclear polarization occures.

The full line in Fig.2 gives the dependence $\mathcal{P}(H_0)$ measured under stationary excitation of heavily doped crystal [*)] $Ga_{0,7}Al_{0,3}As$ for θ =45°. If one excludes the region of small fields, one can see a shift of the Hanle curve, its maximum being at H_0 = - 2 kGs. This value gives the magnitude of H_N.

When the sign of the circular polarization of exciting light is changed a similiar shift in the opposite direction is observed, corresponding to the change of sign of the nuclear field H_N. The shift decreases when θ increases, so that for θ =90° the ordinary Hanle curve has been obtained. This may be readily understood, since the nuclear field is created only by the longitudinal with respect to $\vec{H}_0$ component of the electronic spin and is thus proportional to $\cos\theta$. The distortion of the curve at small fields is due to the fast nuclear spin relaxation in this region, which does not allow the nuclear field to arise.

The dotted line in Fig.2 corresponds to the case when the polarization of the exciting light is modulated between σ^+ and σ^- with a frequency 47 Kc/sec., the alternating intensity of luminescence being detected in σ^+ or σ^- polarization. In these conditions the nuclear polarization cannot to built up and H_N=0. The dotted curve exactly agrees with Eq.2 for H=H_0, the parameters $\mathcal{P}(0)$ and τ being determined from the standart Hanle curve measured for θ =90°.

After the external field is switched on the stationary degree of the luminescence polarization is established during a time of the order of 10-50 sec., such transition processes are represented in Fig.3 for different signs of the external field. Immediatly efter the external field is switched on we have H_N=0. As the nuclear polarization is built up the arising nuclear field shifts the Hanle curve in the direction

[*)] Practically no dependence of the electron spin relaxation time on longitudinal magnetic field is observed in heavily doped p-type samples. In Eq. 2 this dependence is accordingly ignored.

of negative values of H_o. Thus the degree of polarization
increases when $H_o < 0$ and decreases when $H_o > 0$.

The magnitude and direction of the nuclear polarization
may be changed by an applied rf-field. The corresponding
changes in the nuclear field are readily detected by the
polarization of the luminescence. The optically detected NMR
spectrum is present in Fig.4. The sharp drop of polarization
reflects the adiabatic passage of the resonance. The direc-
tion of spins of corresponding nuclei is reversed, so that
the associated nuclear field changes sign. The following
increase of $\mathscr{P}$ is due to relaxation towards the steady state
nuclear polarization.

Fig.4. Optically detected NMR spectrum in conditions of
incomplete Hanle effect. $H_o = -2,7$ kGs.

The incomplete Hanle effect provides, apparently, a
universal means of optical detection of nuclear polarization
and presents rich possibilities for investigation of
nuclear spin processes in semiconductors.

R E F E R E N C E S

1. G.Lampel, Phys.Rev.Lett. 20, 491, 1968.
2. V.L.Berkovits, A.I.Ekimov, V.I.Safarov, ZhETF, 65,346,1973
3. M.I.Dyakonov, V.I.Perel, ZhETF, 65, 362, 1973
4. B.P.Zakharchenya, Proc. XI[th] Int. Conf. on Semicon.
 p.1315, Warsawa, 1972.
5. R.R.Parsons, Canad.J.Phys. 49, 1850, 1971

OPTICAL ORIENTATION IN STRAINED SEMICONDUCTORS

R.I. DZHIOEV, V.G. FLEISCHER, E.L. IVCHENKO,
V.L. VECUA, B.P. ZAKHARCHENYA
A.F. Ioffe Physico-Technical Institute
Leningrad, USSR

Uniaxial stress changes the energy spectrum and
wave function symmetry of electrons in crystals.
Therefore the polarization of recombination
radiation of optically oriented carriers should
change as stress is applied to a semiconductor.
The experimentally observed manifestation of the
stress effect is associated, first, with the
appearance of an optical anisotropy affecting
the polarization of radiation propagating in the
semiconductor (a bulk effect) and, second, with
a change in the interband transition probability
in a stressed crystal (a local effect).

1. OBSERVATION OF STRESS IN EPITAXIAL FILMS

The transverse optical orientation effect was observed in
epitaxial films. In this effect, orientation is detected in the
x-direction perpendicular to the direction of excitation z
(Fig. 1). The orientation appears at the application of a mag-
netic field $\vec{H}$ perpendicular to both these directions as a result
of the spin $\vec{S}_o$ being turned by an angle θ. In this way one can
produce circularly polarized radiation propagating along the
epitaxial film. If the recombination radiation is only weakly
absorbed in the crystal, then in the transverse effect one can
vary the optical path length l of this radiation in the film by
properly displacing the exciting light beam. This provides a
sensitive means of studying the anisotropy of thin semiconductor
films caused by an internal or external stress. Presented in
Fig. 2 are the corresponding dependences of the degree of
circular $\rho_{zx}^{(\sigma)}(l)$ (curves 1 and 2) and linear $\rho_{zx}^{(\pi_1)}(l)$ at 45° to
the y-axis (curves 3 and 4) obtained with a $GaP_{0.3}As_{0.7}$ epi-
taxial film at $77^\circ K$. Curves 1 and 3 refer to the case of
$H_y > 0$, and curves 2 and 4 to $H_y < 0$. The solid curves in
Fig. 2 correspond to the theoretical relationships taking into
account the absorption of recombination radiation. Agreement
with experimental data occurs at the refraction index difference

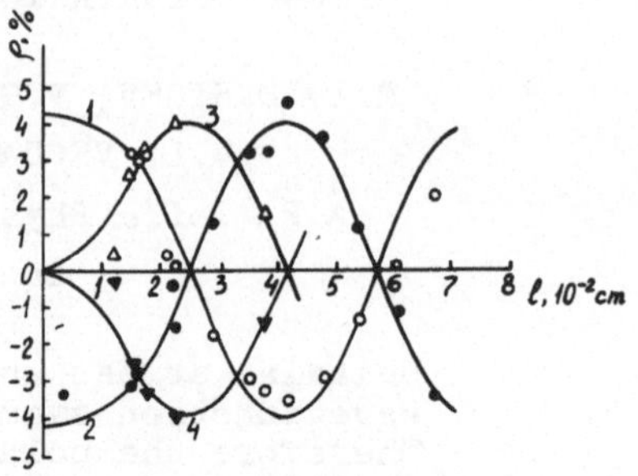

FIG. 1: Experimental set-up (see the text).

FIG. 2: Oscillations of the polarization at moving of the laser beam along the sample.

$n_y - n_z \simeq 10^{-3}$. A theoretical analysis [1] shows that such a value of birefringence in the film corresponds to a relative tensile strain of 0.7×10^{-3} perpendicular to the film surface. Oscillations similar to those presented in Fig. 2 were observed in $Al_xGa_{1-x}As$ epitaxial films and not seen in GaAs single crystals.

The direction of the electron spin precession axis in a transverse magnetic field is determined by the sign of the g-factor. For all the above crystals, the sign of the electronic g-factor turned out to be positive.[+]

2. LOCAL STRESS EFFECT

Uniaxial stress produces splitting of the fourfold-degenerate electronic state at the Γ-point in the valence band into two doubly degenerate levels with the projections of $\pm\frac{3}{2}$ and $\pm\frac{1}{2}$ respectively. The ground state of the hole trapped at the acceptor otherwise undergoes a similar splitting. If the excitation and observation of optical orientation are carried out in the stress direction, the degree of circular polarization of recombination radiation at the frequency $\rho_{zz}^{(\sigma)}(\omega)$ and the degree

[+] For epitaxial samples, the sign of the g-factor was deduced from that of $\rho^{(\sigma)}$ in the first half-period of the oscillations, i.e. at small values of l.

of spin orientation of free electrons P are related by the
following equation:

$$\rho_{zz}^{(\sigma)}(\omega) = P\,\frac{3 - \gamma(\omega)}{3 + \gamma(\omega)} \tag{1}$$

Here $\gamma(\omega)$ is the intensity ratio of transitions at the frequency
ω involving the hole states $\pm\frac{1}{2}$ and $\pm\frac{3}{2}$, accordingly with no
stress applied, $\gamma(\omega) = 1$ and $\rho_{zz}^{(\sigma)} = 0.5\,P$. In the above case
of a stressed epitaxial film (tensile stress along the normal to
its surface), holes occupy at low temperatures predominantly the
$\pm\frac{3}{2}$ level, so that $\gamma < 1$, yielding $\rho_{zz}^{(\sigma)} > 0.5\,P$. Thus application
of stress to a film may result in an increase of $\rho_{zz}^{(\sigma)}$ above the
value of 25% which is the upper limit for crystals of the group
in question. Fig. 3 shows the dependence of $\rho_{zz}^{(\sigma)}$ on the energy
$\hbar\omega$ of the recombination radiation quanta (curve 1) for an epi-
taxial $GaP_{0.3}As_{0.7}$ film at $4.2^\circ K$ obtained in the reflection
arrangement (with excitation and observation carried out through

FIG. 3: Spectral dependence of the intensity J, degree of linear $\rho_{zx}^{(\pi_2)}$ and circular $\rho_{zz}^{(\sigma)}$ polarization of luminescence.

the xy-face in Fig. 1). In this geometry the effect of bi-
refringence is negligible since the path length of light in the
films is $\lesssim 1\ \mu$, the observed increase over the value of $\rho = 25\%$
being associated with the local stress effect. This is con-
firmed by observation of the linear polarization of radiation in
the direction perpendicular to the stress axis (with unpolarized
pumping). Depending on the magnitude of $\gamma(\omega)$, there should be
observed linear polarization along or perpendicular to the stress

axis, its degree being

$$\rho_{zx}^{(\pi_2)} = \frac{1 - \gamma(\omega)}{1 + \frac{5}{3}\gamma(\omega)} \tag{2}$$

Such polarization is indeed observed. The experimental dependence of $\rho_{zx}^{(\pi_2)}$ is shown in Fig. 3 (curve 2). Measurements of ρ were carried out by means of a quartz modulator in the two-channel photon counting arrangement permitting us to attain high accuracy in determining the degree of polarization at low intensities at the edges of the line. Curve 3 in the same figure illustrates the dependence of intensity on $\hbar\omega$. As seen from Fig. 3, within the range 1,80 eV $\gtrsim \hbar\omega \gtrsim$ 1,70 eV the dependences $\rho_{zz}^{(\sigma)}(\omega)$ and $\rho_{zx}^{(\pi_2)}(\omega)$ correlate well, this correlation being in agreement with eqs. (1) and (2). At $\rho_{zx}^{(\pi_2)} = 0.16$ ($\hbar\omega$ = 1.78 eV), we obtain from eq. (2) that $\gamma = 0.64$, which yields at $P_{max} = 0.5$, according to eq. (1), $\rho_{zz\ max}^{(\sigma)} = 32\%$. A dependence of $\rho_{zz}^{(\pi_2)}(\omega)$ similar to curve 2 was observed with a GaAs single crystal subjected to uniaxial compression along the y-axis. One measured here the magnitude of $\rho_{zz}^{(\pi_2)}$ for the light propagating along z and polarized linearly along y or x.

The behavior of linear polarization in this geometry is equivalent to the case of an epitaxial film subjected to tensile stress along the normal to its surface. Note that this stress is not due to the difference in the lattice constants of the solid solution and the GaAs substrate. It was observed in the films of both GaP_xAs_{1-x} and $Al_xGa_{1-x}As$.

REFERENCE

[1] Vecua, V.L.; Dzhioev, R.I.; Zakharchenya, B.P.; Ivchenko, E.L.; Fleischer, V.G.: J. Experim. i Teor. Fiz. <u>66</u>, 1790 (1974).

SPIN-FLIP RAMAN SCATTERING*

S.R.J. BRUECK

Lincoln Laboratory, Massachusetts Institute of Technology
Lexington, Massachusetts 02173, U.S.A.

The polarization selection rules, resonance enhancement and spontaneous linewidth of spin-flip Raman scattering in InSb in the 5 μm near resonance regime are reviewed. The suitability of the InSb spin-flip Raman laser for spectroscopic applications is discussed.

1. INTRODUCTION

The detailed physical properties of spontaneous spin-flip Raman scattering in semiconductors have acquired enhanced significance in conjunction with the development of the spin-flip Raman laser [1-3]. The near resonance regime is of particular interest due to the CW operation of the InSb spin-flip laser at 5 μm [2]. The main features of spontaneous spin-flip scattering discussed in this paper are: (i) polarization selection rules, (ii) cross-section enhancement for photon energies near the InSb energy bandgap and (iii) spontaneous lineshape.

The gain for stimulated spin-flip Raman scattering is given approximately by

$$g \propto \left(\frac{E_g}{E_g - \hbar\omega_i}\right)^2 \frac{(n_\uparrow - n_\downarrow)\,\sigma_T^*}{\Gamma_s}\, c^2\, I_i\, |(\hat{\epsilon}_i \times \hat{\epsilon}_o) \cdot \hat{h}|^2 \tag{1}$$

where $n_\uparrow(n_\downarrow)$ is the population of the lower (upper) conduction band spin state, E_g is the energy gap, $\hbar\omega_i$ the incident photon energy, Γ_s the spontaneous linewidth, σ_T^* the Thompson cross section calculated with the conduction band mass, I_i the incident pump intensity and the final term contains the polarization information. The resonance enhancement is given schematically by the first term in (1). Thus, the three properties discussed here enter directly into the spin-flip laser gain expression as well as influencing other properties of the spin-flip laser such as the tuning characteristics.

* This work was sponsored by the Department of the Air Force.

2. POLARIZATION SELECTION RULES

Detailed measurements of the polarization selection rules for spin-flip Raman scattering have been carried out in the 5 µm near resonance regime for $n = 1 \times 10^{16}$ cm^{-3}, $H = 40$ kG and $T \sim 30$ K [4]. The results of this experiment are given in Table I along with theoretical results of a calculation which includes the detailed InSb band structure. The results within each set of data delineated by the double lines are normalized to the maximum theoretical value within that set and no comparison can be made between these data sets. The slight depolarization measured can be attributed, at least partially, to the effects of the collection optics. These results are in contrast to those reported [5] at a 10.6 µm excitation wavelength where violations of these selection rules were found.

3. RESONANCE ENHANCEMENT

There is a resonant increase in the cross-section for spin-flip Raman scattering and in the gain when the pump photon energy is comparable to the intermediate state (conduction band upper spin level to valence band) energy as is indicated in Equation (1). This resonant enhancement has been studied in the InSb-CO laser system by varying both the incident photon energy and the magnetic field. Figure 1 shows the result of a cross section measurement, corrected for sample absorption, at a fixed magnetic field of 40 kG. The incident photon energy was varied by changing CO laser transitions. There is good agreement between theory and experiment for a variation of over an order of magnitude in the relative cross section. The discrepancy in the vicinity of an incident photon energy of 235 meV may may arise from additional intermediate

INCIDENT PHOTONS

| | | | $\vec{k}_i \perp \vec{H}$ | | | | $\vec{k}_i \parallel \vec{H}$ | |
| | | | $\vec{E}_i \parallel \vec{H}$ | | $\vec{E}_i \perp \vec{H}$ | | $\vec{E}_i = \frac{1}{\sqrt{2}}(\vec{E}_{i+} + \vec{E}_{i-})$ | |
			Experiment	Theory	Experiment	Theory	Experiment	Theory
SCATTERED PHOTONS	$\vec{k}_o \perp \vec{H}$	$\vec{E}_o \parallel \vec{H}$	0.06	0.0	0.69	0.78	0.78	0.78
		$\vec{E}_o \perp \vec{H}$	1.0	1.0	0.12	0.0	0.07	0.0
	$\vec{k}_o \parallel \vec{H}$	$\vec{E}_{o-}$	1.0	1.0	0.04	0.0	<0.1	0.0
		$\vec{E}_{o+}$	0.02	0.0	0.04	0.0	<0.1	0.0

TABLE I: Polarization selection rules-comparison of theory and experiment ($n = 1 \times 10^{16}$ cm^{-3}, $H = 35$ kG, $T \sim 30$ K).

states associated with a Zn acceptor level at this energy. Substantially larger enhancements can be inferred from the very low threshold powers ($\lesssim 5$ mW) observed in low electron concentration material at low magnetic fields for photon energies quite close to the bandgap. Extrapolation of the theoretical curve of Figure 1 to an incident photon energy of 117 meV ($\lambda = 10.6$ µm) yields a relative cross section of 0.1.

Changing the magnetic field varies the electron populations $n_\uparrow$ and $n_\downarrow$ as well as the resonance enhancement. Further effects arise because of the dependence of the resonance enhancement on electron kinetic energy. The result of a measurement of the cross section variation with magnetic field is shown in Figure 2. At low magnetic

FIG. 2: Magnetic field variation of the relative cross section ($n = 1 \times 10^{16}$ cm^{-3}, T$\sim$5 K, $\hbar\omega_i = 230$ meV).

FIG. 1: Resonance enhancement of spontaneous spin-flip Raman scattering as the input photon energy is varied ($n = 1 \times 10^{16}$ cm^{-3}, H$= 40$ kG, T$\sim$30 K).

fields many electrons cannot participate in the spin-flip process due to the equilibrium population of the upper spin state. Above about 25 kG for this electron concentration all of the electrons are in the lowest spin state (the quantum limit) and the population factors are no longer varying. The decrease in relative cross section is due to the variation in the resonance enhancement as higher magnetic fields result in a larger intermediate state energy.

4. LINEWIDTH

Measurements of the linewidth of spin-flip Raman scattering have been made in two geometries, the $\vec{q} \cdot \vec{H} = 0$ collinear geometry ($\vec{q} = \vec{k}_i - \vec{k}_o$ is the scattering wavevector) with both photons propagating normal to the magnetic field (the usual geometry of a spin-flip laser) and the $\vec{q} \cdot \vec{H} \neq 0$, 90^o scattering, geometry with $\vec{k}_i \perp \vec{H}$ and $\vec{k}_o \parallel \vec{H}$. Figure 3 shows the results of this measurement for a electron concentration of 1×10^{16} cm^{-3} at $T = 2$ K. Note the very striking geometry dependence of the linewidth particularly at low magnetic fields.

Two inhomogeneous broadening mechanisms which affect the spin-flip linewidth, non-parabolicity broadening and Doppler broadening are shown schematically in Figure 4 along with the orbital dependence of the spin-flip energy (k_z is the electronic momentum along the magnetic field). Doppler broadening only affects the $\vec{q} \cdot \vec{H} \neq 0$ geometry and accounts for the large linewidth difference. Because of these orbital dependences of the spin-flip frequency, orbital relaxation processes which alter the k_z state of a spin excitation while conserving the total number of spin excitations can significantly affect the lineshape. These effects have been incorporated in a phenomenological theory [4] which involves two collision times: (i) τ_s a time that characterizes the homogeneous spin relaxation and (ii) τ_p a time that characterizes orbital relaxation processes. The results of this lineshape calculation for the $\vec{q} \cdot \vec{H} \neq 0$ geometry ($n = 1 \times 10^{16}$ cm^{-3}, $H = 35$ kG, $\tau_s = 10^{-10}$ sec, $T = 0$ K) are shown in Figure 5 with τ_p as a parameter. In the absence of orbital collisions ($\tau_p = \infty$) the calculated lineshape is highly asymmetric with a sharp peak and a long tail extending towards smaller frequency shifts. As τ_p as decreased the calculated lineshape becomes more symmetric with a smaller total frequency spread and the peak shifts towards an average value of the spin-flip frequency. These dramatic effects on the lineshape can be understood in terms of the concepts of motional narrowing well known in the field of nuclear magnetic resonance. The orbital collisions serve to retard the dephasing of spin-excitations by causing them to perform a random walk in k_z and hence in $\omega_s(k_z)$. In the limit of τ_p collisions rapid compared with the total inhomogeneous frequency spread the calculated lineshape is Lorentzian with a line center given by the average spin

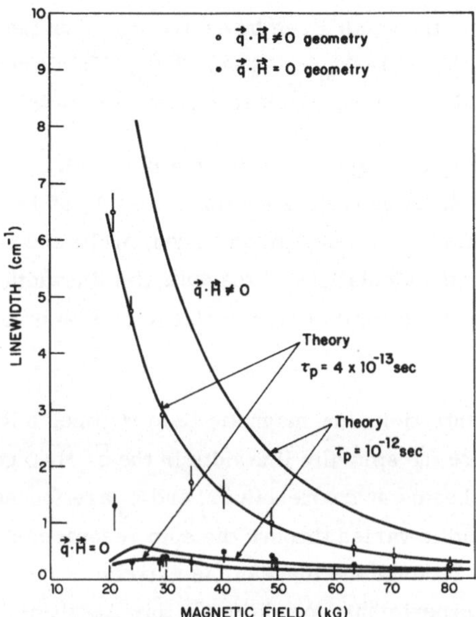

FIG. 3: Comparison of the experimentally observed linewidth with the phonomenological lineshape theory ($n = 1 \times 10^{16}$ cm^{-3} and $\tau_s = 10^{-10}$ sec).

FIG. 4: InSb models used in lineshape calculations: (a) inhomogeneous nonparabolicity broadening and (b) inhomogeneous Doppler broadening.

FIG. 5: Calculated effects of orbital collisions on the lineshape in the $\vec{q}\cdot\vec{H} \neq 0$ geometry. ($n = 1 \times 10^{16}$ cm^{-3}, $H = 35$ kG, and $\tau_s = 10^{-10}$ sec).

frequency $<\omega_s(k_z)>$ and a linewidth given by an average over the inhomogeneous fre-
quency spread; $\Gamma_s = 2\tau_p <[\omega_s(k_z) - <\omega_s(k_z)>]^2>] + 2/\tau_s$. This linewidth is proportional
to τ_p, more rapid orbital collisions result in a narrower linewidth.

The fit between this theory and experiment for the linewidth as a function of magnetic
field is shown in Figure 3. A good fit is obtained for a τ_p of 4×10^{-13} sec which is on
the order of the dc mobility τ. Davies [6] and Yuen, Wolff and Lax [7] have made
more detailed microscopic calculations of the spin-flip linewidth which include addi-
tional effects and point out the limits of applicability of this simple phenomenological
treatment.

In the low carrier concentration, low magnetic field regime, a Raman gain technique
has been used to measure the spin-flip linewidth in the $\vec{q} \cdot \vec{H} = 0$ geometry [8]. In this
technique two CO laser beams at frequencies ω_i and ω_o are focused onto an InSb sam-
ple and the magnetic field is varied through the spin resonance condition $\omega_s = \omega_i - \omega_o$
resulting in gain at ω_o. Because the linewidth is extremely narrow [less than 100 MHz
$(0.003\ cm^{-1})$ for some experimental conditions], this lineshape is not accessible by
conventional spectroscopic techniques. Figure 6 shows the results of this measure-
ment for a sample concentration of $8 \times 10^{14}\ cm^{-3}$ at $T = 2\ K$. The linewidth is now an
exponentially increasing function of magnetic field. Similar but somewhat narrower

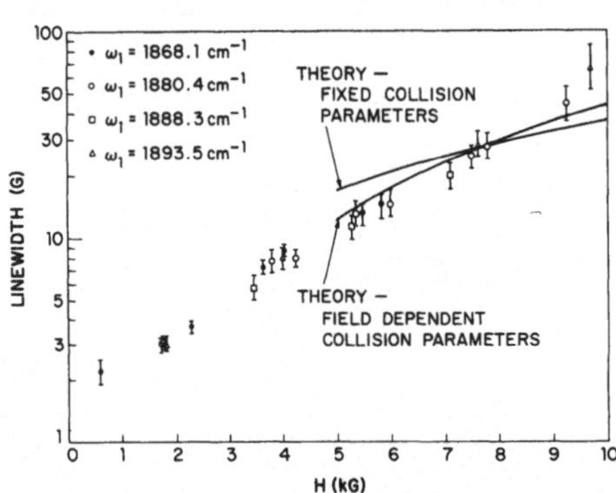

FIG. 6: Linewidth of spin-flip Raman scattering in the low carrier concentration re-
gime ($n = 8 \times 10^{14}\ cm^{-3}$, $T = 2\ K$). The solid curves are the results of the
phenomenological theory with the fixed parameter values $\tau_s = 2 \times 10^{-9}$ sec,
$\tau_p = 4 \times 10^{-12}$ sec and $T_e = 10\ K$ and the field dependent collision parameters
$\tau_s = 1 \times 10^{-8}/H^{3/2}$ sec $(gauss)^{3/2}$, $\tau_p = 3 \times 10^{-13}\ H$ sec/gauss and $T_e = 10\ K$.

linewidths are obtained for $n = 1 \times 10^{14}$ cm^{-3}. A simple extension of the above theory to finite temperature results in an H^2 dependence of the linewidth for a nondegenerate electron gas. The best fit between this theory and the experiment is shown in Figure 6 as well as the results of a modified theory which allows for a magnetic field dependence of the collision times [8]. This gives a somewhat better fit but in the absence of a complete microscopic theory which includes the temperature and energy dependences of these parameters it cannot be used to provide more than a rough correction to the model.

5. DISCUSSION AND APPLICATIONS

In the 5 µm near resonance regime, the CW InSb spin-flip Raman laser has been operated over a wide range of electron concentration ($n = 5 \times 10^{13}$ cm^{-3} to $n = 1 \times 10^{17}$ cm^{-3}) and magnetic field (100 G to 100 kG). The differing properties of spontaneous spin-flip Raman scattering in these parameter regimes lead to different laser behaviors. One characteristic that is particularly important for spectroscopic applications is the fine tuning behavior of the laser. This tuning is characterized by mode jumps as the laser tunes over the longitudinal modes of the InSb cavity. The size of these mode jumps and the tuning rate within a mode are governed by a mode pulling between the spontaneous spin-flip line and the cavity modes [9, 10]. To a good approximation, the tuning rate of a spin-flip laser is given by

$$\frac{d\nu}{dH} \approx \frac{d\nu_s}{dH} \frac{\Gamma_c}{\Gamma_s + \Gamma_c} \tag{2}$$

where $d\nu_s/dH = g\beta$ is the tuning rate of spontaneous scattering, Γ_s the spontaneous linewidth and Γ_c the cavity linewidth. A narrower spontaneous line leads to a higher tuning rate and a smaller mode jump and is therefore better suited to spectroscopic applications. Thus a $n = 10^{15}$ cm^{-3} spin-flip laser (5 mm long) at H = 4 kG has a tuning rate of over 60 MHz/G and mode jumps of only about 200 MHz while a $n = 10^{16}$ cm^{-3} laser at 40 kG has a tuning rate of 16 MHz/G and mode jumps of 0.2 cm^{-1} (6 GHz). An example of the spectroscopic capabilities of a low field spin-flip laser is shown in Figure 7 which shows the Doppler-limited absorption spectrum of the $R_{1/2}(1/2)$ line of NO at 1881 cm^{-1}. The line is fully resolved; the 360 MHz splitting is due to Λ-type doubling.

The spin-flip Raman laser is a useful, relatively high power, tunable laser in the middle infrared. A detailed understanding of spontaneous spin-flip scattering is necessary in order to tailor the output characteristics of the spin-flip laser for specific applications.

FIG. 7: Doppler-limited absorption spectrum of NO $R_{1/2}(1/2)$ line at 1881 cm^{-1} (Torr pressure, 10 cm cell).

This work has been carried out in collaboration with Aram Mooradian and F.A. Blum. Helpful discussions with A.L. McWhorter, R.W. Davies, S. Yuen and P.A. Wolff are greatfully acknowledged.

6. REFERENCES

[1] Patel, C.K.N; Shaw, E.D.: Phys. Rev. Lett. 24, 451 (1970); Phys. Rev. B3, 1279 (1971).

[2] Mooradian, A; Brueck, S.R.J.; Blum, F.A.: Appl. Phys. Lett. 17, 481 (1970).

[3] Allwood, R.L.; Dennis, R.B.; Smith, S.D.; Wherrett, B.S.; Wood, R.A.: J. Phys. C4, L163 (1971).

[4] Brueck, S.R.J.; Mooradian, A; Blum, F.A.: Phys. Rev. B7, 5253 (1973).

[5] Patel, C.K.N. and Yang, K.H.: Appl. Phys. Lett. 18, 491 (1971).

[6] Davies, R.W.: Phys. Rev. B7, 3731 (1973).

[7] Yuen, S.Y.; Wolff, P.A.; Lax, Benjamin: Phys. Rev. B9, 3394 (1974).

[8] Brueck, S.R.J.; Mooradian, A: Optics Commun. 8, 263 (1973).

[9] Brueck, S.R.J.; Mooradian, A.: Appl. Phys. Lett. 18, 229 (1971).

[10] Brueck, S.R.J.; Mooradian, A.: IEEE J. Quant. Electron. (to be published).

STUDY OF MOTT TRANSITION AND OBSERVATION OF STRONG MICROWAVE
INDUCED SPIN-FLIP RAMAN LIGHT SCATTERING IN n-TYPE CdS

R. ROMESTAIN[†], S. GESCHWIND AND G. E. DEVLIN
BELL LABORATORIES, MURRAY HILL, NEW JERSEY USA

P. WOLFF[*]
MASSACHUSETTS INSTITUTE OF TECHNOLOGY, CAMBRIDGE, MASS. USA

Electron delocalization is studied in n-CdS by observing the SFRS linewidths. A new effect is also described wherein simultaneous irradiation of the spins with laser light at frequency ω_L and microwaves at frequency ω_o near the spin resonance results in forward scattering at $\omega_L \pm \omega_o$ which is more than three orders of magnitude greater than spontaneous SFRS.

1. INTRODUCTION

At an excess donor concentration, $n = (N_D - N_A)$ of approximately 2×10^{17} in CdS, the donor orbits have begun to overlap sufficiently so as to mark the onset of a Mott transition. Transport data [1] and NMR Knight shift and relaxation measurements [2] indicate continuing delocalization with increasing concentration and for $(N_D - N_A) > 2\times10^{18}$ it behaves like a metal with a Fermi energy appropriate to its concentration.

2. DELOCALIZATION AS SEEN IN SFRS

In Spin-flip Raman scattering (SFRS) electron delocalization is manifested by a Doppler linewidth due to the motion of the electrons [3]. Since electron collision times, τ_c, are in the range of 10^{-13} -10^{-14} in CdS so that $\Delta\nu_{Dop} \ll 1/\tau_c$, one expects a collisionally narrowed line given by

$$\Delta\nu_{eff} = (\Delta\nu_{Dop})^2 \tau_c = \left(4v_F \lambda \sin\frac{\theta}{2}\right)^2 \tau_c \qquad (1)$$

where θ is the scattering angle, and τ_c may be expressed in terms of the resistivity and effective mass.

† On leave from CNRS, France

* Research sponsored by U. S. Air Force Office of Scientific Research, Air Force Systems Command under Contract/Grant No. AFOSR 71-2010.

2.1 EXPERIMENTAL RESULTS ON SFRS

Table 1 shows the results of SFRS linewidth measurement taken
with a Fabry-Perot on samples with $n = 1.8\times10^{17}$ and 5×10^{17} and
with a double grating monochromator on a sample with
$n = 4\times10^{19}$[†]. Ar^+ laser light at 4880 A° was used.

n	ρ ohm-cm	$\Delta\nu_{eff}$ cm^{-1}	$\Delta\nu_{meas}(cm{-}1)$ $\theta = 90°$ Temp = 2K
1.8×10^{17}	$\sim10^5$	–	0.003
5×10^{17}	0.3	0.11	0.04
4×10^{19}	1.1×10^{-3}	6.2	5.0

TABLE 1. SFRS Linewidth vs. excess donor concentration.

In the specimen with $n = 1.8\times10^{17}$ the observed linewidth of
0.003 cm^{-1} is found to be independent of scattering angle and
as will be described below, reflects the observed EPR linewidth,
i.e. the electrons are localized and there is no Doppler width.
With $n = 4\times10^{19}$, the linewidth is found to have a $\sin^2\theta/2$ de-
pendence and is in good agreement with Eq. 1 in this metallic
regime.

One is not entirely justified in using Eq. (1) for $n = 5\times10^{17}$
as it only applies to completely **delocalized** electrons with Fermi
gas parameters. It is nonetheless interesting that the experi-
mental result is close to the calculated value. More precise
measurements of transport properties and control of sample
homogeneity are required. Variations of $\Delta\nu$ of 50% are found
in different parts of the sample which probably correspond to
a spread of n.

3. MICROWAVE INDUCED CHANGES IN 90° SFRS; POPULATION EFFECTS

The relative intensities of Stokes and anti-Stokes scattering
reflect the unequal population of spin down and spin up popula-
tions respectively. When resonant microwaves are applied to
the sample, an increase in anti-Stokes and decrease in Stokes

† Kindly measured for us by P. Fleury and T. C. Damen.

scattering is observed corresponding to the microwave induced
change in population. As the DC magnetic field is swept through
resonance the variation in SFRS signal reflects the EPR line-
width as shown in Fig. 1 for $n = 1.8 \times 10^{17}$.

FIG. 1. Microwave EPR signal detected by change in intensity of 90° SFRS anti-Stokes line Microwaves were modulated at low frequency and phase sensitive detection was used.

Note that observed 45 gauss linewidth corresponds to the ob-
served SFRS optical linewidth of 0.003 cm^{-1} cited above. If the
DC magnetic field is kept at its peak resonance value and the
microwaves are suddenly switched off, the recovery of the light
signal measures T_1 of the donors which is found to be 5 μsec for
$n = 1.8 \times 10^{17}$, so that the linewidth is not due to T_1.

4. MICROWAVE INDUCED COHERENT FORWARD SFRS

The maximum change in the microwave induced light signal de-
scribed above is equal to half the difference in intensity of
the spontaneous Stokes and anti-Stokes lines. By contrast, if
one observes the microwave induced change in the Raman light in
forward scattering ($\theta = 0°$) with the laser beam close to the c-
axis, more than three orders of magnitude increase in light sig-
nals at $\omega_L \pm \omega_0$ is observed as compared to 90° scattering.

Conversion efficiencies as large as 6% of the laser beam are
seen, as shown in Fig. 2b and 2c. This effect is due to
coherent SFRS from the coherent spin states, or alternatively,
as parametric conversion of the laser light via the transverse
component of the macroscopic magnetic dipole, σ_T, precessing at
microwave frequency, ω_0. The reversal of intensity between
Stokes and anti-Stokes with change in input polarization as seen
in Figs. 2b and 2c is due to phase matching of the dispersion

from ω_L to $\omega_L \pm \omega_o$ by means of the birefringence of CdS.

FIG. 2a. Forward spontaneous SFRS at 2°K with Y(ZX)Y geometry as seen by Fabry-Perot.

FIG. 2b. Microwave Driven Coherent Scattering in Y(ZX)Y geometry. Stokes line is phase matched. Laser beam is a few degrees from c-axis.

FIG. 2c. Microwave Driven Coherent Scattering in Y(XZ)Y geometry.

The above described coherent effect may be seen as follows. Spontaneous SFRS may be described semiclassically as radiation from a Raman dipole given by

$$\vec{D}_R = \langle \downarrow | \vec{\sigma} | \uparrow \rangle \times \vec{E}_L (\alpha e^{-i(\omega_L \pm \omega_Z)\tau + i\varphi} + c.c.) \qquad (2)$$

where σ is the spin operator; $\vec{E}_L, \omega_L$ the incident laser field and frequency respectively; ω_Z the Zeeman frequency; and α a constant involving second order expressions for the dipole operator $e\vec{r}$ between the ground state and excited states, and φ a <u>random</u> phase factor. Radiation in the transverse plane of the dipole in Eq. 2 can be expressed in terms of a <u>spontaneous</u> incoherent differential Raman cross section per center given by

$$\left(\frac{d\sigma}{d\Omega} \right)_{spon} = 4 |\alpha|^2 \frac{(\omega_L \pm \omega_Z)^4}{c^4} \qquad (3)$$

where each donor center radiates with random phase. The microwaves induce coherence, described by σ_T, between spin up and

spin down states such that $\vec{D}_R \sim \alpha\vec{\sigma}_T \times \vec{E}_L\, e^{-i(\omega_L \pm \omega_o)t}$ with well defined phase. These coherent dipoles emit cooperatively in the _forward_ direction only. It can be shown that if phase matching conditions are satisfied then the ratio of scattered power P_s to laser power P_o is given by

$$\frac{P_s}{P_o} = \frac{1}{2}\frac{d\sigma}{d\Omega}\bigg|_{spon} \frac{\lambda_s^2 n^2 \sigma_T^2 L^2}{\varepsilon} \tag{4}$$

where λ_s is the free space optical wavelength, ε the optical dielectric constant, L the interaction length and n the concentration of donors. Substitution of appropriate experimental values into Eq. (4) yields $P_s/P_o \sim 0.08$ in good agreement with observation. It should be noted that this effect is the inverse of a similar parametric infrared mixing in InSb described earlier [5], [6].

As the magnetic field is swept through resonance, the intensity of these forward scattered coherent Raman lines will reflect the EPR signal (σ_T), but it will be orders of magnitude larger than the population effect described in Sec. 3. The microwave saturation properties of this signal with n = 1.8 $\times 10^{17}$ shows that the linewidth is homogeneous with $T_2 \approx$ 4×10^{-9} sec. Thus there is a three orders of magnitude difference in T_1 and T_2 at the onset of the Mott transition. This new microwave-SFRS optical technique should prove a valuable tool in studying electron dynamics in the Mott transition region.

5. REFERENCES

[1] S. Toyotami and K. Morigaki, J. Phys. Soc. Japan 25, 807 (1968).
[2] F. D. Adams, D. C. Look, L. C. Brown, D. R. Locker, Phys. Rev. 4B, No. 7, 2115, (1971).
[3] J. F. Scott, T. C. Damen and P. A. Fleury, Phys. Rev. B 6, No. 10, 3856 (1972).
[4] Y. Yafet, Phys. Rev. 152, 855 (1956).
[5] V. T. Nguyen and T. J. Bridges, Phys. Rev. Lett. 29, 359 (1972).
[6] T. L. Brown and P. A. Wolff, Phys. Rev. Lett. 29, 362 (1972).

QUANTUM EFFECTS IN SPIN-FLIP SCATTERING

B.S. WHERRETT, S. WOLLAND, C.R. PIDGEON, R.B. DENNIS and S.D. SMITH

Department of Physics, Heriot-Watt University, Edinburgh, U.K.

The occupancy of Landau states of finite n and k_z values at
finite temperatures is accounted for in a calculation of the
spin-flip Raman cross-section for conduction electrons in InSb.
The application of the spin-flip Raman laser to electric-
dipole-induced far-infrared generation by resonant difference
frequency mixing is discussed.

1. QUANTUM OSCILLATIONS IN THE RAMAN GAIN

In both the 10µm and 5µm spectral regions, where spin-flip Raman-
laser (SFRL) action has been observed to date, the output power is found
to be highly irregular as a function of applied magnetic field strength
[1, 2]. The 10µm results have been interpreted as being a consequence of
minor magnetic-field-dependent variations in the Stokes absorption
coefficient, emphasized in the output power by the non-linear nature of
the SFRL process. In contrast near 5µm the Stokes losses are slowly
varying and the irregularities, which occur at the lower field strengths,
are to be explained by the quantum oscillations of the Raman gain as the
Landau levels sink beneath the Fermi energy.

The peak Raman gain is, in conventional notation [3],

$$g_p = \frac{16\pi^2 c^2}{\hbar \omega_s^2 \omega_p n_s n_p} \; \frac{I_p}{\Gamma} \left[\left(\frac{e}{mc^2}\right)^2 \; \frac{\omega_s}{\omega_p} \; \Sigma_I \; m^2 |R|^2 \right] \; . \tag{1}$$

The quantum oscillations derive from the summation Σ_I which obeys the
identity

$$\Sigma_I \equiv \Sigma_{n, k_z} \{f(n, k_z, \uparrow) - f(n, k_z, \downarrow)\} \; . \tag{2}$$

$f(n, k_z, \uparrow)$ is the Fermi occupation factor for the spin-up Landau sublevel.

Due to the presence of near-resonant terms in R, for pump photon
energies $(\hbar\omega_p)$ near 5µm, the n and k_z dependence of the energy
denominators is to be accounted for in calculations based on equation
(1). In practise, the term in square brackets is evaluated here. In
the zero temperature limit considered in this section this is equal to
the spontaneous Raman cross-section.

The energy band model used as a basis for the Raman gain calculations

is the "three band" model for InSb in which the magnetic-field and k_z-dependent mixing of the eight conduction and valence basis functions are accounted for [4]. The number of intermediate states to be included in R rises from five for $n = 0$, $k_z = 0$ to thirteen for $n \gg 2$ and finite k_z, all of which allow only (z, -) or (+, z) polarization selection rules [5]. In the energy denominators we employ the parabolic approximation and use magnetic-field independent values of effective masses and g-factors. The errors we introduced are small by comparison with the corrections for finite n, k_z, for field strengths below 30 kG.

Figure (1) shows the calculated Raman cross-sections, using the data $E_g = 0.237$ eV, $\Delta = 0.81$ eV, $E_p = 21.1$ eV [6] and $\hbar\omega_p = 0.235$ eV. In (1a) $n = 0$, $k_z = 0$ cross-sections are included, calculated (i) using the threeband model and simply multiplying the cross-section per electron by the number of active electrons (---) and (ii) including higher bands to the prescription employed in [6] (-·-). Comparing figures (1a) and (1b), it is significant that the closer proximity of the quantum condition ($\hbar\omega_{sf} = E_F$) to the interband Raman resonance for the lower concentration causes a greater maximum possible cross-section. As the spontaneous linewidth decreases with free-carrier concentration this result re-inforces the experimental findings that lower SFRL thresholds can be achieved in low concentration samples.

Experimental measurements of the SFRL output powers [2] are superimposed on figure (1). The agreement in magnetic field positions of the peaks and troughs indicates that the oscillations are indeed due to the movement of Landau levels through the Fermi energy. Employing rate equation theory [7] to transcribe the cross-section calculations to output powers the low field peaks are enhanced with respect to the major peak. That such enhancement is not seen in the SFRL output is attributed to the finite temperature of the InSb sample. From the results of section (2) it may be concluded however that the actual magnitude of the peaks is consistent with a temperature of the order of 10 K.

2. TEMPERATURE DEPENDENCE OF THE RAMAN GAIN

It has been suggested that heating of the Raman-active filament is

responsible for several results obtained in SFRL experiments, particularly
in those carried out under 10.6µm pumping where higher incident powers
are essential. By calculating the number of electronic states which
contribute to scattering (N_S), the role played by the temperature
dependence of the scattering (as opposed to the linewidth and absorption
effects) is investigated.

For 10.6µm pumping R in equation (1) is basically independent of
n, k_z and the Raman gain is proportional to N_S (N_S is the summation
involving fermi factors given precisely by equation (2)). The evaluation
of N_S requires first the calculation of the temperature dependence of
E_F at finite fields. Due to the oscillatory nature of the Landau density
of states, E_F is particularly temperature dependent in the viccinity of
Landau minima and is obtained as the solution of:

$$N = (\frac{eH}{2\pi hc}) \; \Sigma_n \int_{-\infty}^{\infty} \{f(n \; k_z\uparrow) + f(n \; k_z\downarrow)\} \; dk_z \qquad (3)$$

The resulting values for N_S are shown in figure (2).

The experimental results of interest concern scattering from a
9.7×10^{15} cm^{-3} sample, using 10.6µm pumping, where the temperature and
magnetic field dependence of the SFRL output were measured in 1971 by
Smith et. al. [8]: (i) the lowest value of the SFRL threshold occurs
at 45 kG, (ii) no low field quantum oscillations are observed, (iii) the
output power saturates more strongly at high input power levels than is
to be expected from theory, and (iv) if the bulk sample temperature is
increased by external heating a sharp decrease in the SFRL power occurs
between 30 and 50 K. The first two results may be explained purely in
terms of the temperature dependence of N_S by supposing that in the Raman-
active region of the sample the electron temperature is ∿40 K. Using
the absorption loss data of [1] the peak value of the gain to loss ratio
(ie the minimum threshold) would occur at the quantum condition for zero
temperature (25 kG). Also the first quantum oscillation peak would be
above threshold at the highest input power levels used. Results (iii)
and (iv) are not however as easy to interpret. An increase in sample
temperature with increased input power lowers the effective gain-loss
ratio and would look like a heavy saturation but the temperature required
to fit the data would exceed 70 K, which is not consistent with the
position of the peak output power. Also the electron redistribution
required to explain effect (iv) is far in excess of that obtained at the

noted temperature.

3. ELECTRIC DIPOLE INDUCED DIFFERENCE FREQUENCY MIXING

Observations of tunable, far-infrared (FIR) radiation by resonant, difference - frequency mixing in InSb [9] have been explained in terms of the magnetic-dipole strength of the conduction electron spin-resonance transition [10].

Although InSb is essentially centrally symmetric a relatively large conventional second order nonlinear susceptibility is induced by the inversion asymmetry and of particular significance, there is an inversion-asymmetry-induced contribution to the electron spin-resonance transition. The dominant electric-dipole moment M, connecting the two spin levels is linear in k_z and, whilst it leads to a finite absorption proportional to the sum of $|M|^2$ overall k_z electron wave-vectors, the resonant non-linear polarization is directly proportional to M and cancels exactly to zero on summation over positive and negative k_z. We are therefore concerned to calculate the strength of the inversion-asymmetry-induced FIR output by comparison to the magnetic-dipole result. The inversion asymmetry mixes the a and b-set basis functions for InSb in such a manner as to produce an anisotropic contribution to the spin-resonance electric-dipole moment. There is also an isotropic contribution to this matrix element through the spin-orbit correction to the radiation interaction itself. The resulting electric-dipole moment is of order (eH/hc) $P^2 Cm/E_g^2 \hbar$ [11].

Using the empirical estimation for C of 9.3×10^{-11} eV cm [12] the electric-dipole induced mixing is in general two orders of magnitude below that predicted by the magnetization mechanism. Existing emperiments have been carried out under phase matching conditions which are favourable to the polarization selection rules for the magnetization. It is suggested that a magnetic field of ~ 5 kG and a concentration of 4×10^{15} cm^{-3} are suitable conditions for observing electric-dipole induced emission polarized perpendicular to $\underline{H}$ and so testing the magnitude of C. It is further noted that a static electric field in the $\underline{H}$-direction may be used to create an asymmetry in the k_z-distribution of the electrons and remove the cancellation discussed above. A field of the order of 5 V cm^{-1} should be sufficient to give a 1% asymmetry of the electron distribution and produce a k_z-induced electric-dipole mixing of the order of the magnetic-dipole effect.

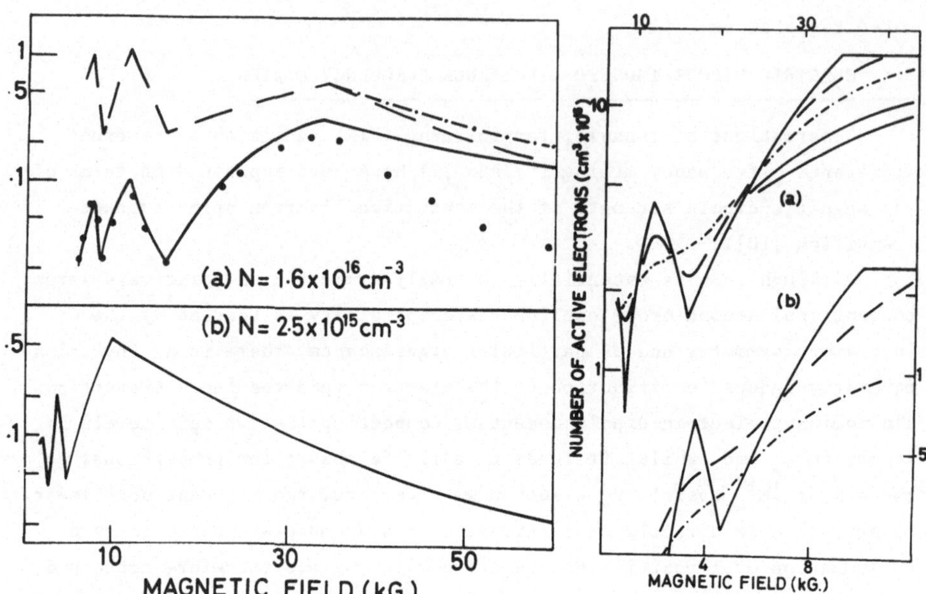

Fig. 1. Quantum Oscillations in the Raman cross-section (see text). Closed circles show relative SFRL experimental outputs.

Fig. 2. Temperature dependence of N_S (a) 1.6×10^{16} cm^{-3} (b) 2.5×10^{15} cm^{-3}, T = 0 —, 10 — , 30K —·—. 50, 70K lower full lines on (a).

[1] Dennis, R.B., Firth, W.J., McNeish, A., Pidgeon, C.R., Smith, S.D., Smith, J.W., Wherrett, B.S., and Wood, R.A., Proc. 11th Int. Conf. on the Phys. of Semiconductors.

[2] Wood., R.A., McNeish, A., Pidgeon, C.R., and Smith, S.D., J. Phys. C. 6, L144 (1973); Smith, S.D., Pidgeon, C.R., Wood, R.A., McNeish, A., and Brignall, N., Vail Conference on Tunable Laser (1973), to be published Plenum Press.

[3] Dennis, R.B., Pidgeon, C.R., Smith, S.D., Wherrett, B.S., and Wood, R.W., Proc. Soc. Lond. A. 331, 203 (1972).

[4] For form of Wavefunctions see Yafet, Y., Phys. Rev., 152, 858 (1966).

[5] Wherrett, B.S., and Wolland, S., to be published.

[6] Pidgeon, C.R., and Brown, R.N., Phys. Rev., 146, 575 (1966).

[7] Wherrett, B.S., and Firth, W.J., IEEE, QE-8, 865, (1972).

[8] Smith, S.D., Allwood, R.L., Dennis, R.B., Firth, W.J., Wherrett, B.S., Wood., R.W. and Hafele, H., Proc 2nd. Int. Conf. on Light Scattering in Solids, p 187, Paris: Flamarion Press (1971), Allwood., R.L., Ph.D. Thesis unpublished.

[9] Van Tran Nguyen, and Bridges, T.J., Phys. Rev. Letts.,29, 359 (1972).

[10] Brown, T.L., and Wolff, P.A., Phys. Rev. Letters, 29, 362 (1972).

[11] Wherrett, B.S., and Pidgeon, C.R., Phys. Rev., B9, 711 (1974).

[12] Pidgeon, C.R., and Groves, S.H., Phys. Rev. 186, 824 (1969).

NEW THEORETICAL RESULTS FOR THE InSb SPIN-FLIP RAMAN LASER

M. H. WEILER, R. L. AGGARWAL and B. LAX[*]
Francis Bitter National Magnet Laboratory[†]
Massachusetts Institute of Technology
Cambridge, Massachusetts 02139 USA

The Stokes output intensity is calculated, taking into
account the spatial dependence of the pump and Stokes
intensities and cavity boundary conditions. Both ana-
lytical and numerical results are obtained for the two
allowed pump polarizations, and are compared with
experiment. The effect of anti-Stokes generation is
discussed.

1. INTRODUCTION

We calculate the Stokes output intensity for the n-type InSb spin-flip Raman
(SFR) laser, which is an efficient source of magnetically tunable radiation
in the region of 10 μm, when pumped with a CO_2 laser[1], or 5 μm, using
a CO laser[2]. Our calculation takes into account the spatial dependence
of the pump and Stokes intensities and the boundary conditions at both ends
of the InSb sample cavity. We obtain the Stokes intensity as a function of
pump intensity and applied magnetic field. Previous calculations [3, 4] have
used mainly traveling-wave techniques, obtaining steady-state solutions by
replacing the absorption coefficients a by $(a - \ln R/L)$, where R is the re-
flectivity and L the sample length. In order to calculate the Stokes output,
we assume that anti-Stokes generation has negligible effect on the direction
and intensity of the Stokes output; we justify this assumption below.

2. PUMP AND STOKES EQUATIONS

Yuen [3] has developed equations for the pump, Stokes and anti-Stokes
electric fields E_p, E_s and E_a and for the electron density matrix operators
$N_1 = \Sigma\, a_1{}^+ a_1$, $N_2 = \Sigma\, a_2{}^+ a_2$ and $M = \Sigma\, a_1{}^+ a_2$, which are averaged over
the electrons states; $a_1{}^+$ creates an electron in the $l = 0$, spin-up conduc-
tion-band ground state, and $a_2{}^+$ creates an electron in the $l = 0$, spin-
down state. Assuming the total electron density is $N_e = N_1 + N_2 =$ constant,
and neglecting anti-Stokes generation, the equations involve only E_p, E_s,
M and $D = N_1 - N_2$. The equations include relaxation terms involving Γ,

[*]Supported by the National Science Foundation.
[†]Also Physics Department, M.I.T.

the spontaneous Stokes linewidth, and $W_{12} = 1/\tau_s$, the spin-flip relaxation rate, with N_{eff} the thermal equilibrium value of D. Into these equations we substitute forward and backwards plane waves in the collinear configuration, with the magnetic field $\vec{H} \parallel \vec{z}$ and the pump and Stokes wave vectors $\vec{k}_p$, $\vec{k}_s \parallel \vec{y}$. We obtain the following equations for the forward (I_{j_1}) and backward (I_{j_2}) pump and Stokes intensities $(j = p, s)$:

$$(I_{p_1})^{-1} \, dI_{p_1}/dy = - (I_{p_2})^{-1} \, dI_{p_2}/dy = - \alpha_p - \mu_s I_s/(1 + \beta I_p I_s) \tag{1a}$$

$$(I_{s_1})^{-1} \, dI_{s_1}/dy = - (I_{s_2})^{-1} \, dI_{s_2}/dy = - \alpha_s + \mu_p I_p/(1 + \beta I_p I_s) \tag{1b}$$

where

$$\mu_j = \frac{16\pi^2 c^2 N_{eff} d\sigma/d\Omega}{\hbar \Gamma n_s n_p \omega_s^2 \omega_j} \,, \quad \beta = \frac{32\pi^2 c^2 \tau_s \, d\sigma/d\Omega}{\hbar^2 \Gamma n_p n_s \omega_s^3 \omega_p} \,, \tag{2}$$

$I_j = I_{j_1} + I_{j_2}$, $\omega_s = \omega_p - \omega_e$ is the Stokes frequency, α_p and α_s are the pump and Stokes absorption coefficients, and $d\sigma/d\Omega$ is the photon cross-section per electron. To obtain numerical solutions, we add to Eq. (1b) small terms representing spontaneous Stokes scattering. Equations (1) are solved by iterative forward and backwards integration, starting with estimated solutions, until the result converges with the boundary conditions satisfied.

3. ANALYTICAL SOLUTIONS

We replace the functions $I_j(y)$ on the right-hand sides of Eqs. (1) by their average values $\langle I_j \rangle = L^{-1} \int_o^L dy \, I_j(y)$ and obtain exponential solutions. Neglecting pump depletion, we obtain the Stokes output intensity

$$I_s = - \tfrac{1}{2} \ln R \langle I_s \rangle = I_{sat} \, (1 - I_{th}/I_p) \tag{3}$$

where the threshold pump intensity is

$$I_{th} = \frac{\hbar \Gamma n_s n_p \omega_s^2 \omega_p}{16\pi^2 c^2 N_{eff} d\sigma/d\Omega} \left(\alpha_s - \frac{\ln R}{L} \right) \left(\frac{\alpha_p L}{1 - R} \right) \left(\frac{1 - Re^{-\alpha_p L}}{1 - e^{-\alpha_p L}} \right) \tag{4}$$

and the saturation Stokes intensity is

$$I_{sat} = \frac{N_{eff} \hbar \omega_s}{4\tau_s} \, L_{eff} \tag{5}$$

where $L_{eff} = L/(1 - \alpha_s L/\ln R)$ is an effective generation length. Other groups have obtained similar results, but with somewhat different factors which come from using $\alpha_j \to \alpha_j - \ln R/L$ and neglecting backwards waves.

Computing $\langle I_p \rangle$ to first order in $\langle I_s \rangle$, we then obtain for I_s

$$I_s = \frac{1}{2} \, I_{sat} \left[I_p + I_d - \sqrt{(I_p - I_d)^2 + 4I_{th} I_d} \right] / I_d \tag{6}$$

where I_d is the pump depletion intensity

$$I_d = \frac{N_{eff} \, h\omega_p}{2\tau_s} \left(\frac{L}{1-R} \right) \left(\frac{1 - Re^{-\alpha_p L}}{1 - e^{-\alpha_p L}} \right) \left[1 - \frac{\alpha_p L (1-R) e^{-\alpha_p L}}{\left(1 - e^{-\alpha_p L} \right) \left(1 - Re^{-\alpha_p L} \right)} \right] \tag{7}$$

If $I_p \gg I_d$ or $I_p \gg I_{th}$, Eq. (6) reduces to Eq. (3) but if $I_p \sim I_d$ even Eq. (6) will not be accurate.

Comparison of the numerical and analytical results shows that Eqs. (4) and (5) for I_{th} and I_{sat}, are extremely accurate, within $\sim 0.3\%$. However, Eq. (3) can be off by $\sim 30\%$ or more when I_p is 2 to 10 times I_{th}, that is, in the region of maximum efficiency. Equation (6) is more accurate, within a few percent depending on the value of I_d.

4. COMPARISON WITH EXPERIMENT

We compare our results with the experiments of Aggarwal et al. [5, 6] for CO_2 laser pumping at 10.59 μm, and a sample with $N_e = 2 \times 10^{16}$ cm^{-3} and 2 cm long. For these parameters we calculate $d\sigma/d\Omega$, N_{eff}, and α_p and α_s to fit our experimental data[7], for the two scattering geometries $y(z, x)y$ and $y(x, z)y$ denoting $k_p(E_p, E_s)k_s$. Substituting these values into Eqs. (4) and (5) we obtain the curves for I_{th} and I_{sat} shown in Fig. 1. Note that I_{th} is lower for the (z, x) polarization but I_{sat} is also lower in this case.

In Fig. 2 we compare our results with those of Aggarwal et al. [5] at 59.1 kOe for the (z, x) polarization. Curve $\underline{a}$ is the numerical results; curve $\underline{b}$ was calculated using Eq. (3); Eq. (6) gives a curve very close to curve $\underline{a}$. Finally, in curve $\underline{c}$, we have integrated the numerical result over a Gaussian pump-beam profile. The numerical threshold agrees with that observed using roughly the measured spontaneous linewidth $\Gamma' = \Gamma/2\pi c = 1.7$ cm^{-1}. Curve $\underline{c}$ approximates the data using a spin-flip relaxation time of $\tau_s \simeq 1.6$ ns rather than the previous estimate, based on an incorrect expression for I_{sat}, of ~ 20 ns.

FIG. 1. Threshold and saturation intensities vs magnetic field, for the two allowed polarizations.

FIG. 2. Stokes output intensity vs input pump intensity, compared with the data in Ref.5 for H = 59.1 kOe.

Finally, in Fig. 3 we compare the Stokes output, from Eq. (6) with a Gaussian pump beam, with that observed by Aggarwal et al. [6] at $I_p \simeq 3 \times 10^6$ W/cm^2. Although there is qualitative agreement, our theory does not account for the strength or shape of the minimum near H $\simeq$ 90 kOe. Measurements of the saturation intensity vs H, for a short sample, would indicate whether τ_s is changing significantly in this region of magnetic field.

FIG. 3. Stokes output intensity vs magnetic field, compared with the data in Ref. 6, for a pump intensity $\sim 3 \times 10^6$ W/cm^2.

5. ANTISTOKES GENERATION

There have been several efforts to include anti-Stokes generation in the
SFR theory [3, 8]. The usual technique [9] is to consider propagation at
small angles $\theta_{s,a} \simeq [(\Delta k - \Delta k_o)\omega_{a,s}/k_p\omega_{s,a}]^{1/2}$, where $\Delta k = 2k_p - k_{sy} - k_{ay}$,
and Δk_o is the collinear phase mis-match $\Delta k_o = (2n_p\omega_p - n_s\omega_s - n_a\omega_a)/c$.
Using indices of refraction for the (z, x) polarization, we obtain

$$\Delta k_o \simeq k_p \left[\omega_{p\ell}^2 (\omega_e^2 + \omega_c^2) + \omega_e^2 (\omega_{LO}^2 - \omega_{TO}^2) \right] / \omega_p^4 \tag{8}$$

where $\omega_{p\ell}$, ω_c, ω_{LO}, and ω_{TO} are the plasma, cyclotron resonance, longitu-
dinal optical phonon and transverse optical phonon frequencies respectively.
In most cases ($N_e \gtrsim 10^{15}$ cm^{-3}, $H \gtrsim 20$ kOe) $\Delta k_o > 0$ and is well above the
peak of the gain curve [9] with $\Delta k = \Delta k_o$ for $\theta_s = \theta_a = 0$, and $\Delta k > \Delta k_o$ for
θ_s, $\theta_a \neq 0$. Therefore we predict 1) the SFR output should peak at
$\theta_s = \theta_a = 0$, and 2) the anti-Stokes intensity $I_a \ll I_s$, in agreement with the
analysis of Wherrett [8]. Consequently, the anti-Stokes generation should
have little effect on the Stokes output.

6. REFERENCES

[1] Patel, C. K. N.; Shaw, E. D.: Phys. Rev. Lett. 24, 451 (1970); Phys.
 Rev. B3, 1279 (1971) and other references cited therein.

[2] Brueck, S. R. J.; Mooradian, A.: Appl. Phys. Lett. 18, 229 (1971).

[3] Yuen, S. Y.; Lax, B.; Wolff, P. A.: Opt. Commun. 10, 4 (1974); see
 also Lax, B.: Proc. 11th Intl. Conf. on the Physics of Semiconductors,
 (Polish Scientific Publishers, Warsaw, 1972), p. 1115.

[4] Dennis, R. B.; Pidgeon, C. R.; Smith, S. D.; Wherrett, B. S.; Wood, R. A.:
 Proc. Phys. Soc. (London) A331, 203 (1972).

[5] Aggarwal, R. L.; Lax, B.; Chase, C. E.; Pidgeon, C. R.; Limbert, D.:
 Appl. Phys. Lett. 18, 383 (1971).

[6] Aggarwal, R. L.; Lax, B.; Pidgeon, C. R.; Limbert, D.: Proc. 2nd Intl.
 Conf. on Light Scattering in Solids (Flammarion Press, Paris, 1971),
 p. 184.

[7] Weiler, M. H.; Aggarwal, R. L.; Lax, B.: Solid State Commun. 14, 299
 (1974).

[8] Wherrett, B. L.: Opt. Commun. 6, 402 (1972).

[9] Shen, Y. R.; Bloembergen, N.: Phys. Rev. A137, 1787 (1965).

CYCLOTRON HARMONICS ABSORPTION AND TEMPERATURE DEPENDENCE OF STIMULATED SPIN-FLIP RAMAN SCATTERING IN n-InSb

R. GRISAR AND H. WACHERNIG

Max-Planck-Institut für Festkörperforschung, Hochfeldmagnet-
labor, B.P. 166, 38042 Grenoble-Cedex, France

New experimental results on the CO_2 laser pumped
stimulated spin-flip Raman scattering (SFR) in
n-InSb show striking spin-optical phonon resonances
at high magnetic fields. The suitability of the
SFR laser for intraband spectroscopy in InSb is
demonstrated. The measured temperature dependence
is ascribed to spontaneous line broadening effects.

1. INTRODUCTION

Three new aspects of stimulated spin-flip Raman scattering (SFR)
in n-InSb, pumped with a pulsed 10.6 µm source, were investi-
gated: The first is the extension of the magnetic field from
10.5 to 14 T. In this range the spin-split energy reaches the
optical phonon energies. The second concerns the intraband ab-
sorption of the scattered light due to cyclotron harmonics in
the medium itself and the third is the temperature dependence
of SFR.

2. EXPERIMENTAL

The CO_2 laser had a peak power of 10 kW and a pulse width of
200 nsec. The experiments were carried out in Voigt configura-
tion and in the collinear scattering geometry (see inset of
Fig.1) on an (100)-oriented, 8 mm thick n-InSb sample with a
carrier concentration of 1.57×10^{16} cm^{-3}. It was mounted in a
variable temperature cryostat which was placed in the 5 cm dia-
meter bore of a 15 T Bitter magnet.

3. RESULTS AND DISCUSSION

For an understanding of all our results the following relation
for the SFR laser intensity I_R is essential:

$$(1) \quad I_R \propto e^{(g_R - \alpha_R)\,l}.$$

α_R is the absorption coefficient for the scattered light, l the sample length and the Raman gain factor g_R is

$$(2) \quad g_R \propto \frac{I_L\, n}{\Gamma}$$

(I_L: pump laser intensity; n: number of scattering electrons in the lowest spin state of the conduction band; Γ: spontaneous Raman line width). The two quantities g_R and α_R critically determine the SFR laser power.

3.1 HIGH MAGNETIC FIELDS

We have measured the spin-flip laser output at constant pump power. Particularly at fields above 10.5 T, where no results

Fig.1: Spin-Flip Raman laser power versus magnetic field. The involved light, phonon and intra-band transition frequencies, respectively, are plotted in the upper part. Arrows indicate the origin of observed minima

were available up to now, a striking new structure appears which is shown in Fig.1. The upper part of Fig.1 serves to identify the origin of these minima. It shows, as a function of magnetic field, the frequencies of the Raman scattered light ω_R and of the intra-conduction band transitions. ω_S is the spin-splitting and ω_C the cyclotron resonance frequency. Also drawn are the frequencies of the CO_2 laser ω_L and of the optical phonons ω_{LO} and ω_{TO}.

The minimum of the SFR power at 9.2 as well as the cutoff at

10.2 T are well known and were ascribed to the absorption at $\omega_R = 2\omega_C$ [1,2] and the tail of the strong absorption at $\omega_R = \omega_C$ [2] near 20 T, respectively. The values of α_R were measured since [3] and are too small to prevent laser action. In this paper we report that even with our relatively low pump power we could operate the SFR laser at higher fields.

An essential new effect leads to a dropoff at 10.5 and 12.5 T, where the spin splitting is in resonance with the TO and LO optical phonons, respectively. At these two fields the spins can relax under $\vec{k}$ - conservation by emission of single phonons. The lifetime of the electrons in the upper spin level, which is the final state in the Raman process, is thus reduced drastically from 10^{-7} sec [4] probably by several orders of magnitude, thereby increasing the spontaneous scattering line width and thus preventing SFR laser action. It is not clear if the coupling of the spin system to the TO phonons at 10.5 T alone can stop the laser action or if the CO_2 laser light absorption peak at 10.4 T affects negatively the scattering conditions. Two absorption peaks for the Raman scattered light at 11.8 and 14 T are indicated in the figure.

These first results demonstrate that spin-flip Raman scattering is a qualified tool to study this new type of spin-phonon coupling.

3.2 INTRABAND SPECTROSCOPY

The Stokes output power at lower fields is shown in more detail in Fig.2. Various absorptions of the Raman scattered light by cyclotron harmonics, indicated in the upper part, lead to dips in the SFR laser power. Parts of this fine structure were already observed previously [2]. Decreasing the pump laser power I_L makes the SFR power more sensitive to small absorptions. We thus could identify cyclotron harmonics up to $7\omega_C$ and phonon-assisted cyclotron harmonics up to $3\omega_C + \omega_{LO}$. The fact that only very weak absorption minima of higher even harmonics appear is due to the selection rules for InSb [5].

By a conventional absorption measurement under similar conditions [3] only harmonics up to $3\omega_C$ could be seen. This comparison

806

Fig.2 Spin-flip Raman laser power versus magnetic field for different pump laser powers in n-InSb at 4 K. Arrows indicate the assignment of the observed minima to intraband absorption of the scattered light.

shows the sensitivity of stimulated SFR for intraband spectroscopy of the scattering material.

3.3 TEMPERATURE DEPENDENCE

Fig.3 shows the relative Stokes output power of the SFR laser versus magnetic field for three different temperatures. Up to 30K

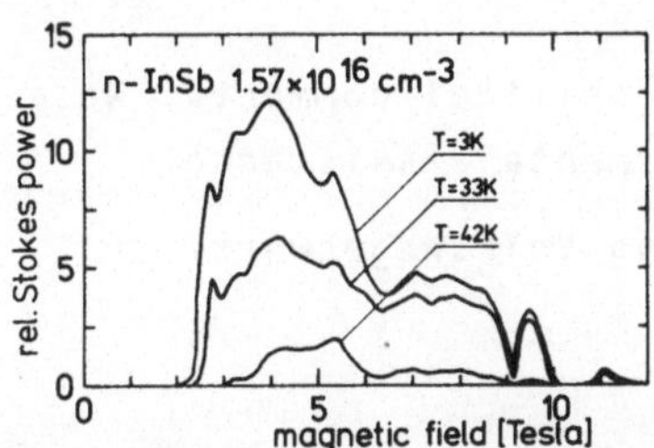

Fig.3: Spin-flip Raman laser power versus magnetic field for different temperatures.

the effect of the temperature is negligible, except below 3 T. Above 30 K the SFR power falls more and more rapidly. The highest temperature at which stimulated SFR could be observed was 47 K, even for the highest fields. At 13 T and 47 K the thermal population of the upper spin level is 0.2%. Obviously this cannot be the reason that the laser action stops.

If we assume that the absorption α_R of the Raman light does not rise significantly with increasing temperature up to 50 K, the only way that the temperature can act on the SFR laser is by broadening the spontaneous Raman line width and thus reducing the gain factor g_R. We measured the threshold pump intensity I_{th} which under the above assumption is proportional to Γ. That is a

consequence of eq.(2).

Fig.4: Threshold pump inten-
sity of the spin-flip
Raman laser versus tem-
perature in n-InSb at
4.2 T.

Fig.4 shows the experimental values for the threshold pump inten-
sity as a function of temperature at a magnetic field of 4.2 T.
The curve is representative for all magnetic fields. Up to 40 K
the threshold pump intensity rises only insignificantly. Between
40 and 45 K it increases drastically by about a factor of ten.

Unfortunately no Raman or infrared data on the temperature depen-
dence of the line width of the spontaneous scattering with a 10.6
μm pump source are available. The detailed data obtained in the
5 μm region [6] with a c.w. CO laser do not exhibit this strong
variation. At present this discrepancy is not fully understood.

4. ACKNOWLEDGEMENT

We wish to thank Prof.K.Dransfeld for critical comments. We are
thankful to M.Berger for expert experimental assistance.

This work was supported by the Stiftung Volkswagenwerk.

5. REFERENCES

[1] Aggarwal,R.L., Lax,B., Pidgeon,C.R. and Limbert,D.,
 Proc. 2nd Int. Conf. Light Scattering in Solids, p.184
 Flammarion Sciences, Paris 1971

[2] Dennis,R.B., Wood,R.A., Pidgeon,C.R., Smith,S.D. and Smith,
 J.W., J.Phys. C 5,L73 (1972)

[3] Weiler,M.H., Aggarwal,R.L. and Lax,B., Sol.St.Commun. 14,
 299 (1974)

[4] Wachernig,H. and Grisar,R., to be published

[5] Bell,R.L. and Rogers,K.T., Phys.Rev. 152, 746 (1966)

[6] Brueck,S.R.J., Mooradian,A. and Blum,F.A., Phys.Rev. B 7,
 5253 (1973)

HOT ELECTRONS

CONTENTS

THEORY OF THE HOT ELECTRON MAGNETOPHONON EFFECT

J.R. BARKER AND B. MAGNUSSON

Department of Physics, University of Warwick, Coventry, U.K.

The basic features and theories of non-ohmic magnetophonon spectra
are briefly reviewed. Numerical calculations for longitudinal and
transverse one phonon m.p. resistivities are presented based on non-
linear quantum kinetic equations in the Landau and crossed field
representations including collision broadening.

1. INTRODUCTION

Ohmic magnetophonon spectroscopy, reviewed in [1], appears to be limited to
temperatures in excess of 40 K by the weak excitation of the relevant phonon
modes. However, the initiation of even moderate electron heating reveals the
appearance of non-ohmic spectra which exhibit markedly different and richer
structures over a lattice temperature regime 4 - 100 K. It has become evident
that the high sensitivity of heated carrier distributions to resonant energy
as well as momentum relaxation is responsible and hitherto unsuspected
processes have been isolated [1]. Unfortunately, the full potential of non-
ohmic m.p. spectroscopy is at present severely limited by the lack of any
comprehensive and quantitative theory. Following a survey of the basic phen-
omena and current theoretical models we report on recent numerical studies
based on the general non-linear quantum kinetic equations for hot electrons.

2. SURVEY OF HOT ELECTRON MAGNETOPHONON STRUCTURES

2.1 GENERAL FEATURES

The full gamut of hot electron m.p. phenomena has been reviewed recently [1].
We concentrate on the low temperature magnetoresistance data. Experimental
spectra present the following features: (i) complex structure: many oscil-
latory (1/B) series present; (ii) poor correlation of extrema positions with
respect to ohmic series; (iii) transverse and longitudinal resistivities tend
to exhibit extrema of the same sign and comparable magnitude in any one
material; (iv) strong dependence of amplitudes on electric field; (v) rogue
extrema, uncorrelated with known phonon processes; (vi) missing series.

2.2 SINGLE L.O. PHONON PROCESSES

The dominant structure in the low temperature (~ 20 K) spectra of n-type InSb
InAs, GaAs, InP, CdTe has been attributed [1] to a resonant capture process
involving an electronic transition from the high density of states region of
a Landau sub-band to a donor level accompanied by the emission of a zone
centre LO phonon. This conjecture is supported by the remarkable fit of the
experimental plots of extremum number versus inverse magnetic field [1] to the
non-orthodox resonance relation $\hbar\omega_L - W(0) \simeq (N+\frac{1}{2})\hbar\omega_c$ where $\hbar\omega_L$ is the LO

phonon energy and $W(B) \simeq W(0) + \frac{1}{2}\hbar\omega_c$ is the magnetic field dependent donor activation energy. The only major uncertainty appears to be an assessment of the correct polaron mass enhancement manifest in the cyclotron frequency ω_c. Two features are puzzling: (i) the transverse and longitudinal resistivities are of similar magnitude and equal sign and display minima in n-InSb, InAs, but maxima in n-GaAs, InP, CdTe; (ii) the failure to observe the orthodox LO phonon series corresponding to $\hbar\omega_L = N\hbar\omega_c$: with InSb and InAs the two series may not be resolved since $W(0)$ is small, but has been tentatively identified in the complex spectra of n-CdTe for $E > 6$ V/cm.

2.3 TWO PHONON PROCESSES

The first m.p. structure to emerge in the III-V's below 20 K as heating onsets is generally attributed to an energy relaxation process involving the emission of two zone edge X point TA phonons (energy $\hbar\omega_{TA}$) of approximately opposite wave-vector in a conventional intra-Landau level resonant transition. Data for n-InSb, GaAs, InP is consistent with m.p. extrema assigned by the relation: $2\hbar\omega_{TA} = N\hbar\omega_c$. A theoretical study of the basic process, its strength and selection rules has been made recently [2, 16] . The mechanism is thought to be very weak but has a lower threshold energy than the LO phonon processes. The 2TA series dissolves at higher field strengths and temperatures as the LO capture series emerges. The simultaneous observation of the 2TA and LO capture series is consistent with a weak 2TA mechanism which does not seriously prohibit electron flow up to the LO capture threshold. Again we have the mystery of a missing series: if the 2TA process goes, we might expect observations of a 2TA phonon assisted capture series which would have an even lower threshold.

2.4 ROGUE STRUCTURE

Almost all spectra display unidentified m.p. structure not necessarily associated with electron-phonon processes. Some features have been tentatively assigned to resonant impact ionisation [1,3] and to resonant excitation of neutral impurities from the ground state to excited states [4]. In addition, unexplained sharp minima or peak splittings have been observed on the very high magnetic field peaks for strong electric fields [1].

2.5 SILICON

Very recently a Fourier transform m.p. spectroscopic study of n-Si [5] reports the resolution of at least 11 different series by applying fields of up to 120 V/cm for $T = 55$ -77 K. Of particular interest is the identification of a TA g-phonon series which at first sight is forbidden by selection rules.

2.6 FIELD DEPENDENCE OF MAGNETOPHONON AMPLITUDES AT HIGHER TEMPERATURES

Here, data is sparse. Curby and Ferry [6] in studies of ρ_{xx} extrema in n-InSb at 77 K for relatively strong fields (1-50 V/cm) report a non-linear shift of the conventional LO series maxima to high magnetic fields with increasing electric field, the amplitudes successively damping out beyond 50 V/cm. Bauer and Kahlert [7], in time resolved measurements on n-InSb (10-77 K), report a non-linear decrease of the normalized amplitudes for ρ_{zz} extrema with increasing field (E > 0.4 V/cm).

3. THEORETICAL MODELS

3.1 THE ELECTRON TEMPERATURE CONCEPT

The most popular interpretations of hot electron spectra rely on the assumption of Maxwellian heating [8,9,19] requiring that the inter-carrier collision frequency considerably exceeds the frequency of electron phonon collisions. Kalashnikov [10] has shown that in this case the electron density matrix may be parametrized by an effective electron temperature T_e, drift velocity v and chemical potential μ. For warm electrons (v∿0) the electron temperature varies as

$$T_e = T\left[1 + \frac{ne^2E^2}{m^*C_e k_B T}\ \Omega_{ep}^{-1}\left(\frac{\omega_\perp}{\omega_c^2 + \omega_\perp^2}\right)\right] \quad ; \ \underset{\sim}{E} \perp \underset{\sim}{B} \tag{1}$$

$$T_e = T\left[1 + \frac{ne^2E^2}{m^*C_e k_B T}\ \Omega_{ep}^{-1}\left(\frac{1}{\omega_\parallel}\right)\right] \quad ; \ \underset{\sim}{E} \parallel \underset{\sim}{B} \tag{2}$$

where C_e is the equilibrium electronic specific heat, and Ω_{ep}, $\omega_\perp$, $\omega_\parallel$ are the energy, transverse and longitudinal momentum relaxation rates (correlation functions). Physically, a dynamic equilibrium is established in which the power gained by electrons from the electric fields is dissipated to the lattice by phonon emmission. At low temperatures, the dominant energy relaxation is assumed to proceed by LO or short wavelength TA phonon emission from high energy electrons. It follows from (1) and (2) that the electron temperature reflects the net energy relaxation rate Ω_{ep}. Provided the Landau structure is well resolved, Ω_{ep} will be enhanced whenever two high density of states regions of the electron spectrum are coupled by the relevant inelastic phonon energy. T_e should then fall to the lattice temperature T: resonant cooling. On the other hand, if ionised impurity scattering dominates the momentum relaxation, $\omega_{\parallel,\perp} \simeq \omega_{ei}^{\parallel,\perp}$, which is generally the case at low temperatures, both the transverse and longitudinal resistivities should maximise at the resonant magnetic fields. Recent numerical calculations for the longitudinal resistivity, based on a similar model but neglecting collision

broadening [9] confirm this general prediction. However, the experimentally observed minima in n-InSb, n-InAs and the missing series phenomenon discussed in §2 are not explicable, and indeed suggest a severe distortion of the electron distribution from Maxwellian form. Furthermore, theoretical studies [17,11]of electron-electron scattering in quantising magnetic fields suggest that the ratio of electron-electron and electron-phonon collision frequencies is smaller than the classical ratio by the large factor $(\hbar\omega_c/k_B T_e)^2$ $\exp(\hbar\omega_c/k_B T_e)$. Thus considerably higher carrier concentrations are necessary for the electron temperature ansatz than is generally assumed. Proper account of screening in high magnetic fields [12] might of course affect this conjecture.

3.2 THE ENERGY DIFFUSION MODEL

Significant progress beyond the effective temperature concept has been achieved by Kurosawa and Yamada [13,14] using the energy-diffusion equation method in which the electron motion is represented at very low temperatures as a Brownian motion in energy space governed by diffusion coefficients derived from the electric field and collision induced particle flows. Here, electrons are considered to diffuse in energy space up to the optic phonon emission threshold ε_o at which they are rapidly returned to the bottom of the band. The mobility is assumed to be limited by ionised impurity scattering, and is derived with the aid of the electron energy distribution function $f(\varepsilon)$ considered to be the same for each Landau sub-band. The latter assumption requires very rapid inter sub-band transition rates, in addition to the diffusion approximation that field plus collision induced transitions should involve very small energy changes except for optic phonon scattering. The divergent undamped state densities are handled by a cut-off procedure. The model predicts a sharp cut-off in $f(\varepsilon)$ at ε_o, and elsewhere $f(\varepsilon)$ is severely distorted from the Maxwellian at even moderate electric field strengths: the distortion being of opposite shape for the transverse and longitudinal field configurations. For transverse fields [13], the average electron energy $\bar{\varepsilon}$ increases and ultimately saturates with increasing electric field, maximising on the m.p.r. condition and leading to minima in the resistivity. With long-itudinal fields [14] $\bar{\varepsilon}$ behaves similarly except that it minimises at resonance but again minima are predicted in the resistivity. Such behaviour is qualitat-ively similar to low temperature InSb data [1], but does not appear to explain the maxima observed in other III-V's.

4. THE QUANTUM KINETIC EQUATION APPROACH

The evident conflict between the existing theoretical models has prompted us to initiate numerical calculations for both transverse and longitudinal non-ohmic resistivities based on rigorous non-linear quantum kinetic equations, incorporating quasi-Lorentzian collision broadening (estimated from ohmic data), for LO phonon m.p.r. in single valley materials. For $\underline{E} \parallel \underline{B}$ we use the Landau basis $(N, k_y, k_z) \equiv \lambda$ and solve the non-linear generalised Boltzmann equation for the electron distribution $f_N(k_z)$. For $\underline{E} \perp \underline{B}$, the crossed field representation [15] is more convenient as the current J_x can be computed from the diagonal elements of the electron density matrix f_λ according to the jump formula:

$$J_x = \tfrac{1}{2}e \sum_{\lambda,\lambda'} (X - X')(f_\lambda R_{\lambda\lambda'} - f_{\lambda'} R_{\lambda'\lambda}) \tag{4}$$

where $X \equiv -(\hbar k_y/eB + m^* E/eB^2)$, $R_{\lambda\lambda'}$ is a field dependent transition rate compounded from electron-phonon and electron-impurity processes whilst f_λ satisfies

$$\sum_{\lambda'} (f_{\lambda'} R_{\lambda'\lambda} - f_\lambda R_{\lambda\lambda'}) = 0 \tag{5}$$

4.1 LONGITUDINAL CONFIGURATION: E ∥ B

Inclusion of even weak collision broadening leads to strong qualitative changes in the electron distribution functions contrasting strongly with the highly singular structure obtained by the generally used zero-damping or cut-off approximations. FIG.1 illustrates the effect for different values of $\gamma = \Gamma/k_B T$, where Γ is the average self-energy. For the warm electron regime, some support is found for a displaced Maxwellian form for $f_N(k_z)$, but in which different temperatures and drift velocities are associated with each sub-band N (FIG.2). However, the distributions are reduced beyond k_o, the LO phonon threshold. At higher fields, the distributions strongly distort, exhibiting carrier pile up near k_o. This behaviour could explain the missing series effect of §2: a strong energy relaxation mechanism with a low threshold will prevent carrier flow to the higher thresholds of other processes. The positions and amplitudes of the m.p. extrema are strongly field dependent, the amplitudes eventually decreasing with electric field as found by Bauer and Kahlert. FIG.3 shows the oscillatory part of ρ_{zz} for InSb in the warm electron region, for weak elastic scattering. We have insufficient data at present to interpret these results as displaying maxima or minima.

FIG.1 Distributions for InSb

FIG.2 Distribution for InSb

FIG.3 m.p. structure in InSb

N indicates ohmic transverse

m.p. extrema positions

4.2 TRANSVERSE CONFIGURATION

We have so far only studied the transverse resistivity in the range 70 –
100 K, for relatively high electric fields (5 – 30 V/cm). Preliminary
results find some qualitative agreement with the experimental data of Curby
and Ferry 6 . Figure 4 shows the first two m.p. peaks in InSb at 77 K,
$N = 5 \times 10^{14}$ cm^{-3}, for two different electric fields.

FIG.4 Non-ohmic spectra in InSb

FIG.5 Distribution functions

Arrows denote the 1st and 2nd ohmic m.p. extremum positions. The maxima
shift gradually to higher magnetic fields and progressively damp out. This
behaviour stems from the highly distorted non-Maxwellian form of the heated
electron distribution. The damping is partially controlled by the distortion
of the collision processes by the total electric field (applied field plus
Hall field) and modifies the energy dependence of the joint spectral density
function for transitions to the form $\pi^{-1}\Gamma_{\lambda\lambda'}/\{[\varepsilon_\lambda-\varepsilon_{\lambda'}+\Delta_{\lambda\lambda'}\pm\hbar\omega_o+eE(X-X')]^2+\Gamma_{\lambda\lambda'}^2\}$
The additional inelasticity $E\Delta X$ is proportional to the momentum transferred
to the scatterer, $\Delta X = \hbar\Delta k_y/eB$, a similar, but less acute effect has been
predicted for zero magnetic field [18]. This mechanism becomes more impor-
tant than collision broadening when $eE\Delta X/\Gamma_{\lambda\lambda'}>1$, where $\Gamma_{\lambda\lambda'}$ is a convolution
of the imaginary part of the initial and final state self energies [18]. In
the present case the critical applied field is about 5 V/cm. FIG.5 illustr-
ates the N=0 and N=1 distribution functions for point of A of FIG.4. Again,
the distributions are level dependent and exhibit a pile up followed by a
sharp fall in the vicinity of the LO phonon threshold k_o, whereas the low
energy portions are strongly heated by rapid scattering in from high energy
states and the inefficiency of acoustic phonon energy dissipation. The
overall distortion is reminiscent of Yamada and Kurosawa's results [13] and
could similarly lead to minima developing in ρ_{xx} when impurity scattering
dominates the momentum relaxation at low temperatures.

5. DISCUSSION

Precision application of hot electron m.p. spectroscopy depends ultimately
on minimising the uncertainties in interpretations of the extrema locations.
Objectively, Fourier spectroscopy [5] appears superior to the subtraction
and double differentiation techniques. There remains however considerable
uncertainty in the conventional polaron mass correction which is adequate
for extrema correlated by the condition $\omega_o = N\omega_c$. Aside from the corres-
ponding capture assisted and two phonon processes, it is not yet clear whether
strongly displaced extrema, which are fabricated from transitions far away
from resonance, are determined by the simple mass correction or by a
functional of the complete dynamical mass $m^*(B)$. Finally, our preliminary
numerical study suggests that hot electron m.p. series are not easily
interpreted in terms of simple periodic oscillations in resistivity arising
from simple physical resonance.

6. REFERENCES

[1] Harper, P.G., Hodby, J. & Stradling, R.A. Rep. Prog. Phys. **36** 1 (1973).

[2] Baumann, K., Acta Physica Austriaca **37** 350 (1973).

[3] Eaves, L. et al. J. Phys. C to be published (1974).

[4] Hoult, R.A., D. Phil. Thesis, University of Oxford (1974).

[5] Portal, J.C., Eaves, L., Askenazy, S., & Stradling, R.A. Sol. St. Comm. (1974).

[6] Curby, R.C., & Ferry, D.K., 11th Semicond. Conf. (Warsaw) 312 (1972).

[7] Bauer, G. & Kahlert, W., private communication (1974).

[8] Pomortsev, R.V. & Kharus, G.J., Sov. Phys.- Sol. St. $\underline{9}$ 1150 (1967).

[9] Peterson, R.L., Phys. Rev. $\underline{B5}$ 3994 (1972).

[10] Kalashnikov, V.P. Physica $\underline{48}$, 93 (1970).

[11] Zlobin, A.M. & Zyryanov, P.S., Sov. Phys. JETP $\underline{31}$ 513 (1970).

[12] Wallace, P.R, J. Phys. C $\underline{7}$ 1136 (1974).

[13] Yamada, E. & Kurosawa, T., J. Phys. Soc. Japan $\underline{34}$ 603 (1973).

[14] Yamada, E. Solid.St. Comm. $\underline{13}$ 503 (1974).

[15] Budd, H.F., Phys. Rev. $\underline{175}$ 271 (1968).

[16] Ngai, K.C. & Ganguly, A.K., Phys. Rev. $\underline{B8}$ 5654 (1973).

[17] Magnusson, B. unpublished (1974).

[18] Barker, J.R., J. Phys. C $\underline{6}$ 2663 (1973).

[19] Zlobin, A.M. & Zyryanov, P.S., Sov. Phys. Uspekhi $\underline{14}$ (1972).

THE EFFECT OF BAND WARPING ON THE ANISOTROPY OF HOT-HOLE DRIFT VELOCITY IN GE

L. REGGIANI, C. CANALI, F. NAVA, G. OTTAVIANI

Istituto di Fisica, Università di Modena, I-41100 Modena, Italy

P. LAWAETZ

Physics Laboratorium III, Technical University of Denmark, DK-2800 Lyngby, Denmark

Time-of-flight measurements in the region of high electric fields in hyperpure Ge ($|N_D - N_A| \simeq 10^{11}$ cm^{-3}) between 40 °K and 77 °K have unambigously evidenced an anisotropic behaviour of the hot-hole drift velocity. Theoretical analysis demonstrate for the first time the direct correlation between the warped shape of the hole energy spectrum and the anisotropy of the hot-hole drift velocity without introducing unknown parameters.

1. INTRODUCTION

In the region of electric fields far away from linear response the hole drift velocity v_d in Ge has been found to exhibit an anisotropic behaviour in accord with the Sasaki effect [1]. The correlation between the v_d anisotropy and the warped shape of the Ge valence band seems to be the physical interpretation of this effect. This correlation may be however masked in principle by the energy and angular dependence of the microscopic scattering mechanisms on the carrier distribution function.

The purpose of this paper is to demonstrate for the first time the direct correlation between the warped shape of the heavy hole energy spectrum and the anisotropy of the hot-hole v_d experimentally measured in Ge between 40 °K and 77°K. Theory supported by the satisfactory agreement with experiments has permitted to obtain a physical insight of the whole phenomenon and to carry on a discussion on/these basic arguments.

i) Correlation among cyclotron resonance experiments, band structure calculation and transport parameters.

ii) Relative importance of acoustic and optical scattering mechanisms.

iii) Role played by the angular dependence of the hole acoustic-phonon

Fig. 1: Field dependence of the hole drift velocity v_d in Ge at 40 and 77 °K. Closed and open circles refer to $\langle 100 \rangle$ and $\langle 111 \rangle$ crystallographic directions respectively.

scattering cross-section.

2. EXPERIMENTS AND THEORY

Two sets of experimental data at 40 °K and 77 °K of the hole drift veloci ty in Ge were carried on as a function of the electric field E applied paral lel to $\langle 100 \rangle$ and $\langle 111 \rangle$ crystallographic directions. The measurments were performed on single crystal of hyperpure Ge ($|N_D - N_A| \simeq 10^{11}$ cm^{-3}) and the experimental apparatus was based on the time-of-flight technique coupled with the Modena University electron gun. The measure uncertainty is within 5%. The experimental results reported in Fig. 1 show a higher v_d for E applied in the $\langle 100 \rangle$ direction than in the $\langle 111 \rangle$ direction. The anisotropy of v_d attains values of about 20% and appears to saturate at the highest applied fields of 10^4 V/cm.

The theoretical analysis makes use of a Monte Carlo technique and is based on a single, warped and parabolic valence band with acoustic and non-polar optical phonon scattering. Accordingly the energy wave-vector relationship is[2]

$$(1) \qquad \epsilon = \frac{\hbar^2}{2m_o} \left\{ A\,k^2 - \left[B^2 k^4 + C^2 (k_x^2 k_y^2 + k_x^2 k_z^2 + k_y^2 k_z^2) \right]^{1/2} \right\}$$

A,B,C being the inverse-mass band parameters and m_o the free electron mass. Their values with other constants used for calculations are reported in Table 1. The warped shape of the heavy mass band[*] evaluated according with

Table 1 — Constants of Ge

Quantity		Value	Unit		
A	[3]	-13.38	—		
B	[3]	- 8.48	—		
$	C	$	[3]	13.14	—
Δ	[4]	0.295	eV		
ϱ	[5]	5.32	gr/cm^3		
θ_{op}	[5]	430	°K		
s_ℓ	[5]	$5.4\ 10^5$	cm/sec		
s_t	[5]	$3.2\ 10^5$	cm/sec		
E_i^o	[pw]	6.5	eV		
$(D_t K)$	[pw]	$9.0\ 10^8$	eV/cm		

pw present work

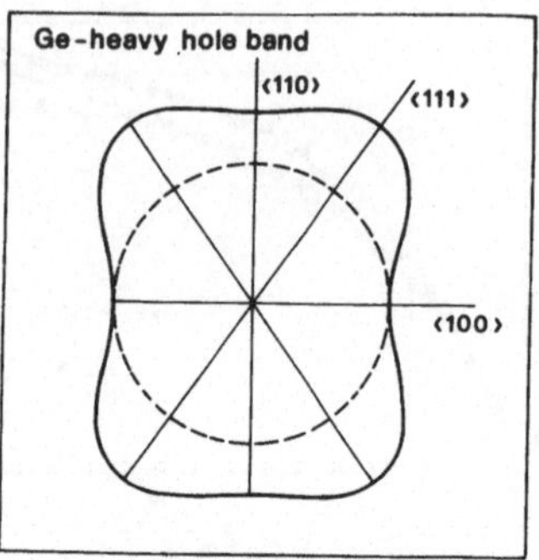

Fig. 2: Ge valence band constant energy contours in the (110) plane. The inner circle is representative of a spherical band.

Eq. 1 is shown in Fig. 2. The reliability of the single band assuption has been controlled by calculating, under the condition of non—degenerate statistics, the ratios n_1/n_h and n_{so}/n_h being n_h, n_1, n_{so} the equilibrium populations of the heavy mass band, the light mass band and the splitt—off band respectively. Values accounting spin—orbit interaction [6] have been found smaller than 0.1 and 0.01 for n_i/n_h and n_{so}/n_h in the energy range of interest in the present study. Furthermore, due to the small inertia of carriers belonging to the light and splitt—off bands, in non—ohmic region the carrier heating should be more pronounced for these bands than for the heavy one. Consequently, a depopulation of these bands in favour of the heavy one is expected with respect to the equilibrium condition. This occurrence further enforce the validity of our model. Following the results of Lawaetz [7] parabolicity also should be a reasonable approximation.

The hole—phonon scattering probability per unit time has been taken as [8]

$$(2) \quad P_{ac} = \frac{(K_B T) E_i^{o2}}{2 V \varrho s^2}\ G(\alpha)\ \delta(\epsilon_s - \epsilon_i) \quad ; \quad G(\alpha) = \frac{1}{4}(1 + 3\cos^2\alpha)$$

$$(3) \quad P_{op} = \frac{\hbar^2 (D_t K)^2}{2 V \varrho K_B \theta_{op}} \left(N_q + \frac{1}{2} \pm \frac{1}{2}\right) \delta(\epsilon_s - \epsilon_i)$$

* It is worthwhile to note that the light and splitt—off bands are characterized by near spherical equienergetic surfaces, thus in principle no anisotropic effects are expected to arise from these bands.

Where subscripts ac, op stand for acoustic and optical, α is the scattering angle, ϵ_i, ϵ_f the energy of the initial and final states, ϱ the crystal density, $s = 1/3\, s_\ell + 2/3\, s_t$ an average sound velocity, V the crystal volume, Θ_{op} the optical phonon energy assumed q independent, N_q the Bose–Einstein distribution function, and E_1^o, $(D_t K)$ the deformation potential parameters. The $\pm$ signs refer to emission and absorption processes. Acoustic scattering mechanism has been considered as usual elastic in the energy equipartition approximation, while the effectiveness of transverse as well as longitudinal modes has been conveyed in the acoustic deformation potential constant [9].

The transport approach, which due to the low carrier concentration $(n \approx 10^{12}\ cm^{-3})$ does not account for the carrier–carrier scattering, follows a time average method. Accordingly, a single carrier suffering a rate of momentum $\hbar \dot{\underline{k}} = e\underline{E}$ has a longitudinal drift velocity v_d [10]

$$(3) \qquad v_d = \lim_{t \to \infty} \frac{1}{eEt} \sum (\epsilon_f - \epsilon_i)$$

where the sum is evaluated over all carrier flight.

3. DISCUSSION

Theoretical results reported in Fig. 1 satisfactory agree with experiments and give evidence for the first time of the anisotropic effects of v_d. Since the inverse-mass band parameters here used have been taken from the most up to date cyclotron resonance data [3] and the values of E_1^o and $(D_t K)$ are in accord with a recent mobility analysis in cubic semiconductors [11], the present theory in the limit of the model used needs no unknown parameters. The relative importance of optical to acoustic scattering mechanisms has been found comparable, the coupling strenght ratio $\dfrac{W_{op}}{W_{ac}} = \dfrac{\hbar s\, (D_t K)}{K_b\, \Theta_{op}\, E_1^o}$ assuming the value 0.96. To test if the angular dependence of acoustic scattering affects the anisotropy of v_d, calculations have been performed by taking $G(\alpha) = 1/2$. Results obtained do not discriminate between isotropic and angular dependent scattering confirming that anisotropy of hot-hole drift velocity originates from the particular warped shape of the Ge valence band structure. Furthermore, the mean energy of holes in excess over thermal

equilibrium has been found to depend on the field orientation, the highest value being for E in the $\langle 100 \rangle$ direction and the lowest for E in the $\langle 111 \rangle$ direction.

4. REFERENCES

[1] Sasaki, W.; Shibuya, M.; Mizuguchi, K. and Hatoyama, G.M.: J. Phys. Chem. Solids 8, 250 (1959)

[2] Dresselhaus, G.; Kip, A.F. and Kittel,C.: Phys. Rev. 98, 368 (1954)

[3] Hensel,J.C. and Suzuchi, K.: Phys. Rev. Lett. 22, 838 (1969)

[4] Baynham, A.C. and Paige, E.G.S.: Proc. Sev.th Int. Conf. Paris (1964) (Academic Press Inc., New York, 1964) p. 149

[5] Conwell, E.M.: High Field Transport in Semiconductors, Supp. 9 (Academic Press, New York and London, 1967) p. 133

[6] Kane, E.O.: J. Phys. Chem. Solids 1, 82 (1956)

[7] Lawaetz, P.: Phys. Rev. 174, 867 (1968)

[8] Costato, M. and Reggiani, L.: Phys. Stat. Sol. (b) 58, 461 (1973)
Due to symmetry there are no apriori arguments for the angular dependence of the hole optical-phonon matrix element [7] , consequently for the present purpose optical-phonon scattering has been chosen isotropic.

[9] Ehrenreich, H. and Overhauser, A.W.: Phys. Rev. 104, 331 (1956)

[10] Fawcett, W.; Boardman, D.A. and Swain, S.: J. Phys. Chem. Solids 31, 1963 (1970)

[11] Costato, M.; Gagliani, G.; Jacoboni, C. and Reggiani, L.: to be published in J. Phys. Chem. Solids

MEASUREMENT OF INTERVALLEY REPOPULATION TIME IN SILICON

C. JACOBONI, C. CANALI, G. OTTAVIANI, A. ALBERIGI-QUARANTA

Istituto di Fisica, Università di Modena, Modena, Italy

The repopulation time of hot electrons between hot and cold valleys has been measured in hyper-pure Si at 8°K. Transit-time measurements with the electric field applied parallel to a $\langle 100 \rangle$ direction show in fact a dependence upon sample thickness, which can be interpreted by taking into account that repopulation takes place exponentially with time. The analysis of the results with Monte Carlo calculations allows determination of the strength of electron scattering with low-energy intervalley phonons.

1. INTRODUCTION

If a high electric field E is applied along a $\langle 100 \rangle$ direction in Si, the two valleys lying along the same direction are "cooler" and more populated than the other four. The electronic transitions necessary to establish this steady-state repopulation are intervalley f transitions (between perpendicular valleys), which require phonons (f-phonons) with equivalent temperature T_f of several hundred °K [1] . If the temperature is sufficiently low, the number of such phonons available to perform f transitions with an absorption process is extremely low. Furthermore, at intermediate fields the average electron energy is much smaller than the f-phonon energy ϵ_f, and emission processes are also very rare. In such conditions the time necessary to reach steady-state repopulation becomes very long. In what follows we shall describe how this time has been measured and microscopically interpreted.

2. EXPERIMENTAL MEASUREMENTS

Transit-time measurements [2] of drift velocity v_d of electrons in hyper-pure Si have been performed at 8°K with E parallel to $\langle 100 \rangle$ and $\langle 111 \rangle$ directions. Several samples with different thicknesses W have been used, and if v_d is evaluated by simply dividing W by the transit time T_R, $v_d(100)$ results strongly dependent upon W. This effect is shown in Fig. 1, where the experimental results obtained for W/T_R in some of the different samples are reported. The continuous line shows the experimental results obtained along

FIG. 1: Drift velocity obtained in samples with different thicknesses.
Dot—dashed line gives the limiting drift velocity for infinite
thickness. The dashed line gives the drift velocity evaluated
with the theoretical repopulation time.

a $\langle 111 \rangle$ direction.

3. INTERPRETATION OF THE MEASUREMENTS

In Fig. 1 it may be seen that the experimental W/T_R with $E /\!/ \langle 100 \rangle$ in the
shortest samples follow the $\langle 111 \rangle$ curve at lower fields and then pass to the
$\langle 100 \rangle$ curve at higher fields. This is due to the fact that electrons do not
have time to reach steady-state conditions, and since anisotropy is
essentially due to valley repopulation $v_d(111)$ is obtained instead of
$v_d(100)$ at lower fields. At higher fields the average electron energy ϵ is
higher, the repopulation time is shorter, and steady-state is approached. In
longer samples the transit time T_R is longer and the effect is shifted to
lower fields.

For a quantitative analysis let $t = 0$ be the time at which the carriers are
created at one side of the sample. Since the intervalley transition rates
for electrons with $\epsilon > \epsilon_f$ are much higher than intravalley rates, while the
time necessary to establish the steady-state electronic distribution func-

tion f_i within each single valley is much shorter than T_R, we may assume that at $t = 0$ the electrons are equally distributed among the valleys and, within each valley, according to f_i. At $t > 0$ the populations $n_c(t)$ and $n_h(t)$ of cold and hot valleys obey the continuity equations

$$dn_c/dt = -n_c P_{ch} + 2n_h P_{hc} \; ; \qquad\qquad dn_h/dt = n_c P_{ch}/2 - n_h P_{hc} \qquad (1)$$

where P_{ch} is the probability per unit time that an electron in a cold valley will jump into a hot valley and inversely for P_{hc}. The solution of Eq.(1) leads to a repopulation exponential with time with a time constant

$$\tau = (P_{ch} + P_{hc})^{-1} \qquad (2)$$

and to an average electron velocity

$$v(t) = 2\, n_c(t)\, v_c + 4\, n_h(t)\, v_h = v_{100} + v_1 \exp(-t/\tau) \qquad (3)$$

where v_c and v_h are the drift velocity in cold and hot valleys; v_{100} is the steady-state drift velocity; v_1 is a constant related to P_{ch}, P_{hc}, v_c and v_h. As we have seen at $t = 0$ v is essentially the same as the drift velocity v_{111} obtained at the same field value, and Eq.(3) may be written as

$$v(t) \approx v_{100} + (v_{111} - v_{100}) \exp(-t/\tau) \qquad (4)$$

The thickness of the sample is covered in a time T_R such that

$$W = \int_0^{T_R} v(t)\,dt = v_{100}\, T_R + v_1\, \tau\, (1 - \exp(-T_R/\tau)) \qquad (5)$$

which for large T_R becomes

$$W = v_{100}\, T_R + v_1\, \tau \qquad (6)$$

This expression can be experimentally tested by directly plotting the sample thickness as a function of the transit time. An example is shown in Fig. 2. The slope of this straight line gives the steady-state drift velocity, v_{100}, reported in Fig. 1 as a dot-dashed line. v_1 can be obtained as $v_{111} - v_{100}$ and then τ is obtained by the extrapolation of Eq.(6) at $T_R = 0$. The resulting repopulation times are shown as function of field strength in Fig. 3.

FIG. 2: Thickness versus experimental transit time.

FIG. 3: Experimental (dots) and theoretical (line) repopulation time.

4. MICROSCOPIC THEORY AND DISCUSSION

A measurement of the repopulation time of electrons in Si when compared with theoretical predictions, leads to important information on the intervalley scattering coupling constants. This information can be maximized if we also take into account the results obtained for the drift velocity as a function of field, field orientation, and temperature [3] .

We have performed Monte Carlo calculations, assuming ellipsoidal valleys with parabolic spectrum and taking into account intravalley acoustic scattering as well as two f and two g intervalley scatterings. Acoustic scattering has been treated with the correct phonon population and energy relaxation. The details of this calculation will be described elsewhere [3] . The interpretation of all the results mentioned above confirms the need for low-energy intervalley phonons [4,5] . Furthermore, if an f-mechanism is active with energy below or equal to the lowest g-mechanism it is not possible to obtain the correct repopulation time without losing most of the anisotropy at high fields. On the other hand a g-phonon with too low an energy ($\lesssim 200°K$) makes the repopulation time much too long (with weak f) or anisotropy at high fields too small (with strong f).

In fitting the results for τ we have given more importance to its lower values since when τ becomes very long other mechanisms (as, e.g. neutral impurity scattering) may contribute to lower its value.

The physical parameters used are: effective masses $m_1 = 0.9163\ m_o$ and $m_t = 0.1905\ m_o$; sound velocity $v_s = 9.037 \cdot 10^5$ cm sec^{-1}; density $\rho = 2.329$ gr cm^{-3}; dielectric constant $\kappa = 11.7$; acoustic deformation potential $E_1 = 9$ eV; equivalent temperatures and coupling constants of intervalley phonons $T_{f1} = 400°$K, $T_{g1} = 300°$K, $D_{f1} = 3 \cdot 10^8$ eV cm^{-1}, $D_{g1} = 2 \cdot 10^8$ eV cm^{-1}. The intervalley phonons with higher equivalent temperatures have no effect in our conditions of temperature and field.

In Fig. 3 the continuous line shows the theoretical results obtained for the repopulation time τ, evaluated by Eq.(2) where P_{ch} and P_{hc} are obtained by the simulated distribution function. The dashed line in Fig. 1 indicates the average velocity of the electrons through a sample with thickness 0.183 mm evaluated by means of Eq.(4) where v_{111} and v_{100} are taken from experiments and τ from the theoretical results.

5. REFERENCES

[1] Conwell, E.M.: High Field Transport in Semiconductors. Academic Press, New York, 1967
[2] Canali, C.; Ottaviani, G.; Alberigi-Quaranta, A.: J. Phys. Chem. Solids 32, 1977 (1971)
[3] Canali, C.; Jacoboni, C.; Nava, F.; Ottaviani, G.; Alberigi-Quaranta, A.: to be published.
[4] Norton, P.; Braggins, T.; Levinstein, H.: Phys. Rev. B 8, 5632 (1973)
[5] Portal, J.C.; Eaves, L.; Askenazy, S.; Stradling, R.A.: private communication.

A.C. CONDUCTIVITY OF SEMICONDUCTORS WITH HOT ELECTRONS

M.S.Kagan, S.G.Kalashnikov, N.G.Zhdanova, P.E.Zilberman

Institute of Radioengineering and Electronics
of the USSR Academy of Sciences, Moscow, USSR

A.c. power gain is shown to be obtainable in a hot
electron semiconductor possessing a positive d.c.
differential conductivity and stable with respect
to the growth of the fluctuations (active behavi-
our). Particular examples of an activity are con-
sidered.

Some finite characteristic relaxation times are needed for a
nonlinear I-V curve to develop in a hot electron semiconduc-
tor. Hence the phase shift arises between the a.c. current
and the voltage. Therefore the average a.c. power delivered
in the semiconductor may become negative even if the slope of
the d.c. I-V curve is everywhere positive. This possibility
was studied recently in ref.$[1,2]$. In principle such mate-
rials may be used as active elements to amplify or generate
the electrical oscillations. However then the problem arises
of the fluctuation stability of such a system. This problem
was not considered in the papers mentioned above. It is shown
in the present paper that the power gain may indeed be obtai-
ned the fluctuation stability being preserved (active behavi-
our). Particular examples of such an activity are considered.
 Consider first a homogeneous unipolar semiconductor con-
taining two types of the traps, 1 and 2, with the field depen-
dent capture coefficients, $c_{1,2}(E)$. The differential conduc-
tivity, $\sigma_d(\omega,k) = dj/dE$, is obtained by solving simultane-
ously the Poisson equation, equation of continuity and the
recombination rate equations. The simplest case is that of
the voltage (of the frequency ω) exciting just the homoge-
neous oscillations with the wave number $k = 0$. Then the im-
pedance $Z(\omega) = L\sigma_d^{-1}(\omega)$, L being the length of the sample.
The fluctuational stability conditions are obtained by stu-
dying the dispersion relation $\omega = \omega(k)$.[1] Active behaviour

1) In such a system the instability is possible even at
$\sigma_d(o) > 0\ [3,4]$.

is to be expected if two conditions are met: a) Re $\sigma_d(\omega) < 0$ at some real values of ω and b) the sign of $\mathrm{Im}\,\omega(k)$ at any real k corresponds to the time decay of the fluctuations. We assume the dielectric relaxation time to be small compared to the recombination times thus neglecting the displacement current. The diffusion and thermoelectric components of the current were neglected as well. Results of such a calculation are shown on the Fig.1. The coordinates ζ , η are defined by

$$\zeta = \frac{\tau_1^{-1}\sigma_{d1}(0) + \tau_2^{-1}\sigma_{d2}(0)}{\sigma(\tau_1^{-1} + \tau_2^{-1})} , \quad \eta = \frac{\sigma_d(0)}{\sigma} .$$

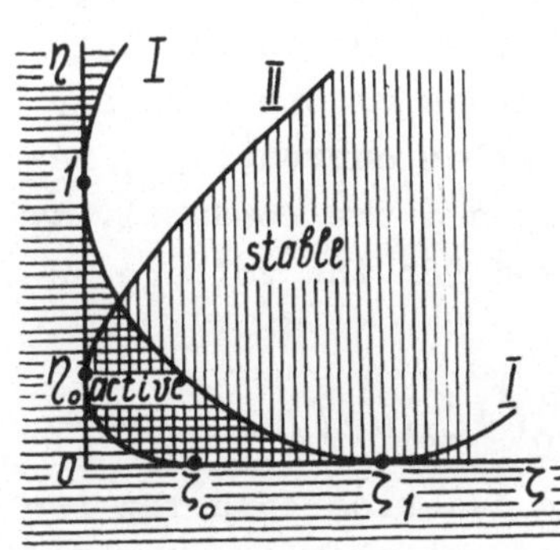

FIG. 1: Stability and activity regions.

Here $\sigma = en_0\mu$ is the d.c.conductivity, the mobility μ being assumed field independent, $\sigma_{d\,1,2}(0)/\sigma$ is the normalised d.c. differential conductivity due to just one kind of traps (1 or 2), $\sigma_d(0)$ is the total d.c. differential conductivity,

$$\tau_{1,2}^{-1} = C_{1,2}(E_0)(N_{1,2} - N_{1,2}^- + n_0) + g_{1,2}$$

the electron life-time due to the traps 1 and 2, $N_{1,2}$ and $N_{1,2}^-$ are the concentrations of the traps and of the trapped electrons, respectively, $g_{1,2}$ is the trap – to band emission probability, E_0 is the static field strength. The region outside of the parabola I (horizontal lines) corresponds to Re $\sigma_d(\omega) < 0$. The parabola II and the part of the ζ–axis ($\zeta > \zeta_0$) separate the stability and instability regions the first of them being shaded by the vertical lines. The instability corresponds either to the NDC ($\eta < 0$) or to the cases considered in ref.[3,4]. The activity region is shaded by the double lines. Here $\sigma_d(0) < \sigma$, thus the sublinear curve is needed.

Analogous calculation was made in the case of the antiblocking contacts ($k \neq 0$). Here again one may have Re $Z(\omega) < 0$ the fluctuational stability being preserved.

Consider now the example of an inhomogeneous active system, namely the sample with a static high field domain, arising due to static NDC of a homogeneous crystal. Exact calculation of $Z(\omega)$ is difficult in this case. To get some idea of the behaviour of such a system we use simplified model. Let a) the field distribution be step-wise, b) the I-V curve contains a complete saturation region and c) the shift of the domain wall exceeds its thickness considerably. The length, l, of the high field region is assumed to obey the following equation

$$\frac{dl}{dt} = \frac{l_s(V) - l}{\tau_l} \, , \tag{1}$$

where the characteristic time τ_1 may depend upon V; $l_s(V)$ is the static value of 1. Only l_s is varied on varying the d.c. bias V_o, while the static values of the high and low field, E_{20}, E_{10} and of the saturation current, j_o, remain constant. For the current density and the conductivity $\sigma_m (m = 1,2)$ in homogeneous regions and for the voltage across the sample we put:

$$j = \sigma_m E_m \, , \ \frac{d\sigma_m}{dt} = g_m - \frac{\sigma_m}{\tau(E_m)} \, , \ V = E_2 l + E_1 (L - l) \, , \tag{2}$$

where g_m = const. Such equations may describe both the recombination and the drift nonlinearity, $\tau(E)$ being the lifetime or the intervalley relaxation time respectively. Let the voltage be $V = V_o + V_\omega \exp(-i\omega t)$, $V_\omega \ll V_o$. Linearising eqs. (1), (2) one obtains the impedance

$$Z(\omega) = \frac{V_\omega}{j_\omega} = \left(1 + \frac{i}{\omega \tau_l}\right)\left(R_2 \frac{1 - i\omega\tau_2}{1 + \beta_2 - i\omega\tau_2} + R_1 \frac{1 - i\omega\tau_1}{1 + \beta_1 - i\omega\tau_1}\right). \tag{3}$$

Here $R_2 = l_s/\sigma_{20}$, $R_1 = (L - l_s)/\sigma_{10}$, $\tau_m = \tau(E_{mo})$, $\beta_m = d \ln \tau_m/d \ln E_{mo}$. The quantities E_{mo} and j_o depend essentially upon the contact conditions [5]. If, in particular, j_o is equal to the peak value, j_p, of the current-voltage curve of the homogeneous sample, then $\beta_1 = -1$ and one obtain from eq.(3):

$$Re Z = R_2 \frac{1 + \beta_2 + \beta_2 \frac{\tau_2}{\tau_l} + \omega^2 \tau_2^2}{(1 + \beta_2)^2 + \omega^2 \tau_2^2} + R_1 \left(1 - \frac{1}{\omega^2 \tau_1 \tau_l}\right). \tag{4}$$

Thus $ReZ < 0$ if the frequency is smaller than some critical value ω_o.

Active behaviour was indeed observed in the samples with static domains of recombination origin. Impendance of Cu-doped compensated Ge was studied at $80^{\circ}K$. Both strong current saturation and the step-wise field distribution were observed in such samples [6]. Fig.2 shows the frequency dependence of $ReY = ReZ^{-1}$. ReY is seen to be negative in some (illumination dependent) frequency region. Self-excitation of the oscillations was observed in the resonant circuit containing such a sample.

Experimentally observed curve $ReZ(\omega)$ was compared with the theory. For antiblocking contacts employed the saturation current was equal to j_p; hence eq.(4) was used. Following [6] the values of τ_m, E_{mo} and R_m were obtained directly. Besides that the condition $(\omega\tau_2)^2 \ll 1$ was fulfilled at $ReZ < 0$. Hence a linear dependence of ReZ upon ω^{-2} was expected in this frequency region according to eq.(4). Such a dependence was indeed observed experimentally. The values of R_1 and τ_1 having been known the time τ_1 could be obtained from the slopes of these lines. Fig.3 shows the τ_1 dependence upon the low field (illumination dependent) conductivity of the sample.

FIG. 2: Frequency dependence of ReY at different low field conductivity 6_o. R - d.c. sample resistance.

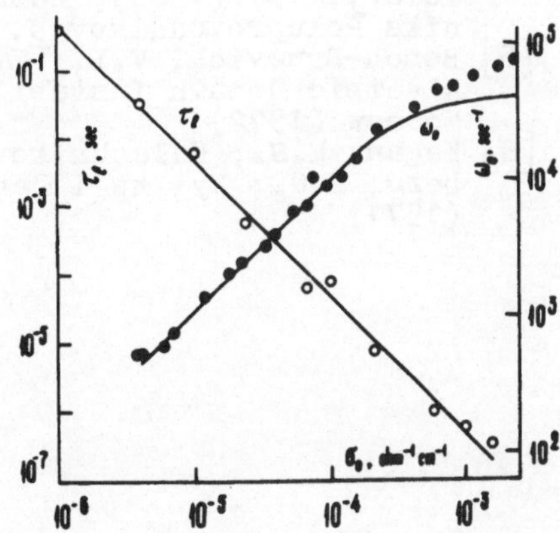

FIG. 3: τ_1 and ω_o versus 6_o.

The experimental values of the higher frequency $\omega_o(\mathrm{ReZ}{=}0)$ are shown in Fig.3 too (the lower frequency is due to the absence of complete current saturation in real cristals, the existence of which was assumed in the calculation). The solid curve was calculated from eq.(4). The value $\beta_2 = 6$ was used. Note however that ω_o is rather insensitive to β_2: under our experimental conditions ω_o varied by a factor of about 3 at $0 < \beta_2 < \infty$.

Thus we come to the conclusion that the activity of the samples containing static domain is due essentially to the domain wall oscillations. This type of an activity is to be expected as well in the Gunn-effect materials where the static domains arises due to the intervalley redistribution of the hot electrons.

The authors are indebted to V.L.Bonch-Bruevich for valuable discussions.

REFERENCES

[1] McGroddy, J.C.; Guéret, P.: Sol. St. Electronics 14, 1219 (1971)

[2] Zilberman, P.E.; Kagan, M.S.; Kalashnikov, S.G.: Fyzika i Technika Poluprovodnikov 7, 1533 (1973)

[3] Volkov, A.F.; Shulman, A.Ya.: Fyzika Tverdogo Tela 11, 3161 (1969)

[4] Bonch-Bruevich, V.L.; Kalashnikov S.G.: Fyzika Tverdogo Tela 7, 750 (1965); Bonch-Bruevich, V.L.: Fyzika i Technika Poluprovodnikov 3, 357 (1969)

[5] Bonch-Bruevich, V.L.; Zviagin, I.P.; Mironov, A.G.: Electric Domain Instability in Semiconductors, "Nauka", Moscow (1972)

[6] Kagan, M.S.; Kalashnikov, S.G.; Kemarskii, V.A.; Landsberg, E.G.: Fyzika i Technika Poluprovodnikov 5, 2041 (1971)

PHOTOCONDUCTIVE PROPERTIES OF CdS SINGLE CRYSTALS WITH DIFFERENT TYPES OF RECOMBINATION DOMAIN INSTABILITIES

K. GERMANOVA, K. MARINOVA, M. MLADENOVA
Faculty of Physics, University of Sofia, Sofia, Bulgaria

Recombination domain instabilities are sensitive
to the properties of the recombination centers.
Results on the correlation between the photocon-
ductive properties and the observed domain in-
stability type in CdS crystals are reviewed.

1. INTRODUCTION

The quantitative characteristics of the different types of
domain recombination instabilities occuring in semi- and
photoconductors depend essentially on the properties of the
recombination centers [1]. However, no sistematic investiga-
tions have been performed so far of the correlation between
the photoconductive properties and the type of domain insta-
bility in the crystals. The present work was aimed at compa-
ring certain parameters of the centers in photoconducting
crystals of CdS which exhibit various types of a recombina-
tion domain instability.

2. EXPERIMENTAL DETAILS

Two basic groups of single crystal platelets of CdS have been
investigated, exhibiting: a) temperature-electric (TE) domains
- static, oscillating [2] and moving [3] ; b) electric (E)
domains, moving in the cathode-anode (ca-domains) and/or anode
-cathode (ac-domains) direction [4,5,6]. Most of the E-domain
samples have been Cu doped after growth in the presence of an
S vapor. The electrically active levels in the samples have
been examined using conventional methods: thermally stimulated
conductivity curves, photoconductivity spectra (Ph), infrared
quenching spectra (IR), temperature dependence of the photo-
conductivity (σ(T)-curves), photoconductivity kinetics [7].
In measuring the photoconductivity spectra, equal incident
light energies have been secured at all wavelengths (the opti-
cal generation rate per unit volume $g=2,4.10^{13}$ $cm^{-3}s^{-1}$ at

$\hbar\omega$=2,42 eV). In measuring the optical quenching spectra the excitation light intensity has been so chosen that the photoconductivity δ_o at maximum photosesitivity be the same for all the samples. The quenching light energy has been the same at any λ . These conditions make it possible to compare the intrinsic and impurity photoconductivities, as well as the degrees of the optical quenching in the different samples.

3. EXPERIMENTAL RESULTS

The experimental results for certain samples of each group of crystals are given in Table 1.

Table 1

cr.	δ_d $\Omega^{-1}cm^{-1}$	$\dfrac{\Delta\delta}{\delta_d}$	$\dfrac{\delta_o}{\delta_i}$	Q%		ΔE_v, eV Ph	IR	$\delta(T)$	N_s cm^{-3}	N_t cm^{-3}
1 TE	4.10^{-10}	10^5	27	67	91	1,01 0,20	1,05 0,20	0,89	5.10^{17}	2.10^{16}
7	8.10^{-11}	10^5	20	62	71	0,96	1,00 0,21	0,84	2.10^{17}	10^{16}
3	8.10^{-11}	10^5	17	64	76	1,07 0,41 0,21	1,05 0,24	0,87 0,25	4.10^{17}	3.10^{16}
2 Ea	3.10^{-10}	10^4	67	38	60	 0,82 0,20	1,03 0,23	0,86 0,54	5.10^{16}	7.10^{15}
6	2.10^{-10}	3.10^4	120	40	67	 0,81 0,20	1,09 0,23	0,88 0,62	6.10^{16}	4.10^{15}
1'	10^{-10}	10^4	150	48	69	1,06 0,20	1,05 0,19	0,86	5.10^{16}	8.10^{15}
5 Eb	5.10^{-11}	7.10^3	4000	28	42	 0,21	1,08 0,21	0,81	9.10^{15}	2.10^{14}
4	2.10^{-11}	2.10^3	1000	10	28	 0,25	1,08 0,23	0,85	8.10^{15}	4.10^{14}
8 Ec	5.10^{-11}	23	10	2	5	 0,29	1,25 0,27			

Ea - ca-domains; Eb - simultaneous occurrence of ca- and ac-domains; Ec - ac-domains; δ_d - dark conductivity; $\Delta\delta = \delta_o - \delta_d$; δ_i - photoconductivity at $\hbar\omega$ =2,42 eV; Q - percent quenching at 0,84 eV and 1,35 eV; N_s and N_t - densities of sensitizing centers and electron traps respectively; trap depth - in the range between 0,20 - 0,65 eV.

4. DISCUSSION

The results in Table 1 show clearly that from the observed type of domain instability one can draw definite conclusions on the photoconductive properties of the crystals and vice versa. Apart from the identity of the energy spectra of the centers, the following well-defined differences have been established: 1) the concentrations of the sensitizing centers and the electron traps are highest in crystals exhibiting TE-domains; 2) samples with TE-domains are more photosensitive than ones with E-domains (cf. the $\Delta\sigma/\sigma_d$ ratio in Table 1) which is related to the higher density of sensitizing centers in TE-domain crystals on the one hand and to the fact that the latter samples have not been Cu doped [8], on the other; 3) TE-domain samples reveal a substantial impurity photoconduction (cf. the σ_o/σ_i ratio in Table 1) as well as a higher degree of optical quenching.

The comparative investigation of the photoconductive properties and the characteristics of recombination domains in both groups of samples enables one to make certain assumption as to the charge state of the sensitizing centers. Namely, in thermal equilibrium the sensitizing centers in E-domain samples are doubly-charged and in TE-domain samples - singly-charged acceptors. This conclusion is based on the following arguments:

i) Two mechanisms of a recombination non-linearity have, for the time being, an experimental verification in photoconducting samples of CdS: the temperature quenching of the photoconductivity (TQP) under the conditions of Joule's heating [2] and the field quenching (FQ) [9]. Comparing the theoretical results in [9] with the experimental data in [5,10] shows that FQ mechanism would explain both the occurrence and the behavior of the E-domains at 300 K and at relatively low critical fields ($F_{cr} \approx 10^3 - 10^4$ V/cm) only if the sensitizing centers are assumed to be double acceptors (Fig. 2).

ii) It is shown both experimentally [3,11] and theoretically [2] that there is an excitation-light intensity threshold for the occurrence of TE-domains. Below this threshold the

FIG. 1: Dependence of the free electron density on the electric field for sample № 1'. The experimental points are denoted by circles. 1 and 2 are theoretical curves derived on the assumption that the sensitizing centers are doubly- and singly-charged acceptors respectively.

FIG. 2: I-V characteristic (curve 1) and theoretical $n(F)$ dependence (curve 2) for sample № 7.

I-V characteristics of the samples are linear (Fig. 3, curve 1). On the other hand, Joule's heating is negligible so that the FQ mechanism could work (provided the appropriate conditions are available) without being masked by the TQP. Curve 2 in Fig. 3 shows the $n(F)$ dependence under the assumption that the sensitizing centers in the crystal are doubly-charged. The critical field for the occurrence of a negative differential conductivity is relatively low - $F_{cr}=7.10^3$ V/cm. However, the observations that no E-domains occur and that the I-V characteristic remains linear at fields up to $1,3.10^4$ V/cm imply that FQ would occur at higher fields. This corresponds to the opinion that the sensitizing centers in TE-domain samples are singly-charged acceptors.

iii) It is shown in [12] that double acceptors (V_{Cd}'') with considerable concentration occur in samples heated in the S

vapor. It could reasonably be assumed that in doping the
samples, which exhibit E-domains, conditions similar to those
described in [12] , are created.
The results in the present work indicate that the systematic
investigation of the recombination domain instabilities could
be a non-trivial and efficient method for studying the cha-
racteristics of various types of impurity centers in semi-
and photoconductors under essentially non-equilibrium con-
ditions [1].

5. REFERENCES

[1] Bonch-Bruevich, V. L.; Zvjagin, I. P.; Mironov, A. G.;
 In: Electrical domain instabilities in semicond.(Publ.
 House Nauka, Moskow, 1972)
[2] Kalashnikov, S. G.; Pustovoit, V. I.; Pado, G. S: Soviet
 Phys. - Semicond. $\underline{4}$, 1255 (1970); $\underline{3}$, 832 (1969)
[3] Germanova, K.: Appl. Phys. $\underline{2}$, 321 (1973)
[4] Böer, K. W.; Wilhelm, W. E.: phys. stat. sol. $\underline{4}$, 237
 (1964)
[5] Germanova, K.; Marinova, K.; Stefanova, M.: phys. stat.
 sol. (a) $\underline{1}$, 421 (1970)
[6] Koepp, R.; Marinova, K.: phys. stat. sol. $\underline{27}$, K117 (1967);
 phys. stat. sol. (a) $\underline{9}$, 491 (1972)
[7] Lashkarev, V. E.; Lyubchenko, A. V.; Sheynkman, M. K.:
 Dokl. Akad. Nauk USSR $\underline{161}$, 1310 (1965)
[8] Bube, R. H.: In: Photoconductivity of Solids (J. Wiley
 and sons, Inc., New-York - London, 1960)
[9] Dussel, G. A.; Böer, K. W.: phys. stat. sol. $\underline{39}$, 375
 (1970); $\underline{39}$, 391 (1970)
[10] Marinova, K.: Comp. rend. l'Acad. Bulg. Sci. $\underline{26}$, 1009
 (1973); Gemanova, K.; Marinova, K.; Mladenova, M.: phys.
 stat. sol. (a) $\underline{14}$, K113 (1972)
[11] Germanova, K.; Koepp, R.; Mir, F.: In: Proc. of the Sym-
 posium on Plasma Phys. and Electr. Inst. in Solids (Min-
 tis Publ. House, Vilnius, 1972); Comp. rend. l'Acad.
 Bulg. Sci. $\underline{26}$, 863 (1973)
[12] Boyn, R.: phys. stat. sol. $\underline{29}$, 307 (1968)

SPONTANEOUS RADIATIVE TRANSITIONS OF HOT ELECTRONS BETWEEN LANDAU-LEVELS IN n-InSb

E. GORNIK, W. MÜLLER and F. KOHL

Institut für Physikalische Elektronik der T.H. Wien
and
Ludwig Boltzmann-Institut für Festkörperphysik,Wien

The recombination radiation from Landau states in impact ionized n-type InSb is analysed with the aid of an n-InSb cyclotron resonance detector. The radiation intensity and emission line width are studied in dependence of the electric and magnetic field intensities and the electron concentration. It is found that the intensity increases linearly with increasing electron concentration, while the line width is proportional to the square root of the total ionized impurity content. The magnetically tunable radiation is applied to measure the Zeeman - splitting of the shallow donors in n-GaAs up to 50 kG.

I. INTRODUCTION

The analysis of the magnetic field dependent recombination emission from impact ionized shallow donors in n-type InSb has, for the first time, given evidence for radiative transitions of hot electrons between Landau-levels [1] .

The energy difference between two Landau-levels is linearly proportional to the applied magnetic field and thus provides a tunable infrared source, which can be magnetically tuned from 25 cm^{-1} to 160 cm^{-1} with magnetic fields between 3 kG and 25 kG. In order to obtain more information about the source the tunable radiation is investigated in dependence of the electron concentration, the electric and magnetic field intensities.

II. EXPERIMENTAL METHODS

The experimental system used consists of two superconducting solenoids, which are positioned in one and the same helium cryostat. One coil (maximum field 30 kG) bears the n-InSb emitter samples, the other (maximum field 50 kG) the detector and analyser sample. The whole system is immersed in liquid helium. The emitter is pulse biased; the detector output is measured either with a lock-in amplifier or a box-car detector. The n-InSb detector sample has a net electron concentration of $5 \times 10^{13} \text{ cm}^{-3}$, an electron mobility of $6 \times 10^5 \text{ cm}^2/\text{Vs}$ at 77 K and typical dimensions of $6 \times 4 \times 0.2 \text{ mm}^3$. The characteristics of the various n-InSb emitter samples covering a wide electron concentration

range are sumerized in Tab.1. The total impurity content is determined from the 77 K mobility [2] . The Zeeman spectroscopy measurements are performed with an epitaxially grown n-GaAs sample with a free carrier concentration and mobility at 77 K of $n=3 \times 10^{14}$ cm^{-3} and $\mu = 7.3 \times 10^4$ cm^2/Vs, respectively.

Table 1:

samples	M3	N13	M6	M4	6F14	S1	S4	S5
$\mu_{77K}(cm^2/Vs) \times 10^{-5}$	5	6.8	6.0	3.7	3.0	2.4	1.2	1.4
$n_{77K}(cm^{-3}) \times 10^{14}$	0.6	0.8	0.5	0.08	6.7	4	14	13
$N_I(cm^{-3}) \times 10^{14}$	7	4.5	5.5	8	21	27	68	62

III. EXPERIMENTAL RESULTS AND DISCUSSION

The emitted intensity due to spontanous radiative transitions between Landau-levels depends on the occupation number of the higher Landau-levels and the radiative transition probability. In an electron temperature model we get for a nondegenerate semiconductor a total emission rate per volume [3,4]

$$(1) \qquad R_{tot} = n \cdot \frac{n e^2 \omega_c^3}{3 \pi \hbar c^3 \varepsilon_o} \; 1^2 \; (1 - e^{-\gamma}) \sum_{N=1}^{\infty} N \cdot e^{-\gamma N} \; ,$$

with $\gamma = \hbar\omega_c/k_B T_e$, T_e the electron temperature, ω_c the cyclotron frequency, n the free carrier concentration, η the refractive index and l the cyclotron orbit radius. The influence of the impurity states and the nonparabolicity is neglected.
The emitted radiation is measured and analysed with a tunable n-InSb cyclotron resonance detector. The results are shown in Fig.1 for various emitter magnetic fields. With increasing magnetic field the double line structure of the detector appears clearly; furthermore, at high magnetic fields, two emission lines are resolved due to spin down and spin up transitions (insert of Fig.1) demonstrating the influence of nonparabolic band structure. The measurements are made under breakdown conditions, and hence the population of the higher Landau-levels increases with increasing electric field. The dependence of the photoconductivity signal (relative response) on the electric field for various emitter magnetic fields is shown in Fig.2 (experimental

Fig.1: Magnetic field dependent photo-signal of an n-InSb detector (B_D) for various magnetic fields and constant emitter current (2.2 mA higher curves, 1 mA lower curves). The insert shows an energy level diagram indicating the absorption (D) and emission (E) transitions.

points). The measurements are compared with calculations of the total emission rate according to eq. (1). The agreement between experiment and theory depends on the choice of the electron temperature on electric field dependence. The resulting dependence is shown in the insert of Fig.2 (full curves) and is compared with values obtained by Kobayashi and Otsuka [5] from cyclotron absorption measurements (circles). Additionally, we used a distribution function of hot electrons in strong magnetic fields computed by Yamada and Kurosawa (YK) [6] : the emission rate saturates at high electric fields but shows too steep an increase on the low electric field side (dashed curve in Fig.2). As a result, the comparison of the different models shows that the electron temperature model gives a better approximation of the experiment for low electric fields (1-5 V/cm), while the distribution function of YK better describes the observed saturation at high electric field.

Fig.2: Comparison of the emission rate according to eq.(1) (full curves) and calculations of Yamada et al. (YK) with experimental values.

The emission line width is determined by the magnetic field dependence of the dominant nonradiative recombination processes; additionally it is influenced by the nonparabolicity of the conduction band. The line width observed in Fig.1 is determined both by emitter and detector. To find the spectral width of the emitted radiation, we used a detector line width resulting from cyclotron absorption measurements [7,8] . Assuming Lorentz profiles for all lines, the emission line width for sample M6 is estimated to be $\Delta B \approx 0.65$ kG for an emitter field of 12 kG and $\Delta B \approx 0.35$ kG for 20 kG.

For 12 kG the resulting emission line is broader than the absorption line presumably according to contributions from the higher spin level; at 20 kG the spin down line is separated, yielding an emission line equal to the absorption line.

Additionally we studied the "Landau-emission" in dependence of the electron concentration. At constant electric (2 V/cm) and magnetic field (12 kG) the integral photosignal increases proportional to the electron concentration in the range from 8×10^{12} cm^{-3} to 4×10^{14} cm^{-3} in agreement with eq.(1) and saturates for higher concentrations. Since the emission line width at 12 kG did not show any uniform tendency in dependence of the electron concentration, we plotted it in dependence of the square root of the total ionized impurity content (Fig.3). The linear dependence on $N_I^{1/2}$ is in good agreement with theoretical calculations of the cyclotron absorption line width for an adiabatic scattering process in a coulomb potential [9,10,11] ; this indicates the same scattering process to be responsible for both the emission and absorption line width. Furthermore the following facts support this assumption:

a) the theoretical [9,10,11] and experimental [8] cyclotron absorption line width in the quantum limit is temperature independent;

b) the measured emission line width remains constant for electric fields between 1 V/cm and 5 V/cm (is independent of the electron temperature as shown by curves for different current in Fig.1);

Fig.3: Dependence of emission line width on ionized impurity content.

c) for a magnetic field of 20 kG the spin split line is separated and the emission line is equal to the absorption line.

In conclusion, our experimental results suggest that the dominant scattering process is adiabatic since nonadiabatic scattering predicts a linear dependence on N_I and a temperature dependence with $T^{-1/2}$.

IV. APPLICATION AS A TUNABLE SOURCE

The magnetically tunable radiation is applied to measure the Zee-
man - splitting of the shallow donors in n-GaAs up to 50 kG. Pho-
toconductivity measurements were performed with the epitaxially
grown n-GaAs samples. Fig.4 shows the photo-
conductive signal in dependence of the mag-
netic field of the emitter sample.
The insert of Fig.4 shows a comparison of
our results (circles)with measurements from
Stillmann et al. [12] and Kaplan et al.[13]
(full curves). For our results the wavenum-
ber of the radiation was determined inclu-
ding nonparabolicity [14] .

Fig.4: Photoconductive signal of an n-GaAs detector
versus emitter magnetic field. A comparison of our
results (circles) with literature [12,13] (full cur-
ves) is shown in the insert.

Acknowledgement

We wish to thank Prof.H.W.Pötzl for his interest in our work and many valu-
able discussion. This work is sponsored by the Fond zur Förderung der wis-
senschaftlichen Forschung. We are grateful to Mr.E.Swiggart (N.R.L., Washing-
ton) for supplying pure n-InSb material and to Dr.Zschauer, Siemens München,
for the epitaxial GaAs.

Literature

[1] Gornik E., Phys.Rev.Lett. 29 (1972), 595
[2] Kranzer D., unpublished thesis T.H.Vienna (1972)
[3] Robinson L.C., Physical Principles of Far-infrared Radiation Vol.10,
 Academic Press (1973)
[4] Stern F., Solid State Physics, Vol.15; F.Seitz and D.Turnbull ed.,
 New York, Academic Press (1967)
[5] Kobayashi L., Otsuka E., Proc. 11. Int. Conf. on the Physics of Semi-
 conductors, Warshow 1972, p. 903
[6] Yamada E., Kurosawa T., Proc. Int. Conf. on the Physics of Semiconduc-
 tors, Moscow 1968, p. 805
[7] Apel J.R., Pöhler T.O., Westgate T.R., Joseph R.I., Phys.Rev.B4,
 436 (1971)
[8] Kaplan R., Mc Combe B.D., Wagner R.J., Solid State Comm. 12, 967 (1973)
[9] Kawabata A., J.Phys.Soc.Jap. 23, 999 (1967)
[10] Shin E.H., Argyres N.P., Lax B., Phys.Rev. B7, 3572 (1973)
[11] Lodder A., Fujita S., Physics Letters 46A, 381 (1974)
[12] Stillman G.E., Wolfe C.M., Dimmock J.O., Solid State Comm.7, 921 (1969)
[13] Kaplan R., Kinch M.A., Scott W.C., Solid State Comm. 7, 883 (1969)
[14] Johnson E.J., Dickey D.M., Phys.Rev. B1, 2676 (1970)

HOT ELECTRON EFFECTS IN SILVER HALIDES IN QUANTUM LIMIT

K. KAJITA

Department of Physics, University of Tokyo
Hongo, Tokyo, 113 Japan

T. MASUMI

Department of Pure and Applied Sciences, University of Tokyo
Komaba, Tokyo, 153 Japan

Hot electron effects in quantum limit were investigated by observing photoconductivity in pure AgCl and AgBr crystals at 4.2K up to 13kV/cm and 52kOe using a fast-pulse technique with blocking electrodes. It is recognized that the hot electron effect in strong magnetic fields occurs above the critical field E_H related to the magnetic field H through the equation $eE_H r_c = \hbar\omega_c$, where r_c and ω_c are the cyclotron radius and frequency for electrons. These results are discussed in terms of the electron excitation from the N=0 Landau level to the N=1 level abruptly enhanced above E_H.

1. INTRODUCTION

There have been a large number of investigations on the hot electron effect in semiconductors such as Ge, Si and InSb [1]. Conductivity measurements in semiconductors at low temperatures and extremely high electric fields, however, usually encounters some difficulties such as heating of crystals and impact ionization of impurity electrons. Few reliable data are available on the transport phenomena in extremely high electric fields in semiconductors. On the other hand, in ionic crystals such as alkali and silver halides, neither heating of crystals nor impact ionization occurs on application of high electric fields because of their highly insulating characters. Thus, ionic crystals are believed to be suitable materials for the survey of extremely high electric field effects in the transport phenomena of photo-electrons. Hot electron effects in ionic crystals at high electric fields were studied previously by one of the authors on pure silver halides [2]. Here, we report on the hot electron effect in strong transverse magnetic fields in ionic crystals.

2. EXPERIMENTAL METHODS AND RESULTS

Transient photoconductivity measurements were performed in pure AgCl and AgBr crystals cut out of zone refined ingots using a fast-pulse technique with blocking electrodes in electric and magnetic fields up to 13kV/cm and 52kOe, respectively. Several pure AgCl crystals having the value of electron mobility between 8,000cm^2/V·sec and 50,000cm^2/V·sec at 4.2K were used to clarify the effects of impurities on the hot electron effect. Pulsed light of duration of 2μsec from an xenon flash tube or of duration of 10nsec from an AVCO dye laser at λ=3695Å were used for excitation of electrons. Further details of experimental methods can be found in the reference [3]. Experimental results here are obtained at liquid He temperatures where the "quantum limit" conditions are fulfilled at higher magnetic fields. (e.g.; At 52kOe, the Landau energy gap is about 14meV (16K) in AgCl with the effective mass of electron polaron m_p=0.43m_0.)

FIG.1 illustrates the electric field dependences of the electron drift velocity v_d parallel to the electric field E in a pure AgCl crystal at 4.2K and fixed magnetic fields in which the electron mobility μ_0 at low electric fields is about 38,000cm^2/V·sec. At zero magnetic field, the Ohmic relation between v_d and E are found to be valid below 30V/cm. Above 30V/cm, v_d is approximately proportional to $E^{1/2}$ [2]. The "effective temperature" of electrons in this region is considered to be higher than the lattice temperature. Application of a transverse magnetic field H reduces v_d. In a strong magnetic field, the v_d-E curve consists of two regions below and above a critical electric field E_H. Below E_H, v_d increases rather linearly with increasing E, whereas, above E_H, it increases abruptly up to the value at zero

FIG.1 : The drift velocity of electrons v_d versus the electric field E. Arrows indicate the critical field E_H.

magnetic field. The critical field E_H denoted by arrows in FIG.1 moves towards higher fields as the magnetic field increases. It is noted that v_d at E_H are nearly constant in various magnetic fields. Several experimental runnings on different crystals clarified the fact that E_H is determined only by H and is independent on the value of μ_0 in the crystal.

FIG.2 shows the curves of v_d versus H at fixed E. i) At sufficiently <u>low electric fields</u>, the portion of these curves at low H agrees with the

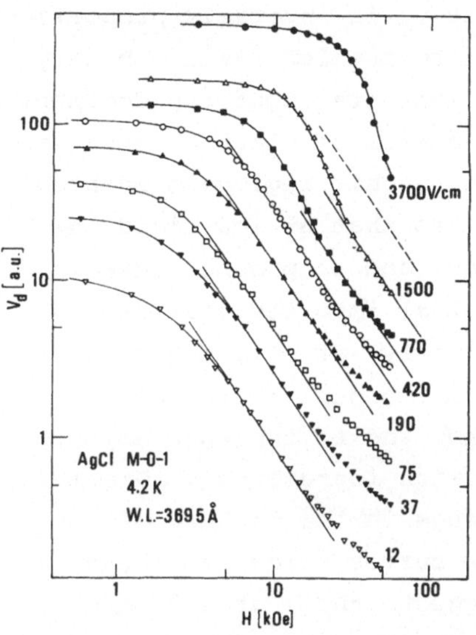

FIG.2 : The drift velocity of electrons v_d versus the magnetic field H.

usual v_d-H relation in the transient case, $v_d(H)=v_d(0)/[1+(\omega_c\tau)^2]$, where ω_c and τ are the cyclotron frequency and the scattering relaxation time of electrons, respectively [4]. At high H, v_d is found to be proportional to H^{-1}. ii) At a <u>strong electric field</u>, where warming of electrons is observed at H=0, the v_d-H curve has a form different from that in the case i). At a moderately high H, v_d begins to decrease abruptly until the critical field in H above which v_d is proportional to $H^{-3/2}$.

Data obtained on two crystals at various magnetic fields are shown together in FIG.3. Here, the quantity defined by $\gamma=\mu_0\cdot v_d$ is plotted on the ordinate as a function of the normalized field $\varepsilon=eEr_c/\hbar\omega_c \propto E\cdot H^{-3/2}$ on the abscissa, where r_c is the cyclotron radius of an electron in the lowest Landau level. Since μ_0 is inversely proportional to the rate of electron scattering in low H region while v_d in high H region is proportional to the rate of scattering, the value of γ reflects the degree of similarity between two types of scattering centers in both regions. Two points should be emphasized here: First, when $\varepsilon<1$, all the experimental points lie on a single line with slope of about 1.0; namely, v_d is proportional to $E\cdot H^{-3/2}/\mu_0$ in this region. The fact

that v_d is inversely proportional to the low field mobility μ_0 indicates that the dominant mechanism of electron scattering in this region is identical to that at low field region, and is probably due to neutral impurity scattering. Secondly, at $\varepsilon > 1$, the curves begin to depart from each other and to increase abruptly with increasing ε. Drastic change in the feature of the $\gamma-\varepsilon$ curves below and above $\varepsilon = 1$ suggests that, at $\varepsilon = 1$, the electron system changes to a new phase. It is most probable that above $\varepsilon = 1$, the "elec-

FIG.3 : $\gamma = \mu_0 \cdot v_d$ as a function of $\varepsilon = eEr_c/\hbar\omega_c$, where μ_0 is the electron mobility at low fields and r_c and ω_c are the cyclotron radius and the frequency for electrons.

tron temperature" exceeds the energy of the Landau-splitting. It was also found that, where $\varepsilon \gg 1$, v_d becomes independent on the crystals indicating that the mechanisms of electron scattering in this region is not due to neutral impurities but to other mechanisms independent on crystals, probably due to "acoustic phonons". Qualitatively similar data were obtained in AgBr crystals in which the situation was found somewhat complicated because both electrons and holes are mobile.

3. DISCUSSIONS

Based on these data, we derive the following scheme on the mechanism of warming of electrons in strong magnetic fields. First, in the region $\varepsilon < 1$, the "electron temperature" is considered to remain at the lattice temperature. In AgCl, the rate of energy relaxation of electrons due to phonon scattering is considered to be too fast to induce the hot electron effect in this region. Secondly, when ε exceeds 1, the energy of the lowest Landau level (N=0) at x=0 becomes comparable to that of the first excited state (N=1) at $x=r_c$ as illustrated in FIG.4, where x is the center of cyclotron motion of electrons. An electron in the N=0 state at x=0 can easily be scattered in this region to the

848

N=1 state at $x=r_c$, then to the N=2 state at $x \simeq 2r_c$ and so on. As these electrons cannot dissipate the energy supplied from the electric field, they become hot. The electron drift velocity at $\varepsilon > 1$ increases anomalously with the increase in scattering due to phonons caused by the rise in the "electron temperature". Because the scattering due to phonons is considered to be dominant at high "electron temperatures", the electron drift velocity becomes independent on crystals in the region where $\varepsilon >> 1$. Accordingly, we believe that the hot electron effects in quantum limit due to the enhanced excitation between the adjacent Landau levels at extremely high electric fields have been observed, for the first time, in silver halides.

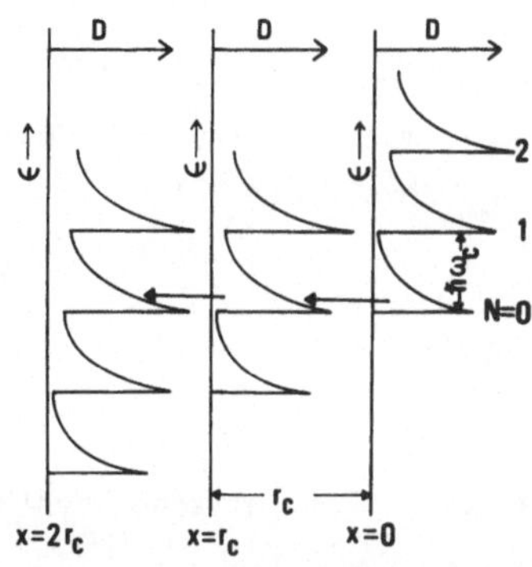

FIG.4 : Schematic pictures of the density of states of electrons D(ε) in a magnetic field at $\varepsilon=1$. The center of cyclotron motion of electrons is denoted by $x=0$, r_c and $2r_c$.

The authors are indebted to A, Muramatsu for his preparing the crystals. This work was supported by the Grants-in-Aid for Scientific Research A and for Special Project Research, and also in parts by the Fuji Photo Company.

4. REFERENCES

[1] Conwell, E.M.: "High Field Transport in Semiconductors" edited by Seitz,F.; Turnbull,D.; Ehrenreich,H. (Academic Press, New York, 1967)
[2] Masumi,T.: Phys. Rev. <u>159</u>, 761 (1967)
[3] Borders,J.A.; Hodby,J.W.: Rev. Sci. Instr. <u>39</u>, 722 (1968)
[4] Brown,F.C.: in "Polarons and Excitons" edited by Kuper, C.G.; Whitfield,G.D. (Oliver and Boyd Ltd., Edinburgh and London, 1963) p.323

RESONANT COOLING OF HOT ELECTRONS IN QUANTIZING MAGNETIC FIELDS IN n-InSb

W. RACEK and H. KAHLERT

Ludwig Boltzmann Institut für Festkörperphysik, and Institut für Angewandte
Physik der Universität Wien, A-1090 Vienna, Austria

G. BAUER

I. Physikalisches Institut der RWTH Aachen, Aachen, Fed.Rep.Germany

P. KOCEVAR

Institut f. Theoretische Physik der Universität Graz, Graz, Austria

Measurements of the magnetophonon effect in n-InSb under hot
electron conditions are reported for B∥j and B⊥j. Besides the
normal LO series additional structure is detected at magnetic
fields where spin-flip transitions are expected. For B∥j, the
amplitudes of the oscillations at 10 K decrease with increasing
electric field and increase with increasing impurity content of
the samples. At 4.2 K and 10 K a considerable deviation from
the exponential dependence of the amplitudes on harmonic number
is observed. Measurements of the Hall coefficient reveal, that
the carrier concentration between 4.2 K and 20 K depends on
the electric and magnetic field.

1. INTRODUCTION

By application of moderate electric fields E it is easy to raise the mean
energy of the electron ensemble in III-V compounds at liquid helium tempe-
ratures. In this range, elastic scattering dominates the mobility, but the
emission of acoustical and optical phonons determines the rate at which the
electrons transfer energy to the lattice. If a quantizing magnetic field is
applied, the electron-lattice interaction will vary in an oscillatory way
with increasing magnetic field B and the mean electron energy will exhibit
a corresponding variation. Whenever the energetic distance between Landau
levels equals a characteristic phonon energy, resonant emission of phonons
allows a decrease of the mean electron energy. The reappearance of the mag-
netophonon (MP) effect under hot electron conditions in several III-V com-
pounds has been attributed to this resonant cooling process [1,2] . A time-
resolved measurement of the amplitudes of the 2TA [1] series revealed, that
in fact the time dependent increase of the mean electron energy can be mea-
sured rather directly [3]. Besides the main LO [1] series belonging to spin-
conserving transitions, a spin flip series had been detected in the hot-elec-
tron photoconductivity in quantizing magnetic fields in n-InSb [4] .

This paper deals with measurements of the hot electron MP effect in the B∥j and B⊥j configuration in electric fields up to 2 V/cm. We report data concerning the dependence of the oscillatory amplitudes of the LO series on the impurity content, on the electric field strength and on the harmonic number N. Measurements of the Hall effect give evidence, that the magnitude of the extrema most probably is not only caused by mobility changes but is also determined by variations of the carrier concentration n.

2. EXPERIMENTAL METHODS

Samples of n-InSb with carrier concentrations ranging from $0.9-5.8 \times 10^{14} \mathrm{cm}^{-3}$ and mobilities of $2.76-5.2 \times 10^{5} \mathrm{cm}^{2}/\mathrm{Vsec}$ (77K) were investigated in the temperature range from 4.2 to 77 K. Voltage pulses of 1 μsec duration and a repetition rate of 50 Hz were applied and a differentiating technique was used which allowed a time resolved detection of the MP oscillations [5] . The measurements were performed with a 50 Ω source and load impedance. A careful analysis was necessary to accurately deduce the sample resistance from the voltage drop.

3. RESULTS AND ANALYSIS

Resonant transitions between Landau levels are to be expected whenever their energetic separation is equal to a charcteristic phonon energy $\hbar\omega_i$ (2TA,LO)

$$\varepsilon_{L+N} - \varepsilon_L = \hbar\omega_i \quad \text{or} \quad \varepsilon_{L+N} - \varepsilon_L = \hbar\omega_i - \varepsilon_i(B) \tag{1}$$

where ε_L is the energy of a Landau level with quantum number L taking into account nonparabolicity of the conduction band, and $\varepsilon_i(B)$ is the donor binding energy. In Fig. 1 the resonance magnetic fields are plotted vs harmonic number N for transitions ending at the L = 0 Landau level.

FIG. 1: B^{-1} vs harmonic number N; full curves spin conserving transitions, dashed curves: spin flip transitions; dash-dotted curves: transitions from Landau levels to donor ground states calculated by using Larsens[6] $\varepsilon_i(B)$ dependence. A band edge effective mass of 0.0139 m_o, an energy gap of 0.235eV and a band edge g-factor of -52 was used. For $\hbar\omega_{LO}$ and for $\hbar\omega_{2TA}$ 23.9 meV and 10.3 meV were substituted, respectively.

At temperatures higher than 11K, Stradling and Wood [7] and Akselrod et.al.[8] have observed 2TA and LO transitions. At 4.2K additional structure is observable. The calculated peak positions of Fig. 1 together with experimental recordings of $d^2\rho/dt^2$ for the B||j configuration are shown in Fig. 2. Especially at T = 4.2 K the contribution of the transitions to the impurity states rather than to the fundamental Landau level accounts for the noticeable shift of the main series or even forms a separate series. At low magnetic fields the LO impurity series coincides with the $N^- - 0^+$ spin-flip positions. At 10K already the adjacent peaks tend to merge.

FIG. 2: $d^2\rho_{||}/dt^2$ vs magnetic field B.

a) 2TA series, E = 180 mV/cm, t = 75ns, T = 10 K
b) 2TA series, E = 440 mV/cm, t =200ns, T =4.2 K
c) LO series, E = 880 mV/cm, t =200ns, T =4.2 K.
The arrows (pointing up: LO, pointing down:2TA) indicate the following transitions:
N : N - 0 ; fundamental series
N' : $N^-- 0^+$; spin-flip series
N'' : $N^+- 0^-$; spin-flip series
N_i : donor series

Since in the temperature range investigated ionized impurity scattering is the limiting process for momentum relaxation, it is of particular interest to study the dependence of the positions and amplitudes of the MP oscillations on the total impurity content. The dependence of the normalized oscillation amplitudes on carrier concentration $n = N_D - N_A$ and on E is shown in Fig. 3. Both curves illustrate the importance of ionized impurity scattering on the longitudinal momentum relaxation in quantizing magnetic fields. In order to demonstrate the difference of hot electron MP oscillations between low lattice temperatures (T < 20 K) and T = 77 K, Fig. 4 shows the normalized oscillation amplitudes vs harmonic number for various E and T. At 4.2 K, in contrast to 77 K, there is a reversed dependence of $\Delta\rho/\rho_0$ on N for low electric fields and short times after the beginning of the pulse and thus for low mean electron energies. As E and T are increased and the mean electron energy raises accordingly, the plot $\Delta\rho/\rho_0$ approaches the normal exponential behaviour.

FIG. 3: Relative amplitudes of the N=4 extremum (fundamental LO series) vs initial electric field E_O and carrier concentration n for B||j configuration.

FIG. 4: Relative amplitudes of the LO series
vs harmonic number for B∥j;
open circles: T=77K, E_o=700mV/cm, t=50ns
squares : T=77K, E_o=900mV/cm, t=50ns
triangles : T=10K, E_o=400mV/cm, t=50ns
full circles: T=42K, E_o=200mV/cm, t=35ns
(the dashed line connecting the triangles
is only included for clarity).

At 77 K the relative amplitudes of the hot MP

oscillations even increase with increasing E.

The curves indicate that at low lattice tempera-

tures and low energy gain from the electric field, where transitions to im-

purity states seem to be of importance, the anomalous behaviour of the hot

electron MP amplitudes is found.

Results of the measurements in the B⊥j configuration are difficult to in-

terpret because of their complexity. In Fig. 5 data for three different ini-

tial electric fields are presented. While at low electric fields and cor-

responding low mean electron energy we find in agreement with Stradling and

Wood [7] and with Akselrod et al. [8] that the main LO series exhibits mi-

nima in the magnetoresistance, at high electric field there is either a con-

siderable shift of the minima with magnetic field or even a change to maxi-

ma. A shift of the extrema towards higher magnetic fields could be explained

by transitions not terminating at L = 0 but at the L = 1,2 ... Landau levels.

However, in an intermediate electric field region the extrema nearly vanish,

thus suggesting that really a change of sign occurs.

FIG. 5: $-d^2\rho_\perp/dt^2$ vs B; LO series at T=10K,

a) E_o = 180mV/cm, t = 25 ns
b) E_o = 1430mV/cm, t = 10 ns
c) E_o = 2190mV/cm, t = 800 ns

The arrows have the same meaning as in Fig. 2.

As already discussed by Stradling and Wood [7],

the question arises whether the MP-oscillations

at low temperatures are caused by mobility vari-

ations alone or are also influenced by a change

of the carrier concentration of mobile carriers.

We have performed Hall measurements under nonohmic conditions in the inter-

esting range of electric and magnetic fields, sampling at 800 ns after the

rise of the pulse. It is well known that at helium temperatures in n-InSb

of comparable carrier concentration magnetic freezeout occurs.

Our ohmic Hall measurements showed that also at T = 10 K the carrier con-
centration changes in magnetic fields of a few kG. The evaluation of the
ohmic and non-ohmic data yielded the dependence of n on B shown in Fig. 6.

FIG. 6: Dependence of n on B under ohmic
(open circles) and nonohmic (full circles,
E = 200mV/cm) conditions; T = 10 K.

The results show that a monotonuous change of
n with B in the electric field region occurs
where hot electron MP oscillations are found.
In addition we have found an oscillatory
variation of the hot-electron Halleffect with B with the period of the LO
series superimposed to this monotonuous decrease. An oscillatory variation
of the hot electron Halleffect in n-GaAs was also found by Wood et al. [9] .
Thus our results indicate that a theory of the hot electron MP effect at
low temperatures applicable to n-InSb should not only consider a resonant
change of the mobility with B but has also to consider a change of the
carrier concentration with E and B.

4. ACKNOWLEDGEMENTS

We thank Prof. K. Seeger, Prof. P. Grosse and Prof. K. Baumann for their
interest in this work. It was supported by Ludwig Boltzmann Gesellschaft
and Fonds zur Förderung der wissenschaftlichen Forschung, Vienna.

5. REFERENCES

[1] For a review see: Harper, P.G.; Hodby, J.; Stradling, R.A.:Rep.Prog.Phys.
 36, 1 (1973)
[2] Yamada, E.: Proc.Int.Conf.Phys.Semiconductors, Warsaw 1972, p. 274
 Peterson, R.L.: Phys.Rev. B 5, 3994 (1972)
[3] Racek, W.; Bauer, G.; Kahlert, H.: Phys.Rev.Letters 31, 301 (1973)
[4] Morita, S.; Takano, S.; Kawamura, H.: Solid State Commun. 12,175 (1973)
[5] Bauer, G.; Erkinger, E.; Kahlert, H.; Racek, W.: J.Phys. E 6, 186(1973)
[6] Larsen, D.M.: J.Phys.Chem.Solids 29, 271 (1968)
[7] Stradling, R.A.; Wood, R.A.: J.Phys.C 3, 2425 (1970)
[8] Akselrod,M.M.;Lugovykh, V.P.; Pomortsev, R.V.; Tsidil'kovskii, I.M.:
 Sov.Phys.Solid State 11, 81 (1969)
[9] Wood, R.A.; Stradling, R.A.; Molodyan, I.P.: J.Phys. C 3, L 154 (1970)

ELECTRON HEATING BY NON-UNIFORM ELECTRIC FIELD

S. AŠMONTAS, J. POŽELA, K. REPŠAS, O. VASILEC

Institute of Semiconductor Physics of the Academy of
Sciences of the Lithuanian SSR, Vilnius

Current-voltage characteristics of asymmetrically neck-
ed samples cut from uniform single crystal of german-
ium become asymmetric at high electric fields. Current
instabilities due to impact ionization are observed. A
dc electromotive force arises in such samples in a
large microwave field. These phenomena are caused
by the build up of a static high field domain when
charge carriers are heated by non-uniform electric field.

1. INTRODUCTION

Investigation of electron heating in non-uniform electric field shows
that current-voltage characteristic of asymmetrically necked samples
becomes asymmetric at high electric fields and dc electromotive
force, called bigradient emf, arises [1] . These phenomena were
explained by the hot electron diffusion which is asymmetric as two
different gradients of temperature are present [2] . This report deals
with similar bigradient phenomena under conditions when ohmic
currents dominate over diffusion. Results on microwave oscillation
at dc fields above the threshold for impact ionization are also given
and discussed.

2. THEORY

Let us consider distributions of electric field and electron density
caused by a steady flow of ohmic current along an asymmetrically
necked biconical sample (Fig. 1(a)). A one-dimensional approach with
diffusion neglected is sufficient to describe the main features of the
phenomena of interest at low electron densities and small cone
angles. If the electric field at the contacts $E(0) = E_0$ is supposed
to be a non-heating one (mobility $\mu(E_0) = \mu_0$), it follows from
Poisson's and continuity equations:

$$\frac{1}{S(x)}\frac{d}{d(x)}\left[S(x)E(x)\right] = -\frac{eN_d}{\varepsilon_r \varepsilon_0}\left[\frac{\mu_0 E_0 S_0}{\mu(E)E(x)S(x)}-1\right] , \qquad (1)$$

where S(x) is the cross-section of the sample and n(x) is the
electron density at distance x, N_d is the donor density, $\mathcal{E}_r$ and $\mathcal{E}_0$
are the dielectric constants.

FIG. 1: A schematic view of:
(a) sample configuration,
(b) distributions of the
electric field and (c),(d)
of electron densities
along the sample.

An analytic solution of equation (1) was found assuming that mobility
of electrons is constant up to a certain field E_c ($\mu(E) = \mu_0$) and
the drift velocity is saturated $V_d(E \geqslant E_c) = \mu_0 E_c$ at the higher fields.
The solution is illustrated in Fig. 1(b) by curve 1 when contact A
is the anode and by curve 2 for the opposite direction of the
current. A high field domain at the neck of the sample due to space
charge accumulation arises; its shape and dimensions depend on
the current value and direction. Fig. 1(c) and 1(d) show the electron
density distributions along the sample. The current-voltage charac-
teristics for both directions of the current are presented in Fig. 2
for the sample with: $\mathcal{E}_r = 16$, $N_d = 2 \cdot 10^{14}$ cm^{-3}; $S_0/S_1 = 10^4$;
$r_1 = 5 \cdot 10^{-3}$ cm; $\mu_0 = 3900$ cm^2/V·sec; $E_c = 1,7$ kV/cm; $\measuredangle_1^0 = 45°$;
$\measuredangle_2^0 = 12°$.

Asymmetry with $(I_1 > I_2)$ is observed at voltages corresponding to sufficient deviations from Ohm's law.

FIG. 2: Calculated (solid lines) and measured (points) I-V characteristics of the asymmetrically necked sample.

3. EXPERIMENTAL RESULTS AND DISCUSSION

The asymmetrically necked samples were cut from germanium single crystal. Ohmic contacts were formed at the ends. Fig. 2 (points) shows the current-voltage characteristics measured with 10 ns voltage pulses at room temperature for the $10\,\Omega$ cm n-Ge samples with the following dimensions: $S_1 = 8 \cdot 10^{-5}$ cm^2; $S_0 = 0,2$ cm^2; $\alpha_1^0 = 45^\circ$; $\alpha_2^0 = 12^\circ$. It is noteworthy that the measured I-V characteristics confirm the main theoretic predictions: they are asymmetric in the non-ohmic region and there is no current saturation. At about 500 V a steep increase in the current I_1 starts, however a similar behaviour is not observed to I_2 at this voltage. Estimation of the electric field in the domain for case 1 gives 10^5 V/cm. Therefore, the current increase may be ascribed to impact ionization [3]. In case 2 the same estimation predicts impact ionization at higher voltages.

Fig. 3 shows the current oscillogramm due to a higher voltage. Impact ionizations and carrier transit across the necked part of the sample are supposed to be responsible for the observed oscillations.

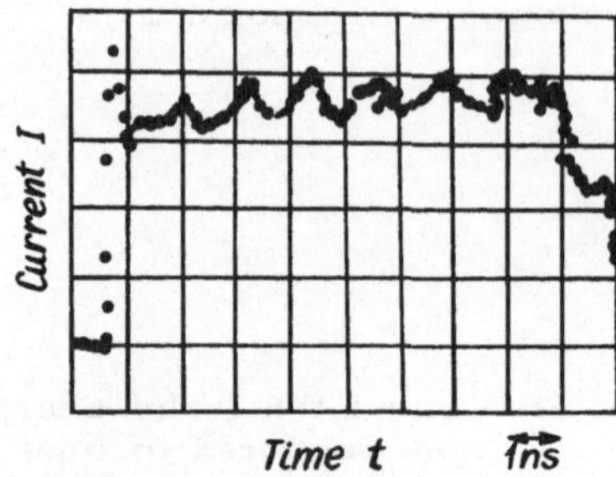

FIG. 3: Current oscillograph trace.

The samples of a similar shape subjected to microwave power exhibit the bigradient emf with its positive sign on the contact B of Fig. 1(a).

FIG. 4: The dependence of the bigradient emf upon microwave power.

Fig. 4 shows experimental dependence of the emf on the incident microwave power of the X band.

The authors wish to thank A. Matulionis and L. Subačius for their helpful advice and assistance.

4. REFERENCES

[1] Ašmontas S.; Požela J.; Repšas K.: Lietuvos fizikos rinkinys 11, 243 (1971)
[2] Ašmontas S.; Repšas K.: Lietuvos fizikos rinkinys 13, 545 (1973)
[3] Sze S. M.: Physics of Semiconductor Devices. New York: Wiley, 1969.

ACOUSTO- AND PIEZOELECTRIC EFFECTS

CONTENTS

ELECTRIC FIELD ECHOES AND STORAGE
IN PIEZOELECTRIC SEMICONDUCTORS

R. L. Melcher and N. S. Shiren

IBM Thomas J. Watson Research Center

Yorktown Heights, New York, U.S.A.

The nonlinear interaction between propagating strain
waves and uniform electric fields in piezoelectric
semiconductors gives rise to a variety of interesting
phenomena. These can be separated into parametric back-
ward wave processes and acoustically induced charge
transfer storage effects. The latter are analogous
to the storage of phase holograms.

1. INTRODUCTION

Since the initial discovery of an anomalous "echo" phenomenon by
Popov and Krainik [1] and Kessel et al. [2], the effect has been observed
in a wide variety of piezoelectric crystals (see Shiren, et al. [3] for a
list of references). In this talk we review the progress made in the
past year and one half on understanding the "echo" phenomena in piezo-
electric semiconductors.

All the phenomena discussed here are consequences of the nonlinear
interaction between propagating strain (elastic) waves and applied, uniform
electric fields. The observed effects can be separated into parametric
backward wave processes and acoustically induced charge transfer storage
effects. Although the microscopic origins of the two effects are quite
closely related, the former is purely dynamic in character whereas the
latter is essentially static. Both effects arise from electric field-
assisted detrapping of electrons bound in shallow impurity states near
the conduction band. The probability for field-assisted tunneling from
bound states is given by [4]:

$$P = A(\varepsilon_T, |E|)e^{-\frac{B(\varepsilon_T)}{|E|}}, \qquad (1)$$

where the prefactor, A, is dependent upon the shape of the trap potential
and B is proportional to the three-halves power of the trap depth ε_T.
Crucial to the understanding of the electric field-strain wave interactions
of interest here is the dependence of P upon the absolute value of the
total field, E, present in the crystal.

In a typical experiment a radio or microwave frequency electric field pulse is applied to the crystal at time $t = 0$. The first pulse, E_1, generates via the surface piezoelectric effect a forward-propagating strain wave. At time $t = \tau$, a second pulse E_2 is applied to the sample. The nonlinear interaction between the field E_2 and the strain S_1 (which was generated by E_1 at time $t = 0$ and is still propagating in the crystal at time $t = \tau$) is the origin of both parametric and storage effects. At the time τ the total field in the crystal may be written:

$$E = K^2 E_1 \cos (\omega_1 t - k_1 x) + E_2 \cos \omega_2 t, \tag{2}$$

where the first term is the piezoelectric field which accompanies the propagating strain wave of frequency ω_1 and wave vector k_1 and the second term is the field at frequency ω_2 applied at $t = \tau$. The different harmonic components contained in P when this total field is used in Eq. (1) are the source of the nonlinearities which give rise to the parametric and storage effects.

2. PARAMETRIC BACKWARD WAVE PROCESSES

When $\omega_2 = \omega_1$, $|E|$ and consequently P contain harmonic components at the frequency $2\omega_1$. Therefore, the conduction band charge density, n, contains a component $n_2 \cos 2\omega_1 t$. The drift current $J = q\mu nE$, where q is the electronic charge and μ the mobility, contains the harmonic backward wave component $\cos (\omega_1 t + k_1 x)$. From the equation of charge continuity the space charge density is found also to have a backward wave component. This backward wave space charge gives rise to a backward propagating strain wave. At time $t = 2\tau$, this backward strain wave reaches the surface of the crystal and, via the inverse piezoelectric effect, generates an electric field signal into the receiver tuned at frequency ω_1. In addition, other harmonic components of P contribute to this process.

The backward wave process involves the wave vector reversal of the forward-propagating strain wave. Therefore, all loss of phase coherence of the forward wave due to elastic scattering events is recovered by the backward wave when it reaches the crystal surface to form the echo at $t = 2\tau$.

3. STORAGE PHENOMENA

When $\omega_2 = \omega_1$, a static harmonic component of P exists which varies in space as $\cos k_1 x$. Through the diffusion or drift (in the presence of a

dc electric field) currents, an inhomogeneous space charge density can be built up. This space charge causes an inhomogeneous redistribution of the trapped charge density into a cosine charge "grating" which can persist for long times (the order of months at 4.2 K). The charge grating can be driven at time $t = T \gg \tau$ by a uniform electric field $E_3 \cos \omega_1 t$ to generate forward and backward strain waves. The backward strain wave generates at time $t = T + \tau$ an "echo" in the receiver. Similarly, the forward-propagating strain wave generated at $t = T$ by the field E_3 is coherent with the grating at $t = T + \tau$ and causes the grating to radiate an "echo". These two means of driving the charge grating can be separated experimentally.

The charge grating can be looked upon as the holographic image of E_1 with E_2 considered to be the reference beam [5].

4. REFERENCES

[1] Popov, S. N; Krainik, N. N.: Fiz. Tver. Tela <u>12</u>, 3022 (1970) [Sov. Phys. Solid State <u>12</u>, 2440 (1971)].

[2] Kessel, A. R.; Safin, I. A.; Gol'dman, A. M.: Fiz. Tver. Tela <u>12</u>, 3070 (1970) [Sov. Phys. Solid State <u>12</u>, 2488 (1971)].

[3] Shiren, N. S.; Melcher, R. L.; Garrod, D. K.; Kazyaka, T. G.: Phys. Rev. Letters <u>31</u>, 819 (1973).

[4] Haering, R. R.: Can.J. Phys. <u>37</u>, 1374 (1959).

[5] Amodei, J. J.; Staebler, D. L.: RCA Rev. <u>33</u>, 71 (1972).

LIGHT DIFFRACTION STUDIES OF CdS PHONON MASERS

J. VRBA and R.R. HAERING
Department of Physics
University of British Columbia
Vancouver, B. C., Canada

The ultrasonic field of acoustoelectric oscillators (phonon masers) is investigated by means of light diffraction. Contrary to earlier beliefs, it is found that phonon masers normally oscillate in "off-axis" modes. The nature of the off-axis oscillation is explained on the basis of the angular variation of the phonon maser parameters and of the reflection and conversion coefficients at the phonon maser boundaries. Good agreement between experiments and theory is found.

1. INTRODUCTION

The interaction between acoustical phonons and conduction electrons in piezoelectric semiconductors can, under the influence of a static electric field, lead to ultrasonic amplification [1], [2]. This amplification may be used for the generation of ultrasonic oscillations if feedback is provided in the amplifier. Acoustical feedback may be achieved by cutting and polishing the piezoelectric semiconductor in the shape of a plane-parallel cavity. The early concept of such an oscillator (phonon maser) was as follows: A d.c. electric field of sufficient magnitude is applied perpendicular to the cavity walls and the ultrasonic waves are assumed to be reflected back and forth inside the cavity, propagating perpendicularly to the cavity walls. The ultrasonic wave propagating in the direction of the electron drift (forward direction) is amplified and the ultrasonic wave propagating in the direction antiparallel to the electron drift (reverse direction) is attenuated. White [1] recognized that the forward gain can exceed the reverse loss and therefore there exists a possibility of net round trip gain. In the actual oscillator, the lattice ultrasonic loss and the loss suffered by the ultrasonic wave at the crystal boundaries must also be taken into account. The analysis of such systems was first given by Gurevich and Laikhtman [3] and the effect was first successfully demonstrated by White and Wang [4], who called these systems acoustoelectric oscillators. We prefer to call them phonon masers because they are in fact acoustic analogues of solid state masers or lasers.

Most of the experimental studies of phonon masers have been based on the

observation of the electric signals which always accompany the ultrasonic wave. Because the ultrasonic waves are the primary feature of the phonon maser, it is desirable to study these waves directly. We have done this using diffraction of light by the ultrasonic waves as a tool. Before proceeding with the discussion of our results, we briefly review the general features of light diffraction by ultrasonic waves.

Consider an ultrasonic running wave propagating through a solid medium. The corresponding variation of the strain produces a variation of the index of refraction, which represents a phase grating for the incident light in the Raman-Nath limit [5]. The diffraction pattern consists of many symmetrical diffraction spots arranged along a straight line parallel to the direction of ultrasonic phase propagation. By measuring the deflection angle and the optical frequency of a given diffraction spot we can determine the ultrasonic wavelength and frequency. If ultrasonic waves propagating in different directions are present in the crystal, we obtain a two-dimensional diffraction pattern [5].

2. OFF-AXIS OSCILLATIONS

The nature of the ultrasonic field in the phonon maser will be demonstrated for a maser which is cut with its cavity normal parallel to the crystallographic c-axis (c-axis phonon maser). The behaviour of phonon masers oriented differently (we have investigated phonon masers cut with their cavity normals at angles 15°, 30°, 45°, 60°, 75°, and 90° to the c-axis) exhibit the same general features but, for symmetry reasons, the interpretation of the c-axis phonon maser is most straight forward. If we _assume_ that the ultrasonic waves in a c-axis phonon maser propagate perpendicular to the cavity walls, we expect that only longitudinal waves will be present (the shear wave piezoelectric coupling constant is zero for this orientation). Therefore we expect that the diffraction pattern will consist of spots aligned along a straight line parallel to the c-axis.

The result of a diffraction experiment performed on a c-axis CdS phonon maser driven by voltages such that $v_s < v_d < v_\ell$ (v_s = shear velocity, v_d = drift velocity, v_ℓ = longitudinal velocity) is shown in Fig. 1a and the experimental configuration is shown in Fig. 1c. The diffraction pattern is clearly two-dimensional [6] and corresponds to "off-axis" waves propagating at an angle to the c-axis. Detailed analysis indicates that the spots A and A' correspond to longitudinal waves propagating at about 45° to the

FIG. 1: Diffraction experiment for c-axis phonon maser. (a) diffraction pattern when $v_s < v_d < v_\ell$; (b) diffraction pattern when $v_\ell < v_d$; (c) experimental arrangement.

c-axis and the spots B, B' and C, C' correspond to shear waves propagating at angles of about 20° and 70° to the c-axis, respectively. Only when the driving voltage is increased so that $v_d > v_\ell$ do we observe the expected one dimensional diffraction picture corresponding to longitudinal waves propagating parallel to the c-axis (Fig. 1b).

To explain the experimental results it is necessary to consider the effect of the phonon maser cavity walls and the angular dependence of ultrasonic gain and loss [7]. In the following treatment the maser cavity is assumed to be an infinite plane-parallel slab with parallel walls acting as free boundaries for ultrasonic waves.

The effect of the cavity walls: When a shear wave polarized in the plane of incidence is incident at an oblique angle onto a cavity wall, it is partially reflected as a shear wave of the same polarization and partially converted into a longitudinal wave [8]. The shear wave angle of incidence θ_s is equal to the shear angle of reflection and the propagation angle θ_ℓ of the converted longitudinal wave is related to θ_s by the acoustical form of Snell's law (see Fig. 2). It is therefore sufficient to describe the system

FIG. 2: Reflection and conversion of ultrasonic waves at the phonon maser cavity walls

of associated shear and longitudinal waves inside the phonon maser slab by the angle θ_s alone. For a longitudinal wave incident at an oblique angle

onto a boundary the situation is similar. Assuming that the amplitude conversion coefficients, $r_{s\ell}$ and $r_{\ell s}$, and the amplitude reflection coefficients, r_{ss} and $r_{\ell\ell}$, are given by the square root of the corresponding energy reflection coefficients and using the isotropic free surface formula, we can write

$$r_{\ell\ell} = r_{ss} = \left| \frac{\sin2\theta_s \, \sin2\theta_\ell - \left(v_\ell/v_s\right)^2 \cos^2 2\theta_s}{\sin2\theta_s \, \sin2\theta_\ell + \left(v_\ell/v_s\right)^2 \cos^2 2\theta_s} \right| \tag{1}$$

$$r_{\ell s}^{\,2} = r_{s\ell}^{\,2} = 1 - r_{ss}^{\,2} \tag{2}$$

A plot of Eqs. 1 and 2 is shown for CdS in the upper part of Fig. 3. At small θ_s the conversion coefficient increases with increasing θ_s, reaches maximum for $\theta_s \simeq 25°$ and then quickly drops to zero as θ_s approaches the critical angle.

FIG. 3: Reflection and conversion coefficients as a function of θ_s (upper part). The piezoelectric coupling constants for shear (χ_s) and longitudinal (χ_ℓ) waves as a function of θ_s (lower part). The experimentally observed oscillation angle, $\theta_s \simeq 20°$, is denoted by a dashed line. Shear wave angle, θ_s, and longitudinal wave angle θ_ℓ are connected by Snell's law.

The angular dependence of gain and loss: In piezoelectric semiconductors the ultrasonic attenuation constant, Γ, has the form [2]

$$\Gamma = \frac{\chi q_0 \omega \tau \gamma}{2\left[\left(1 + q_0^2 R^2\right)^2 + \left(\gamma\omega\tau\right)^2\right]} \tag{3}$$

where χ is the piezoelectric coupling constant, τ is the conductivity relaxation time, q_0 and ω are the wavevector and the frequency of the ultrasonic wave, R is the Debye screening length, and

$$\gamma = 1 - \frac{v_d}{v} \tag{4}$$

is the parameter describing the effect of d.c. electric field on the attenu-
ation constant Γ. For $v_{d/v} > 1$ we get $\gamma < 0$ and Γ then corresponds to
amplification. We assume that for off-axis waves the gain or loss of the
ultrasonic waves is still described by Eq. 3 provided that the magnitude of
the electric field is replaced by its projection on the direction of propa-
gation and provided that the angular dependence of all parameters is proper-
ly taken into account. The parameter most affecting the magnitude of Γ is
the piezoelectric coupling constant χ. Its angular variation due to crystal
anisotropy is displayed as a function of θ_s in Fig. 3 for both shear and
longitudinal waves in CdS.

On the basis of the above discussion the qualitative nature of the wave pro-
pagation in the oscillating phonon maser can already be demonstrated. First
assume that an ultrasonic wave is propagating nearly perpendicular to the
cavity walls (no conversion). If the phonon maser oscillates, the gain in
the forward direction must be sufficiently large to overcome both the non-
electronic losses and the electronic losses in the reverse direction. The
requirement of high forward gain can be relaxed if the electronic loss in
the reverse direction can be diminished. This can be achieved by off-axis
oscillation. Consider a system of shear and longitudinal waves at $\theta_s \simeq 20°$
(the experimentally observed angle) propagating inside the maser cavity.
For this angle of propagation $\chi_s \neq 0$, $\chi_\ell \simeq 0$, and the conversion co-
efficient $r_{s\ell} \simeq 0.8$. (The approximate oscillation angle is marked by a
dashed line in Fig. 3). If the applied d.c. electric field is such that
$v_s < v_d < v_\ell$, the shear wave in the direction of gain is amplified. At
boundary I (see Fig. 2) this wave is partially reflected as a shear wave
and partially converted into a longitudinal wave, both propagating in the
reverse direction. The shear wave in the reverse direction is strongly
attenuated, while the longitudinal wave is very little affected because
$\chi_\ell \simeq 0$ and also $v_d < v_\ell$. Hence the longitudinal wave reaches boundary II
in Fig. 2 without substantial loss of energy. At boundary II both shear and
longitudinal waves are reflected and converted mainly into the shear wave
propagating in the direction of gain. (It is shown below that the amplitude
of the lingitudinal wave created at boundary II is small.) Thus due to the
off-axis propagation and the resulting mode conversion at the boundaries the
effective net round trip gain is increased and occurs at a lower threshold
d.c. field. In the forward direction most of the acoustical energy is
stored in the shear wave which is being amplified. In the reverse direction,
when the shear wave is being attenuated, the acoustical energy is partly

carried by the longitudinal wave, which is very little affected by the
d.c. field.

The exact oscillation angle θ_s is the result of a compromise between the
projection of the electric field, the angular dependence of the coupling
constants χ_s and χ_ℓ, and the angular dependence of the reflection and con-
version coefficients. For different orientations of the phonon maser cavity
with respect to the crystallographic axes, the angle θ_s is different [9].

A more precise picture of the off-axis propagation can be obtained when a
solution for the infinite layer solid ultrasonic waveguide [10] is gene-
ralized for an active medium [7]. The exact solution of the problem is
quite complicated but useful physical approximations can be made. It is
found [7] that the ultrasonic waves in the phonon maser are quite accurately
approximated by "modes" (not normal modes) M_1 to M_4 (see Fig. 4) or by com-
binations of these modes. In Fig. 4 the ultrasonic waves associated with

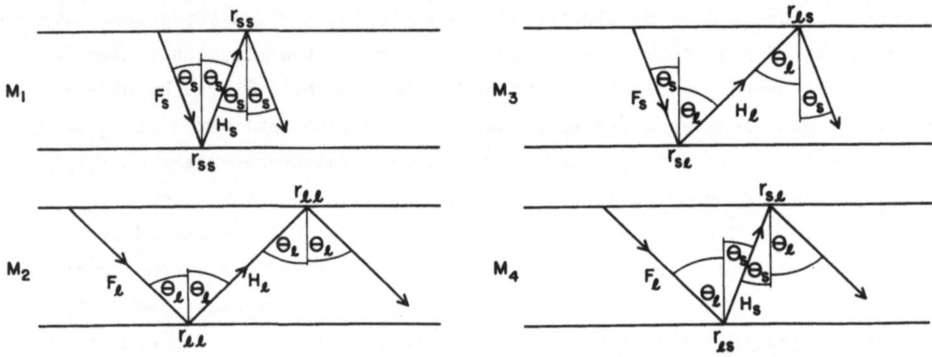

FIG. 4: Ray diagrams of important modes of propagation in the phonon maser
cavity.

these modes are represented by corresponding ray diagrams. The symbols F
and H denote the change of the ultrasonic amplitude in the forward and re-
verse directions, and the subscripts ℓ or s denote longitudinal or shear
waves. In c-axis phonon masers for conductivities $\sigma \gtrsim 10^{-4}$ $(\Omega\text{cm})^{-1}$ and
drift velocities such that $v_s < v_d < v_\ell$ the calculation of θ_s based on
mode M_3 alone gives very good agreement with experiment. For
$\sigma < 10^{-4}$ $(\Omega\text{cm})^{-1}$ both M_1 and M_3 must be considered. For drift velocities
$v_d > v_\ell$, mode M_2 propagating at $\theta_\ell = 0°$ is the solution to the problem.

The above analysis accounts only for the diffraction spots A, A', B, and B'
in Fig. 1. The finite lateral size of the phonon maser allows for additio-
nal shear waves created by conversion at the end of the phonon masers.

These additional shear waves diffract light into the spots C and C'. The
agreement between theory and experiment is good [6].

3. FREQUENCY SPECTRA

In the above theoretical analysis we were concerned only with the existence
of the off-axis oscillations and it was possible to neglect the phase con-
ditions inside the maser cavity. The details of the frequency spectrum
and the distribution of the propagation angles can be obtained by incorpo-
rating the phase conditions into the calculations. When this is done it is
found that oscillations develop for frequencies and propagation angles such
that the interference between modes M_1 and M_3 is constructive [11]. That
this is the condition for the initiation of oscillation can be seen from
the following argument. Let us again assume $v_s < v_d < v_\ell$ and let us in-
vestigate a shear wave propagating in the direction of gain. The energy of
this shear wave, is divided between shear and longitudinal waves at boun-
dary I. At boundary II the energy is either returned back to the shear wave
via the modes M_1 and M_3 (solid lines in Fig. 2), or is dissipated, or
"leaked" into longitudinal waves created at boundary II (dashed lines in
Fig. 2). Because the longitudinal wave in the forward direction is not
amplified, we expect that the oscillations first develop when this longi-
tudinal "leak" is minimized. This happens when M_1 and M_3 interfere con-
structively [11]. The condition for the constructive interference of the
M_1 and M_3 predicts a discrete set of propagation angles and oscillation fre-
quencies. For illustration, the theoretically predicted and experimentally
observed spectra in a c-axis phonon maser are shown in Fig. 5. The oscil-
lation frequencies consist of groups of lines which are nearly multiples of
three times the fundamental shear maser cavity frequency.

FIG. 5: (a),(b) Experimental, frequency
spectrum of c-axis phonon maser
($L \simeq 3.4 \times 10^{-2}$ cm); (c) theo-
retical spectrum (maximum
stiffening); (d) multiples of
fundamental shear cavity fre-
quency; (e) scale, in units of
fL, where f is the oscillation
frequency and L is the maser
thickness.

It is interesting to note that in b-axis phonon masers (c-axis parallel to
the cavity walls) whose thickness is less than about 10^{-2} cm, oscillation
frequencies which are multiples of three times the fundamental shear cavity

frequency, are also observed [12]. An explanation based on a nonuniform distribution of electric field inside of the phonon maser cavity was offered [13]. We propose an alternate explanation based on the frequency spectrum of the off-axis oscillations. It has been demonstrated that the b-axis phonon maser can oscillate in an off-axis mode [6, 7]. If the waves in a thin b-axis phonon maser propagate at large angles to the cavity normal, then the 3rd harmonic character of the spectrum is quite natural. Preliminary experiments now in progress support this explanation.

4. CONCLUSION

By analyzing the angular dependence of the ultrasonic loss coefficient and the reflection and conversion coefficients at the free boundaries we were able to show that, as the applied d.c. voltage is increased, c-axis phonon masers first oscillate in an off-axis mode. The threshold voltage for this off-axis mode is comparable to that for phonon masers cut in the shear orientation (90° or b-axis phonon masers). The experimental and theoretical analysis of phonon masers cut in a variety of orientations indicates that this off-axis behaviour represents the usual situation, because it normally corresponds to the condition of maximum net gain.

5. REFERENCES

[1] White, D.L.: J. Appl. Phys. $\underline{8}$, 2547 (1962)

[2] Gurevich, V.L.: Sov. Phys. Semicond. $\underline{2}$, 1299 (1969)

[3] Gurevich, V.L.; Laikhtman, B.D.: Sov. Phys. Solid State $\underline{7}$, 2603 (1965)

[4] White, D.L.; Wang, W.C.: Phys. Rev. $\underline{149}$, 628 (1966)

[5] Vrba, J.; Haering, R.R.: Can. J. Phys. $\underline{51}$, 1341 (1973)

[6] Vrba, J.; Haering, R.R.: Can. J. Phys. $\underline{51}$, 1359 (1973)

[7] Vrba, J.; Haering, R.R.: Can. J. Phys. $\underline{51}$, 1350 (1973)

[8] Nadeau, G.: Introduction to Elasticity, Holt, Rinehart and Winston, New York (1964)

[9] Vrba, J.; Haering, R.R.: Can. J. Phys. (1974) in print

[10] Redwood, M.: Mechanical Waveguides, Pergamon Press, New York (1960)

[11] Vrba, J.; Haering, R.R.: to be published

[12] Marshall, F.G.: Electronics Lett. $\underline{6}$, 382 (1970)

[13] Butcher, P.N.; Janus, H.M.: J. Phys. C: Solid State Phys. $\underline{5}$ 567 (1972)

COMPARISON OF EXPERIMENTAL AND THEORETICAL RESULTS OF GAIN SATURATION IN ACOUSTOELECTRIC DOMAINS [+]

H. KABELKA and H. KUZMANY

L. Boltzmann Institut für Festkörperphysik and Institut für Angewandte Physik der Universität Wien, A-1060 Wien, Austria

Experimental results of the gain saturation in acoustoelectric domains obtained by Brillouin scattering are compared with nonlinear theories for strong carrier bunching and parametric conversion. The agreement is satisfactory, particularly for the theory of parametric conversion and for the early stages of domain formation. However, for the fully developed domain, considerable deviation from all the theories considered is obtained.

1. INTRODUCTION

Brillouin scattering (BS) has turned out to be the most powerful experimental technique to study the development of acoustoelectric (ae) domains [1,2]. In such experiments usually the intensity of scattered light is observed as a function of the acoustic frequency f and the distance x of the domain from the cathode. On the other hand, in nonlinear theories, the gain saturation is investigated as a function of the ae potential $\bar{\psi} = e\varphi/k_B T$ [3,4], or as a function of the acoustic strain S or acoustic energy density ϕ [5]. Therefore gain saturation should be determined experimentally also as a function of ϕ if a direct comparison between experiment and theory is desired.

2. EXPERIMENTAL TECHNIQUES AND THEORETICAL MODELS

BS experiments on ae domains have been performed with a standard experimental setup [1,2] and crystals from the Eagle Picher Corp. and Airtron Corp. for CdS and ZnO, respectively. In all experiments a transverse sample geometry was used with samples having dimensions of about 0.4 x 0.18 x 0.03 cm (ZnO) and 0.5 - 0.9 x 0.2 x 0.03 cm (CdS). BS was observed along the propaga-

[+]This work was supported by the Ludwig Boltzmann Gesellschaft and the Fonds zur Förderung der wissenschaftlichen Forschung.

tion of the domain for the frequency spectrum around the frequency of maximum gain f_m. The depletion of the incident light was corrected according to [6], but this correction was smaller than 20% in all cases. From the increase of the relative intensity of scattered light, I/I_o, the "local" gain α_1 was determined for all frequencies as a function of x and, simultaneously, as a function of I/I_o itself. In the case of GaAs experimental results of Spears [6], which were analysed in exactly the same way as mentioned above, have been used.

The experimental results have been compared with three theoretical models which will be assigned in the following as the strong bunching model (sb-model) [3], the parametric conversion model (pc-model) [5], and the random fluctuation model (rf-model) [4].

In the sb-model $\bar{\varphi}$ was used as a parameter for which the nonlinear gain coefficient $\alpha_{nl}(\bar{\varphi})$, the acoustic energy density $\phi(\bar{\varphi})$ and the amount of carrier bunching were determined. The latter was then used to find the local electric field. Starting from thermal equilibrium the gain saturation α_{nl}/α_o' was obtained as a function of ϕ as shown in Fig.1 (inner scale). This calculation can be done exactly only for one acoustic frequency which was chosen as f_m. α_o' is an effective amplification factor acting before gain saturation starts. It is not very different from the linear amplification factor α_o and is constant over a reasonable range of ϕ.

For the pc-model the theory by Conwell and Ganguly [5] has been extended to larger values of S and the gain saturation has been calculated by determining the amount of acoustic intensity which is downconverted into subharmonic frequencies in a similar way as described in [7]. However, in the present case, the full cascade of downconversion over the whole spectrum was calculated by solving numerically the system of nonlinear equations for collinear and phasematched sound propagation of the form

$$(1) \quad \phi_N(x) = \phi_{NL}(x) - 2\phi_{N,N+1}(x) + 2\phi_{N-1,N}(x)$$

In Eq.(1) ϕ_N, ϕ_{NL}, $\phi_{N,N+1}$ and $\phi_{N-1,N}$ mean the local energy density of the N-th subharmonic frequency, the value it would ob-

tain according to a linear theory, the amount which is deliver-
ed to its subharmonic wave N+1 and the amount it gains from the
wave N-1, respectively. For the comparison with the experiment
the local nonlinear gain α_{nl} of Φ_N was determined for all sub-
harmonic frequencies starting from f_m and was plotted as a
function of Φ_N in a similar way as for the sb-model.

3. COMPARISON OF EXPERIMENTAL AND THEORETICAL RESULTS

FIG. 1: Experimental re-
sults (points, outer
scale) and theoretical
values (lines, inner
scale) for the gain sa-
turation in ae domains
in CdS for the two mo-
dels indicated.

FIG. 2: Similar diagram
as in Fig.1 but for
GaAs.

In Fig.1 experimental results for CdS
are compared with calculations accord-
ing to the sb-model and the pc-model.
The theoretical result for the pc-mo-
del was fitted to the points by using
a suitable factor of proportionality
between Φ and I/I_o. Obviously, the pc-
model gives a much better fit than the
sb-model does. A similar result is
shown in Fig.2 for GaAs. The pc-model
still gives a better fit to the expe-
rimental points. However, the differ-
ence between the two models is not as
strong as for CdS. The good agreement
of the pc-model is a strong argument
for its validity at the early stages
of the domain. It also suggests to use
it for a calculation of the whole spe-
ctrum of the domain according to Eq.
(1). The result is shown in Fig.3 for
CdS. The curves are as calculated,
starting at the cathode with a thermal
equilibrium flux. Best fit to the re-
sults of light scattering was obtain-
ed by a proper position of the scale
for Φ to that for I/I_o. The distance from the cathode at which
strong gain saturation should occur theoretically for 2 GHz,
coincides within 0.25 mm with the experimental result. Also the
initial gain of the subharmonic frequency is represented very

well (The linear gain of f_m is used to determine the effective parameter of activity). However, the subharmonic wave saturates at remarkably lower intensities than predicted theoretically which might be due to carrier bunching. This effect is obviously still more pronounced for the second subharmonic wave (0.5 GHz). Results for ZnO are similar to those for CdS.

Fig. 3: Development of the energy density in ae domains for various frequencies (lines, inner scale) and for the light scattering intensity (points, outer scale) with distance x from the cathode.

4. THE MODEL OF RANDOM ACOUSTIC FLUCTUATION

In this model which is due to Bonch Bruevich [4] the acoustic flux is assumed to cause random fluctuations in the potential energy of the lattice atoms resulting in a generation of allowed levels in the forbidden energy gap. Since the density of the levels increases with ϕ and all levels below the Fermi level

are filled the concentration of free carriers decreases with increasing ϕ. Accordingly, f_m is shifted towards lower frequencies and a spectral distribution in the domain similar to experimental results is expected. Fig. 4 shows the development of ϕ for two frequencies as calculated by starting from thermal equilibrium for ZnO (full lines). Though the gain saturation at 4 and also at 1 GHz is very similar to the experimental results (broken lines) the position of the lower frequency is not well represented.

FIG. 4: Development of two acoustic waves in an ae domain in ZnO as calculated by the rf-model (full line, inner scale) and as observed experimentally by BS (broken line, outer scale).

5. LIGHT SCATTERING EXPERIMENTS ON THE AE OSCILLATOR

Light scattering experiments in the near infrared (λ = 1.15 μ)
have been performed for the ae oscillator. Following the same
idea as described above, the intensity of scattered light was
used as a measure for $\overline{\varphi}$. Thus, particularly for the conditions
of current saturation, this intensity could be compared with
calculated values of $\overline{\varphi}$ which cause a complete current satura-
tion for various values of carrier concentration and applied
field [9]. Very good agreement was observed and detailed re-
sults will be published elsewhere.

The authors acknowledge Prof. Seeger's continual interest in
the progress of this work.

6. REFERENCES

[1] Meyer, N.I.; Jorgensen, M.H.: Festkörperprobleme 10, 21
 (1970)
[2] Kuzmany, H.: phys. stat. sol. September 1974
[3] Wonneberger, W.: phys. stat. sol. 40, 127 (1970)
[4] Bonch-Bruevich, V.L.: JETP Lett. 15, 392 (1972)
[5] Conwell, E.; Ganguly, A.K.: Phys. Rev. B 4, 2535 (1971)
[6] Spears, D.L.: Phys. Rev. B 2, 1931 (1970)
[7] Kuzmany, H.: J. Appl. Phys. 44, 4385 (1973)
[8] Wonneberger, W.: Z.Naturforsch. 26a, 1625 (1971)
[9] Hess, K.; Kuzmany, H.: phys. stat. sol. (a) 9, 253 (1972)

INTENSE LASER MODULATION OF ACOUSTOELECTRIC INTERACTIONS[+]

J. K. WAJDA and RALPH BRAY

Physics Department, Purdue University
West Lafayette, Indiana 47907, USA

A strong modulation of acoustoelectric current and
phonon intensity in n-GaAs by 3000 W, Q-switched YAG
laser pulses is reported. The response to the light
pulse is greatly dependent on the stage in the evolu-
tion of the acoustoelectrically amplified domain. The
interpretation of these effects is based on the photo-
excitation of electrons from deep traps, greatly in
excess of the dark population of $\sim 7\mathrm{x}10^{14}$ cm^{-3} at 300 K.

1. INTRODUCTION

Narrow domains of intense acoustic flux can be generated in semiconduct-
ing n-GaAs by acoustoelectric amplification of the piezoelectrically active
$\langle 110 \rangle$ fast-TA phonons from the thermal background [1]. We have found such
domains to exhibit very strong response to extrinsic light at 1.06 μ. Of
particular interest are the diverse forms of the photoconductive effect at
different stages in the life of the domain and the possibility of rapidly
modulating the acoustoelectric interactions. In addition, the photoconductive
response permits the very simple detection of the spatial extent and propaga-
tion of the domains which augments previous probing techniques [1]. By the
same process, domains can provide a scan of the light distribution in the
sample.

2. EXPERIMENTAL PROCEDURES AND RESULTS

The photoconductive studies were made at 300 K with n-GaAs samples of
typical dimensions 2 x 0.2 x 0.1 cm^3, and with carrier concentration
$n \cong 7 \times 10^{14}$ cm^{-3}. A Q-switched YAG laser supplied 200 nsec-wide pulses at
power levels up to $\sim$ 3000 watts. The samples were illuminated either locally,
transverse to the current, or longitudinally, with the uniformity of excita-
tion limited by an inherent absorption coefficient of $\sim$ 1.5 cm^{-1} [2].

[+]This work was supported by the National Science Foundation (GH-33344X
and MRL Program GH-33574A1).

Fig. 1 shows voltage and current pulse traces representing the acoustoelectric effect in the dark and the superimposed photoconductive spikes due to the short light pulses. The dark current and voltage remain flat (ohmic) during the initial growth stage of the acoustic domain. After an incubation time of ~ 1.7 μsec, the domain grows enough to cause the large drop in current due to strong momentum transfer from the electrons to the amplified phonons. After the external voltage pulse is shut off, the current persists as momentum is transferred from the phonons back to the electrons. The decay of the persistent current reflects the attenuation of the acoustic flux in the domain [3].

The magnitude and sign of the photoconductive effect is very dependent on the stage of evolution of the acoustoelectric domain and on the light intensity. Transverse illumination in a spot of about 0.01 cm^2 in the ohmic regime produces relatively small current modulation at t_1 in Fig. 1 (b). The photoconductive signals shown at t_2, t_3, and t_4 are all for light focussed directly in the domain. At t_2, excitation in a mature domain gives an extremely strong response. The much larger effect in the domain is due in part to its concentration of the voltage applied to the sample. This regime is most suitable for probing the spatial extent of the domain. The response is also positive and still quite appreciable at t_3 early in the persistent current stage, but becomes negative at t_4 when the acoustic flux is much weaker. Throughout, the current responds very rapidly to the pulse illumination, with only small aftereffects.

In Fig. 2, the dependence of the photoconductive signals on light intensity in longitudinal geometry is compared for the different stages of the acoustoelectric effect. The accompanying changes of acoustic intensity in the domain, 250 nsec after the peak of the YAG pulse, were determined by Brillouin scattering measurements. A pulsed Hg lamp, filtered to give 0.98 μ radiation, was used for this purpose [3, 4]. The results shown in Fig. 3 represent relative acoustic intensities at 0.8 GHz, well below the frequency of maximum net gain ($f_m \approx 3$ GHz) in these samples [4].

A puzzling question was how extrinsic excitation in the domain could so strongly affect the current, increasing it by almost an order of magnitude. To determine whether the effect was peculiar to the domain, we checked the ordinary photoconductivity in small (0.2 x 0.2 x 0.05 cm^3), uniformly illuminated samples, and found close to an order of magnitude, sublinear increase

FIG. 1: (a) voltage pulse; (b) current pulse with superimposed photo-
conductive spikes (dashed lines). The circled area is shown
enlarged in (c) and (d).

FIG. 2: Current as a function
of YAG intensity at 4
different times as in-
dicated in Fig. 1 (b).

FIG. 3: Brillouin scattering
signal in the domain,
taken with pulsed Hg
lamp 250 nsec after
the peak of YAG laser
pulse.

in conductivity. These effects were found in both oxygen-doped and oxygen-free samples, obtained from Monsanto. Such large, positive effect for weakly absorbed, extrinsic light at 300 K suggests electron excitation from deep traps, capable of providing excess carrier concentrations greater than 5×10^{15} cm^{-3}. Very large concentrations of deep traps have been reported in semi-insulating material [5]. Residual absorption (1-2 cm^{-1}) near 1.06 μ has been reported [6] for semiconducting GaAs, and attributed to deep traps.

3. ANALYSIS AND DISCUSSION

The various observations can be analyzed in terms of the photoexcited excess electron population and its modulation of the acoustoelectric gain coefficient $\alpha = \alpha_o \gamma$. In small signal theory, $\gamma = (v_d/v_s -1)$ where v_d is the electron drift velocity, v_s is the sound velocity; α_o is the frequency-dependent, piezoelectric coupling coefficient [3, 4]. The rise in n increases the piezoelectric screening and thereby shifts the frequency dependence of α_o. Thus, α_o is decreased at $f < f_m$. The more complicated variation of γ with n is described below. Our discussion is for photoexcitation localized in the domain, and the voltage held constant across the sample. However, the qualitative aspects of the phenomena are not sensitive to these conditions.

Intense light incident on an embryonic domain at t_1 (in the ohmic regime) is observed to extend the incubation time by the duration of the light pulse because of the shorting out of the local electric field and the acoustoelectric gain.

When the light is incident at t_2 in a mature domain, the photocurrent increases rapidly with light intensity (Fig. 2), in fact even more rapidly than in our small test samples (not shown). This indicates that at the available light intensity, not only n but also v_d is increased in the domain. This conclusion is in agreement with an analysis [7] based on the accoustoelectric equations describing flux growth and current modulation [8]. Note that at the same time the phonon intensity is increased (Fig. 3), indicating that the gain α has become higher. This increase in α is attributed to the factor γ, which apparently is more responsive to the increase in v_d in the domain than to the simultaneous decrease in domain field, as observed by voltage probe measurements.

The persistent current j_p is flat for very strong flux and given [3] by nev_s. At t_3, j_p increases with the carrier concentration. Since the

881

acoustic flux now has to drive more electrons, it must attenuate faster, as is indeed observed in Fig. 3. As seen in Fig. 2, the current does not increase indefinitely with n; ultimately the acoustic intensity is insufficient to drive all electrons with $v_d = v_s$. The eventual decrease in accousto-electric current with increasing n becomes apparent at much lower light intensity when the acoustic flux is relatively weak. This is seen at t_4, where the acoustic flux is already greatly attenuated (and the spectrum of frequencies present is centered much below f_m). Here the current drops during the light pulse mainly because α_o is reduced for the low frequency flux as a result of stronger piezoelectric screening. The reduced acoustoelectric interaction should permit the acoustic flux to live longer, and this is observed in curve t_4 of Fig. 3.

The demonstrated possibilities of using intense light for modulation of acoustoelectric interactions should be important for studying basic effects of strong piezoelectric fields. For example, the effect of the partial screening out of the piezoelectric field by excess carriers on the broadening of the intrinsic absorption edge [9], is presently being studied.

4. REFERENCES

[1] Meyer, N. I.; Jorgensen, M. H.: Festkörperprobleme 10, 21 (1970)

[2] We are indebted to D. Abramsohn for this measurement.

[3] Palik, E. D.; Bray, R.: Phys. Rev. B3, 3302 (1971)

[4] Spears, D. L.: Phys. Rev. B2, 1931 (1971)

[5] Shah, J.; Yacoby, Y: Phys. Rev. 174, 932 (1968), which also contains earlier references.

[6] Bois, D.; Pinard, P.: Phys. Stat. Sol. (a) 7, 85 (1971)

[7] Wajda, J. K.; Bray, R.: to be published elsewhere.

[8] Bray, R.: IBM J. Res. Develop. 13, 487 (1969)

[9] Garrod, D. K.; Bray, R.: Proc. 11th Intern. Conf. Phys. Semicond. (Warszawa, 1972), p. 1167

MAGNETIC SEMICONDUCTORS

CONTENTS

MODULATED REFLECTANCE SPECTRA OF EUROPIUM CHALCOGENIDES

K-E. LÖFGREN, T. TUOMI and T. STUBB

Semiconductor laboratory

Technical research centre of Finland

SF - 02150 OTANIEMI, Finland

Electroreflectance and wavelength modulated reflectance spectra have been measured in EuO and EuSe single crystals in the energy range of the E_1 reflectance peak. The studied peak is a $4f^7(^8S_{7/2}) \rightarrow 4f^6(^7F_J)5dt_{2g}$ transition in Eu^{2+}. At 77 and 85 K the spectra show at least five extrema in the energy range from 1.24 to 2.00 eV for EuO and from 2.00 to 2.53 eV for EuSe. The fine structure observed in the derivative spectra can be interpreted as originating from a multiplet splitting of the excited 7F_J state.

1. INTRODUCTION

The optical spectra of the magnetic and semiconducting europium monochalcogenides show two relatively narrow peaks, E_1 around 2 eV and E_2 around 4 eV [1]. They are associated with the $4f^7 \rightarrow 4f^6 5d$ transition, where the 5d state is split by the crystal field into t_{2g}- and e_g- bands. Reflectance measurements made with circularly polarized light have revealed a fine structure in the E_1 and E_2 peaks consisting of three to four peaks, respectively, near the magnetic ordering temperature.
Recent magnetic field-modulated reflectance measurements on EuS [2] and electric field-modulated reflectance measurements on EuSe [3] have shown a multiplet structure of the E_1 peak even at temperatures well above the ordering temperature. This fine structure has been interpreted as originating from the multiplet splitting of the $4f^6(^7F_J)5dt_{2g}$-state with J = 0,1,2,3,4,5,6.

2. EXPERIMENTAL PROCEDURE

The measured crystals were grown from a metal-rich solution. The wavelength modulated reflectance (WMR) spectra were measured on a freshly cleaved surface of a single crystal. The

electroreflectance (ER) spectra of EuO were measured with a co-
planar electrode configuration. The electrodes were gold con-
tacts vacuum deposited on the cleaved plane. The distance bet-
ween the electrodes was 0.5 mm and the crystal was about 0.3
mm thick. The ER spectra of EuSe were, on the other hand, measu-
red with a dry sandwich method and the crystal in liquid nitro-
gen [3]. The measurements were performed at 77 K(EuSe), 85 K(EuO)
and 300 K. The EuO crystal was placed on a coldfinger of a
cryostate cooled with liquid nitrogen. An unpolarized beam of
light from a quartz-prism monochromator fell at near normal in-
cidence on to the crystal.
The wavelength modulation was achieved with a glass plate vibra-
ting at 21 Hz inside the monochromator. The electric field modu-
lation was achieved with a beat-frequency oscillator. For the
ER-measurements the lock-in amplifier was tuned to the double
frequency of the applied field.

3. EXPERIMENTAL RESULTS AND DISCUSSION

Both EuO and EuSe had a structureless E_1 peak in the reflectance
spectrum. At room temperature the ER- and WMR-measurements did
not give any significant fine structure.
At 85 K, on the other hand, a multiplet structure appears, in
the WMR-curves after correcting for the modulated incident in-
tensity $\Delta I_o/I_o$. Figure 1 shows both the reflectance spectrum
and the modulated reflectance spectra of EuO in the same photon
energy region. The ER-spectra were measured with three diffe-
rent electric field strengths, all exhibiting at least five
peaks at the following photon energies: (1.24), (1.28), 1.34,
1.47, 1.61, 1.73 and 1.91 eV. The WMR-spectrum shows structure
at (1.35), 1.45, 1.57, 1.73 and 2.00 eV. The values in paren-
theses are somewhat uncertain values. As can be seen from
figure 1 the field strength has an influence on the different
peaks changing their relative strengths.
At 77 K a splitting of the E_1 peak appears in the spectra of
EuSe [3]. The ER-spectrum exhibits peaks at 2.00, 2.10, 2.21
and 2.41 eV whereas in the WMR-spectrum there are peaks at
2.00, 2.18, 2.40 and 2.53 eV.

FIG. 1:
(a) A reflectance spectrum, (b) wavelength modulated reflectance spectrum, and (c) electroreflectance spectrum of EuO. The ER-spectrum was measured with different modulating field strengths at 107.5 Hz and the lock-in amplifier was tuned to 215 Hz. At the maximum the signal was 3 mV. The zero level refers to the 5 kV/cm curve, the other two have been shifted vertically.

The observed fine structure is similar to that found in the absorption spectrum of Eu^{2+} in EuF_2 [4] where the final state is a $4f^6(^7F_J)5de_g$ state. The fine structure found in the modulated reflectance spectra from EuO at 85 K and EuSe at 77 K can be explained with the multiplet splitting of the final state $4f^6(^7F_J)5dt_{2g}$ as in EuS.

Acknowledgements - The authors wish to thank Henrik Stubb for the EuO single crystals and T.B. Reed for the EuSe single crystals.

4. REFERENCES

[1] Pidgeon,C.R.; Feinleib,J.; Scouler,W.J.; Dimmock,J.O.; Reed,T.B.: IBM j. res. develop. 14, 309 (1970)

[2] Güntherodt,G.; Wachter,P.: AIP conf. proc. no. 10 p. 1284 and Surf. sci. 37, 288 (1973)

[3] Löfgren,K-E.; Tuomi,T.; Stubb,T.: Sol. state commun. (in press)

[4] Freiser,M.J.; Methfessel,S.; Holtzberg,F.: J. appl. phys. 39, 900 (1968)

THERMO-REFLECTANCE SPECTRA OF MAGNETIC SEMICONDUCTORS: EuO, EuS, EuSe AND EuTe

T. MITANI and T. KODA

Department of Applied Physics
University of Tokyo, Tokyo, Japan

Thermo-reflectance (TR) spectra are measured on
EuO, EuS, EuSe and EuTe. A doublet structure
observed in near U.V. region is ascribed to the
Wannier exciton associated with the direct band
edge, $\Gamma_{1c}-\Gamma_{15v}$. Splitting of this doublet at
the ferromagnetic phase transition of EuS and
EuO is interpreted as due to the spin splitting
of these non-localized bands. Uniqueness of the
TR measurements for the study of magnetic semi-
conductors is emphasized.

1. INTRODUCTION

Coexistence of the localized magnetic and non-localized band-
like electrons endows many interesting features to the magnetic
semiconductors. The purpose of this work is to reveal fine
structures in the optical spectra of Eu-chalcogenides by modu-
lation technique, thereby providing information on the inter-
relation between the localized and non-localized electrons in
magnetic semiconductors. The thermo-reflectance (TR) measure-
ments have been proved to be especially useful for this purpose
[1]. Here we report further details of the TR spectra of Eu-
chalcogenides.

2. EXPERIMENTAL

Reflection spectra were measured on cleaved or polished surfaces
of Eu-chalcogenide single crystals. Samples, about 2x2x0.1 mm
in size, were glued on a thin stainless steel sheet, which was
used as a heater by passing a pulsed current with frequency of
10 Hz. The temperature modulated reflectivity $\Delta R/R$ was measured
in the visible and U.V. region. The sample temperature was
determined optically from the temperature shift of the unmodu-
lated reflectivity with an accuracy of about $0.05^{\circ}C$. Details of
the experimental methods are to be described elsewhere [2].

3. EXPERIMENTAL RESULTS

Fig. 1 illustrates the TR spectra observed in the paramagnetic
region. The unmodulated absorption spectra reported by Busch et

al [3] are also shown for comparison.

The observed TR spectra exhibit several groups of signals, which can be classified into three groups, A,B,C, as denoted in Fig.1. By lowering temperature, each group shows a marked increase in intensity. We shall not describe all of the features of these signals here, but present only the typical examples of the C signals, which demonstrate the magnetic ordering effect on the TR spectra in these magnetic semiconductors most prominently. In the paramagnetic region, the C signals appear as a doublet, C_1 and C_2, as seen in Fig.1. (The assignment is somewhat ambiguous in EuTe, where an extra sharp line, marked * in Fig.1, appears between C_1 and C_2, but, judging from the observed features, this line appears to be of a different origin from the C doublet.) With decreasing temperature, intensities of the doublet components are remarkably enhanced, as shown in Fig.2 for EuS and EuSe. Near the magnetic phase transition temperature, temperature modulation ΔT of as small as $0.05 \sim 0.01^\circ C$ is sufficient to observe TR signal $\Delta R/R \approx 10^{-3}$.

By reducing temperature further, these doublet components exhibit characteristic critical behaviour: near the ferromagnetic ordering temperature of EuS, the doublet suffers an abrupt change of the spectral shapes, and at the same time, splits into multiple components below T_c. We plotted, in Fig.3, the peak (maximum or minimum) positions of the C signals of EuS against temperature. The original doublet, which slightly shifts to lower energy with decreasing temperature about T_c, suddenly splits at T_c, clearly indicating that the electronic levels involved are under a direct influence of the ferromagnetic spin ordering in EuS. A similar, or even more remarkable, splitting is also observed in EuO.

On the other hand, the behaviour of the C doublet in EuSe in considerably different from those in EuS and EuO: the intensities and spectral shapes of the doublet components are changed rather discontinuously at around 4K, but no overall splitting of the doublet occurs as observed in EuS and EuO. The behaviour of the corresponding signals in EuTe is essentially the same as that of EuSe. Application of an external magnetic field (15.6KG) at 6.9K was found to cause splitting of the C components of EuSe.

4. DISCUSSIONS

As for the interpretation of the observed TR signals, the fairly
broad A group signals are evidently ascribed to the well known
$4f$-$5d(t_{2g})$ transitions which have been subjected to a number of
investigations hitherto. The fine structures and their temper-
ature dependences which are not described here in detail can be
accounted for in terms of the magnetic exciton model proposed by
Yanase and Kasuya [4].

On the other hand, the C group signals, as described before, can
not be considered to arise from the localized $4f$ electron tran-
sitions, but are believed to be ascribable to the Wannier ex-
citons associated with the direct band edge, Γ_{1c}-Γ_{15v}. This
assertion is based on the following facts: (1) The doublet
separations, C_1-C_2, are nearly equal to the spin-orbit splittings
of the exciton peaks observed in the corresponding Ba-chalcogen-
ides having the same valence electrons as Eu-chalcogenides.
(2) The spectral shapes of the C components are well explained
by assuming damped Lorentz oscillators as a phenomenological
model for excitons (see Fig. 4), while observed narrow spectral
shapes cannot be accounted for by the $4f$ multiplet transitions
nor by the interband transitions without an exciton effect.
(3) The observed splitting of the doublet in EuS and EuO at the
ferromagnetic phase transition can be successfully analyzed by
the exciton model. By assuming temperature dependent spin split-
ting of the Γ_1 conduction and Γ_{15} valence bands and introducing
the spin-orbit and electron-hole spin-exchange interactions, the
calculated exciton levels can reproduce the observed behaviour
of the C signals in EuS and EuO quite well [5]. This fact,
together with the fact that such splitting is absent at the
antiferromagnetic phase transition of EuSe and EuTe, is believed
to be the most convincing support for the proposed interpretation.
Interpretation of the B signals is less clear at present and waits
for further study.

5. CONCLUSIONS

Usefulness of the TR measurements on magnetic semiconductors has
been established by this work. Its uniqueness lies in the fact
that the effective modulation mechanism is primarily due to the

fluctuation in the magnetic spin order, unlike the thermal
modulation of the phonon populations as usually the case in the
temperature modulation spectra of non-magnetic semiconductors.
In view of this result, it is very tempting to observe the modu-
lation spectra of these Eu-chalcogenides by directly affecting
the spin system alone by means of a weak A.C. magnetic field.
Such efforts are now under way.

ACKNOWLEDGEMENTS

We gratefully thank Dr. T.B. Reed (for EuO, EuS and EuSe),
Dr. T. Nakayama (for EuS), and Prof. E. Hirahara (for EuTe), for
kindly supplying us with their single crystal samples.

REFERENCES

[1] Mitani, T.; Koda, T.: Phys. Letters 43A, 137 (1973)
[2] Mitani, T.; Koda, T.: Solid State Phys. (in Japanese) 9,
 July (1974), (in press)
[3] Busch, G.; Guntherodt, G.; Wachter, P.: J. Physique 32,
 C1-924 (1971)
[4] Yanase, A.; Kasuya, T.: J. Phys. Soc. Japan 25, 1025 (1968)
[5] Mitani, T.; Koda, T.: to be published.

FIG.1 : Thermo-reflectance (in the paramagnetic region) and abs-
orption (R.T.,Ref.[3]) spectra of Eu-chalcogenides

FIG.2 : Critical behaviours of the C signals in EuS and EuSe across the magnetic phase transition temperatures

FIG.3(left) : Splitting of the C doublet of EuS at the ferromagnetic ordering temperature

FIG.4(right): Observed and calculated TR spectra for the C doublet in EuSe (Two adjacent damped Lorentz oscillators are assumed in the calculation.)

POSITIVE MAGNETORESISTANCE DUE TO CONDUCTION-BAND SPLITTING IN THE Eu-CHALCOGENIDES

Y. SHAPIRA and R. L. KAUTZ

Francis Bitter National Magnet Lab.,[*] MIT, Cambridge, Mass. 02139 USA

T. B. REED[**]

Lincoln Laboratory, MIT, Lexington, Mass. 02137 USA

A positive magnetoresistance observed in n-type EuS, EuSe, and EuTe samples is attributed to the effect of spin splitting of the conduction band on nonmagnetic scattering.

Usually the magnetoresistance in a magnetic semiconductor is negative and is caused by a decrease in spin-disorder scattering or, sometimes, by an increase in the number of conduction electrons [1]. In contrast, we have observed a positive magnetoresistance in n-type EuS, EuSe, and EuTe single crystals with electron concentrations $n \sim 10^{19}$ cm^{-3}. This behavior is attributed to spin splitting of the conduction band, induced by the magnetic field $\vec{H}$.

Representative data for the H-dependence of the resistivity ρ are shown in Fig. 1. As H increases from zero, ρ first increases, reaches a maximum at H_{max}, and decreases at higher H. This behavior is observed only in a

FIG. 1. Dependence of the resistivity ρ for an EuTe sample on external (applied) magnetic field H_{ext}, at 64.4 K.

[*]Supported by the National Science Foundation.
[**]The Lincoln Laboratory portion of this work was sponsored by the Dept. of the Air Force.

limited temperature range well above the magnetic ordering temperature. For temperatures T above and below this range, ρ decreases monotonically with increasing H. The T-dependence of $(\Delta\rho/\rho)_{max} = [\rho(H_{max}) - \rho(0)]/\rho(0)$ for several EuTe and EuS samples is shown in Fig. 2. The T-dependence of H_{max} for the same samples is shown in Fig. 3. Similar experimental data for EuSe are presented in [2,3].

FIG. 2. T-dependence of $(\Delta\rho/\rho)_{max}$ $\equiv [\rho(H_{max}) - \rho(0)]/\rho(0)$ for samples EuTe 3B ($\underline{n} = 2.6 \times 10^{19}$ cm^{-3}), EuTe 1B ($\underline{n} = 8.4 \times 10^{18}$ cm^{-3}), and EuS 4B ($\underline{n} = 1.3 \times 10^{19}$ cm^{-3}). The insert shows the T-dependence of the zero-field resistivity $\rho(0)$ for sample 3B.

FIG. 3. T-dependence of H_{max} for the three samples in Fig. 2.

In many nonmagnetic semiconductors one observes a classical positive magnetoresistance with a magnitude $(\Delta\rho/\rho) \sim (\mu H)^2$, where μ is the mobility. However, the present results cannot be attributed to this usual effect because: 1) in some cases the observed positive magnetoresistance is larger than $(\mu H)^2$ by a factor $10^3 - 10^4$; 2) for a given H, the observed positive magnetoresistance is sometimes larger at temperatures where μ is lower; 3) the classical magnetoresistance usually depends on the direction of the electrical current $\vec{I}$ relative to $\vec{H}$, whereas our data for $\vec{I} \parallel \vec{H}$ are nearly the same as for $\vec{I} \perp \vec{H}$ (see Fig. 1). We attribute the positive magnetoresistance to spin splitting of the conduction band, as outlined in the following paragraphs.

As H increases from zero the localized spins of the Eu^{++} ions become increasingly aligned, and the s-f (or d-f) interaction between these spins and the spin of a conduction electron splits the conduction band into two spin subbands separated by an energy $\delta(H)$[1]. This splitting is much larger than the usual splitting (of order $\mu_B H$, where μ_B = Bohr magneton) in nonmagnetic semiconductors. With increasing δ, the concentration n^+ of electrons in the lower (+) subband increases at the expense of the concentration n^- in the higher (-) subband (see Fig. 4). For very large δ, all the electrons are in the (+) subband.

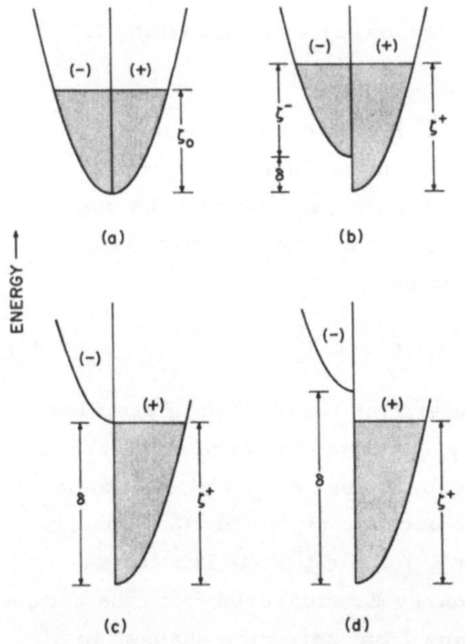

FIG. 4. Schematic diagram showing the distribution of conduction electrons between two spin subbands separated by an energy $\delta(H)$. It is assumed that the statistics is degenerate and that $n = n^+ + n^-$ = const. a) $\delta = 0$. b) $\delta < \zeta^+$. c) $\delta = \zeta^+$. d) $\delta > \zeta^+$.

In the temperature range where the positive magnetoresistance is observed the total resistivity is due to both magnetic scattering (spin-disorder scattering) and nonmagnetic scattering (mostly ionized-impurity scattering but also acoustical-phonon scattering). Within a factor of two or so, the resistivity ρ_{im} due to the impurities alone can be estimated from the residual resistivity at low T and high H. For our EuTe samples, $\rho(140 \text{ kOe}, 4.2 \text{ K})/\rho(0, 77 \text{ K}) \cong 0.8$, which indicates that for the temperatures of interest ρ_{im} is a significant part of the total resistivity ρ.

In the usual treatments of magnetoresistance in a magnetic semiconductor, the resistivity ρ_{NM} due to the nonmagnetic scattering mechanisms is presumed to be H-independent, and is regarded as a constant background. Since the resistivity ρ_M due to magnetic scattering decreases with increasing H[1], the net magnetoresistance in such treatments is negative. The key feature of our model is that ρ_{NM} is not independent of H. Under certain conditions (apparently satisfied in our samples) ρ_{NM} increases as $\underline{n}^+/\underline{n}^-$ changes with increasing H. This increase of ρ_{NM} can more than offset the concomitant decrease in ρ_M. The result is a net positive magnetoresistance at low H. At high H, where all the electrons are in the (+) subband, ρ_{NM} is constant and the total resistivity decreases with increasing H because ρ_M continues to decrease.

The physical reasons why a change in $\underline{n}^+/\underline{n}^-$ leads to a change in ρ_{NM} are most easily explained when the statistics of the conduction electrons is degenerate. Consider Fig. 4 and assume for the moment that the resistivity is due only to nonmagnetic scattering. Since the statistics is degenerate, the resistivity for any δ is given by

$$\rho^{-1} = (e^2/m^*)\left[n^+ \tau(\zeta^+) + n^- \tau(\zeta^-)\right], \tag{1}$$

where $\tau(\zeta^+)$ and $\tau(\zeta^-)$ are the electron relaxation times at the Fermi level for the (+) and (-) subbands, respectively, $\underline{e}$ is the charge of an electron, and $\underline{m}^*$ is the effective mass. Even though we assume $\underline{n} = \underline{n}^+ + \underline{n}^-$ to be constant, ρ will change with increasing δ because $\tau(\zeta^+)$ and $\tau(\zeta^-)$ change. Two factors contribute to the change of $\tau(\zeta^+)$ and $\tau(\zeta^-)$: (i) The relaxation time $\tau(E)$ depends on the electron energy E, measured from the bottom of the subband. Therefore, as δ increases from zero, the changes in ζ^+ and ζ^- affect $\tau(\zeta^+)$ and $\tau(\zeta^-)$. As an example, for acoustical-phonon scattering $\tau(E) \sim E^{-1/2}$ so that at high fields ($\underline{n}^+ \gg \underline{n}^-$) the resistivity is higher than at zero field because $\tau(\zeta^+)$ has decreased. ii) For ionized-impurity scattering, τ depends not only on E but also on the impurity screening radius $\underline{r}_o$. For degenerate statistics

$$r_o^{-2} = 4\pi e^2 N(\zeta)/\mathcal{K}, \tag{2}$$

where $\mathcal{K}$ is the dielectric constant, and $N(\zeta)$ is the density of states at the Fermi level (summed over all subbands). Equation (2) leads to an increase of $\underline{r}_o$ with increasing $\underline{n}^+/\underline{n}^-$, which tends to increase ρ. Detailed calculations[3] show that in our samples ρ_{NM} should increase with increasing

$\underline{n}^+/\underline{n}^-$ when the statistics is degenerate or nearly degenerate.

It can be shown that in the limit of classical statistics, ρ_{NM} is independent of $\underline{n}^+/\underline{n}^-$ because the average energy of a(+) or (-) electron, as well as $\underline{r}_o$, are independent of $\underline{n}^+/\underline{n}^-$. However, ρ_M decreases with increasing H even in the limit of classical statistics. For our samples the transition from degenerate to classical statistics occurs at temperatures of order 10^2 K. Therefore, for temperatures well above 100 K the negative magneto-resistance due to ρ_M is dominant and the positive magnetoresistance disappears (see Fig. 2). The disappearance of the positive magnetoresistance at low T is due to a different cause. As T decreases towards the magnetic ordering temperature, ρ_M increases and ultimately becomes much larger than ρ_{NM}. The negative magnetoresistance due to ρ_M is then larger than the positive magnetoresistance due to ρ_{NM}. This explanation is supported by the data for sample EuTe 3B in Fig. 2. With decreasing T, the magnitude of the positive magnetoresistance for this sample falls rapidly near 20 K, which is where the zero-field resistivity $\rho(0)$ increases rapidly due to an increase in ρ_M.

The model outlined above gives the correct order of magnitude for H_{max}, and the qualitative dependence of H_{max} on T and $\underline{n}$[3]. For degenerate and nearly degenerate statistics, H_{max} is approximately the field where $\delta = \zeta^+$ because at this point the transfer of electrons into the (+) subband is essentially complete (see Fig. 4). As T decreases, a given magnetic field leads to a stronger alignment of the Eu^{++} spins, which in turn leads to a larger δ. Therefore, for a lower T the condition $\delta = \zeta^+$ is satisfied at a lower H. Also, for a sample with a higher $\underline{n}$ the Fermi energy is higher and (for a fixed T) the condition $\delta = \zeta^+$ is satisfied at a higher H. These conclusions agree with the data in Fig. 3.

A more detailed account will be published later [3].

REFERENCES

[1] Haas, C.: CRC Crit. Rev. Solid State Sci. $\underline{1}$, 47 (1970); Phys. Rev. $\underline{168}$, 531 (1968); IBM J. Res. Dev. $\underline{14}$, 282 (1970).

[2] Shapira, Y.; Kautz, R.L.; Reed, T.B.: Phys. Lett. $\underline{A47}$, 39 (1974).

[3] Shapira, Y.; Kautz, R.L.: to be published. Shapira, Y.; Foner, S.; Oliveira, N.F., Jr.; Reed, T.B.: to be published.

ELECTRONIC STRUCTURE AND MAGNETIC ORDER OF THE ISOELECTRONIC COMPOUNDS EuS AND GdP.

G. GÜNTHERODT[*] and P. WACHTER

Laboratorium für Festkörperphysik, ETH Zürich, 8049 Zürich, Hönggerberg, Switzerland.

$Eu^{2+}S^{2-}$ and $Gd^{3+}P^{3-}$ are isoelectronic magnetic semiconductors which are ferro-and antiferromagnetic, respectively. The electronic structure of these compounds is derived from optical reflectivity and a KK analysis. The separation in energy between the $4f^7$ state and the 5d conduction band amounts to 1.65 eV and 7 eV for EuS and GdP, respectively. This difference is the main reason for the two types of magnetic order in both compounds.

1. INTRODUCTION

The rare earth (RE) monochalcogenides and monopnictides form a very interesting group of materials: the crystal structure is of the simple rocksalt type, generally, the compounds exhibit some kind of magnetic order, divalent RE monochalcogenides are insulators or semiconductors, trivalent RE monochalcogenides are metals and the trivalent RE monopnictides show an ambivalent conducting character (1-8). In this paper we want to compare the electronic structure of EuS and GdP as derived from the Kramers-Kronig (KK) analysis of reflectivity of single crystals. These materials have been selected because GdP is the first member of the pnictide series where the growth of large and well defined single crystals has now become possible (8,9), and EuS and GdP are isoelectronic. In both compounds the spectroscopic ground state of the cation is $4f^7$ ($^8S_{7/2}$) and the magnetic moment per cation is 7 μ_B. In both materials one expects a valence band formed by $3p^6$ orbitals of the anion and empty conduction bands of 5d and 6s character. However, EuS is a ferromagnetic insulator while GdP is an antiferromagnet and,so far, exhibits metallic conductivity. It is the aim of this paper to clarify this different behavior by deriving an energy level scheme for both compounds.

[*] Present address: IBM T. J. Watson Research Center

2. EXPERIMENTAL METHODS AND RESULTS OF THE KK ANALYSIS

The reflectivity of polished single crystals of EuS and GdP has been measured at room temperature for photon energies from 0.03 eV to 12 eV. The samples were situated either in high vacuum or inert gas atmosphere. The reflectivity has been analysed in terms of the optical constants by means of the KK relation. The procedure has been described elsewhere (10). As a result of this analysis we show in Fig.1 the real and imaginary part of the dielectric function versus photon energy of EuS and GdP. On first sight it is obvious that the dielectric function in EuS is much more structured than in GdP. Another remarkable difference between both spectra is the fact that ε_2 for EuS tends to zero for photon energies less than the energy gap of 1.65 eV, whereas ε_2 in GdP becomes infinite for small energies. In addition, ε_1 levels off to $\varepsilon_\infty = 4.7$ for small energies in EuS, while ε_1 tends to minus infinity in the case of GdP. This behavior is typical for an insulator in the case of EuS and for a material with high electron concentration for GdP. Superpositions of intra-and interband transitions in GdP as evident in ε_2, result in a coupling of plasmons and interband transitions. Thus we have to decompose ε_1 and ε_2 into contributions from free and bound electrons. The basic assumption is that the minimum of ε_2 at 0.65 eV in Fig.1 corresponds to the onset of interband transitions and above this energy the contributions of free carriers to ε_2 are negligible. The resulting decomposition of ε_1 into the contribution ε_1^f of free and ε_1^b of bound carriers is shown in Fig.2. For small energies ε_1^b becomes constant and $\varepsilon_\infty = 1 + \varepsilon_\infty^b = 11.2$. This value characterizes the sample in the absence of free electrons.

Fig. 1: Dielectric function of
EuS and GdP at 300 K

On the other hand ε_1^f describes the sample in the absence of inter-
band transitions. Obviously there is a shift of the true, unper-
turbed plasma resonance ($\varepsilon_1^f=0$) of the conduction electrons from
1.53 eV to 0.52 eV ($\varepsilon_1=0$) due to interband transitions. Similar
shifts have been observed in the Gd chalcogenides and in LaS (7).

3. DISCUSSION

The discussion of optical transitions and the electronic structure
in the case of EuS relies on an energy level scheme first pro-
posed by Methfessel (2), which has been confirmed by Wachter (3)
on the basis of a purely ionic crystal model using the thermo-
chemical Haber-Born cycle. In these models the 5d conduction band
is subject to the cubic crystal
field and splits in two rather
narrow subbands of t_{2g} and e_g
symmetry. This implies that the
matrix elements of an optical
transition into these bands are
only weakly energy dependent and
thus maxima in ε_2 correspond to
peaks in the joint density of
states. Consequently, the two
most prominent peaks in ε_2 at
2.2 eV and 4.3 eV have been assign-
ed to transitions from the $4f^7$
level into the 5d subbands. Then

Fig. 2: Decomposition of ε_1
into free and bound con-
tributions ε_1^f and ε_1^b.

the crystal field splitting turns out to be 2.1 eV. Making use of
low temperature photoconductivity (3, and ref.cit.there), magneto-
optics above and below T_c(3), photoemission (11) and specific ex-
periments concerning the bottom of the conduction bands (12) we
can give an energy level scheme of EuS as sketched in Fig. 3.
This model has been discussed at length by Wachter (3) and Gün-
therodt (10). The main point is that the energy separation of the
$4f^7$ states and the bottom of the 5d band is only 1.65 eV and that
the 4f states are above the p-derived valence band of sulfur.

In the case of the RE monopnictides the only existing proposal

for a relative energy level scheme has been given by Wachter (3),
again on the basis of a Haber-Born-cycle. It follows from this
scheme because the Gd ion in GdP is in a trivalent state that op-
tical transitions from the $4f^7$ level to states above the Fermi ni-
veau E_F are to be expected in the vacuum UV region, similarely as
in GdS (7). Thus we assign the two peaks of ε_2 at 6.9 eV and 8.7
eV in Fig.1 to transitions from the $4f^7$ level of Gd into the cry-
stal field split $5dt_{2g}$-and $5de_g$ subbands, respectively.Analogously
we conclude that the peaks at 1.45 eV and 3.25 eV correspond to
transitions from the p-derived valence band of phosphorus into the

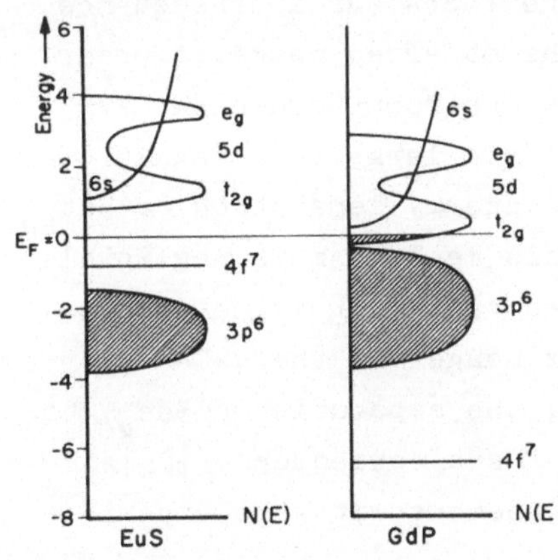

Fig.3: Energy level scheme
of EuS and GdP

crystal field split 5d subbands.
This assignment is confirmed by
the observation of an energy dif-
ference of 1.8 eV between both sets
of peaks of ε_2, a value which then
corresponds to the crystal field
splitting of the 5d band. The ma-
ximum of ε_2 at 3.25 eV is presum-
ably broadened by transitions
from the valence band into the 6s
conduction band, extending over a
much wider energy range. Hence we
deduce, the $4f^7$ levels in GdP are
positioned about 7 eV below E_F (Fig.3). The onset of interband
transitions p→E_F has been determined to be 0.65 eV. Since the 5d
conduction bands are not expected to be parabolic it is difficult
to determine the energy difference E_F-bottom of the band. We only
can say that an energy gap is less than 0.65 eV, if it exists at
all. Because of this small energy gap it is not surprizing that
the slightest deviation from stoichiometry, the presence of im-
purities or Schottky defects in cation and anion sublattice will
result in an excess of carriers. In GdS and GdP we expect the 5d
conduction band to be lowest (7,8). We thus can use the same m*
as in GdS which has been determined to be m*=1.2m (7). The true
unperturbed plasma resonance in GdP at 1.53 eV (Fig.2) will then
provide us with the free carrier concentration which for this

sample is $N=2\cdot10^{21}cm^{-3}$. The lowest carrier concentration in GdP which we have measured is $N=9.8\cdot10^{20}cm^{-3}$ which corresponds to 0.046 free carriers per Gd ion.

Darby and Taylor (13) nave invoked the RKKY interaction for the exchange mechanism of the RE monopnictides, assuming 1.4 carriers per RE ion. Also the work of Birgenau et al (14) suggests between 1 and 2 free carriers per RE ion in e.g. PrP. It is quite evident that these numbers are an order of magnitude too large compared with the measured carrier concentrations in GdP. Using $N=10^{21}cm^{-3}$ and the formalism of the RKKY theory, one computes GdP to be ferromagnetic in contradiction with the facts. As a consequence, the RKKY interaction cannot explain the observed magnetic order.

It is well established (3) that the Anderson-Goodenough exchange interaction $J_1=b_1^2J_{fd}/2S^2U_1^2$ yields a large ferromagnetic exchange constant in EuS because U_1, the energy separation 4f-5dt$_{2g}$, is small (Fig.3). On the other hand this mechanism is negligible in GdP due to a much larger U_1. Instead, antiferromagnetism in GdP can be explained by a 180^o superexchange via the valence orbitals of the anions $J_2=b_2^2/U_2$ with U_2, the separation p-5de$_g$, being less for GdP than for EuS (Fig.3). As a consequence $|J_2|>J_1$ in GdP, whereas $J_1>|J_2|$ for EuS in agreement with the experimental observation.

4. REFERENCES

(1) Busch G., J. appl. Phys. <u>38</u>, 1386 (1967)
(2) Methfessel S., Mattis D., Handbuch Phys. Bd 28/1 (1969)
(3) Wachter P., CRC Crit. Rev. Solid State Sciences <u>3</u>,189 (1972)
(4) Junod P., Menth A., Vogt O.,Phys.condens. Matter <u>8</u>,323(1969)
(5) McGuire T.R., Gambino R.J., Pickart S.J., Alperin H.A.,
 J. appl. Phys. <u>40</u>, 1009 (1969)
(6) Jayaraman A., Narayanamurti V., Bucher E., Maines R.G.,
 Phys. Rev. Letters <u>25</u>, 368 (1970)
(7) Beckenbaugh W., Evers J., Kaldis E., Güntherodt G., Wachter
 P., J. Phys. Chem. Solids (in print) (1974)
(8) Güntherodt G., Kaldis E., Wachter P., Phys.Rev.Lett.(In pr.)
(9) Kaldis E., J. Crystal Growth (in print) (1974)
(10) Güntherodt G., Phys. condens Matter (in print) (1974)
(11) Eastman D.E., Kuznietz M., J. appl. Phys. <u>42</u>, 1396 (1971)
(12) Schoenes J., Wachter P., Phys. Rev. B (April) (1974)
(13) Darby M.I., Taylor K.N.R., Phys. Letters <u>14</u>, 179 (1965)
(14) Birgenau R.J., Bucher E., Maita J.P., Passell L., Turberfield
 K.C., Phys. Rev. <u>8</u>, 5345 (1973)

MAGNETIC FIELD-INDUCED RAMAN SCATTERING IN EuSe[*]

V. J. TEKIPPE[†], R. P. SILBERSTEIN[†], M. S. DRESSELHAUS[†]

Department of Electrical Engineering
and Center for Materials Science and Engineering
Massachusetts Institute of Technology, Cambridge, Mass.02139,USA

R. L. AGGARWAL

Francis Bitter National Magnet Laboratory[*]
Massachusetts Institute of Technology, Cambridge, Mass.02139,USA

We report here several new features associated with
the magnetic field, temperature, and laser excitation
energy-dependence of the magnetic field-induced Raman
scattering effect in EuSe. New features include the
persistence of this line above T_C and the observation
of two energy ranges where resonant enhancement occurs.

A detailed study is in progress of the magnetic field-induced
Raman scattering observed at the zone center LO phonon frequency
in EuSe[1] and relates to other work in the europium chalcogen-
ides[2,3,4]. Our study emphasizes the temperature and magnetic
field dependence of the scattering intensity for T near the mag-
netic critical temperature, T_c=4.6K. In addition we present the
results for the dependence of this scattering intensity on the
laser excitation energy in the range from $\sim$1.9eV to 2.7eV, which
is the regime of the E_1, E_1' and E_1'' reflectivity maxima of EuSe[5].
The Raman results reported here were obtained in the back scat-
tering geometry and in the Voigt configuration with the electric
vector of the incident light, $\vec{E}_i$, perpendicular to the applied
magnetic field, $\vec{H}$. Laser excitation energies in the range from
$\sim$1.9eV to 2.7eV were provided by an Ar^+ ion laser and a dye la-
ser pumped by it. Magnetic fields up to 10.5kOe were obtained
from a nine-inch Varian magnet. The low temperature range used
in these measurements was provided by a Super Vari-Temp Dewar and
the temperature was monitored by a carbon resistor thermometer.
Two distinct types of first-order Raman scattering have been
reported in EuSe: i) a broad line extending from about the TO

*Supported by the National Science Foundation
†Visiting Scientist, Francis Bitter National Magnet Laboratory

phonon frequency (127 cm^{-1}) to beyond the LO phonon frequency
(182 cm^{-1}) and ii) a sharp line at the LO phonon frequency[1].
For T near T_c, it is possible to observe both types of scat-
tering simultaneously as shown in Fig.1. For H=0 and at T∿5K we

FIG. 1: First-order Raman spectra of EuSe at T∿5K for H=0, 4kOe,
8kOe and 10.5kOe, using laser excitation at 2.41eV
(λ_i=5145 Å) with $\vec{E}_i \perp \vec{H}$ and $\vec{E}_s \perp \vec{H}$ or $\vec{E}_s \parallel \vec{H}$. The inten-
sities have been corrected for the instrument function
and are all plotted on the same linear scale.

observe a broad unpolarized line in the region from ∿130 cm^{-1} to
∿190 cm^{-1}. The intensity of this broad line increases as T in-
creases, but its shape remains relatively unchanged up to T∿300K.
As a magnetic field is applied at T∿T_c, the intensity of the
broad line decreases as shown in the right side of Fig.1. At
the same time, however, a sharp line is observed to emerge from
the broad line at the LO phonon frequency. This sharp line is
observed when the electric vector of the scattered light, $\vec{E}_s$, is
perpendicular to $\vec{H}$ as shown in the left side of Fig.1.
Both types of Raman scattering described above are anomalous

since symmetry considerations forbid such scattering in face
centered cubic EuSe. However, when the spin structure is ex-
plicitly included in the symmetry analysis, these symmetry re-
strictions are relaxed. In the paramagnetic phase, the random
ordering of the spins removes both the translational and in-
version symmetry of the system. With the loss of inversion sym-
metry, Raman scattering from odd-parity phonon modes is no lon-
ger strictly forbidden. With the loss of translational symmetry,
phonons from throughout the Brillouin zone can contribute to the
scattering. In the ferromagnetic phase, the spin ordering re-
stores the translational symmetry but not necessarily the in-
version symmetry and thus Raman scattering from zone center
phonons could become dominant.

This "spin order-disorder" argument provides a qualitative in-
terpretation of the experimental results for both types of scat-
tering. It accounts for the observation of the broad-line scat-
tering for $T > T_c$ and its quenching either by lowering T below T_c
or upon application of a magnetic field at $T \sim T_c$ [1]. The results
for the sharp-line scattering can be understood with the help of
the magnetic phase diagram[6] given as an insert in Fig.2. This
insert shows that ferromagnetic ordering is achieved only by the
application of a magnetic field. In this connection, Fig.1
shows the emergence of sharp-line scattering at the zone center
LO phonon frequency for $H > 0$. The intensity of this scattering
increases monotonically with increasing magnetic field, indic-
ative of the effect of the applied field toward increasing the
ferromagnetic order. The lack of ferromagnetic spin alignment
for $H = 0$ and $T < T_c$ may account for the absence of the sharp-line
scattering phenomenon in this regime. The magnetic phase dia-
gram also indicates that in the presence of internal magnetic
fields in excess of $\sim 2kOe$, the crystal will gradually change
from ferromagnetic to paramagnetic ordering as a function of
temperature. Thus the spin order-disorder argument predicts
that the intensity of the sharp line would be quenched as T goes
from 2K to $T > T_c$. This is in qualitative agreement with the T
dependence of the sharp-line intensity at $H = 10.5kOe$ (see Fig.2).
Similar results are obtained at lower values of $\vec{H}$.

FIG. 2: Peak intensity (arbitrary units) of the sharp line in
EuSe vs. T at H=10.5kOe and laser excitation energy of
2.41eV (λ_i=5145 Å). The insert shows the magnetic
phase diagram of EuSe (after Griessen et al. Ref.[6]).

With the limited range of available laser excitation energies,
Tsang et al.[1] saw large resonant enhancement effects in the
intensity of the sharp-line scattering. Our work extends their
excitation energy range and, in addition to corroborating their
observation of a strong resonant enhancement near 2.4eV, we find
a region of weaker resonant enhancement near 2.1eV. This is
shown in Fig.3 where the peak intensity of the sharp-line scat-
tering observed at T~2K and H=10.5kOe at several different ex-
citation energies is compared to the peak intensity observed at
2.41eV (λ_i=5145 Å) under the same conditions. The magneto-
reflectivity data of Pidgeon et al.[5,7] also show structure in
the same energy range; however, it is difficult to make a de-
tailed comparison because of the difference in experimental
conditions. The study of the excitation energy-dependence of
the sharp-line scattering is continuing along with the measure-
ment of the magneto-reflectivity under the same experimental
conditions.

FIG. 3: Peak intensity of the sharp line in EuSe relative to that at 2.41eV (λ_i=5145 Å) vs. excitation energy at H=10.5kOe and T∿2K.

In addition to the broad-line scattering reported previously in EuS[1,3] we have recently observed the sharp-line scattering at the LO phonon frequency in this material[4]. Although the scattering exhibits many characteristics similar to those observed in EuSe, there are significant differences which are consistent with the predictions of a spin order-disorder mechanism.

REFERENCES

[1] Tsang, J. C.; Dresselhaus, M. S.; Aggarwal, R. L.; Reed, T. B.: Phys. Rev. B <u>3</u>, 984 (1974); B <u>3</u>, 997 (1974).
[2] Holan, G. D.; Webb, J. S.; Dennis, R. B.; Pidgeon, C. R.: Solid State Commun. <u>13</u>, 209 (1973).
[3] Schlegel, A.; Wachter, P.: Solid State Commun. <u>13</u>, 1865 (1973).
[4] Tekippe, V. J.; Silberstein, R. P.; Dresselhaus, M. S.; Aggarwal, R. L.: to be published.
[5] Pidgeon, C. R.; Feinleib, J.; Schouler, W.; Hanus, J.; Dimmock, J. O.; Reed, T. B.: Solid State Commun. <u>7</u>, 1325(1969).
[6] Griessen, R.; Landolt, M.; Ott, H. R.: Solid State Commun. <u>9</u>, 2219 (1971).
[7] Pidgeon, C. R.; Feinleib, J.; Reed, T. B.: Solid State Commun. <u>8</u>, 1711 (1971).

CHARGE COMPENSATION IN A-SITE SUBSTITUTED $CuCr_2S_4$ SPINELS

H. GÖBEL, L. TREITINGER, H. PINK, W. K. UNGER, E. BAYER

Forschungslaboratorien der Siemens AG, München, F.R.G.

Materials in the system $Cu_{0.5+x}In_{0.5-x}Cr_2S_4$ have
have been prepared with $0.1 \geq x \geq -0.1$. At x=0, the
free holes causing $CuCr_2S_4$ to be a metallic ferro-
magnet with a Curie temperature of 420 K are com-
pensated and the material becomes an antiferromagne-
tic semiconductor. By conductivity, susceptibility,
and magnetization measurements the magnetic exchange
interaction in these materials was found to be con-
ductivity controlled.

1. INTRODUCTION

Among the ferromagnetic chalcogenide spinels, $CuCr_2S_4$ is one
of the most interesting ones because of its high Curie tem-
perature (420 K) and p-metallic conductivity. The high con-
centration of free holes is caused by the monovalence of
Cu leaving one hole per molecule, when placed on the A-site
of the spinel structure. The ferromagnetism is due to indi-
rect exchange interaction between the localized Cr-3d-spins
via the antiparallel spin polarization of the free holes in
the Cu-σ-band. The schematic band structure of this model
given by J.B. GOODENOUGH /1/ is shown in Fig. 1.

Fig. 1: Schematic band struc-
ture of $CuCr_2S_4$ after /1/ and
this work.
(CB: conduction band, VB:
valence band, E_F: Fermi ener-
gy)

In this work, we report the substitution of monovalent Cu by
trivalent In with the aim to control the concentration of

free holes and to study by this way conductivity controlled ferromagnetism in Cu-Cr-spinels.

2. EXPERIMENTAL

The specimen preparation has been described recently /2/. In the range $-0.1 \leq x \leq +0.1$, X-ray diffraction investigations showed no second phase present. The composition was determined approximately from the lattice constants which increase about linearly with increasing In content.

3. RESULTS

3. 1 ELECTRICAL CONDUCTIVITY (TRANSPORT PROPERTIES)

Figure 2 shows resistivities that room temperature measured by a four point method on pressed bars or on hot pressed and polished

Fig. 2: Room temperature resistivity ρ and activation energy E_A as a function of x determined from the lattice constant a_0 by assuming linear dependence $a_0(x)$.

discs of representative specimens. The sharp maximum at the composition $x = 0$ is caused by a transition from p-type to n-type conductivity as indicated by the change of sign of the Seebeck voltage. The resistivity of these compensated specimens is about 9 orders of magnitude larger than that of $CuCr_2S_4$.

The same figure shows the activation engergies determined from the high temperature curves of the resistivities. The activation energy behaves similar to the resistivity. At the

compensated composition it reaches a maximum with 0.93 eV
leading to an energy gap width of 1.86 eV.

By measuring the optical transmission, it was found that be-
sides the states contributing to the conductivity there are
localized states tailing into the energy gap.

3.2 MAGNETIC PROPERTIES

Figure 3 shows in its top part the Curie temperature deter-
mined by measuring the temperature dependence of the magne-
tic hysteresis. Near the composition x = 0, there is a tran-

Fig. 3: Curie temperature T_C, Neél
temperature T_N, and paramagnetic
Curie temperature θ vs. x and
vs. lattice constant a_o.

Fig. 4: Magnetic moment m per
formula unit at 4.2 K vs. applied
field B for different spinel com-
positions

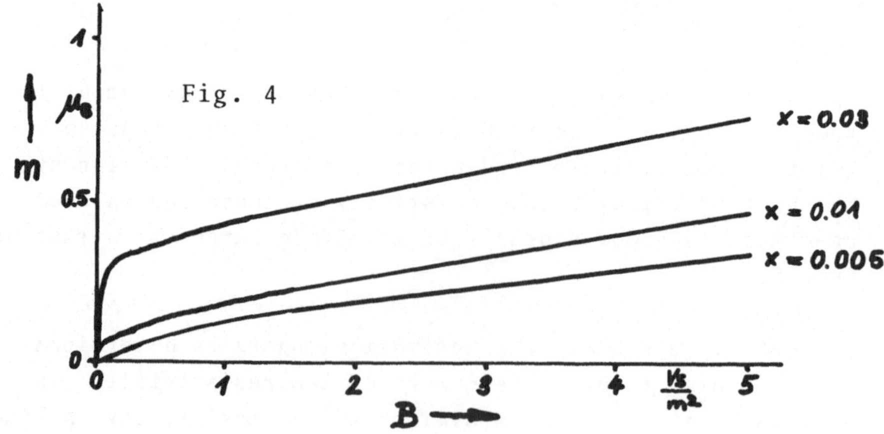

sition from ferromagnetism to antiferromagnetism, the speci-
mens still being p-type. All n-type materials investigated
are antiferromagnets.

The bottom part of Fig. 3 shows the paramagnetic Curie tem-
peratures determined from susceptibility measurements. The
paramagnetic Curie temperature shows a pronounced minimum
at the composition of compensated specimens.

The similarity of Figs. 2 and 3 confirms that with growing
carrier concentration the paramagnetic Curie temperature
rises independent of the type of the carriers.

Fig. 5: Extrapolated moments of magnetization $m_{B=0}$ and effective moments m_{eff} from susceptibility measurements per formula unit

Fig. 6: Magnetic orde-
ring temperatures
T_C, T_N, θ vs. room tempe-
rature resistivity of
$Cu_{0.5+x} In_{0.5-x} Cr_2S_4$

To examine the nature of the magnetic transition, we have
carried out magnetization measurements using a vibrating
sample magnetometer with a super conducting coil producing
magnetic fields up to 7 Vs/m^2.
Figure 4 shows the results of these magnetization measure-
ments carried out at the temperature of 4.2 K. In contradic-
tion to the behaviour of $CuCr_2S_4$ which is saturated at a
field of about 1 Vs/m^2, the In-substituted specimens show
no saturation at magnetic fields as high as 5 Vs/m^2 indica-
ting the diamagnetic In to change the spin arrangement when
substituted for Cu. The same observation was made by
F.K. LOTGERING for the In-substituted $CdCr_2S_4$ /4/.

By extrapolating the magnetization curves to B = 0, we got
the values of magnetization shown in Fig. 5 as a function
of the composition. It can be seen this extrapolated magne-
tization value decreases with composition indicating again
the transition from ferro-to antiferromagnetism.

The effective magnetic moment m_{eff} calculated from the slope
of the susceptibility curves has a maximum for compensated
samples. The maximal value of m_{eff} is (5.55 $\pm$ 0.1) μ_B for
x =0 which is close to 5.45 μ_B, the spin-only value of
Cr^{3+} ions. For x $\neq$ 0, m_{eff} is smaller than 5.55 μ_B because
of the larger concentration of mobile charge carriers which
reduce the net magnetic moment of the Cr ions.

4. DISCUSSION

Our experiments confirm the initially shown level scheme
given by J.B. GOODENOUGH /1/. Since the ferromagnetism
vanishes in the semiconductive region, the valence band edge
must be formed by the spin polarized Cu - 6- band. In addi-
tion, the dependence of the activation energy on the Cu con-
tent points to this level scheme.

Figure 6 shows the dependence of the magnetic ordering tem-
perature on the resistivity indicating that the magnetic ex-
change interaction is mainly controlled by the concentra-
tion of free carriers. In this figure we have inserted the

corresponding results of K. MIYANTANI et al. /6/ on the mixed system $CuCrSe_{4-x}Br_x$ in which similar charge compensation effects exist. In this system, a full compensation could not be reached because of preparative difficulties /7/.

The strong conductivity control of ferromagnetism in Cu-Cr-spinels opens the possibility of influencing magnetic properties of the spinels by charge carrier injection.

5. REFERENCES

/1/ Goodenough, J.B.: J. Phys.Chem.Sol. <u>30</u>, 261 (1969)
/2/ Unger, W.K.; Göbel, H.; Treitinger, <u>L.</u>; Pink, H.:
 Intern. J. Magnetism (1974)
/3/ Pinch, H.L.; Woods, M.J.; Lopatin, E.: Mat.Res.Bull <u>5</u>,
 425 (1970)
/4/ Lotgering, F.K.; van der Steen, G.H.A.M.: Solid Commun.
 7, 1827 (1969)
/5/ Lotgering, F.K. van Stapele, R.P.: J. Appl. Phys. <u>39</u>
 417 (1968)
/6/ Miyatani, K; Minematsu, K; Wada, Y; Okamoto, F; Kato,K;
 Baltzer, P.K.: J. Phys.Chem. Solids <u>32,</u> 1429 (1971)
/7/ Pink, H; Unger, W.K; Schäfer, H; Göbel, H.: Appl. Phys.
 (July or August 1974)

EXTRAORDINARY HALL EFFECT IN THE MAGNETIC SEMICONDUCTOR $Sn_{1-x}Mn_xTe$

M. ESCORNE, A. GHAZALI and P. LEROUX-HUGON

Laboratoire de Physique des Solides, CNRS,

92190 Meudon-Bellevue, France.

We have studied the extraordinary Hall effect (EHE) in the
magnetic semiconducting solid solutions $Sn_{1-x}Mn_xTe$ over the
ranges of carrier concentrations : $3.10^{20} < p < 8.10^{20}$ cm-3, and
Mn concentrations : $2.5 < x < 7.5$ at.%. We have calculated the
skew scattering contribution to the EHE arising from both spin
independent and spin dependent potentials ; we have also estimated
the current renormalization contribution. Good agreement has been
found between the calculated values of extraordinary Hall
coefficient and the experimental ones over the whole range of
parameters studied.

1. INTRODUCTION.

The EHE, i.e. that part of the Hall effect which in a magnetic
conductor depends on magnetization rather than on magnetic induction, is a
typically spin dependent transport phenomenon. Luttinger [1] has shown that
the EHE arises from the interband matrix elements of the spin-orbit coupling.
The study of this effect which would be vanishingly small for free electrons
requires a knowledge of the actual band structure. A detailed analysis of
the EHE [1,2] allows one to separate different contributions the relative
importance of which varies with carrier concentration and scattering
mechanisms. Finally theoretical [3], and experimental [4] results have
pointed out that, in a semiconductor, the EHE may be defined with a small
number of band parameters namely the ratios $\alpha = \epsilon_F/\epsilon_g$ and $\beta = \Delta/\epsilon_g$ (ϵ_F Fermi
energy, ϵ_g energy gap, Δ spin-orbit energy). It is obviously of interest to
study the EHE in a magnetic semiconductor where the band parameters and the
scattering parameters are fairly well known.

$Sn_{1-x}Mn_xTe$ solid solutions are small gap, highly degenerate, p type
semiconductors with a rather large spin-orbit coupling energy ($\epsilon_g = 0.27$ eV,
$p \sim 10^{20} - 10^{21}$ cm^{-3}, $\Delta = 0.55$ eV). Indirect exchange through carriers
couples the magnetic moments carried by the Mn ions and leads to a
ferromagnetic ordering at low temperatures and to spin polarization of the
carriers ; for the studied range of carrier concentrations p, the Curie
temperature T_c (°K) $\simeq 1.4$ x (at.%) depends only slightly on p.
Monocrystalline samples have been grown in which the Mn concentration and
the carrier concentration are varied indepedently, and the EHE is easily

detectable [5]. $Sn_{1-x}Mn_xTe$ then constitutes a very suitable material for the study the EHE. Upon varying (i) the magnetic parameters, (ii) the Fermi energy and (iii) the scattering parameters, detailed comparisons between experiment and theory are possible. The band structure is rather intricate ; however, a simple nonparabolic band scheme allows one to account for the electronic properties (effective masses, mobility).

2. ESTIMATION OF THE EHE.

Let us consider first the <u>skew scattering</u> contribution to the EHE. It arises from the fact that, when the spin-orbit coupling is properly included in the wave functions, the transition probability $W_{kk'}$, associated with a scattering event contains a term which is antisymmetric with respect to $\vec{k}$ and $\vec{k}'$ [3]. Our band scheme will be the one of InSb with the parameters ϵ_g and Δ appropriate to the matrix SnTe, the multivalleys character being taken into account.

In $Sn_{1-x}Mn_xTe$, at low temperatures, three different mechanisms take part in the scattering processes namely : (i) scattering by native impurities (Sn vacancies). They are specified by their potential - U_I. We will assume their concentration N_I to be equal to holes one. (ii) scattering by the spin independent part - U_M of the Mn impurities potential. N_M is their concentration. (iii) scattering by the s-d like interaction between the carrier's spin $\vec{s}$ and spin $\vec{S}$ of Mn ions : - $J\,\vec{s}.\vec{S}$. For the sake of simplicity, we assume these potentials to be constant within the unit cell volume Ω and vanishing outside.

Once the wave functions are known, one may calculate (to second Born approximation) that part of the transition probability which is responsible for the skew scattering [3]. A straightforward extension of this calculation allows one to include the three scattering contributions into the Hall resistivity expression ρ_H^{sk}, using Boltzmann equation. We have retained only the lowest order terms :

$$\rho_H^{sk} = \frac{m_c\,m_d^{\,2}}{4(6\pi^2)^{1/3}e^2\hbar^5 p^{1/3}}\left[\begin{array}{l} 8/5(2\sqrt{2}bc - b^2)[N_I(U_I\Omega)^3 + N_M(U_M\Omega)^3]\,\dfrac{J\Omega}{2\epsilon_F} \\[2mm] + (2\sqrt{2}bc + b^2)\,(U_M\Omega)^2\,J\Omega \end{array}\right] \times \frac{M}{g\mu_B} \quad (1)$$

where m_c and m_d are the conductivity and density of states masses, M the magnetization per unit volume and the factors :

$$2\sqrt{2}bc \overset{+}{\underset{-}{}} b^2 = \frac{2}{3} \frac{\alpha\beta}{(1 + 2\alpha)(1 + \beta)} \left(2 \overset{+}{\underset{-}{}} \frac{\beta}{2\beta + 3}\right)$$

specify the orbital and spin wave functions mixing. It is worth making some comments on (1).

(a) ρ_H^{sk} is a linear function of M. For the spin independent potentials, this M dependence is due to the spin polarization $(p\!\uparrow - p\!\downarrow) \propto JM$; while, the exchange potential is itself linear in JM and the $p\!\uparrow$ and $p\!\downarrow$ contributions add up. The latter statement means that the exchange term remains finite even for a negligible spin splitting.

(b) As, in our case, the relative spin polarization uses to be small (< 20 %), the exchange contribution to ρ_H^{sk} is larger than the other ones although the exchange scattering is responsible only for a small part of the (longitudinal) resistivity ρ.

(c) The sign of ρ_H^{sk} depends on the sign of scattering potentials. One verifies, for an exchange interaction under the form $- J\vec{s}.\vec{S}$ with $J > 0$, that the exchange part of ρ_H^{sk} has the very sign of the ordinary Hall effect (OHE) ; the same is true for a spin independent attractive potential.

Let us now turn to the <u>current renormalization</u> contribution to the EHE. Using an effective Hamiltonian, Nozières and Lewiner [2] have recently given a simple and comprehensive discussion of this contribution. They have shown that :

$$\frac{\rho_H^{sk}}{\rho_H^{ren}} = \frac{\pi \, p \, U\Omega}{\Gamma} \tag{2}$$

where $\Gamma = \hbar/\tau$ (τ is the relaxation time) ; U is a spin independent potential. This result is not readily valid if several spin dependent and spin independent scattering potentials are operating. Nevertheless we propose to estimate ρ_H^{ren} from :

$$\frac{\rho_H^{sk}}{\rho_H^{ren}} = \frac{\pi}{\hbar e^2} \frac{m_c}{\rho} \left[\frac{N_I(U_I\Omega)^3 + N_M(U_M\Omega)^3}{N_I(U_I\Omega)^2 + N_M(U_M\Omega)^2}\right] \tag{3}$$

where the structure of (2) is maintained. One will verifies that ρ_H^{ren} and the OHE have opposite signs.

3. EXPERIMENTAL RESULTS.

Fig. 1 - Hall resistivity as a function of magnetic field at different temperatures for sample 3.

Fig. 2 - (ρ_H/H) versus (M/H) plot of the data of sample 3.

We have studied solid solutions in the range $2.5 < x < 7.5$ at.%. The homogeneity of the samples have been controlled and the Mn concentration measured by electron microprobe analysis. The carrier concentration has been varied by annealing Sn-electroplated samples. Transport properties have been measured using an a.c. bridge with a lock-in detector. Magnetization has been measured on the same samples with a vibrating sample magnetometer. From this data, we extract the magnetic (T_c) and scattering (U_I, U_M, J) parameters.

The Hall resistivity has been measured as a function of magnetic field (0 - 17 KOe) for different temperatures (1.5 - 20K). Typical data are displayed on the figure 1 : obviously OHE and EHE add up with the same sign and similar magnitude. Following our previous discussion, the total Hall resistivity is readily splitted into two terms :

$$\rho_H = R_o H + R_e M \qquad (4)$$

where R_o and R_e are respectively the ordinary and extraordinary Hall coefficients. This separation is easily achieved in a ρ_H/H versus M/H diagram as given in figure 2. From this plot, we see that, for $T < T_c$ ($T_c = 7.5$ K in this case), points corresponding to different M and T values

lie on the same straight line, this implies a constant R_e as predicted by our analysis. It can be noticed that, in the paramagnetic range, the EHE values are still appreciable, as theoretically predicted, but with a deviation from the linear behaviour.

Experimental data together with theoretical results for a number of samples are given on the following table :

Sample	x (at.%)	T_c (K)	p $(10^{20}cm^{-3})$	μ (cm^2/Vs)	$(10^{-8}\Omega cm/gauss)$			
					R_e^{sk} cal	R_e^{ren} cal	R_e^{tot} cal	R_e exp
1	7.4	10.5	8.6	19	5.41	− 3.56	1.85	2.52
2	7.4	9.2	3.3	90	3.54	− 1.74	1.80	2.30
3	5.4	7.5	5.5	67	4.18	− 1.41	2.77	2.78
4	2.5	3.5	5.4	108	3.53	− 0.76	2.77	2.67

4. DISCUSSION AND CONCLUSION.

The spanned range of Mn and carrier concentrations allows the magnetic properties (T_c) and the scattering efficiency (μ) to vary between wide limits. Nevertheless the discrepancy between calculated and experimental values of R_e (see table) never exceeds 25%. This is a gratifying result, especially if the involved approximations are considered. Let us point out that for the two first samples, x being the same but p differing by a factor of about 3, EHE are nearly equal, as are the EHE of the two last samples which have the same p but different x. As seen on the table, rather different values of R_e^{sk} and R_e^{ren} compensate each other so that the algebraic sum does not vary very much from sample to sample.

As a conclusion, our study emphasizes the interest of magnetic semiconductors in the study of the EHE : varying the relevant parameters, we have been able to show the importance of different contributions to this effect. The experimental data are fitting in closely with theoretical results.

5. REFERENCES.

[1] Luttinger, J.M. : Phys. Rev. 112, 739 (1958).
[2] Nozières, P. ; Lewiner, C. : J. Physique 34, 901 (1973).
[3] Leroux-Hugon, P. ; Ghazali, A. : J. Phys. C 5, 1072 (1972).
[4] Chazalviel, J.N. ; Solomon, I. : Phys. Rev. Lett. 29, 1676 (1972).
[5] Ghazali, A. ; Escorne, M. ; Rodot, H. ; Leroux-Hugon, P. : AIP Conf. Proc. 10, 1374 (1972).

CRITICAL SCATTERING AND BAND CONDUCTION
IN NEARLY STOICHIOMETRIC FeO

I. BALBERG

The Racah Institute of Physics, The Hebrew University,
Jerusalem, Israel

We have observed a resistivity anomaly in the
vicinity of the Néel temperature T_N of Fe_xO
where $0.99 \leqslant x \leqslant 1.00$. The critical exponents of the
temperature derivative of the resistivity were
found to be 0.4 ± 0.1, both above and below T_N,
this in accord with the theory of critical spin
fluctuations. The present results have also
enabled an estimation of the holes mobility μ_o
in the present material. The values obtained
$5 \leqslant \mu_o \leqslant 10$ cm^2/V sec, provide new evidence for band
type conduction in the transition metal monoxides.

1. INTRODUCTION

In recent years significant progress was made in the understand-
ing of transport properties in solids at the vicinity of criti-
cal points [1,2]. Of these properties, resistivity anomalies,
due to critical spin fluctuations were most extensively studied.
However, quantitative comparisons of the observed anomalies with
theory, such as determination of critical exponents, were carried
out for a few metals [1-3] but not for semiconductors. In this
study we have examined the resistivity anomaly in nearly stoichio-
metric Fe_xO ($0.99 < x < 1.00$) and we have determined the critical
exponents for this anomaly. Since the conduction mechanism in
the transition metal monoxides is still debated [4,5] and the
contribution of critical scattering is still unexplored, we have
used our observation to estimate the carriers mobility and to
determine the conduction mechanism in FeO. This method for
mobility estimation has not been used before.

2. EXPERIMENTAL

The Fe_xO material used for the present measurements has a lattice
constant $a_o = 4.326$ Å. Using the known relation [6] between
a_o and x we have determined the value of x. From the

magnetic susceptibility of this material we found that the mag-
netic Néel temperature is about 190°K, that the Curie-Weiss
temperature, θ, is about 180°K and that the susceptibility, χ,
in the vicinity of the Néel point is about 10^{-2} emu/mole.
Details of the sample preparation and the magnetic properties of
the material will be presented elsewhere [7].

For the resistivity measurements the pallets prepared were cut
into bars of 8x2x1 mm^3. Four indium contacts were soldered to
the bars. The resistivity was measured using a standard four
probe technique and the temperature was monitored by a copper-
constant and thermocouple. The voltage between the two voltage
probes versus the thermocouple voltage was displayed on a
recorder. The measurements were taken in the temperature range
$78 \leqslant T \leqslant 300°K$.

3. RESULTS AND DISCUSSION

In Fig. 1 the curve ρ_1 shows the temperature dependence of the
resistivity as recorded over the temperature range $165 \leqslant T \leqslant 215°K$.
Below 185°K and above 189°K the resistivity decreases with in-
creasing temperature while between these temperatures the

FIG. 1: Recorded trace of
the resistivity dependence
on temperature in nearly
stoichiometric Fe_xO
(curve ρ_1) and the
expected dependence in the
absence of critical pheno-
mena (curve ρ_2).

resistivity increases with temperature. This anomalous resist-
ivity increase in the vicinity of the Néel temperature is
associated with critical phenomena and must be due to an intrin-
sic effect since changes such as in the dimensions of the sample
in the critical region, will cause an effect that is two orders
of magnitude smaller than the observed one [7]. In Fig. 1 we
have also added curve $\rho_2(T)=\rho_0\exp(0.09/KT)$ which represents the
expected temperature dependence of the resistivity if no criti-
cal phenomena were present. Here 0.09 eV is the resistivity
activation energy below 165°K, k is the Boltzman constant and
$\rho_0=\rho_1(160°K)\exp(-0.09/k\cdot160)$.

The resistivity anomaly observed here is similar to the one ob-
served in the antiferromagnetic semiconductors MnTe and NiO. In
these materials it was found that the anomaly is dominated by
critical scattering [8,9]. In view of this and the analysis to
be given below we assume that this is also the origin of the
present anomaly. It is thus useful to consider the quantity
$P\equiv(\rho_1-\rho_2)/\rho_2$ since under this assumption $P=\tau_0/\tau_s$ where τ_s is the
mean free time due to critical scattering while τ_0 is due to all
other interactions. Thus for comparison of the present results
with theory, we have derived dP/dT. The dependence of this
quantity on temperature is shown in Fig. 2. It is seen that

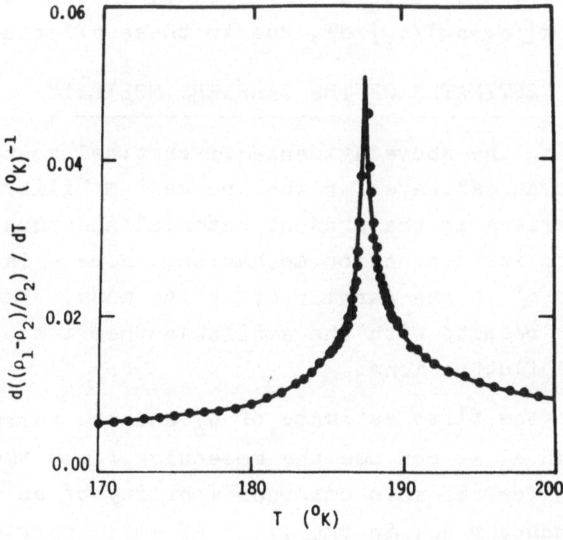

FIG. 2: The
temperature
derivative of
the normalized
resistivity
$(\rho_1-\rho_2)/\rho_2$

this derivative is positive over the whole temperature range and its maximum is at $T_N = 187.5°K$.

The results of Fig. 2 were found to be expressible in the power low form

$$dP/dT \propto \varepsilon^{-\lambda} \tag{1}$$

over the range $10^3 \leqslant \varepsilon \leqslant 10^{-1}$ [7]. Here $\varepsilon = |T-T_N|/T_N$ and λ, the critical exponent, is 0.4±0.1 both above and below T_N. The corresponding theoretical prediction is that $\lambda = 1/3$ [7,10]. This good agreement is an indication for the importance of scattering by short range fluctuations in antiferromagnetic semiconductors.

The fact that in the present system as in some other systems [11] the observed critical exponents are somewhat larger than those predicted by the available theories [7,10,12] may be attributed to dynamic effects [1]. In the nearly stoichiometric Fe_xO dynamic effects seem to be more important than in previously studied materials since τ_o is relatively short in this material and thus cannot be assumed to be large compared to the lifetime of the critical fluctuations [1,2]. That dynamic effects may lead to an increase in the critical exponents is apparant from the expected dependence of $d(1/\tau_s)/dT$ on ω, the fluctuations frequency [10]. This is in view of the dynamic scaling prediction [13] that $\omega \propto \xi^{1/2}$ where ξ is the correlation length that is given by $\xi = \xi_o \cdot \varepsilon^{-2/3}$. Hence, an enhancement of the divergence of $d[(\rho_1 - \rho_2)/\rho_2]/dT$, due to these effects, is expected.

4. ESTIMATES OF THE CARRIERS MOBILITY

Using the above evidence for critical scattering we are trying to get an estimate for the "normal" mobility $\mu_o = e\tau_o/m^*$ of the carriers in the present material and thus to get some insight into its conduction mechanism. Here e is the electronic charge and m^* is the carrier effective mass. The idea is to compare our results with the available theories on scattering by critical fluctuations.

For the first estimate of μ_o and the corresponding mean free path ℓ_o we can use the molecular field theory of Haas [14] who got for the spin disorder mobility of an antiferromagnetic semiconductor μ_s, in the limit of small carrier-wave number, the

relation:

$$\mu_s = C\left[m^{*\,5/2}\,J^2\,(kT_N)^{3/2}\,a_o^6\,x\right]^{-1} \tag{2}$$

where C is a constant and J is the exchange constant of carrier-ion spin-interaction. In transition metal monoxides a value of J=0.5eV [14,15] and a value of $m^*=4m_o$ [4], where m_o is the free electron mass, seem to be good estimates. Using the above given values for all the parameters in eq.(2) we get $\mu_s=12$ cm^2/V sec. The effect of critical scattering will be sampled by the resistivity when $\mu_s \approx \mu_o$ and thus μ_o is of the order of 10 cm^2/V sec. From this and the value of the thermal velocity at T_N we also obtain $\ell_o \approx 10$Å.

Another estimate can be achieved by considering the coherence of the critical scattering. If two spins separated by a distance R are strongly correlated they will not scatter coherently if $R \gg \ell_o$. As ξ, the correlation length, approaches ℓ_o the contribution of the short range, specific heat type fluctuations, becomes important [1,7,12]. Since all our evidence indicates that in the present case the scattering is dominated by the short range fluctuations we can expect that the onset of the resistivity anomaly will be when $\xi \approx \ell_o$. Experimentally, the anomaly is observed for $\varepsilon \lesssim 0.1$ (see Fig. (1)). In the Orenstein-Zernike approximation the value of ξ, for an fcc antiferromagnet of the second kind can be given by [16]:

$$\xi = (a_o/\sqrt{6})\varepsilon^{-1/2} \tag{3}$$

provided $T_N \approx \theta$ as in the present case. Substitution of $\varepsilon=0.1$ in eq.(3) yields $\xi=4.5$Å. Using the exponent predicted by scaling theory, i.e. 2/3 rather than 1/2 in eq.(3) yields $\xi=6.3$Å. Thus ℓ_o is expected to be of the order of 5Å, in good agreement with the estimate derived above by using the theory of Haas.

The above relatively high value for the mobility indicates that in the nearly stoichiometric material, the conduction is dominated by band conduction or by large-polaron conduction [4,5]. Since we have a p-type material [17] the estimated mobility is that of holes. This is to be contrasted with the n-type material, the low mobility and the defect hopping conduction in the non-stoichiometric material [17]. It seems thus that the defect hopping gives way to band like conduction as

stoichiometry is approached in Fe_xO.

5. REFERENCES

[1] P.P. Craig and W.I. Goldburg, J. Appl. Phys. 40, 964 (1969) (References)
[2] R.D. Parks, in Magnetism and Magnetic Materials - 1971 AIP Conference Proceedings No. 5, edited by C.D. Graham, Jr. and J.J. Rhyne (AIP, New York, 1972) p. 630, (References).
[3] R.A. Craven and R.D. Parks, Phys. Rev. Letters 31, 383 (1973).
[4] A.J. Bosman and H.J. Van Daal, Adv. in Phys. 19, 1 (1970).
[5] D. Adler and J. Feinleib, Phys. Rev. B2, 3112 (1970).
[6] B. Heutschel, Z. Naturforsch 25A, 1996 (1970).
[7] I. Balberg, J.S. Helman and H.L. Pinch, to be published.
[8] J.D. Wasscher, A.M.J.H. Seuter and C. Hass, Proc. Int. Conf. on Semiconductor Physics, Paris, (1964) p.1269.
[9] I.G. Austin, A.J. Springthorpe, B.A. Smith and C.E. Turner, Proc. Phys. Soc. 90, 157 (1967).
[10] Y. Suezaki and H. Mori, Prog. Theor. Phys. 41, 1177 (1969).
[11] M.P. Kawatra, J.A. Mydosh and J.I. Budnik, Phys. Rev. B2, 665 (1970).
[12] M.E. Fisher and J.S. Langer, Phys. Rev. Letters 20, 665 (1968).
[13] B.I. Halperin and P.C. Hohenberg, Phys. Rev. 177, 952 (1969).
[14] C. Hass, Phys. Rev. 169, 531 (1968).
[15] R.W. Cochrane, F.T. Hedgcock and J.O. Ström-Olsen, Phys. Rev. B8, 4262 (1973).
[16] P.G. de Gennes, J. Phys. Chem. Solids 6, 43 (1958).
[17] I. Bransky and D.S. Tannhauser, Physica 37, 547 (1967), and Trans. of the Metallurgical Soc. of AIME, 239, 75 (1967).

ACKNOWLEDGEMENTS

The author is indebted to Y. Goldstein and J.S. Helman for helpful discussions.

THEORY OF SPIN-DEPENDENT PHONON RAMAN SCATTERING IN MAGNETIC SEMICONDUCTORS

H. KAMIMURA

Department of Physics, University of Tokyo,
Bunkyo-ku, Tokyo, Japan

N. SUZUKI

Faculty of Engineering Science, Osaka University,
Toyonaka, Osaka, Japan

A general theory is developed for the spin-dependent phonon Raman scattering in magnetic semiconductors. The integrated Raman intensity is expressed as $S(T)=(n_0+1)|R+M<S_0 \cdot S_1>|^2$. Three types of temperature variation of intensity are predicted in ferro- and antiferro-magnetic cases, depending on values of R/M. It is shown that the mechanism of variation of the d electron transfer with lattice vibrations is essentially important in producing the large temperature variations. The theory is successfully applied to explaining experimental results in $CdCr_2S_4$.

1. INTRODUCTION

Recently the effects of magnetic ordering on the first order phonon Raman scattering in magnetic semiconductors have attracted both experimental and theoretical interests [1-7]. In this paper, first we develop a phenomenologinal theory of the spin-dependent phonon Raman scattering and derive a general formula for the integrated Raman intensity which markedly depends on magnetic ordering through the spin correlation function. Secondly, possible microscopic mechanisms of the spin-dependent Raman process are investigated. Particularly, two types of spin-dependent scattering mechanisms, variation of the d electron transfer with lattice vibration and that of the non-diagonal exchange interaction, are considered. Finally the present theory is successfully applied to explaining the phonon Raman scattering experiment in ferromagnetic semiconductor $CdCr_2S_4$ (T_c=84.5K) [1].

2. DERIVATION OF A GENERAL FORMULA

Here we derive a general formula for the phonon Raman scattering in magnetic semiconductors in which the valence band is full

while the conduction band is empty and the ground state of a
transition metal ion is orbitally non-degenerate.

We try to express the Raman scattering cross section in terms
of the spin-dependent polarizability tensor. For this purpose,
we write down a general form of polarizability tensor as follows:

$$\tilde{\alpha} = \tilde{\alpha}_o + \sum_{j,\rho} \tilde{A}_{j\rho} S_{j\rho} + \text{---} + \sum_{i,j} \tilde{P}_{ij} (\vec{S}_i \cdot \vec{S}_j) + \text{---} . \tag{1}$$

Then we expand the above $\tilde{\alpha}$ in powers of ionic displacements.
In doing this, we take into account the most important terms of
$\tilde{\alpha}$; spin-independent term $\tilde{\alpha}_o$ and scalar type interaction $\tilde{P}_{ij}$ with
the nearest neighbor spin pair i,j. Thus the polarizability
tensor $\tilde{\alpha}_p$ which is associated with the one phonon scattering is
expressed in terms of the phonon creation and destruction opera-
tors $b^+_{q,p}$ and $b_{q,p}$ as follows:

$$\tilde{\alpha}_p = \tilde{\alpha}_{p1} + \tilde{\alpha}_{p2}, \tag{2}$$

$$\tilde{\alpha}_{p1} = \sum_{q,p} \sum_{R} C(\vec{q},p) [\vec{\nabla}_R \tilde{\alpha}_o \cdot \vec{e}(\vec{q},p)] e^{-i\vec{q}\cdot\vec{R}} (b^+_{q,p} + b_{-q,p}),$$

$$\tilde{\alpha}_{p2} = \sum_{q,p} \sum_{R} \sum_{i,j} C(\vec{q},p) [\vec{\nabla}_R \tilde{P}_{ij} \cdot \vec{e}(\vec{q},p)] e^{-i\vec{q}\cdot\vec{R}} (b^+_{q,p} + b_{-q,p}) (\vec{S}_i \cdot \vec{S}_j).$$

The differential Raman scattering intensity is calculated from
the well known formula

$$d\sigma/d\omega d\Omega = (\omega^4/c^4)\frac{1}{2\pi} \int_{-\infty}^{\infty} dt \; e^{i(\omega-\omega_o)t} <\tilde{\alpha}_p(0) \cdot \tilde{\alpha}_p(t)> . \tag{3}$$

Inserting (2) into (3) we can obtain the integrated intensity of
one phonon Raman scattering in the following form:

$$S(T) = (n_o+1) |R + M<\vec{S}_o \cdot \vec{S}_1>/S^2|^2 . \tag{4}$$

Here R and M are the temperature independent constant associated
with the first derivative of $\tilde{\alpha}_o$ and $\tilde{P}_{ij}$, respectively.

In ferromagnetic case, noticing the general behavior of the
nearest neighbor spin correlation function $<\vec{S}_o \cdot \vec{S}_1>$, we can
classify the reduced integrated intensity $I(T)$, which is defined
as $S(T)$ derived by the phonon population number (n_o+1), into
three types as shown in Fig.1. In antiferromagnetic case we can
also obtain the three patterns similar to that in Fig.1.

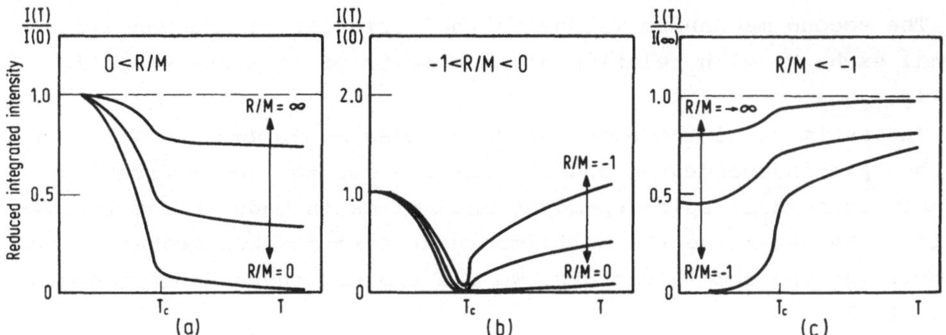

FIG. 1: Three patterns of temperature variation of the reduced integrated intensity I(T)/I(0).

3. MICROSCOPIC MECHANISMS

Though there may be various mechanisms which give rise to the spin-dependent polarizability $\tilde{\alpha}_{p2}$ in eq.(2), we consider two different kinds of mechanisms which are expected to be dominant. The first one comes from variation of the d electron transfer energy with relative displacements of ions. As shown in Fig.2a, this is the third order perturbation: (i) First an incident photon is absorbed and a d electron in a transition metal ion i is excited to the conduction band. (ii) Then a d electron in a neighboring metal ion j transfers to the i-th ion by emitting or absorbing one phonon by the d electron-phonon interaction H_{d-p}

$$H_{d-p} = \sum_{i,j} \sum_{R} \vec{\nabla}_R t_{ij} \cdot \delta\vec{R}\, d_i^+ d_j. \tag{5}$$

(iii) Finally the virtually excited electron in the conduction band recombines with the j-th ion, returning the electronic system to its ground state and a scattered phonon is emitted.

FIG. 2: Microscopic mechanisms of the spin-dependent process.

The second mechanism is due to the variation of the non-diagonal exchange with relative displacements of ions, as shown in Fig. 2b.

The estimate of the order of magnitudes of the Raman tensors for the spin-independent mechanism (R) and for the above first (M_I) and second (M_{II}) spin-dependent mechanisms is made in the following: By retaining the most important term in each mechanism and omitting the common factors, we can express these Raman tensors as follows:

$$R \propto \Xi p_k^2/\Delta E_k^2, \quad M_I \propto z(a \cdot \nabla_R t) p_k'^2/\Delta E_k'^2, \quad M_{II} \propto z(a \cdot \nabla_R J_{nd}) p_u^2/\Delta E_u \Delta E_e.$$

For R, the most typical values are $p_k \simeq \hbar/a$, $\Delta E_k =$ 2 to 4eV (the band gap) and electron-phonon interaction energy in a conduction band $\Xi \simeq$ 1eV. For the estimation of M_I, we consider a magnetic semiconductor in which the absorption band edge corresponds to the strong charge transfer transition from or to the d level. For the charge transfer transition, $p_k'^2 \simeq 10^{-1} p_k^2$, while $\Delta E_k' \simeq \Delta E_k$. Further $a \cdot \nabla_R t \simeq a \cdot t/a \simeq$ 1eV. Thus $M_I/R = 10^{-1}$ to 1 for the number of nearest neighbors z=6. For M_{II}, $p_u \simeq p_k$, $\Delta E_u \simeq$ 10eV, $E_e =$ 2 to 3eV. Further $a \cdot \nabla_R J_{nd} \simeq a \cdot J_{nd}/a = 10^{-3}$ to 10^{-2}eV. Then $M_{II}/R = 10^{-2}$ to 10^{-1}.

From the above estimation, it seems that, among the spin-dependent phonon Raman tensors, the first mechanism M_I is most likely to become the same order of magnitude as R and is the most important to produce the spin-dependent scattering.

4. APPLICATION TO $CdCr_2S_4$

Here we apply the present theory to $CdCr_2S_4$. All of the Raman active modes Γ_1^+, Γ_{12}^+ and 3 Γ_{25}^+ are observed in $CdCr_2S_4$ [1]. In applying the theory, first we investigate the selection rule concerning the spin-dependent Raman tensor M. From the symmetry consideration, we obtain the following tensor form of M for each Raman active mode:

$$M(\Gamma_1^+) = \begin{pmatrix} -16(2A+B)/\sqrt{6} & 0 & 0 \\ 0 & -16(2A+B)/\sqrt{6} & 0 \\ 0 & 0 & -16(2A+B)/\sqrt{6} \end{pmatrix},$$

$$M(\Gamma_{12}^+) = \begin{pmatrix} (A-B)/\sqrt{3} & 0 & 0 \\ 0 & (A-B)/\sqrt{3} & 0 \\ 0 & 0 & -2(A-B)/\sqrt{3} \end{pmatrix}, \begin{pmatrix} -(A-B) & 0 & 0 \\ 0 & (A-B) & 0 \\ 0 & 0 & 0 \end{pmatrix},$$

$$M(\Gamma_{25}^{+}) = \begin{pmatrix} 0 & 0 & 0 \\ 0 & 0 & -2C \\ 0 & -2C & 0 \end{pmatrix}, \begin{pmatrix} 0 & 0 & -2C \\ 0 & 0 & 0 \\ -2C & 0 & 0 \end{pmatrix}, \begin{pmatrix} 0 & -2C & 0 \\ -2C & 0 & 0 \\ 0 & 0 & 0 \end{pmatrix}. \tag{6}$$

From the cubic symmetry of the crystal and the non-degenerate ground state of a Cr^{3+} ion, we have A$\simeq$B. Thus M almost vanishes for Γ_{12}^{+} mode and this is consistent with the observed fact.

Now we compare the calculated results on temperature variation of reduced integrated intensity with observed ones. In doing this, for the spin correlation function in eq.(4) we use that calculated by Callen for $CdCr_2S_4$ [8]. Then the parameter R/M is determined from the ratio of the observed intensity at T =0K to that at about $T=2T_c$. In Fig.3, the theoretical curve of I(T)/I(0) is shown for the Γ_{25}^{+} mode (A line) in $CdCr_2S_4$ as well as the experimental points. The agreement between theory and experiment is excellent. The calculated results for other modes also reproduce the experimental curves very well.

FIG. 3: Comparison of the theoretical result (solid line) with the experimental plots (open circles) for A line (Γ_{25}^{+} mode) of $CdCr_2S_4$.

5. REFERENCES

[1] Steigmeier, E.F.; Harbeke, G.: Phys. kondens. Materie 12, 1 (1970).
[2] Shepherd, I.W.: Phys. Rev. B5, 4524 (1972).
[3] Suzuki, N.; Kamimura, H.: Solid State Commun. 11, 1603 (1972).
[4] Suzuki, N.; Kamimura, H.: J. Phys. Soc. Japan 35, 985 (1973).
[5] Meixner, A.E.; Dietz, R.E.; Rousseau, D.L.: Phys. Rev. B7, 3134 (1973).
[6] Anastassakis, E.; Burstein, E.: J. Phys. C5, 2468 (1972).
[7] Sokoloff, J.B.: J. Phys. C5 2482 (1972).
[8] Callen, E.: Phys. Rev. Lett. 20, 1045 (1968).

LIGHT SCATTERING IN THE VICINITY OF PLASMA
FREQUENCIES IN PYRITE-TYPE MAGNETIC SEMICONDUCTORS

E. ANASTASSAKIS,[*†] B. PAPANICOLAOU[†] AND C. H. PERRY[*]

Department of Physics, Northeastern University

Boston, Massachusetts 02115, U.S.A.

We report the results of an experimental study of surface
Raman scattering by $q \approx 0$ optical phonons, using excitation
frequencies in the vicinity of the plasma frequency ω_p. The
observed scattered intensity in $CuSe_2$ ($\omega_p \approx 1.71$ eV) is
found to follow closely the fifth power of an appropriately
defined skin depth, in agreement with theory. Deviations
from this dependence are attributed to resonance enhancement
caused by interband electronic transitions in the region of
interest.

1. GENERAL

Most of the experimental studies of Raman scattering (RS) from opaque mate-
rials have been made in semiconductors, using lasers in the visible region.
RS from metallic surfaces has been reported only for a limited number of
materials and is due to their low scattering efficiency [1]. In non-metallic
compounds there are situations where normally weak RS efficiency may be
substantially enhanced if the incident excitation frequency ω_i is chosen to
be in the vicinity of a characteristic energy gap in the material. This
type of resonance between incident photons and selective interband elec-
tronic transitions has been extensively studied in recent years [2]. Accord-
ing to the theoretical treatment of RS from metallic surfaces of Mills et al.
[3] the changes in the dielectric function due to R-active optical phonons
include contributions from intraband as well as interband electronic transi-
tions. Thus, in principle it should be possible to create resonance condi-
tions for RS even in metals provided that ω_i is chosen to be in the appro-
priate region of the spectrum. To our knowledge no such measurements have
been reported. In metals, however, there exists a distinct possibility of
enhancing the low RS efficiency if ω_i is chosen to be close to the plasma
frequency ω_p. The reason for this is that close to ω_p metals exhibit a
minimum in the reflectivity and a maximum in the skin depth δ; the inte-
grated RS efficiency can be shown [3] to vary as δ^5. Accordingly, the RS
intensity should exhibit a noticeable maximum close to ω_p.

*Supported in part by the National Science Foundation and the Office of
 Naval Research.
†Supported by the Research Corporation.

We have explored the situation discussed above, by studying the surface RS in the transition metal dichalcogenide compound $CuSe_2$. We have observed an appreciable increase in the RS intensity as ω_i approaches ω_p.

It might be attempted to describe the present enhancement as due to a "resonance" between ω_i and ω_p. However in metals the electron plasma does not perform the same physical function as that of a resonant intermediate electronic state in insulators and semiconductors. Here, the "resonance" mainly improves the impedance matching between the metal and the vacuum as ω_i approaches ω_p. This means that close to ω_p a higher percentage of the incident intensity enters the material ($\sim \delta^2$) and a higher percentage of the RS intensity produced inside the material is emitted into the vacuum ($\sim \delta^2$). The combined effect of impedance matching ($\sim \delta^4$) is further supplemented [4] by a coherence factor ($\sim \delta$). Thus, the net dependence of the RS efficiency on δ^5 is justified [3].

The present type of enhancement is unique in metals; the free-carrier concentration may be such that the corresponding plasma frequencies lie in the visible or near UV region of the spectrum where most of the lasers operate. Inelastic RS by free-carrier collective excitations (plasmons) in semiconductors has been observed and studied in the past [5]; here the relatively few carriers yield plasma frequencies mostly in the far infrared. Clearly, the mechanism of RS by phonons in metals with ω_i close to ω_p is fundamentally different than that of RS by plasmons in semiconductors.

2. OPTICAL PROPERTIES OF $CuSe_2$

Copper diselenide crystallizes according to the pyrite structure (point group T_h). Although it is often listed as a magnetic semiconductor its optical and electronic behavior more closely resembles that of a material with metallic character. The real (n) and imaginary ($\varkappa$) parts of the complex refractive index, $\hat{n}$, have been measured by Bither et al. [6]. The region of interest is reproduced in Fig. 1. We have also calculated and included in Fig. 1 the real ($\varepsilon_1 = n^2 - \varkappa^2$) and imaginary ($\varepsilon_2 = 2n\varkappa$) parts of the dielectric constant. The values of n and $\varkappa$ have been used to compute reflectivity values, based on the

FIG. 1: Optical constants of $CuSe_2$ (Ref. [6]).

expression $R = |(\hat{n}-1)/(\hat{n}+1)|^2$ which is valid for normal incidence only [7]. The curve which results from the fitting of these values with a classical model is shown in Fig. 2.

The classical skin depth $\delta_{CL} = \lambda_o/2\pi\varkappa$ is defined [3, 7] as the distance over which the fields decay by a factor of e. (λ_o is the wavelength in vacuum.) δ_{CL} does not necessarily describe the actual skin depth, for which one also needs to take into consideration the electron mean free path ℓ. Thus, in the regime of the anomalous skin effect [8] it can be shown that $\delta_{AN} \approx \ell^{1/3}\delta_{CL}^{2/3}$, assuming that the anomalous and classical conductivities are related by $\sigma_{AN} \approx (\delta_{CL}/\ell)\sigma_{CL}$. In our case we do not have sufficient information to support this approximation, and we will simply assume that $\delta_{eff} = \beta\delta_{CL}^{\alpha}$, where $0 < \alpha \leq 1$, and $\beta > 0$. A plot of δ_{CL}^5 and $\delta_{AN}^5/\ell^{5/3}$ vs. ω_i is shown in Fig. 2. Both curves have been normalized to their values at 2.40 eV where $\delta_{CL} \simeq 633$Å. Clearly, these functions exhibit a maximum at (1.95 ± 0.01) eV; this is somewhat higher than $\omega_p = (1.71 \pm 0.01)$ eV which, by definition, corresponds to $n = \varkappa$ or $\varepsilon_1 = 0$, Figs. (1,2). As expected the reflectivity exhibits its minimum also at $\omega = 1.95$ eV.

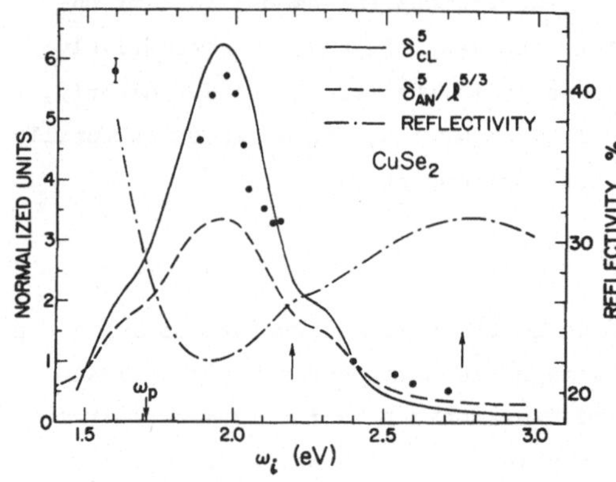

FIG. 2: Calculated and observed intensities (all normalized at 2.40 eV) as a function of the excitation frequency. The incident radiation is polarized ∥ to the surface. The scattered radiation is unpolarized. The reflectivity curve is included for comparison. The interband electronic transitions are indicated by the vertical arrows.

3. EXPERIMENTAL RESULTS AND DISCUSSION

The R spectra of CuSe$_2$ have been reported elsewhere [9]. We restrict the present study to the behavior of the phonon at ~ 260 cm^{-1} which exhibits the strongest RS intensity. We ignore the adjacent mode at ~ 270 cm^{-1} which causes an apparent asymmetry of the band at 260 cm^{-1}. Such an asymmetric band-shape should not be confused with that of a band describing RS by odd-parity phonons in metals. According to Mills et al. this possibility does exist in metals. Within the experimental limitations our measurements have not revealed any RS activity of this type. The experimental details are

essentially those described previously [9]. The same samples were used at 300 K. The peak intensities were measured and normalized against CaF_2. The excitation frequencies were provided by an Ar^+ laser and a continuously-tunable Dye laser. The incident radiation was polarized either $\parallel$ or $\perp$ to the plane of scattering. The detected radiation was unpolarized. According to the selection rules of Mills et al. only the components of the incident and scattered field parallel to the surface participate in the RS process. Thus, a $\perp$ incident field has the correct polarization relative to the surface but is also reflected more. On the other hand, a $\parallel$ incident field is not completely parallel to the surface but is reflected less, especially if the angle of incidence is close to the "critical" angle θ_c. It can be shown [7] that $\sin^3\theta_c \simeq \varkappa/2$, provided that $n^2 + \varkappa^2 \gg 1$. Since this condition is met in our region we obtain $\theta_c = 60^\circ \pm 5$. Although the ratio of the RS intensities observed for the $\parallel$ and $\perp$ incident fields was ~ 2 (a ratio of 1.75 was obtained from the optical constant data), a $\parallel$ incident field was used as it consistently gave a better signal to noise ratio. The CaF_2-normalized intensities were further normalized at 2.40 eV. The resulting points $I_\perp(\omega_i)$ are shown in Fig. 2 within the error bar indicated (this error bar applies to the theoretical curves as well). Although insufficient laser power did not allow us to complete the measurements over the entire range, the overall behavior is clear. A $\ell nI_\perp$ vs. $\ell n\delta^5_{CL}$ plot allowed us to determine an approximate value for the exponent α introduced earlier as $\alpha \simeq 0.86$. This indicates that δ_{eff} is "slower" than δ_{CL} ($\alpha = 1$) but "faster" than δ_{AN} ($\alpha = 2/3$). An exact value for δ_{eff} requires knowledge of β, and this cannot be determined from the present analysis.

There are two regions in Fig. 2 where the experimental results appear to deviate from the δ^5_{eff} dependence, i.e., in the neighborhood of 2.15 eV and at $\omega_i > 2.40$ eV some insight into these anomalies is provided by fitting the reflectivity curve with a classical dispersion model and a free-carrier Drude term [10]. The results of the fit (Fig. 2) indicate the presence of additional oscillators.[†] **Their frequencies (in eV) in the region of interest** are 1.35, 2.20, 2.75 with oscillator strengths 0.27, 0.03, 2.22 and damping constants 0.18, 0.09 and 0.44 respectively. Such oscillators may be attributed to $T_{2g} \Rightarrow E_g$ electronic transitions [6].

[†]Our far infrared reflectance measurements (0.01-0.10 eV) **were** completely dominated by the free carriers and no information about the infrared-active phonons could be extracted.

In view of these results it is reasonable to associate the shoulder observed at ~ 2.15 eV with the onset of resonant RS due to a weak and narrow interband transition at 2.20 eV, and the enhancement at $\omega_i > 2.40$ eV to broad and not so weak interband transitions centered about 2.75 eV. Separation of the relative contributions to the total dielectric behavior indicates that the classical plasma frequency alone with no electronic transitions present would occur at ~ 5.40 eV.

In summary, we have studied the RS intensity in $CuSe_2$ as a function of ω_i in the vicinity of the effective plasma frequency ω_p. The experimental results are found to be consistent with the frequency dependence of δ_{eff}^5, where δ_{eff} is proportional to $\delta_{CL}^{0.86}$. The behavior of the reflectivity in the region of interest is indicative of underlying weak interband electronic transitions; this is further substantiated by the RS enhancements observed. The present study clearly demonstrates the feasibility of RS as an effective experimental technique for the study of metallic surfaces. It can be extremely informative and well within the reach of conventional experimental means, if appropriate physical mechanisms are invoked.

ACKNOWLEDGMENTS

We wish to thank Drs. T. A. Bither and R. J. Buchard for providing the $CuSe_2$ samples, Mr. D. Duchardt for his analysis of the optical constants, and Mrs. Gertrude Tang for excellent typing.

4. REFERENCES

[1] Feldman, D. W.; Parker, J. H.; Askin, M.: Phys. Rev. Lett. 21, 607 (1968); also, in: "Light Scattering Spectra of Solids." Ed. G. Wright, Springer-Verlag: New York, 1969, page 389.
[2] Cardona, M.: Surf. Sci. 37, 100 (1973), and references therein.
[3] Mills, D. L.; Maradudin, A. A.; Burstein, E.: Ann. Phys. 56, 504 (1970).
[4] Loudon, R.: Journal de Physique 26, 677 (1965).
[5] Pinczuk, A.; Brillson, E.; Burstein, E.; Anastassakis, E.: in "Light Scattering in Solids." Ed. by M. Balkanski, Flammarion:Paris, 1971, page 115 and references therein. Also Phys. Rev. Lett. 27, 317 (1971).
[6] Bither, T. A.; Bouchard, R. J.; Cloud, W. H.; Donohue, P. C.; Siemons, W. J.: Inorg. Chem. 7, 2208 (1968).
[7] Born, M.; Wolf, E.: *Principles of Optics* (Pergamon Press, 1965), Ch. XIII.
[8] Wooten, F.: "Optical Properties of Solids," Academic Press, N.Y. 1972.
[9] Anastassakis, E.: Solid State Commun. 13, 1297 (1973).
[10] Geick, R.; Hakel, W. J.; Perry, C. H.: Phys. Rev. 148, 824 (1966).

MAGNETOOPTICAL INVESTIGATIONS OF THE 2-DIMENSIONAL FERROMAGNET $(CH_3NH_3)_2CuCl_4$.

H. AREND, J. SCHOENES and P. WACHTER

Laboratorium für Festkörperphysik, ETH Zürich, 8049 Zürich
Hönggerberg, Switzerland

The optical absorption and Faraday rotation of
$(CH_3NH_3)_2CuCl_4$ has been investigated in the spec-
tral range between the absorption edge and 0.5 eV
and for temperatures between 300 K and 4 K. From
the Faraday rotation in weak magnetic fields and
for $T > T_c = 8.9$ K one can deduce the critical ex-
ponent of the susceptibility. A cross-over region
close to T_c can be found, where the system changes
from 3-dimensionality to 2-dimensional Ising.

1. INTRODUCTION

Materials of the kind $(C_nH_{2n+1}NH_3)_2M^{2+}X_4$, with n=0, 1...10, M^{2+}
being a doubly positively charged 3d-metal (mostly Cu or Mn) and
X a halogen, are known since about 10 years (1). These materials
have been investigated magnetically, with specific heat measure-
ments, by means of neutron diffraction, optically and with EPR
(for a review article see ref. 2). The preparation of large enough,
well defined and perfect single crystals now also permits magneto-
optic measurements. First experiments have been performed with
$(CH_3NH_3)_2CuCl_4$. In this ferromagnetic compound it is the Cu^{2+}
ion with $3d^9$ electrons, which has a magnetic moment of 1 μ_B,
corresponding to S = 1/2.
This compound is termed a 2-dimensional ferromagnet, because the
crystalline structure consists of "magnetic" $CuCl_2$ layers, which
are separated by 2 layers of nonmagnetic material, $(CH_3NH_3Cl)_2$.
The Cu-Cu distance between adjacent layers is about twice the
distance within a layer (1). Since the magnetic superexchange
strongly depends on the distance between the magnetic ions, we
have an intralayer exchange J and a much smaller interlayer
exchange J', so that $J' \simeq 10^{-3} J$ (3).
From the theoretical point of view 2-dimensional ferromagnets
are of great interest, because Mermin and Wagner (4) rigorously
proved that a 2-dimensional Heisenberg model cannot have a

long range magnetic order at finite temperatures. However, Stanley
and Kaplan (5) have shown, that a 2 dimensional Ising model may
possess a finite magnetic ordering temperature.

2. OPTICAL ABSORPTION AND ELECTRONIC TRANSITIONS

In the spectral range between the absorption edge and 0.5 eV trans-
mission measurements on cleaved single crystals were possible. For
comparison they have been performed also on $(CH_3NH_3)_2MnCl_4$ = MAMC.
As can be seen in Fig. 1 $(CH_3NH_3)_2CuCl_4$ = MACC at 300 K has an ab-
sorption peak at 0.8 µm with a shoulder at about 0.95 µm. At low
enough temperatures the peak and the shoulder can be clearly resol-
ved into 2 maxima at 0.75 µm (1.65 eV) and 0.90 µm (1.37 eV). Upon
cooling the absorption edge shows very strong thermochromic effects,
inasmuch as a large blue shift is observed. Between 100 K and 300 K

it amounts to $-1.35 \cdot 10^{-3}$ eV/degree. No
shift of the absorption edge or the 2
absorption maxima has been observed
on cooling below T_c. Important quali-
tative conclusions concerning the elec-
tronic transitions can be drawn by com-
paring the absorption of MACC and MAMC,
where it is observed that the 2 absorp-
tion peaks shown in Fig. 1 are absent
in MAMC. Therefore they must be rela-
ted with the metal ion. We have to con-
sider that the Cu^{2+} ion has a ground
state 2D which splits in the crystal
field. In contrast to this the Mn^{2+}

Fig.1 Optical density of MACC ion has a 6S ground state which is
not influenced by the crystal field. Consequently all intra 3d
transitions of the Mn ion are spin flip transitions with only a
very weak oscillator strength. Arend et al. (6) have shown that
the crystal structure of MACC can best be described by a perovskite
lattice, where the central Cu ion is surrounded octahedrally by 6 Cl
ions. Thus the point symmetry appears cubic in first order. There is,
however, also a tetragonal distortion of the Cl octahedron, where
the c/a ratio is larger than 1. Optical transitions can thus be ex-

pected from the ground state Γ_{t3} to Γ_{t1}, Γ_{t4} and $\Gamma_{t5}^{1,2}$ states. The two latter transitions correspond to the observed absorption maxima at 0.9 µm and 0.75 µm, respectively. The transition Γ_{t3} to Γ_{t1} is expected in the infrared beyond 4 µm, where it coincides with molecular vibrations and cannot be clearly identified at 300 K.

3. MAGNETOOPTICAL MEASUREMENTS

The crystals could only be cleaved perpendicular to the c-axis, so the direction of light propagation was parallel to the c-axis. The magnetic field has been applied also parallel to the c-axis. In Fig. 2 we show the wavelength dependence of the Faraday rotation Θ in an applied magnetic field of 1.25 kOe at a temperature above and below T_c. The wavelength range covers the region of the absorption edge (charge transfer) and the crystal field split d-d transitions.

Fig.2 Faraday effect of MACC

In Fig. 3 we show the Faraday rotation of MACC for a fixed wavelength of 0.76 µm in function of temperature for a very small field of about 30 Oe and a larger field of 6 kOe. The high field curve exhibits the typical behavior of a ferromagnet with a tendency to saturation for temperatures towards zero K. Of more interest is the curve in low applied fields. Here we observe a pointed maximum at T_c. In the Faraday configuration the applied field is perpendicular to the plane of the crystal, so that due to the demagnetizing factor N the internal field becomes much smaller than 30 Oe for $T \simeq T_c$. On the other hand the drop of the Faraday rotation for temperatures below T_c indicates that the susceptibility is less than 1/N. If we assume that MACC is ferromagnetic in agreement with ref. 2,3, we must interpret these results in such a way that the c-axis is the direction of hard magnetization and that the spins lie in the ab-plane. Since the Faraday rotation Θ is proportional to the susceptibility χ, we can deduce the latter from

our measurements. Because of demagnetizing effects the measured
susceptibility (Faraday rotation) has to be normalized according to
$\chi \sim \Theta /(\Theta_{T_c} - \Theta)$ (7).

4. DISCUSSION

The discussion shall only be concerned with the 2 or 3 dimension-
ality of MACC. The dimensionality of a material is best expressed
by means of critical exponents in the vicinity of T_c. Since we ob-
tain the susceptibility from our Faraday rotation measurements it
is reasonable to use γ as the critical exponent. Thus $\chi_o \sim \varepsilon^{-\gamma}$
($\varepsilon>0$, $H_{int} \rightarrow 0$) with $\varepsilon = (T/T_c)-1$. We have shown in the last chapter
that our experimental arrangement closely fulfils the condition

Fig.3 Faraday effect of MACC

$H_{int} \rightarrow 0$ in the critical temperature
region. In Fig. 4 we have plotted
in a double log scale the suscepti-
bility versus ε. We clearly observe
two regions with $\gamma = 1.73 \pm 0.05$
and $\gamma = 1.35 \pm 0.1$.
Over the larger portion of the tem-
perature scale the critical expo-
nent γ agrees to within experimen-
tal error with the exponent derived
for a 2 dimensional Ising model
($\gamma = 1.75$). Indeed, the presence of
a small magnetic ab-anisotropy (3)
makes an Ising model most adapted
to the experimental system. At tem-
peratures closer to T_c we observe a region with a reduced critical
exponent. Considering theoretical models (e.g. see ref.2) we realize
that γ for 3 dimensional systems is always less than for 2 dimensio-
nal systems (with the exception of the spherical model). Thus the
experimentally observed decrease of γ from 1.73 to 1.35 indicates a
"crossover" from 2 dimensionality to 3 dimensionality close to T_c.
However, a clear distinction between 3d Heisenberg ($\gamma=1.40$), 3d XY
planar ($\gamma=1.33$) or 3d Ising ($\gamma=1.25$) is not possible due to experi-
mental resolution.
In any real system there will exist a magnetic interlayer exchange

J', although it might be quite small compared to the intralayer exchange J. For MACC $J' \simeq 10^{-3} J$ (3). But it is obvious that for $T \simeq T_c$ it is the interlayer exchange J' which is determining the magnetic behavior, resulting in ferro-or antiferromagnetism. Thus, inevitably for T appraching T_c and no matter how small J' compared with J, the 3 dimensionality of any real system will manifest itself. Such a "cross over" region as we have observed, and which is also described in ref. 2 has been predicted by Stanley et al.(8). A second "cross over" region from 2 dimensional Ising to 2 dimensional Heisenberg due to magnetic anisotropy could not be observed i n our experiments. Instead γ approaches one for higher temperatures, as demanded by molecular field theory.

Fig.4 Critical exp.γ of MACC

5. REFERENCES

(1) Willet R.D., J. Chem. Physics 41, 8 (1964)
(2) De Jongh L.J., Miedema A.R., Advances in Phys. 23, 104 (1974)
(3) De Jongh L.J., van Amstel W.D., Miedema A.R., Physica 58, 277 (1972)
(4) Mermin N.O., Wagner H., Phys. Rev. Letters 17, 1133 (1966)
(5) Stanley H.E., Kaplan T.A., Phys. Rev. Letters 17, 913 (1966)
(6) Arend H., Hoffmann R., Waldner F., Solid State Commun. 13, 1629 (1973)
(7) Huang C.C., Pindak R.S., Ho J.T., Solid State Commun. 14, 559 (1974)
(8) Liu L.L., Stanley H.E., Phys. Rev. Letters 29, 927 (1972)

EXCITONS

CONTENTS

MAGNETO-OPTICS OF FREE AND BOUND EXCITONS IN CdTe

K.CHO[§], W.DREYBRODT[+], P.HIESINGER, S.SUGA and F.WILLMANN[++]

Max-Planck-Institut für Festkörperforschung, Stuttgart

Through the analysis of the magneto-reflectance of the 1s excitons arising from the Γ_6-Γ_8 band pair, the effect of the $\vec{k}$-linear term on the exciton states with exchange splitting between the J=1 and J=2 states has been clearly demonstrated. The existence of the electron hole exchange splitting is also shown in the case of (D^0X) bound exciton complex through the analysis of the magneto-luminescence spectrum including the two-electron transitions.

§1. INTRODUCTION

Many zincblende type semiconductors have a direct gap at the Γ_8 valence and Γ_6 conduction band edges. The excitons originating from this pair of bands are highly degenerate because of the four fold (Γ_8) and the two fold (Γ_6) degeneracies of the electron bands. This high degree of degeneracy is reflected also in bound exciton complexes. Although there are many interesting aspects arising from this degeneracy, we concentrate in this paper on the electron hole exchange interaction and the effect of the $\vec{k}$-linear term in the Γ_8 bands. Through the analysis of the magneto-optical measurements, CdTe turns out to be a typical material which shows the above effects clearly.

§2. FREE EXCITON

In the framework of the effective mass formulation, the exciton states are described in terms of the following material constants : the mass (m_e) and the g-factor (g_c) of the conduction electron, the Luttinger parameters[1] (γ_1, γ_2, γ_3, κ, q) for the Γ_8 valence bands including the coefficient for the $\vec{k}$-linear term (K_ℓ), the coefficients for the two types of the electron hole exchange interaction[2] (Δ_1, Δ_2), and the dielectric constant (ϵ_0) which screens the Coulomb interaction. In spite of the many independent

§ On leave of absence from the Institute for Solid State Physics, The University of Tokyo.
+ Present address: Fachbereich Physik, Universität Bremen.
++ Present address: Siemens AG, München

parameters, the magneto-optical behavior of the 1s excitons can
be clearly discribed in the low field region by a symmetry argu-
ment connected with a degenerate perturbational treatment
which is an extension of the scheme proposed by Altarelli and
Lipari [4] (AL). If we neglect the magnitude of the photon wave
vector, the splitting pattern linear in magnetic field is given

Fig.1 Linear splitting pattern.
Dotted lines are not observable.
Positive g-values are taken as
an example. The Γ_4-Γ_3 splitting
is due to the Δ_2 exchange inter-
action.

in Fig.1. The existence of the exchange splitting gives a differ-
ent pattern from that of AL who neglected Δ_1 and Δ_2. The consid-
eration of the diamagnetic part gives various patterns including
further splitting, shift and anticrossing effect[3], but it does
not affect the selection rule. For example, even in the case of
the lowest symmetry, $\vec{H}$//<110>, we can see at most four σ lines
and four π lines. This is a general result in the framework of
the effective mass formulation for the excitons with vanishing
translational wave vector, because we have considered all the
possible effects which are allowed by symmetry. On the other
hand, the measurements of the
magneto-reflectance [5] gives the
splitting pattern in Fig.2. Here
the existence of the exchange
splitting is explicitly demon-
strated, and all the eight mem-
bers of the 1s exciton are re-
solved. The lines 1,2, and 3 are
the main structures in the σ_+, σ_-
and π spectra, respectively. At
the converging point (H=0) of the
lines 4∿8, we observe a small dip
in the prominent reflectance line.
Its existence is known for a long

Fig.2 Fan chart of the
reflectance minima.

time[6] without any interpretation. The intensities of the lines 4∿8 grow up with the magnetic field. One can trace the lines 4, 6 and 7 down to the zero field. But the lines 5 and 8 appear only at finite fields. A tentative fitting by means of the above-mentioned theory using appropriately chosen band parameters[5] gives the splitting pattern in Fig.3. The correspondence between the experimental and theoretical curves at low fields is the following :

1 : $|1,1\rangle$, 2 : $|1,-1\rangle$,
3 : $|1,0\rangle$, 4 : $|2,-2\rangle$,
5 : $|2,-1\rangle$, 6 : $|2,0\rangle$, 7 : $|2,2\rangle$, 8 : $|2,1\rangle$.

Fig.3 Energy levels of exciton (Theory). The vertical line at 4 T indicates the validity limit of the perturbation theory.

Although the energy positions are well reflected, there is an essential discrepancy between the experimental and the theoretical results: The states $|2,\pm2\rangle$ are observed in the $\sigma_\mp$ spectra, respectively, instead of the π spectrum. In the theoretical curves, the states $|2,\pm2\rangle$ are <u>very weakly dipole allowed in the π spectrum</u>, but they are <u>completely forbidden in the σ spectrum</u>.

In order to overcome the difficulty, we consider the effect of the $\vec{K}$-linear term combined with the finite photon wave vector by extending the theory of Mahan and Hopfield[7] (MH). The additional perturbation term due to this effect is given by

$$H' = \hbar K_\ell \left[K_x \{ (J_y^2 - J_z^2) J_x \} + \text{cycl. perm.} \right] \{ m_o / (\gamma_1 m_e^* + m_o) \} \quad , \quad (1)$$

where K_x is the x-component of the photon wave vector $\vec{K}$ and J_x, J_y and J_z are the components of the effective angular momentum operator for the J=3/2 hole states[1]. The x, y and z axes are parallel to the three <100> crystal axes, respectively and $\{AB\}=$ (AB+BA)/2. By using the j-j coupling scheme defined in [3] as the basis, the effect of H' is described in the following matrix forms for the magnetic field along the [110] crystal axis. The suffices F and V stand for the Faraday ($\vec{K}//[110]$) and the Voigt

$(\vec{K}//[1\bar{1}0])$ geometries, respectively.

$$
\begin{array}{c}
\begin{array}{cccccccc}
|2,2\rangle & |2,1\rangle & |2,0\rangle & |2,-1\rangle & |2,-2\rangle & |1,1\rangle & |1,0\rangle & |1,-1\rangle
\end{array}\\
\begin{array}{c}
\langle 2,\ 2| \\ \langle 2,\ 1| \\ \langle 2,\ 0| \\ \langle 2,-1| \\ \langle 2,-2|
\end{array}
\left(
\begin{array}{ccccc|ccc}
0 & -\sqrt{3} & 0 & -3\sqrt{3} & 0 & 1 & 0 & -9 \\
 & 0 & \sqrt{2} & 0 & -3\sqrt{3} & 0 & \sqrt{8} & 0 \\
 & & 0 & \sqrt{2} & 0 & \sqrt{6} & 0 & \sqrt{6} \\
 & & & 0 & -\sqrt{3} & 0 & \sqrt{8} & 0 \\
 & & & & 0 & 9 & 0 & -1
\end{array}
\right) \times N_F \quad (2)
\end{array}
$$

$$
\left(
\begin{array}{ccccc|ccc}
-\sqrt{6} & 0 & 1 & 0 & 0 & 0 & 1 & 0 \\
 & \sqrt{24} & 0 & 0 & 0 & -\sqrt{18} & 0 & -\sqrt{2} \\
 & & 0 & 0 & -1 & 0 & -\sqrt{54} & 0 \\
 & & & -\sqrt{24} & 0 & \sqrt{2} & 0 & \sqrt{18} \\
 & & & & \sqrt{6} & 0 & 1 & 0
\end{array}
\right) \times N_V \quad (3)
$$

where the diagonal block for $|1,\pm1\rangle$ and $|1,0\rangle$ is omitted because it is always zero, and

$$N_F = (-i\sqrt{3}/16)\hbar K_\ell K \{m_o/(\gamma_1 m_e^* + m_o)\} , \qquad (4)$$

$$N_V = (-\sqrt{6}/16)\ \hbar K_\ell K \{m_o/(\gamma_1 m_e^* + m_o)\} . \qquad (5)$$

Note that the quantization axis for the basis is parallel to the magnetic field $(//[110])$, that is, one excites $|1,\pm1\rangle$ states in the $\sigma_\pm$ and $|1,0\rangle$ in the π spectra, respectively. From (2) we understand the preferential mixing of the $|2,\pm2\rangle$ states into the $|1,\mp1\rangle$ states, respectively, in the Faraday configuration, and $|2,0\rangle$ state into the $|1,0\rangle$ state in the Voigt configuration, in accordance with the experimental observation. This mixing mechanism between the $J=1$ and $J=2$ exciton states also explains the zero field anomaly in the reflectance spectrum.

With respect to the magnitude of this effect one should stress that it is large enough to induce the structures, but that it is not so large as to produce extra observable splitting due to the finite matrix elements in the diagonal block for the $J=2$ states. A simple estimate based on the second order perturbation with respect to the exciton-photon interaction shows that for $\hbar K_\ell \sim 4 \times 10^{-9}$ eV cm (which corresponds to the slope of the Γ_8 valence bands by 3.5×10^{-9} eV cm along the $<001>$ direction) one gets 5% of the normalized oscillator strength for $|2,-2\rangle$ state in the low field σ_+ spectrum. The above-mentioned value of K_ℓ is not unrea-

sonable in comparison with the value for CeS estimated by MH, considering the much larger spin-orbit interaction in the valence bands of CdTe.

It is interesting to compare the case of MH (the 1s exciton states of the B series in CdS) and ours: In their case the spin triplet forbidden state has no degeneracy and they neglected the exchange splitting in contrast to our case with five fold degeneracy of the J=2 exciton states and the existence of the exchange interaction. The main arguments for the effect of the $\vec{k}$-linear term associated with the finite photon wave vector are the polarization dependence and the line shape analysis at zero field in the case of MH, and the selective mixing of the forbidden states with the allowed states in a magnetic field in our case. Thus MH put stress on the splitting of the two transversal exciton states, but we emphasize the feature of the preferential relaxation of the selection rule.

A similar reflectance anomaly at zero external field is known in the $Z_{1,2}$ exciton peak in CuBr. By means of the magneto-optical measurement this anomaly is ascribed to the Γ_4 spin triplet states of the 1s exciton [8]. From the analogy with CdTe, it is interesting to study this anomaly in a high magnetic field using polarized light for various directions of the magnetic field. If the anomaly is again due the $\vec{k}$-linear term, one will be able to find similar peculiarity of the selective mixing between the forbidden and allowed states. The only difference in this case is the splitting of the J=2 states into $\Gamma_4 + \Gamma_3$ due to the Δ_2 exchange interaction [8].

Uniaxial stress measurements will also be interesting in connection with $\vec{k}$-linear term because this is the very point of difference in symmetry properties between the stressed zincblende and the wurtzite crystals. In the uniaxially stressed zincblende structure the $\vec{k}$-linear term exists in both of the split bands coming from the Γ_8 bands. If we denote the two series of the excitons originating from these two valence bands as A and B series in analogy with the wurtzite case, both the A and B excitons can show the reflectance anomaly in contrast to the real wurtzite case where only the B exciton can have the effect of the $\vec{k}$-linear term. (The $\vec{k}$-linear effect from the Γ_7 conduction

band is quite negligible[9].)

§3. (D^oX) BOUND EXCITON

According to the standard model of the bound exciton complexes
[10], the ground state of the neutral donor-exciton complex is
composed of the two electrons with antiparallel spins and a hole.
Thus only the hole spin is responsible to the energy splitting
of the complex in a magnetic field. Therefore the splitting
pattern has nothing to do with that of the free exciton. This
model implies that the electron-electron exchange interaction is
predominantly large in comparison with the other spin-spin inter-
actions. In order to realize this situation, the orbital wave
functions of the two electrons must overlap to a considerable
amount. If this overlap is very small, however, the situation
changes drastically with respect to the spin multiplicity. We
can consider the case where the electron hole exchange is much
larger than the electron-electron exchange interaction. In this
case we first classify the states according to the angular momen-
tum of the exciton, and then introduce the coupling between the
donor spin and the exciton angular momentum as a small perturba-
tion. The splitting pattern of these states is very similar to
that of the free exciton except for the superposition of the
donor spin splitting. At low field where the magnetic splitting
energy is comparable to the electron-electron exchange splitting,
there can be a complicated situation, but this can be masked
within the line width of the emission spectrum. The only qualita-
tive difference is the possible existence of the spin flip re-
combination process where the exciton radiatively recombines
with the simultaneous spin flip of the donor electron. The inten-
sity of this transition depends on the relative magnitude of the
electron-electron exchange splitting with respect to the magnetic
splitting of the sub-levels of the complex. At higher fields the
spin flip process becomes negligible. This situation is analogous
to the Paschen-Back effect known in the atomic physics[11]. In
the remaining part of the paper, we demonstrate that this case
is realized in the very shallow (D^oX) bound exciton complex in
CdTe.

Fig.4 shows the luminescence
spectrum below the exciton ab-
sorption edge[†] (E_T is the re-
flectance minimum due to the J=1
exciton states.). The details of
the experiment are described
elsewhere[12]. The lines (D^+X)
and (A^OX) are assigned to the
radiative recombinations from
the ionized donor-exciton and
the neutral acceptor-exciton
complexes, respectively. The

Fig.4 Luminescence spectrum of
bound excitons at zero field.

main part of our interest is the lines A, B, A' and B'. The
upper and the lower parts of the figure are relatively shifted
by 10.5 meV. This energy shift agrees well to the known value
(10.6 ± 0.4 meV) of the energy difference between the 1s and the
2p states of a donor in CdTe[13]. The replicating aspect of the
lines strongly suggests that the lines A and B are due to the
recombination of an exciton bound to a neutral donor, and that
the lines A' and B' are due to the Auger processes where the
exciton recombines with the simultaneous excitation of the donor
electron to the 2p state. As the origin of the two lines A and B
, we can easily exclude the possibility of the different donor
species: If we have two species of donors, the energy difference
between A and B should be due to the difference in the donor
potential. Since the effect of the donor potential is larger for
an electron in the neutral donor state than for an exciton in
the (D^OX) bound complex, the transition energies from the 1s to
2p states for the two species of donors should be different by
more than the A-B interval. Since the lines (A, A') and (B, B')
coincide very well by the same energy shift of 10.5 meV, both
A and B must originate from the same species of donor. Further-
more the intensity ratio of the line A and B obeys the thermal
Boltzmann law with an activation energy of 0.5 meV between 2.7

† At and above the absorption edge also, we observe very inter-
esting structure due to the lower (LP) and the upper (UP) polar-
iton branches. (See the insert of Fig.4) By increasing the ex-
citation intencity by ∿240 times the UP luminescence becomes
predominant (curve I) with a remarkable high energy tail in com-
parison with the case of the low excitation (curve II).

and 20 K. This fact also supports the single species model, because for the establishment of the thermal equilibrium for two species of donors one needs excitation processes of the bound excitons at least to the free exciton states (with the activation energy of ~3 meV) which occur very rarely at low temperatures.

The Zeeman pattern of the lines A and B is shown in Fig.5. The close resemblance of this pattern to that of the free exciton (Fig.2) should be noticed. It suggests that the lines A and B result from the exchange splitting of the bound exciton. The spin flip transitions are not observed in our measurement. Furthermore the Zeeman pattern of the lines A' and B' can be well understood as the superposition of the pattern in Fig.5 and that of the 2p (and 2s) states of the donor electron[12]. Thus it is clear that the scheme explained in the first part of this section is realized in the (D$^{\text{o}}$X) bound exciton complex.

Fig.5 Fan chart of the (D$^{\text{o}}$X) lines.

REFERENCES

[1] J.M.Luttinger: Phys. Rev. 102 1030 (1956)
[2] G.L.Bir; G.E.Pikus; L.G.Sublina and D.L.Fedorov: Sov. Phys.-Solid State 12 926 (1970)
[3] K.Cho; S.Suga; W.Dreybrodt and F.Willmann: to be published
[4] M.Altarelli and N.O.Lipari: Phys. Rev. 7B 3798 (1973)
[5] W.Dreybrodt; K.Cho; S.Suga and F.Willmann: to be published
[6] B.Segall and D.T.F.Marple: Physics and Chemistry of II-VI Compounds, ed. N.Aven and J.S.Prener, North Holland Company, Amsterdam, (1967)
[7] G.D.Mahan and J.J.Hopfield: Phys. Rev. 135 A428 (1964)
[8] C.Wecker; M.Certier and S.Nikitine: J. Physique 32 63 (1971)
[9] J.J.Hopfield: J. Appl. Phys. Suppl. 32 2277 (1961)
[10] D.G.Thomas and J.J.Hopfield: Phys. Rev. 128 2135 (1962)
[11] E.U.Condon and G.H.Shortley: The Theory of Atomic Spectra, Cambridge
[12] S.Suga; W.Dreybrodt; F.Willmann; P.Hiesinger and K.Cho: to be published in Solid State Commun.
[13] D.R.Cohn; D.M.Larsen and B.Lax: Phys. Rev. B6 1367 (1972)

NON-COULOMB POTENTIALS IN IMPURITY AND BOUND-EXCITON PROBLEMS

B. Hönerlage, U. Rößler, U. Schröder, D. Strauch

Fachbereich Physik, Universität Regensburg, 84 Regensburg

Fed. Rep. Germany

Detailed investigations of the ground and excited
states of impurities and bound-exciton complexes
and their variation with the chemical impurity in
II-VI compounds were recently carried out. In order
to explain these data theoretically we split the
Hamiltonian, describing the impurity or the bound
exciton complex into one part H_o that does not
depend on the chemical impurity and a second
part $(p \cdot W)$ that considers the chemical impurity
by a parameter p.
Some consequences of this ansatz are discussed for
the impurity and the bound-exciton problem. In
particular we consider the influence of electron-
phonon coupling.

1. INTRODUCTION

In theoretical investigations of the bound-exciton problem
only the Coulomb potential of the charged impurity has
been considered so far, while the influence of the chemical
nature of the impurity was not taken into account [1]. It
is known, however, that in some crystals the binding energy
E_{BX} of the exciton to the complex depends linearly on the
chemical impurity (Haynes' rule). Recent measurements for
n-doped CdS [2] and ZnSe [3] yield the linear relation

$$E_{BX} \propto (E_{2p} - E_{1s}) \qquad (1)$$

where E_{2p} and E_{1s} are the 2p and 1s energies of the donor,
respectively. In addition, the linear relation

$$(E_{2p} - E_{2s}) \propto (E_{2p} - E_{1s}) \qquad (2)$$

for the impurity states was found in these experiments.

2. CHEMICAL SHIFT

As the Coulomb potential of the impurity does not describe

the chemical shift given by equations (1) and (2), we search
for the most general form of the impurity potential, that
accounts for the chemical nature of the impurity. For this
purpose we write the Hamiltonian for the bound-exciton
problem

$$H_{BX} = H^o_{BX} + p \cdot W_{BX} \tag{3}$$

and for the impurity problem

$$H_{imp} = H^o_{imp} + p \cdot W_{imp} \quad . \tag{4}$$

In these equations H^o_{BX} and H^o_{imp} are Hamiltonians that
depend on the host lattice only, but include the Coulomb
potential and even non-Coulomb potentials, that do not
depend on the chemical nature of the impurity. The latter
is considered in one and the same parameter p for a given
impurity, while W_{imp} and W_{BX} are again independent of the
impurity but otherwise quite general potential operators.
Eq. (3) is a generalization of an idea of Baldereschi
who explained the linear relation (1) by assuming W_{imp} to
be a square-well potential [3]. By treating $p \cdot W_{imp}$ as a
perturbation of the 1s, 2s, and 2p states of H^o_{imp} one
obtains the results [4]

$$E_{BX} = A' + B' \cdot (E_{2p} - E_{1s}) \tag{5}$$

and

$$E_{2p} - E_{2s} = A + B \cdot (E_{2p} - E_{1s}) \quad . \tag{6}$$

Thus, the chemical shift as expressed in the experimental
results of eqs. (1) and (2) can be explained by assuming
much more general impurity potentials than the simple square
well potential of Baldereschi.

3. NON-COULOMB POTENTIALS

The constants A and A' are defined by the equations

$$(E^o_{2p} - E^o_{2s}) = A + B \cdot (E^o_{2p} - E^o_{1s}) \tag{7}$$

and

$$E^O_{BX} = A' + B' \cdot (E^O_{2p} - E^O_{1s}) , \qquad (8)$$

where B and B' can be expressed by expectation values of W_{imp} and W_{BX}. Without specifying these potentials it is possible to obtain the values of A, B, A', and B' from the experimental data of [2, 3]. As the eigenvalues of the hydrogen model for the impurity problem of H^O_{imp} do not fulfil eq. (7) we conclude that non-Coulomb potentials must be considered in H^O_{imp}. For polar semiconductors as ZnSe and CdS it is likely that electron-phonon interaction influences the impurity states. To test this assumption we apply the bound-polaron theory of Engineer and Tzoar [5]. The results for the hydrogen model (HM) and for the bound-polaron model (PM) are given in Table 1 together with

$$\Delta = \frac{E^O_{2p} - E^O_{2s} - A}{B(E^O_{2p} - E^O_{1s})} \qquad (9)$$

which according to eq. (7) must be equal to 1.

TABLE 1

	A [meV]	B	$E^O_{2p}-E^O_{2s}$ [meV]		$E^O_{2p}-E^O_{1s}$ [meV]		Δ	
			HM	PM	HM	PM	HM	PM
ZnSe	-3.63	0.202	0.	0.97	21.75	23.13	0.83	0.98
CdS	-5.17	0.232	0.	0.94	24.3	26.26	0.92	0.997

From Table 1 one can conclude that other possible extensions of H^O_{imp} (e.g. many-body corrections) should be small compared to the influence of the electron-phonon coupling.

4. BOUND-EXCITON PROBLEM

Using the experimental values for A' and B' and the values for $E^O_{2p} - E^O_{1s}$ from Engineer and Tzoar [5] we can derive E^O_{BX} from eq. (8). The results are given in Table 2.

TABLE 2

		A' [meV]	B'	E^O_{BX} [meV]
ZnSe	ionized donor	-11.68	0.89	8.91
	neutral donor	- 0.25	0.22	4.84
CdS	ionized donor	-12.79	0.78	7.69
	neutral donor	0.58	0.29	8.11

Obviously, all theories for the bound-exciton problem which neglect the chemical shift should be compared with these values E^O_{BX}.

5. REFERENCES

[1] Schröder, U.: Festkörperprobleme XIII, Pergamon-Vieweg-Verlag, 171 (1973) and references therein

[2] Nassau, K.; Henry, C.H.; Shiever, J.W.: Proc. 10th Int. Conf. Phys. Semicond., Cambridge, Mass., p. 629 (1970)

[3] Merz, J.L.; Kukimoto, H.; Nassau, K.; Shiever, J.W.: Phys. Rev. B6, 545 (1972)

[4] Hönerlage, B.; Rößler, U.; Schröder, U.: to be published

[5] Engineer, M.H.; Tzoar, N.: Phys. Rev. B5, 3029 (1972); Phys. Rev. B8, 702 (1973). The values for E^O_{2s} and E^O_{2p} were obtained by linear extrapolation from the data of this reference.

EXCITONS IN DEGENERATE AND NON-DEGENERATE SEMICONDUCTORS §

JOHN D. DOW

Department of Physics and Materials Research Laboratory

University of Illinois, Urbana, Illinois, U. S. A. 61801

An exact closed-form solution of a model for excitons in degenerate semiconductors is presented. The optical absorption and luminescence spectra reduce to the Coulomb exciton absorption and luminescence (for both bound and continuum excitons) in the limit of vanishing free conduction electron density. In the high density limit, the free electron absorption and emission spectra are obtained. As the free conduction electron density decreases from infinite density, resonance-like structures appear and emerge from the continuum, forming bound exciton states. These continuum resonances should be observable in the absorption and emission spectra of degenerate semiconductors. As the model semiconductor becomes decreasingly degenerate, the optical spectra change continuously, with no evidence of an abrupt phase transition. The relation of this model to theories of x-ray spectra, Mott transitions, and enhanced interband transitions in free-electron metals is discussed.

1.INTRODUCTION

Recent interest in highly-excited semiconductors, the optical properties of solids, and metal-insulator transitions provides motivation for examining a simple model of optical absorption and luminescence by semiconductors containing mobile charges. The simplest model of such a system treats the free charges as a medium characterized by a screening wavevector k_s. (1).

2.THE MODEL

In the effective mass approximation, the optical absorption and emission spectra are given by Elliott's exciton theory (2):

$$S(\omega) = C \sum_{\nu} |\psi_\nu(0)|^2 \delta(\hbar\omega - E_{gap} - E_\nu) F_S . \tag{1}$$

§ Research supported by the U. S. National Science Foundation under grants NSF-GH-33634 and NSF-GH-39132.

Here C is proportional to the square of a Brillouin-zone-center transition matrix element, and is taken to be a constant; we have assumed a narrow hole band at an energy $-E_{gap}$ with respect to the bottom of the parabolic conduction band; and we have neglected the effects of Fermi sea rearrangement during the optical transition. The Pauli factor F_S is different for absorption and emission, and may be calculated for any given temperature and screening wavevector, using an appropriate screening theory (<u>e.g.</u>, Debye-Hückel or Thomas-Fermi theory). For example, at 0 °K for absorption we have $F_S = \theta(\hbar\omega - E_{gap} - E_F)$, where $\theta(x)$ is the unit step function and E_F is the Fermi energy. Since we are primarily interested in the energy-distribution of the oscillator strength, and not whether a particular part of the lineshape is observable in absorption of emission data, we shall loosely refer to Eq.(1) with F_S set equal to unity as the "absorption".

The quantity $\psi_\nu(0)$ is the overlap of the electron's envelope wavefunction with the localized hole-envelope, and is obtained by solving an effective-mass Schrödinger equation with the (Hulthen) screened electron-hole interaction: (3)

$$V(r) = \frac{-(e^2\pi^2 k_s/6\varepsilon_0)}{\exp(k_s r\pi^2/6) - 1} \; \stackrel{\sim}{\sim} \; -(e^2/\varepsilon_0 r)\exp(-k_s r),$$

a form insignificantly different from that of the usual screened potential. The advantage of the Hulthen form is that it permits an exact, analytic solution of the s-wave Schrödinger equation, with the result (for continuum states):

$$\text{"Absorption"} = C(\alpha^2/4\pi^2 kRa^2)(k\alpha a^2/2 - \text{Im}\{\Psi(b_+ + 1)$$
$$+ \Psi(b_- + 1)\})^2 \, |\Gamma(1 - 2ik/\alpha)/\{\Gamma(b_+)\Gamma(b_-)\}|^2, \qquad (2)$$

where we have
$$\alpha = \pi^2 k_s/6, \qquad k = \{2m(\hbar\omega - E_{gap})/\hbar^2\}^{1/2}, \qquad \text{and}$$
$$b_\pm = -ik/\alpha \pm (2\alpha^{-1}a^{-1} - k^2/\alpha^2)^{1/2},$$

with $\Gamma(z)$ and $\Psi(z)$ being the gamma and digamma functions.

Eq.(2) has been evaluated in Figs. (1) and (2), where quantities are expressed in terms of the exciton radius

Fig. 1: "Absorption", defined in Eq.(2), for various values of screening wavevector k_s.

Fig. 2: "Absorption", defined in Eq.(2). There is a bound exciton with energy somewhere in the shaded region for each value of the screening wavevector k_s.

$a = \hbar^2 \varepsilon_0 / me^2$ and Rydberg $R = e^2 / 2\varepsilon_0 a$. Observe that in the extreme metallic limit ($k_s \to \infty$) the square-root-of-energy free-electron result is recovered, whereas in the insulator limit ($k_s \to 0$) the Elliott exciton result is obtained, including a Rydberg series of bound states. As $k_s a$ decreases from ∞ to 2, we see that the absorption becomes less free-electron-like and approaches the exciton continuum absorption. For $1.4 < k_s a < 2$, a continuum resonance forms in anticipation of the bound 1s exciton state to emerge for $k_s a \lesssim 1.2$. The continuum resonance's oscillator strength decreases as the binding energy of the bound state increases (the bound states are not shown in Fig.2) until for $k_s a \simeq 0.4$ a weaker resonance begins building up for the 2s exciton, which also emerges from the continuum as $k_s a$ decreases. This situation repeats itself until, for $k_s = 0$, all the bound excitons have emerged from the continuum.

The continuum resonances should be observable in the emission spectra of degenerate semiconductors. (Note that for $k_s a > 0.8$, virtually all the structure in the "absorption" corresponds to energies that would lie below the Fermi energy; hence these structures should manifest themselves primarily in emission spectra.)

3. DISCUSSION

One unanticipated result of these calculations is the weak dependence of the absorption on conduction electron density n, which in a Thomas-Fermi model, is proportional to $(k_s a)^6$. Following the qualitative reasoning that led Mott to suggest the metal-insulator transition named after him (4), we originally guessed that the absorption should be approximately free-electron-like for $k_s a > 1$ and should change rather abruptly to exciton-like absorption for $k_s a < 1$. No such abrupt change occurs; the model shows no evidence of a Mott transition.

A feature omitted from this model is the many-electron relaxation effect thought responsible for the Kondo effect, and, until recently, believed to be the cause of the x-ray threshold anomalies in metals. That such effects are important

in the optical spectra of solids is doubtful; nevertheless systematic studies of the absorption and luminescence spectra of semiconductors with controlled impurity-content could very well provide definitive information on the subject (5). Indeed, a principal purpose of this paper is to stimulate and to encourage such experiments, which might best be carried out using wavelength-modulation spectroscopy.

Finally we note that these calculations indicate a significant enhancement of interband transitions by the electron-hole interaction, and therefore may provide an explanation of the anomalous oscillator strength of the interband transition in Na (6). A detailed and quantitative theory of the effect for Na, however, would have to account for the finite valence-band mass and the central-cell corrections to the screened electron-hole interaction.

4. REFERENCES

1. Gay, J. G.: Phys. Rev. $\underline{B4}$,2567(1971).
2. Elliott, R. J.: Phys. Rev.$\underline{108}$,1384(1957).
3. See, $\underline{e.g.}$, Sitenko, A. G.: $\underline{\text{Lectures in Scattering Theory}}$ (Pergamon, New York, 1971).
4. Mott, N. F.: Rev. Mod. Phys.$\underline{40}$,677(1968).
5. For a review, see J. D. Dow, Comments on Solid State Physics (in press).
6. Overhauser, A. W.:Phys. Rev. $\underline{156}$,844(1967).

EXCITONS IN A BOUNDED MEDIUM:
LINESHAPE STUDIES OF CdS REFLECTANCE

F. EVANGELISTI,[o+] A. FROVA[+] and F. PATELLA
Istituto di Fisica, Università di Roma, Rome, Italy

Reflectance spectra of the CdS excitonic lines
are interpreted in terms of a surface "dead layer"
whose depth is directly correlated to the cor-
responding orbital sizes. Spatial dispersion plus
interference fully account for the observed line-
shapes. The effect of surface treatments is also
investigated.

It is known that the reflectance structures of discrete lines of
Wannier excitons are generally different from those of a classic-
al oscillator. A characteristic feature is a different lineshape
for the various levels of the series, even within a given semi-
conductor. If the transverse-longitudinal splitting is large,
deviations from classical behavior are introduced by spatial dis
persion effects. This was first pointed out by Hopfield and
Thomas [1], who accounted for the behavior of the n=1 level in
CdS, provided an exciton-free surface region was postulated (the
"dead layer"). This layer is due to the impossibility for the ex
citon to have its center of mass closer to the surface than an
exciton radius, without severe distortion. A calculation [2] in-
clusive of image forces does actually predict ionization of the
exciton for distances below the exciton Bohr radius. Hopfield and
Thomas [1], did not discuss the poor reproducibility of the exper-
imental results and made no analysis of structures other than the
ground level. It is the aim of this paper to study samples with
different surface conditions, to prove that, in ideal cases, the
Hopfield and Thomas approach is capable of accounting for all the

(o) Also Istituto di Matematica, Università di Camerino, Italy.
(+) Gruppo Nazionale di Struttura della Materia del Consiglio Na-
zionale delle Ricerche.

features observed in the exciton series, if a dead layer of depth
proportional to the orbital size of the exciton state considered
is introduced. Experimental nonreproducibility will then be ascri
bed to further perturbations which may be present at the surface,
e.g. built-in fields. The work is motivated by earlier studies in
GaAs and InP [3], where the application of surface fields caused
a drastic spectral change in the n=1 level, due to exciton ioni-
zation and creation of an extrinsic dead layer. The experiment
conclusively showed that interference across this layer is the
major source of anomalous behavior, as spatial dispersion in tho-
se materials is relatively unimportant.

Our CdS samples were in the form of high-resistivity platelets
[4], presumably Class I according to the Gross-Novikov classif-
ication [5] . The starting surface condition is likely to imply
a layer of excess Cd, giving rise to a weak surface field [6].
Some samples were subsequently heat-treated in a vacuum of 2x
10^{-6} torr. For "moderate" treatment, which corresponds to evapo-
ration of the excess Cd, the bands should tend to flatten; for
"heavy" treatment, simultaneous evaporation of Cd and S causes
the development of a strongly doped surface layer. The demarka-
tion line between the two cases is not clearcut and the occurr-
ence of one or the other process is best monitored by the charac
ter of the exciton spectral behavior. In all cases, normal-incid
ence reflectance was measured from a plane containing the c-axis,
with light polarized perpendicular to it. The measurements were
taken with the samples immersed in superfluid helium ($\sim$ 1.8 K),
after a chamber pre-evacuation at 10^{-3} torr.

Fig. 1 shows typical reflectance spectra, all normalized to a val
ue of 0.31 at 4890 Å (see details in the caption). Ignoring the
B-exciton, the main features can be summarized as follows:

i) Only some virgin crystals exhibit the longitudinal-exciton spi
ke, first reported in [1]. The n=2 structure has variable amplit-
ude, without an apparent correlation with the spike.

FIG.1: Experimental reflect_ance spectra corresponding to: A) as-grown sample B 1045, supplied by Boer [4]; B) as-grown sample 505, supplied by Al_bers and Roessler [4]; C) moderately treated 505 (temperature cycling: 2' from R.T. to 790 K, 2' at 790 K, fast cooling); D) heavily treated 505 (cy_cling: same as C ex_cept 950 K).

ii) Moderate treatment of sample 505 enhances the n=2 structure and bears out a dip at n=3 (see magnified box).

iii) The n=2 lineshape is always reversed with respect to n=1, a behavior consistent with many II-VI semiconductors.

iv) Heavily treated sample 505 shows a completely upset spectrum, overall attenuation of the structured components and no evidence of excited states.

As opposed to possible alternative explanations, such as split ex_citon [7], or changes in damping, all the observed features are uniquely accounted for by a varying dead-layer depth. To support this view, we have calculated the reflectance in terms of interfe_rence across the dead layer with inclusion of spatial dispersion. Interference occurs between the structureless beam reflected at the helium—dead-layer boundary and the light returning from the dead-layer—bulk boundary. If the dead-layer depth l (corresponding to a round-trip phase delay $\theta = 4\pi l/\lambda_m$) is used as a parameter, virtually any kind of lineshape is obtained. We shall report only those relevant to (a) moderately treated sample 505/1, correspond_ing to the most ideal surface condition (it will be clear below

FIG.2: Calculated spectra for chosen values of the dead-layer depth l. Arrows give the positions of the longitudinal and transverse exciton. Paramenters used: exciton total mass=0.9 m_o, polarizability =0.0125, background dielectric constant =8.1 and broadening =0.1 meV.

that measurement of a large n=2 signal and detection of the n=3 level point to a nearly field-free surface); and (b) heavily treated sample 505/2, where a depletion layer has developed. The curves of interest are shown in Fig.2. For l=0 (top), discrepancies with any of the experimental spectra are readily apparent. For l≠0, best agreement with sample 505/1 is obtained for dead-layer depths

$$l=70 \text{ Å} \quad (n=1) \qquad l=300 \text{ Å} \quad (n=2) \qquad l=980 \text{ Å} \quad (n=3) \tag{1}$$

as shown by the central curves in the figure. Finally, a depth l=190 Å has to be assumed to account for the heavily treated sample (bottom). Our discussion is done strictly in terms of lineshapes, as theory does not fully account for the attenuation observed in the n=2 and 3 levels. This is because we have considered a perfectly transparent dead layer, which is incorrect. We should indeed expect gradual extinction of the beam reflected from the dead-layer—bulk interface when the former gets deeper and deeper. This explains for instance the attenuated structures of sample 505/2. Let us try to correlate the observed dead-layer depth with the exciton size. Lacking a detailed knowledge of the exciton confinement effect, in our calculations we had to assume zero exciton polarizability over a uniform layer with sharply defined boundaries. Likewise, for every exciton state we take as a reference figure the average diameter

$$\langle d \rangle = 3\, n^2\, r_B \qquad (r_B = \text{exciton Bohr radius}),$$

which is, for CdS,

$$\langle d_1 \rangle = 95\ \text{Å} \qquad \langle d_2 \rangle = 380\ \text{Å} \qquad \langle d_3 \rangle = 850\ \text{Å} .$$

Comparison with values in (1) indicates that the exciton size is dominant in setting the depth of the dead layer, with a one-to-one correspondence. However, if the exciton-free layer were just of intrinsic origin, each exciton level would have a reproducible characteristic lineshape, which is not observed. We conclude that in general defects or surface-like states must affect the dead layer to some extent. Heavily treated samples give strong evidence of this effect.

The picture drawn here has general significance. It applies to GaAs and InP [3], and it can be shown to hold for several other materials. Other effects, such as the lineshape sensitivity to white light illumination, can be explained on the same basis.

The authors are grateful to R. Generosi and S. Rinaldi for assistance during the experiment and to Prof. G. Chiarotti for useful discussions and encouragement.

REFERENCES

[1] Hopfield ,J.J.; Thomas, D.G.: Phys. Rev. 132, 563 (1963)
[2] Deigen, M.F.; Glinchuk, M.D.: FTT 6, 3250 (1963) (Transl. Sov. Phys.-Solid State 5, 2377 (1964))
[3] Evangelisti, F.; Fischbach, J.U.; Frova,A.: Phys. Rev. B, Feb. 1974
[4] The authors are deeply indebted to Prof. K.W. Boer and to Drs. W.A. Albers, Jr. and D.M.Roessler for supplying the material.
[5] Gross,E.F.; Novikov, B.V.: J. Phys. Chem. Solids 22, 87 (1961)
[6 Bragagnolo, J.A.; Wright, C.; Boer, K.W.: to be published Bragagnolo, J.A.; Boer, K.V.: to be published
[7] Brodin, M.S.; Kritskii, A.V.; Strashnikova, M.I.: Zh. Eksp. Teor. Fiz. 10, 217 (1969) (Transl. JEPT 10, 136 (1969))

ON THE THEORY OF EXCITON STARK EFFECT

M.M.DENISOV and V.P.MAKAROV

P.N.Lebedev Physical Institute, Academy of Sciences of the
USSR, Moscow

It is shown that the terms$\sim \mathscr{E}_\alpha k_\beta$ in the exciton Hamiltonian
are responsible for some peculiarities in the Stark effect.
Calculations for dipole excitons in CuCl investigated experi-
mentally in various works are presented.

1. INTRODUCTION

It is well known that the terms$\sim \mathscr{H}_\alpha k_\beta$ in the exciton Hamil-
tonian, where $\vec{\mathscr{H}}$ is the magnetic field and $\vec{k}$ is the wave vector
(WV), are responsible for the effect of magnetic field inver-
sion (see, for example, $[1,2]$). In the present paper we ana-
lyse the effect of terms$\sim \mathscr{E}_\alpha k_\beta$ in the exciton Hamiltonian,
where $\vec{\mathscr{E}}$ is the electric field, on the Stark effect. It is
turned out that the appearance of some Stark components in
the optical spectrum is due to the terms$\sim \mathscr{E}_\alpha k_\beta$. This conclu-
sion is proposed to be verified on dipole excitons in CuCl
where the Stark effect was investigated experimentally
(see, for example, $[3]$).

2. ENERGY AND WAVE FUNCTION OF CRYSTAL IN ELECTRIC FIELD

The interaction with electric field is considered as per-
turbation without using any exciton model. The calculation
made within the framework of the usual perturbation theory
in the limit of uniform field $\vec{\mathscr{E}}$ and with accuracy of terms

$\sim \vec{\mathcal{E}}^2$ in energy and $\sim \vec{\mathcal{E}}$ in wave function (WF) gives the following results: the WF of the crystal ground state is

$$|0;\vec{\mathcal{E}}> = |0> + \sqrt{\frac{(2\pi)^3}{V_0}} \sum_\lambda \frac{e\vec{\mathcal{E}}\cdot\vec{R}_{\lambda 0}|\lambda 0>}{E_0 - E_\lambda(0)} \ ; \tag{1}$$

the WF of the exciton state with the WV $\vec{k}$ is

$$|j\vec{k};\vec{\mathcal{E}}> = \sum_{j'=1}^{s} a_{j'}^{(j)}(\vec{k}) \left\{ |j'\vec{k}> + \sum_\lambda{}' \frac{e\vec{\mathcal{E}}\cdot\vec{R}_{\lambda j'}(\vec{k})|\lambda\vec{k}>}{E_{j'}(\vec{k}) - E_\lambda(\vec{k})} \right. +$$

$$\left. + \sqrt{\frac{(2\pi)^3}{V_0}} \frac{e\vec{\mathcal{E}}\cdot\vec{R}_{0j'}|0>}{E_{j'}(0) - E_0} \delta(\vec{k}) \right. , \tag{2}$$

$$\sum_{j''=1}^{s} a_{j''}^{(j)*}(\vec{k}) \, a_{j''}^{(j')}(\vec{k}) = \delta_{jj'} . \tag{3}$$

$a_{j'}^{(j)}(\vec{k})$ together with exciton energy are solutions of the set of equations which at small $\vec{k}$ is

$$\left[E_j^{exc}(\vec{k};\vec{\mathcal{E}}) - E_{j'}^{exc}(\vec{k})\right] a_{j'}^{(j)}(\vec{k}) = \sum_{j''=1}^{s} \left(e\vec{\mathcal{E}}\cdot\vec{R}_{j'j''} + \mathcal{E}_\alpha \mathcal{E}_\beta S_{j'j''}^{\alpha\beta} + \right.$$

$$\left. + \mathcal{E}_\alpha k_\beta S_{j'j''}^{\alpha\beta} \right) a_{j''}^{(j)}(\vec{k}). \tag{4}$$

Here $|0;\vec{\mathcal{E}}>$ and $E_0(\vec{\mathcal{E}})$ ($|0>$ and E_0) are the WF and the energy of the ground state at $\vec{\mathcal{E}}\neq 0$ (at $\vec{\mathcal{E}}=0$); $|\lambda\vec{k};\vec{\mathcal{E}}>$ and $E_\lambda(\vec{k};\vec{\mathcal{E}})$ ($|\lambda\vec{k}>$ and $E_\lambda(\vec{k})$) are the WF and the energy of the exciton state at $\vec{\mathcal{E}}\neq 0$ (at $\vec{\mathcal{E}}=0$); $E_\lambda^{exc}(\vec{k};\vec{\mathcal{E}})=E_\lambda(\vec{k};\vec{\mathcal{E}})-E_0(\vec{\mathcal{E}})$ and $E^{exc}(\vec{k})=E_\lambda(\vec{k})-E_0$; $(-e)$ is the electron charge; V_0 is the volume of the crystal; s is the number of exciton states $|j\vec{k}>$ with the WV $\vec{k}$, having at $\vec{\mathcal{E}}=0$ equal or almost equal energies; the summation over twice repeated indices $\alpha,\beta=x,y,z$ is implied; the matrix elements (ME) $\vec{R}_{\lambda 0}=\sqrt{(2\pi)^3/V_0}<\lambda|\hat{\vec{R}}|0>$, where $\hat{\vec{R}}=\sum\hat{\vec{r}}$ is the operator of spatial coordinates of all N_0 electrons of the crystal; the ME

$$\vec{R}_{\lambda\lambda'}(\vec{k}) = \frac{(2\pi)^3}{V_0}\frac{\partial}{\partial\vec{k'}}\left[\langle\lambda\vec{k}|\sum e^{i(\vec{k}-\vec{k'})\cdot\vec{l}}|\lambda'\vec{k'}\rangle - N_0\,\delta_{\lambda\lambda'}\,\delta(\vec{k}-\vec{k'})\right]_{\vec{k'}=\vec{k}} \quad (5)$$

and $\vec{R}_{\lambda\lambda'} = \vec{R}_{\lambda\lambda'}(o)$. Note that $\vec{R}_{\lambda\lambda'}(\vec{k}) = \left[(2\pi)^3/V_0\right]\langle\lambda\vec{k}|\hat{\vec{R}}|\lambda'\vec{k}\rangle$ only at $\lambda\neq\lambda'$. So in contrast to 4 , no physically meaningless values $\langle\lambda o|\hat{\vec{R}}|\lambda'o\rangle$ appear in our results. The expressions for the ME $S_{jj'}^{\alpha\beta}$ and $\tilde{S}_{jj'}^{\alpha\beta}$ are not written because their symmetry properties but not their concrete forms are important for further consideration.

From (4) and taking into account that the WV $\vec{k}$ changes it's sign and the electric field $\vec{\mathcal{E}}$ doesn't change it under the time reversal operation $\hat{\theta}$ it follows that $\vec{R}$ is the vector and $\hat{\theta}\hat{\vec{R}}\hat{\theta}^{-1}=\hat{\vec{R}}$, $\hat{S}^{\alpha\beta}$ is the symmetrical tensor of the second rank and $\hat{\theta}\hat{S}^{\alpha\beta}\hat{\theta}^{-1}=\hat{S}^{\alpha\beta}$, and $\hat{\tilde{S}}^{\alpha\beta}$ is the second rank tensor and $\hat{\theta}\hat{\tilde{S}}^{\alpha\beta}\hat{\theta}^{-1}=-\hat{\tilde{S}}^{\alpha\beta}$. Therefore the ME $\vec{R}_{jj'}$ and $S_{jj'}^{\alpha\beta}$ are real values and $\tilde{S}_{jj'}^{\alpha\beta}$ are imaginary values, in particular the diagonal ME $\tilde{S}_{jj}^{\alpha\beta}=0$. As a result of it the reversal of the direction of observation $(\vec{k}\rightarrow -\vec{k})$ does not effect the Stark components. But some Stark components which are absent with $\tilde{S}_{jj'}^{\alpha\beta}\,\mathcal{E}_\alpha k_\beta$ neglected appeare due to effect of $\tilde{S}_{jj'}^{\alpha\beta}\,\mathcal{E}_\alpha k_\beta$.

3. THE STARK EFFECT FOR DIPOLE EXCITONS

Here using the cubic crystal CuCl (T_d) as an example we consider more detail the Stark effect taking into account the term $\sim\mathcal{E}_\alpha k_\beta$ in (4) and the splitting of exciton levels into levels of longitudinal (LE) and transverse (TE) excitons. Choosing coordinate axes $[x,y,z]$ along the cubic axes of the crystal we denote the WF of dipole excitons by $\vec{\psi}=[\psi_x,\psi_y,\psi_z]$.

There can be the level of paraexciton (PE) near the level of dipole (ortho) excitons. We denote the WF of PE at $\vec{k}=0$ by ψ_p. In crystals of group T_d $\vec{\psi} \sim \Gamma_5$ and in CuCl $\psi_p \sim \Gamma_2$. In this case $s=4$ (see (4)).

It is possible to obtain from the symmetry that all nonvanishing ME in CuCl (T_d) are as follows

$$R^z_{xy}=R^x_{yz}=R^y_{zx}=R, \quad \vec{R}_{jj'}=\vec{R}_{j'j}; \quad S^{xx}_{xx}=S^{yy}_{yy}=S^{zz}_{zz}=S_1, \quad S^{yy}_{xx}=S^{zz}_{xx}=S^{zz}_{yy}=S^{xx}_{yy}=$$

$$S^{xx}_{zz}=S^{yy}_{zz}=S_2, \quad S^{xy}_{xy}=S^{yz}_{yz}=S^{zx}_{zx}=S_3/2, \quad S^{xx}_{pp}=S^{yy}_{pp}=S^{zz}_{pp}=S_4, \quad S^{\alpha\beta}_{jj'}=S^{\alpha\beta}_{j'j};$$

$$\tilde{S}^{xy}_{xy}=-\tilde{S}^{yx}_{xy}=\tilde{S}^{yz}_{yz}=-\tilde{S}^{zy}_{yz}=\tilde{S}^{zx}_{zx}=-\tilde{S}^{xz}_{zx}=i\tilde{S}_1, \quad \tilde{S}^{yz}_{xp}=-\tilde{S}^{zy}_{xp}=\tilde{S}^{zx}_{yp}=-\tilde{S}^{xz}_{yp}=\tilde{S}^{xy}_{zp}=-\tilde{S}^{yx}_{zp}=$$

$$i\tilde{S}_2, \quad \tilde{S}^{\alpha\beta}_{jj'}=-\tilde{S}^{\alpha\beta}_{j'j}, \tag{6}$$

where all constants R, S and $\tilde{S}$ are real values.

The WF of LE $\psi_{\parallel}$ and TE $\psi_{\perp 1}$ and $\psi_{\perp 2}$ can be written in the form [5]

$$\psi_{\parallel}=\vec{n}\cdot\vec{\psi}, \quad \psi_{\perp 1}=\vec{e}\cdot\vec{\psi}, \quad \psi_{\perp 2}=(\vec{e}\times\vec{n})\cdot\vec{\psi}, \tag{7}$$

where $\vec{n}=\vec{k}/k$ is the direction of light propagation and $\vec{e}$ is the polarization vector of light. At $\vec{\mathcal{E}}=0$ the optical transition from the ground state $|o\rangle$ is allowed only to the exciton state $|\psi_{\perp 1}\rangle$.

In general, the formulaes (1)-(4) and (6)-(7) allow to find energies and intensities of the Stark components at any directions $\vec{n}$, $\vec{e}$, and $\vec{\mathcal{E}}$.

As an example we consider two cases where the terms $\sim \mathcal{E}_\alpha k_\beta$ in (4) are important in the Stark effect.

1) $\vec{\mathcal{E}} \parallel [001]$, $\vec{n} \parallel [1\bar{1}0]$, $\vec{e} \parallel [001]$. In this geometry two Stark

are invisible and two-corresponding to longitudinal and one
of transverse excitons-appear in the spectrum:

$$E_1^{exc}(\vec{\zeta})-E_\perp^{exc}\cong S_1\zeta^2, \quad f_1(\vec{\zeta})\cong f; \quad E_\parallel^{exc}(\vec{\zeta})-E_\parallel^{exc}\cong -eR\zeta+S_2\zeta^2,$$

$f_\parallel(\vec{\zeta})\cong(\tilde{S}_1\zeta k/E_{\parallel\perp})^2 f$, where f is the oscillator strength of
the exciton transition at $\vec{\zeta}=0$ and $E_{\parallel\perp}=E_\parallel-E_\perp$.
2) $\vec{\zeta}\parallel[001]$, $\vec{n}\parallel[1\bar{1}0]$, $\vec{e}\parallel[110]$. In this geometry also only two
of four Stark components-corresponding to one transverse ex-
citon and paraexciton-appear in the spectrum:

$$E_1^{exc}(\vec{\zeta})-E_\perp^{exc}\cong eR\zeta+S_2\zeta^2, \quad f_1(\vec{\zeta})\cong f; \quad E_p^{exc}(\vec{\zeta})-E_p^{exc}\cong S_4\zeta^2, \quad f_p(\vec{\zeta})=$$

$(S_2\zeta k/E_{p\perp})^2 f$, where $E_{p\perp}=E_p-E_\perp$.
The excitation of LE in the first case and PE in the second
case is associated with the terms $\sim\zeta_\alpha k_\beta$ in (4). We think that
it is interesting to verify this conclusion experimentally.
Note also the evident possibility to obtain from experiments
of this kind the values of splittings $E_{\parallel\perp}$ and $E_{p\perp}$.

4. REFERENCES

[1] Agranovich, V.; Ginzburg, V.: Kristallooptika s uchetom pro-
stranstvennoi dispersii, Nauka, 1965, in Russian.
[2] Knox,R.: Sol. St. Phys., Suppl. 5, 1 (1963)
[3] Deiss, J.; Daunois, A.: Surface Science 37, 804 (1973)
[4] Cherepanov, V.; Druzinin, V.; Kargopolov,J.; Nikiforov, A.:
Fiz. Tverd. Tela 3, 2987 (1961)
[5] Denisov, M.; Makarov, V.: Phys. Stat. Sol. (b) 56, 9 (1973)

STARK EFFECT ON THE ENERGY BANDS OF CuCl AND Cu_2O

A. DAUNOIS, J.L. DEISS, S. NIKITINE

Laboratoire de Spectroscopie et d'Optique du Corps Solide
Groupe de Recherche n° 15 du CNRS
5, rue de l'Université, 67000 Strasbourg

France

The electroreflectance of single oriented
crystals of CuCl and Cu_2O has been studied in
the region of the exciton transitions. The
field-induced reflectivity changes are shown to
be the superposition of the quadratic field
effect of the excitons and of the Stark effect
of the bands of the crystal. The Stark effect on
the bands (linear in CuCl, quadratic in Cu_2O)
shows a pronounced anisotropic dependence on the
directions of the field and light polarization.

1. INTRODUCTION

The changes of the optical properties of semiconductors in an
external electric field are usually explained by the Franz-
Keldysh effect on band to band transitions or by the quadratic
field effect on the excitons. Both effects result from the
influence of the applied field on the relative motion of elec-
trons and holes in the photoexcited crystal. But, the optical
properties of a crystal depend also on its particular one-elec-
tron band structure, and the electric field may then remove the
degeneracy of the bands associated with the considered inter-
band or excitonic transitions. This Stark effect of the bands
has been discussed theoretically by Enderlein et al. [1] in
diamond type crystals, but very little experimental work has
been done so far. In this paper, experiments showing the exis-
tence of such a Stark splitting of the bands in cubic crystals
(CuCl and Cu_2O) are reported and discussed.

2. EXPERIMENTAL RESULTS

The transverse electroreflectance (ER) of single oriented crys-
tals of CuCl and Cu_2O has been investigated at 85°K, in the
region of the fundamental exciton transitions. Linearly polari-

zed light and a sensitive modulation technique have been used. The use of such a differential technique (synchronous detection) allows the possible detection of weak splittings or shifts of the exciton lines, and especially, it allows the separation of ER signals with different field dependences (linear or quadratic field effects).

The reflectance spectra of CuCl in the near UV and of Cu_2O in the blue and violet region reveal two sharp peaks. These two peaks correspond to 1S excitons, which belong to two different series, as a result of the spin-orbit splitting of the higher valence band (VB).

In presence of an applied field, the observed ER structures are explained by the well-known quadratic field effect on the Wannier excitons, giving rise to a shift or a broadening (or both together) of the exciton lines. These isotropic envelope effects on the 1S excitons in CuCl and Cu_2O have been studied and discussed in detail elsewhere [2]. However, for different directions of the field $\vec{E}$ and polarization vector $\vec{\varepsilon}$, these ER spectra exhibit pronounced polarization effects [3]. In order to analyze this anisotropic dependence of the ER response, the electric field has been applied respectively along the crystallographic axes 001, 110 and 111 of the crystals.

In CuCl, the ER spectra are characterized by the existence of two distinct electrooptic effects, linear and quadratic, which

Fig. 1: Linear electrooptic effect in CuCl for $\vec{E}$ // 001.

Fig. 2: Linear electrooptic effect in CuCl for $\vec{E}$ // 110.

are separated by blocking the lock-in amplifier at the same or at the double frequency of the applied field. The linear electrooptic effect represented in figures 1 and 2 for two different orientations of the field, shows a strong anisotropic dependence whereas the quadratic electrooptic effect (envelope effect) is totally isotropic.

When $\vec{E} \, /\!/ \, 001$ (fig. 1), the ER structures are totally polarized; they are present for $\vec{\varepsilon} \, /\!/ \, 110$ and nearly absent for $\vec{\varepsilon} \, /\!/ \, 001$. For $\vec{E} \, /\!/ \, 110$ (fig. 2), the linear ER structures are symmetrically reversed when $\vec{\varepsilon}$ is parallel to $11\sqrt{2}$ and $11\overline{\sqrt{2}}$, respectively. They are practically absent for $\vec{\varepsilon}$ parallel or perpendicular to the field. Similar results have been obtained by Mohler [4], who analysed the ER data by a Kramers-Kronig analysis.

In Cu_2O, which has a center of symmetry, no linear electrooptic effect is observed in the ER response. However, the study of the ER spectra of the blue and violet series reveals polarization effects in addition to the quadratic field effect of the 1S excitons.

For $\vec{E} \, /\!/ \, 001$ and $\vec{E} \, /\!/ \, 110$, the ER spectra can be separated into two distinct signals differing in position and intensity for the two directions $\vec{\varepsilon} \, /\!/ \, \vec{E}$ or $\vec{\varepsilon} \perp \vec{E}$ (fig. 3). For $\vec{E} \, /\!/ \, 111$, the two ER signals are not separated and they have practically the same intensity (fig. 4). This separation of the ER structures in figure 3 into two distinct contributions with polarized light results probably from a splitting of the 1S exciton lines in the

Fig. 3: Electroreflectance of Cu$_2$O for $\vec{E} \, /\!/ \, 001$.　　Fig. 4: Electroreflectance of Cu$_2$O for $\vec{E} \, /\!/ \, 111$.

applied field. At the field strengths used (50 kV/cm), the separation is weak (about 4 cm^{-1}), but pronounced polarization effects are observed on the amplitude of the ER signals (fig.3). The measurements have been performed with single crystals which had been annealed at 1100°C and then cooled very slowly (two days), to avoid internal stress effects. In some crystals, which were not annealed, greater splittings had been observed.

3. DISCUSSION - CONCLUSION

The electroreflectance structures observed in CuCl and Cu_2O are principally due to the quadratic field effect of the excitons (envelope effect) [5]. The detailed study of these ER spectra shows a pronounced anisotropic dependence on the directions of field and light polarization. However, as the field effect on the 1S exciton envelope is totally isotropic, it cannot explain the polarization effects observed experimentally. But, as these 1S excitons are built from degenerate valence and conduction bands, the electric field may remove this electronic band degeneracy. Consequently, the Stark splitting of the bands should show up in the ER spectra together with the envelope effects of the excitons.

In CuCl, a non-centrosymmetric crystal, these two different field effects are easily separated on account of their different field dependences, linear or quadratic. The linear field effect results from an anisotropic shift of the two 1S excitons. This shift is due to a linear Stark splitting of the Γ_7 and Γ_8 VB participating in the formation of these excitons [4]. The strong angular dependence of this linear Stark effect observed in the ER spectra of CuCl is well explained by the following expression :

$$\frac{\Delta R}{R} \propto (\alpha\beta'\gamma' + \beta\gamma'\alpha' + \gamma\alpha'\beta') E$$

where $(\alpha\beta\gamma)$ and $(\alpha'\beta'\gamma')$ are the direction cosines of $\vec{E}$ and $\vec{\varepsilon}$ respectively. This expression has been deduced from the change of the dielectric tensor developed according to a power law of the applied field.

In Cu_2O, a centrosymmetric crystal, the polarization effects

in the quadratic ER response result probably from a splitting of the 1S excitons, itself due to a splitting or shift of the bands [6]. Since this splitting is the same for the blue and violet series, it is then due to a splitting of the higher CB Γ_8^- (Γ_{12}^- without spin), which is common to both series. The experimental splitting scheme (doublet for $\vec{E} \parallel 001$ and $\vec{E} \parallel 110$, singlet for $\vec{E} \parallel 111$) is in agreement with the Stark splitting predicted for the Γ_{12}^- state by group theory. Now, in Cu_2O, such a Stark splitting of the Γ_{12}^- CB can only be a quadratic field effect, as the field mixes only band states of different parity. Such a field-induced mixing between the Γ_{25}^+ VB and Γ_{12}^- CB explains both the magnitude and multiplicity of the observed Stark splitting. In the relatively low fields applied (up to 50 kV/cm) this splitting is in fact very weak, but it is believed that the polarization effects observed in the ER spectra of Cu_2O are due to a degeneracy removal of the Γ_8^- conduction band.
Finally, the existence of a Stark effect on the energy bands of CuCl and Cu_2O shows that this effect should not be neglected in the interpretation of the ER spectra of a solid.

4. REFERENCES

[1] Enderlein, R.; Keiper, R.; Tausendfreund, W.: Phys. Stat. Sol. 33, 69 (1969)

[2] Deiss, J.L.; Daunois, A.: Surf. Sci. 37, 804 (1973)

[3] Mohler, E.: Phys. Stat. Sol. 29, K55 (1968)
Daunois, A.; Deiss, J.L.; Nikitine, S.: C.R. Acad. Sci. (France) 268, 977 (1969)

[4] Mohler, E.: Phys. Stat. Sol. 38, 81 (1970)

[5] Blossey, D.F.: Phys. Rev. B 2, 3976 (1970); 3, 1382 (1971)
Dow, J.D.; Lao, B.Y.; Newman, S.A.: Phys. Rev. B 3, 2571 (1971)

[6] Brahms, S.: Phys. Stat. Sol. (b) 51, 509 (1972)

WAVELENGTH MODULATION SPECTRA OF EXCITONS IN Cu_2O

T. ITOH[†] and S. NARITA

Department of Material Physics, Faculty of Engineering Science,
Osaka University, Toyonaka, Osaka, Japan

A study of exciton absorption spectra of Cu_2O in a
wide temperature range of $4.2K \sim 300K$ has been done by
a wavelength modulation technique. The line shapes of
the exciton spectra are analysed on the basis of
Toyozawa's theory. The band masses are determined
by the aid of the impurity associated transition.

1. INTRODUCTION

A large number of experimental studies on Cu_2O have been carried
out by various methods. The experiments have stimulated the
theoretical activity. Theoretical progress on the exciton ab-
sorption line shape associated with exciton-phonon interaction
has been performed by Toyozawa[1,2,3]. However, very interest-
ing parts of his results are mostly related to the exciton-sta-
tes in high temperature regions where the analysis of the line
shapes of conventional spectra becomes seriously difficult. The
wavelength derivative spectroscopy is more advantageous for the
study: we can make clearer structures of lines and can exclude
the background of spectra more easily. Moreover, by this tech-
nique we can distinguish weak and uncertain structures concealed
in the background in the conventional spectra.

2. WAVELENGTH DERIVATIVE SPECTRA AND STRUCTURE ASSIGNMENTS

The present wavelength modulation spectroscopy has been prac-
tised by the technique of vibrating a mirror near the exit-slit
of a diffraction grating monochromator. Figure 1 shows wave-
length modulation absorption spectra of Cu_2O at various tem-
perature. Here, the symbols D_n^Y and D_n^G represent the structures
associated with the direct transitions from the ground state to
the nth excited states in the yellow Y and green G series. (In
the case of $n=1$, D_n means the transition to the 1s state and in
the case of $n \geq 2$, D_n indicates the transition to the np state).
$D_n^{Y'}$ corresponds to the transition to the ns state. I_a and I_e

[†]New Address: Faculty of Science, Tohoku Univ. Sendai, Japan.

Fig. 1: Wavelength derivative spectra of Cu₂O at various temperatures.

indicate the indirect transitions associated with 110cm^{-1} phonon absorption and emission.

3. INDIRECT TRANSITIONS

Let us consider exciton indirect transitions to the 1s state of the yellow series. From the symmetry consideration of the relevant states and phonons in Cu_2O given by Elliott[4] and by the aid of the Raman scattering[5] and luminescence[6] data, it can be concluded that we have possibility to observe the indirect transitions associated with phonons of $\Gamma_{25}^-(\sim90\text{cm}^{-1})$, $\Gamma_{12}^-(\sim110\text{cm}^{-1})$, $\Gamma_{15}^{-(1)}(\sim150\text{cm}^{-1})$, $\Gamma_{2}^-(\sim 350\text{cm}^{-1})$, and $\Gamma_{15}^{-(2)}(\sim660\text{cm}^{-1})$. The structures A, I_e, B, C, and I_e' indicated in Fig.2 correspond to the phonon assisted indirect transitions to the yellow 1s state. The structure I_e' is associated with the $\Gamma_{15}^{-(2)}(\sim660\text{cm}^{-1})$ phonon, and A, B, and C are newly observed indirect edges associated

Fig. 2: Wavelength derivative spectrum (W.D.S.) of absorption coefficients α at 4.2K around the indirect edge I_e, indicating several indirect transition edges.

with phonons with energies, $88\,\mathrm{cm}^{-1}(\Gamma_{25}^{-})$, $151\,\mathrm{cm}^{-1}(\Gamma_{15}^{-(1)})$, and $359\,\mathrm{cm}^{-1}(\Gamma_{2}^{-})$, respectively. Though the $\sim150\,\mathrm{cm}^{-1}$ phonon assisted transition has already been discussed by Compaan et al.[7], our derivative spectrum at the energy shows a peak-like line shape reflecting the step-like figure of $E^{1/2}$ energy dependence in the conventional spectrum, which corresponds with an allowed transition, while their spectrum has $E^{3/2}$ dependence indicating a symmetry forbidden transition. All the other derivative spectra of the yellow 1s indirect edges also show peak-like line figures.

4. YELLOW 1s DIRECT EXCITON LINE

The structure which appears midway between I_a and I_e in Fig.1 corresponds to the direct transition to the yellow 1s exciton state. According to Toyozawa's theory[2], this line differs from other lines in the exciton-phonon interaction mechanism from the reason that there exist no states with the same energy in other excited states associated with phonon scattering of K=0 exciton, and the line figure is strongly asymmetric with a tail toward the high energy side. From his theory, the half width of the line is expected to show a quadratic temperature dependence in the high temperature region. Figure 3 shows the experimental results of the temperature change and the agreement with theory is satisfactory.

Fig. 3: Temperature dependence of half width of the yellow 1s direct line. It has a quadratic temperature dependence in high temperature region.

5. nTH DIRECT EXCITON LINES

Toyozawa[1] has theoretically derived that nth direct lines ($n \geq 2$) have asymmetric Lorentzian line figures. Therefore, the functional line shapes of the energy derivatives

are given by

$$\ell(W) = -[A(W^2 - 1) + W]/(W^2 + 1)^2, \qquad (1)$$

where A is the degree of asymmetry and $W = h\nu - h\nu_t/\Gamma$, ($h\nu$: the photon energy, $h\nu_t$: the transition energy and Γ/h: the damping frequency). We searched the best fits of the theoretical derivative curves to the experimental ones by assuming the linear backgrounds and taking A as a variable parameter. An example at 4.2K is shown in the inset of Fig.4. In the fitting process, we determine the values of A, $h\nu_t$, and the half width H. From the values of $h\nu_t$ of both the series, we obtain the Rydberg constants, the exciton reduced masses, and the series limit energies E_g as functions of temperature. Toyozawa[1] predicted that the half width H in the Lorentzian region is proportional to temperature except for very low temperatures. The present experiment shows that H has almost linear temperature dependence above ∿100K. If we assume that only a single phonon takes part in the phonon broadening effect, the half width of the 2p yellow line can be represented by $H_{2p}^Y = 23 \cdot \coth(110\text{cm}^{-1}/2kT)$ (cm^{-1}), suggesting that the 110cm^{-1} phonon has a predominant role in Cu_2O crystals. Similarly, the temperature dependences of the half width of the green 2p line, the Rydberg constants and the band gap energies of both the series show the important role of the 110cm^{-1} phonon in Cu_2O.

Fig. 4: Comparison of the temperature dependence of the half width of the yellow 2p line with theory. The inset shows the best fit of a theoretical asymmetric Lorentzian line figure (A=-0.38 and H=23cm⁻¹) to the experimental energy derivative spectrum of the yellow 2p direct exciton at 4.2K.

6. IMPURITY ASSOCIATED TRANSITION AND BAND MASSES

The broad peak E seen in Fig.1 has different height from sample to sample.

Thus, it is reasonable to assign the structure E to a transition associated with impurity. The fact that the temperature shift of E is almost equal to that of the green series limit suggests that E is closely related both to the heavy hole band and to the conduction band. For such a transition, two cases are possible: one is the case of the initial state being some ionized acceptor state and the final state the conduction band (transition A) and the other is the case of the former being the heavy hole band and the latter the ionized donor state (transition B). Referring to the calculation by Johnson[8], the maximum spectral height is known to be proportional to b^2 in transition A and b^{-2} in transition B, (b= the electron band mass m_e^*/ the heavy hole band mass m_{hh}^*). From the above consideration and b<1, E is probable to be associated with transition B. In the same calculation the following relation is derived:

$$E_g^{(G)} - h\nu_t^{imp.} = [E_g^{(G)} - h\nu_p^{imp.}]/(1-0.19b) \; [=E_{D.2p}], \quad (2)$$

where $h\nu_t^{imp.}$ is the transition energy, $h\nu_p^{imp.}$ peak position and $E_g^{(G)}$ the green series limit. Assuming a hydrogen-like impurity level, the binding energy of the 2p state is given by

$$E_{D.2p} = m_e^* . e^4/(2^2 . 2\hbar^2 . \epsilon^2). \quad (3)$$

Then we have

$$E_g^{(G)} - h\nu_p^{imp.} = (1.19 - 0.19 m_e^*/\mu^{(G)}) \cdot m_e^* \cdot e^4/(8\hbar^2 \cdot \epsilon^2), \quad (4)$$

where $[\mu^{(G)}]^{-1} = m_e^{*-1} + m_{hh}^{*-1}$. By solving this quadratic equation for m_e^*, and using the exciton reduced mass at 4.2K for the yellow series $\mu^{(Y)} \approx 0.39 m_0$ and for the green series $\mu^{(G)} \approx 0.58 m_0$, we obtain $m_e^* \approx 0.99 m_0$, $m_{\ell h}^* \approx 0.64 m_0$ (the light hole band mass) and $m_{hh}^* \approx 1.40 m_0$ for 4.2K.

7. REFERENCES

[1] Toyozawa, Y.: Prog. Theor. Phys. 20, 64 (1958).
[2] Toyozawa, Y.: Prog. Theor. Phys. 27, 89 (1962).
[3] Toyozawa, Y.: J. Phys. Chem. Solids 25, 59 (1964).
[4] Elliott, R. J.: Phys. Rev. 124, 340 (1961).
[5] Yu, P. Y.; Shen, Y. R.; Petroff, Y.: Solid State Commun. 12, 973 (1973)
[6] Petroff, Y.; Yu, P. Y.; Shen, Y. R.: Phys. Rev. Letters 29, 1558 (1972)
[7] Compaan, A.; Cummins, H. Z.: Phys. Rev. B6, 4753 (1972)
[8] Johnson, E. J.: "Semiconductors and Semimetals" Vol.3, p. 153 (edited by Willardson, R. K. and Beer, A. C., Academic Press, New York, 1967).

INVESTIGATION OF EXCITON STATES IN A HIGH MAGNETIC FIELD BY THE ADIABATIC METHOD

NUNZIO O. LIPARI
Xerox Corporation, Webster, N. Y. 14580

M. ALTARELLI*
Department of Physics, University of Illinois
Urbana, Illinois 61801

An adiabatic method which simultaneously takes into account the electron-hole Coulomb interaction and the valence-band structure of cubic semiconductors is used to interpret high-resolution magneto-optical data. The adiabatic potentials are calculated explicitly for all the exciton states of interest and used to establish the magnetic field range of validity of the method for diamond and zincblende semiconductors. The determination of band parameters of InSb and GaAs from experimental data is briefly discussed.

1. INTRODUCTION

Recently [1], it has been shown that high resolution interband magneto-optical data can be unambiguously interpreted only by simultaneously taking into account the electron-hole Coulomb interaction and the complicated valence-band structure of cubic semiconductors. This was accomplished in the framework of the adiabatic method, which reduces the problem to a system of coupled one-dimensional Schrödinger equations involving effective adiabatic potentials, which are here calculated explicitly for all the exciton states of interest. The range of validity of this method is then discussed in detail, focusing on the possible complications arising from the complex Laudau-level structure of the valence band. The specific example of the analysis of the magneto-absorption data of Johnson [2] for InSb is discussed to assess the accuracy of determination of valence band parameters. The case of GaAs is also briefly discussed.

2. ADIABATIC POTENTIALS

The one-dimensional adiabatic potentials $V_{n\ell}(z)$ are given by [1]

$$\frac{V_{n_i \ell_i}(z)}{\sqrt{\gamma}} = \frac{-2n_i!}{(n_i+|\ell_i|)!} \int_0^\infty d\rho \; \frac{\rho^{|\ell_i|} e^{-\rho} \left[L_{n_i}^{|\ell_i|}(\rho) \right]^2}{\sqrt{2\rho+\gamma z^2}} \tag{1}$$

*Supported by ARO-DA-HC04-74-C-0005

where n_i and ℓ_i are quantum numbers, as defined in Ref. 1, $L_{n_i}^{|\ell_i|}$ is a generalized Laguerre polynomial, and $\gamma = e\hbar H/2\mu_0 c R_0$ with H the magnetic field in the z direction, R_0 the effective Rydberg and $1/\mu_0 = (1/m_e) + \gamma_1$ [1]. In Ref. 1, the matrix elements of $V_{n\ell}(z)$ were calculated analytically in a gaussian basis set, thus allowing an accurate computation of the exciton levels. In order to discuss the range of validity of the adiabatic method and the accuracy of the gaussian expansion approach, we need to calculate explicitly the potentials. This is accomplished by assuming the completeness of the gaussian set and writing

$$V_{n_i \ell_i}(z) = \frac{\sum_{m,j,k} S_{mj}^{-1} \left[V_{n_i \ell_i} \right]_{jk} c_k e^{-\alpha_m z^2}}{\sum_m c_m e^{-\alpha_m z^2}} \tag{2}$$

where c_i is an arbitrary set of coefficients defining a test function, and S_{ij} is the overalap matrix of the nonorthogonal basis set. In Fig. 1 the results for the most relevant potentials, V_{00}, V_{01} and V_{02} are displayed. For comparison, the exact analytical result for V_{00} is also shown and the excellent agreement provides a good test of the method and gives us confidence in the results obtained for the other potentials.

FIG. 1: Adiabatic potentials $V_{n\ell}(z)$ in units of $\mathrm{Ryd}\sqrt{\gamma}$ as a function of $\sqrt{\gamma}\,z$, computed as described in the text. The dots indicate values of $V_{00}(z)$ as computed from its exact analytical expression.

3. RANGE OF VALIDITY OF THE THEORY

In order to discuss the range of validity of the adiabatic method, let
us recall the basic approximations involved. In the standard Landau
level theory [3], it is shown that the Hamiltonian is decoupled into
manifolds of up to four interacting levels. When the Coulomb interaction
is turned on, further couplings are introduced. In the adiabatic approxi-
mation, for high values of the magnetic field, one neglects this extra
coupling between different manifolds. Since the coupling via the Coulomb
interaction of manifolds with different quantum number ℓ (i.e., with
different angular momentum component in the field direction) is
identically vanishing, only the neglect of the coupling for different n's
has to be discussed. A typical interband Landau level scheme for a
given ℓ is shown in Fig. 2. The criterion for the validity of the
adiabatic approach is that the energy separation between the level
of interest and those of different manifolds be large in comparison with
the depth of the effective adiabatic potential. Due to the presence of
heavy-and light-hole ladders, degeneracies or near degeneracies are
possible for particular values of the field. Generally, this affects only
highly excited levels, and not those which are most relevant experi-
mentally; however, when the present method is used to analyze experi-
mental data, it is desirable, in order to achieve a good determination
of the band parameters, to include as many of the observed levels as
possible. In these cases, it is necessary to exert some caution, and
it is profitable to look at the Landau level structure for the field
under consideration, and make sure that there are no higher levels close
to the ones assigned to the experimentally observed structure.

Consider, as an example, the analysis [1] of magneto-absorption data [2]
of the InSb with $\gamma \sim 52$. In Figure 3, the Landau levels are shown, and
it is seen that, although some of the excitonic levels considered are
associated with higher Landau levels, due to the high value of γ no
sizeable complication is present.

FIG. 2: Interband Landau levels
of Ge with $\ell = 0$, for H $\|$ (110),
$\gamma = 11.52$ (H = 72 kG), in effective
units, relative to the zero field
energy gap. The levels are
labeled by the quantum number n.
The dotted curve under the n = 0
level represents the depth of the
$V_{00}(z)$ potential.

FIG. 3: Interband Landau levels
of InSb, $\ell = -2$, for H $\|$ (110),
$\gamma = 52.78$ (H = 39.1 kG) in effective
units, with respect to the zero
field energy gap (average with
respect to conduction band spin).
The solid levels are those con-
sidered in Ref. 1. Dotted curve
as in Fig. 2.

In the case of materials for which very high γ's are not attainable,
as GaAs, one is on the contrary obliged to include only the lower
lying levels. In the analysis of the magnetoreflectance data at $\gamma = 7.29$,
used to determine the value of the κ parameter [4], it is possible to
use the adiabatic method with reasonable confidence only for low-lying
exciton series (Fig. 4).

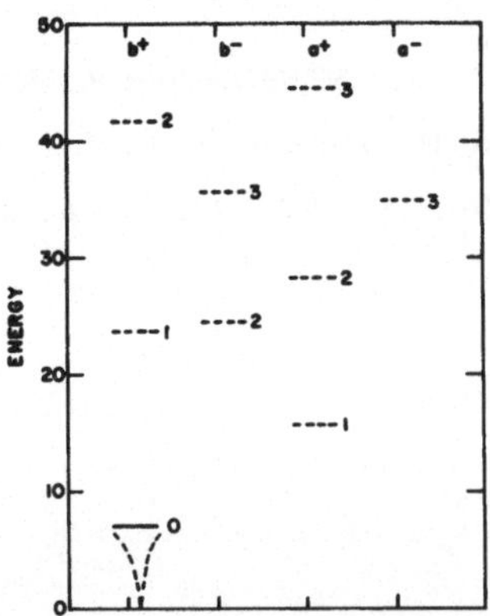

FIG. 4: Same as Fig. 3, for GaAs, $\ell = 0$, $\gamma = 7.29$ (H = 215kG). Solid line is the level considered in Ref. 4.

In summary, a prescription to establish the range of validity of the adiabatic method was given, and used to discuss the accuracy of the theoretical interpretation of high field magneto-optical data.

5. REFERENCES

[1] Altarelli, M.; Lipari, N. O.; Phys. Rev. B 9, 1733 (1974).

[2] Johnson, E. J.; Phys. Rev. Letters 19, 352 (1967).

[3] Luttinger, J. M.; Phys. Rev. 102, 1030 (1956).

[4] Altarelli, M.; Lipari, N. O.; Dingle, R.; Bull. Am. Phys. Soc. 19, 245 (1974), and to be published.

HIGH-PRESSURE BEHAVIOUR OF
THE EXCITON AND THE ABSORPTION EDGES IN Ga Se

J. M. BESSON and K. P. JAIN [x]

Laboratoire de Physique des Solides, associé au C. N. R. S.

Université de Paris VI - 75230 Paris Cedex 05, France

A. KUHN

Laboratoire de Luminescence II, équipe associée au C. N. R. S.

Université de Paris VI - 75230 Paris Cedex 05, France

From a study on the optical absorption edge of Ga Se under
hydrostatic pressure up to 14 kilobars, it is shown that the in-
direct gap energy decreases with pressure faster than the di-
rect edge and the exciton level associated with it. Increasing
overlap of the indirect edge and the direct exciton causes en-
hanced interference between the exciton line and the continuum
of structure due to indirect transitions, which depletes the ex-
citon strength under pressure.

Absorption coefficient studies of Ga Se at room temperature and pres-
sure have revealed a peak due to an exciton which is associated with the
direct edge (1). The results have also led to the proposal that the lowest
edge is an indirect one close (about 50 milli eV) to the direct edge (2). Re-
cent work on the absorption edge under hydrostatic pressure lends support
to this conjecture (3).

In this paper, we present data on absorption coefficient, α , of Ga Se at
pressures up to 14 kilobars in the region of the band gap. Our results con-
firm the existence of an indirect gap at 1. 970 eV close to the direct edge
at 2. 030 eV. The strength of the exciton peak decreases as the pressure
is increased. This can be understood in terms of the pressure coefficients

[x] Permanent adress : Physics Department, Indian Institute of Technology,
New Delhi - 29, India.

of the two gaps. For the direct edge, $dE_g^d / dP = -4.2 \times 10^{-6}$ eV/bar while for the indirect edge $dE_g^i/dP = -11.5 \times 10^{-6}$ eV/bar. The sign and magnitude of these pressure coefficients can be understood in terms of the known band structure of Ga Se (3).

Bridgman-grown, ε -type samples were illuminated parallel to the c-axis (E $\perp$ c polarization).

The behaviour of the absorption edge under hydrostatic pressure is given in the figure. The significant features on increasing the pressure are :

- The oscillator strength of the exciton line decreases and disappears for large pressures. The pressure coefficient is :

$$d E_{exc} / dP = (-4.2 \pm 0.3) \times 10^{-6} \text{ eV/bar.}$$

- The low-energy tail becomes less steep. As shown in the insert (b), at high enough pressures (above 8 kilobars), the effect of indirect transitions becomes apparent.

The damping of the exciton under pressure has been discussed before elsewhere (3). We assign this effect to increasing interference of the discrete exciton line with a continuum due to indirect interband electronic transitions. As the pressure is increased, the ratio of the exciton (discrete) and indirect interband (continuum) transition amplitudes decreases. This leads to increasing interference between the two sets of states and the corresponding damping of the exciton.

It seems most natural to analyse the high-energy part of the spectrum in terms of direct allowed and forbidden transitions. The simplest way of doing this is to convolute the room-pressure absorption curve for the bound and continuum exciton states near the M_o edge with a Lorentzian, which includes lifetime effects (4). A fit of the theory and the experiment below 2.10 eV gave : $E_g^d = 2.033$ eV, R = 30 milli eV and $\Gamma = 20$ milli eV, where R is the exciton Rydberg and Γ the phenomenological width parameter. Above 2.10 eV, the calculated values of α for direct allowed transitions are lower than the experimental ones. The difference between the two is attributed to the contribution of direct forbidden transitions,

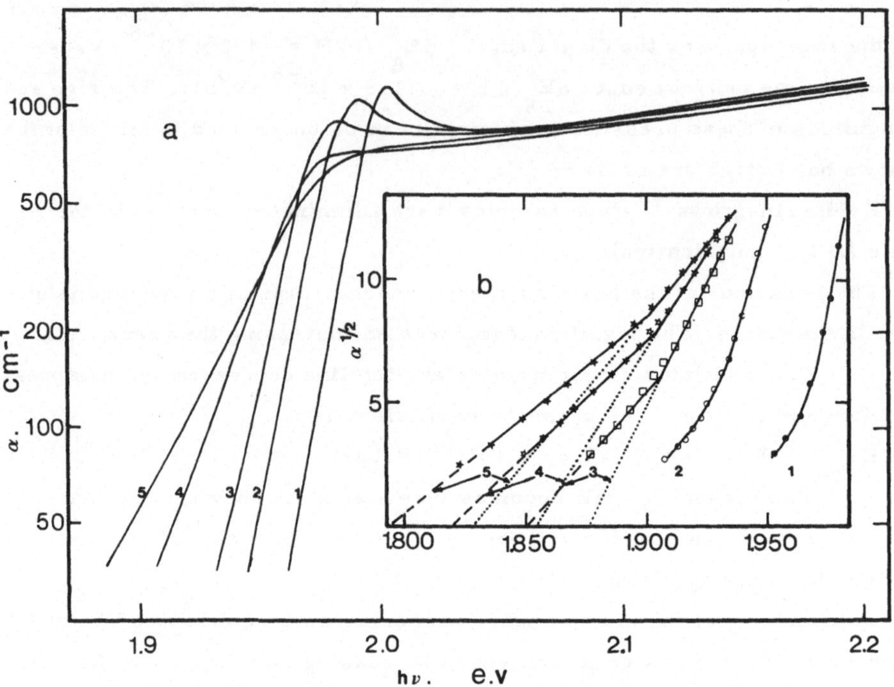

a : Absorption coefficient of Ga Se at 300 K vs. photon energy.
1 : 1 bar - 2 : 2300 bars - 3 : 4400 bars - 4 : 8000 bars -
5 : 12000 bars.

b : Square root of absorption coefficient vs. photon energy.
1 : 1 bar 2 : 5300 bars 3 : 9400 bars 4 : 12200 bars
5 : 14400 bars. At low pressures (curves 1 and 2), the indirect
edge is not separated from the tail of the exciton peak. At higher
pressures (curves 3, 4 and 5), $\alpha = A (h\upsilon - E_g^i \pm h\upsilon_{ph})^{-2}$.
Dashed lines : plus sign (absorption). Dotted lines : minus sign
(emission) $-h\upsilon_{ph} = 16$ milli eV.

since the difference between the two extrapolates, at high energies, to
a $(h\upsilon - E_g^d)^{3/2}$ law.

Finally, the behaviour of the low-energy tail at high pressures gives
evidence for an indirect edge : this portion of the curve was analysed in
terms of allowed transitions which, close to the edge, must dominate the
indirect forbidden contribution, by analogy with the direct edge.

In insert (b), the square root of the absorption coefficient clearly ex-

hibits two straight portions corresponding to absorption and emission of a phonon. The midpoint of the intercept of those lines with the abcissa axis gives an estimate of E_g^i at each pressure. Extrapolation to zero pressure gives $E_g^i = 1.970 \pm 0.010$ eV, and $\dfrac{dE_g^i}{dP} = -11.5 \pm 1 \times 10^{-6}$ eV/bar. This is completely consistent with the expected value for E_g^i (5) and $\dfrac{dE_g^i}{dP}$ (3). Furthermore, the phonon which appears to be responsible for indirect transitions is the 16 milli eV one (135 cm^{-1}), in consonance with Khelladi's results (5).

We conclude by stating that the study of absorption edge of Ga Se under hydrostatic pressure has established the existence of an indirect edge some 60 milli eV below the direct one at room pressure. This contention is strengthened by the appearance of interference effects which damp the exciton. Lastly, the pressure coefficient of the direct and indirect edge can be understood in terms of band-structure of Ga Se.

REFERENCES.

(1) See for instance: Bassani; F., Greenaway, D. L., and Fisher, G., in Physics of Semiconductors : Proceedings of the 7th International Conference, edited by M. Hulin (Academic Press, New York, 1964).

(2) Aulich, E. ; Brebner, J. L. ; Mooser, E. : Phys. Status Solidii, 31 - 129 - (1969).

(3) Besson, J. M.; Jain, K. P. and Kuhn, A. : Phys. Rev. Letters, 32, 936 (1974).

(4) Balzarotti, A. and Piacentini, M. : Solid State Commun., 10, 421, (1972).

(5) Khelladi, F. Z. : Ph. D. thesis, Université de Paris VI, unpublished (1961).

ORIGIN OF ANOMALIES IN THE EXCITON REFLECTION SPECTRA
OF CdS CRYSTALS

I. BROSER and R. BROSER

Fritz-Haber-Institut der Max-Planck-Gesellschaft, Berlin-Dahlem

and

Institut für Festkörperphysik der Technischen Universität Berlin

Germany

Reflection spectra of CdS-crystals have been studied
under different geometric conditions in the exciton
region. Several possibilities for the origin of
anomalies, due to thin exciton free surface layers,
to spatial dispersion and to unisotropy, are discussed.

1. INTRODUCTION

Reflection spectra in the exciton region of semiconducting
crystals frequently show anomalies such as narrow peaks (spikes)
which can in principal be understood by taking into account
simultaneously spatial dispersion effects [1] and exciton free
thin layers at the surface [2]. There exist, however, till now
several contradictions in literature as to the quantitative
correlation of theory with experimental observations for
different materials under different conditions [3-5] . Using CdS
as a model substance, a detailed study of exciton reflection
spectra, including variations of damping, surface layers and
angles of incidence has been performed. We report the main
results of this work and show that a unique interpretation for
the origin of the different types of reflection anomalies can
be given.

2. NORMAL INCIDENCE TO THE $(11\bar{2}0)$-FACE AND $E\perp c$

From as grown crystals two types - with and without a spike at
ω_L - can be selected. A striking effect is the observation that
reflection spectra of crystals with a spike are altered in wide
limits by the action of intense highly absorbed additional
light. Preexcitation enlarges, simultaneous excitation dimin-
ishes the magnitude of the spike. A treatment of such crystals
with chemical or mechanical means causes the total disappear-

ance of the reflection peaks (fig. 1). The reflection spectra
of such cleaned crystals can no longer be affected by addi-
tional excitation of light.

FIG. 1:
Reflection
spectra of a CdS
crystal for nor-
mal incidence
and $E \perp c$ at
4,2 K.
a) Crystal sur-
face as grown
b) crystal sur-
face cleaned

To correlate this behaviour with the existing theoretical and
experimental work in literature the postulation of two differ-
ent types of surface layers seems to be necessary: The first is
created even in ideal crystals according to the assumptions of
THOMAS and HOPFIELD [2], by total reflection of excitons at the
surface and has a constant thickness of less than 50 $\overset{o}{A}$. As can
be seen from fig. 2, such a layer causes a spike only for
crystals with very small damping factors ($\Gamma < \cdot 10^{-5}$ eV).

FIG. 2:
Calculated reflec-
tion spectra in-
cluding spatial dis-
persion, surface
layers and damping.
Normal incidence,
$\omega_0 = 2,5529$, $m^x = 0,9$,
$\mathcal{E}_o = 8,1$, $4\pi\alpha_o = 0,01$
(notations accord-
ing to [2])

The second type of a layer can be created by all kinds of effects by which excitons are annihilated, i.e. which give rise to a great damping factor in the surface region. One such effect will be the existence of a high surface field, which might be caused by applying a voltage across the crystal [6] or by impurity induced space charges. These space charges can be enlarged by filling impurity or surface states with electrons or holes (preexcitation by light), surface fields will be destroyed by the existence of free carriers (simultaneous excitation of light). As can be seen from fig. 2, thicker layers are less sensitive to the damping factor in the bulk and can therefore be observed also for less perfect crystals. There is no simple relationship between the width of the peak and the magnitude of the damping factor. Γ as well as ℓ of an exciton-free layer can be different for different excitonic states even for the same crystal.

3. OBLIQUE INCIDENCE TO THE $(11\bar{2}0)$-FACE AND $E_{\pi} \perp c$

For this geometry (E_π = electric field vector parallel to plane of incidence) spatial dispersion effects can result even for a layer free crystal in a reflection spike [7,8], which disappears, however, already for damping constants $\Gamma < 10^{-6}$ eV. For crystals with no spike at normal incidence (i.e. $\Gamma > 10^{-6}$ eV) one should not expect to observe such an anomaly (fig. 3a - crystal B 109). Thus, the question arises why PERMAGOROV et alt. [3] have found spikes under oblique incidence. The answer might be drawn from fig. 3a - crystal B 108. Here one sees that small layer-type spikes are strongly enhanced at great angles of incidence. To distinguish between these two types of reflection anomalies an independent determination of the damping constant seems to be desirable.

4. OBLIQUE INCIDENCE TO THE $(11\bar{2}0)$-FACE, E_{π} AND c IN SAME PLANE

It is well known [9], that for this geometry for zero angle of incidence there is no possibility to excite the $A_{n=1}$ exciton in a hexagonal CdS crystal. If, however, the angle is increased [10], a mixed longitudinal-transverse mode exists. While the absorption of such "longitudinal excitons" can be observed only for

FIG. 3: Reflection spectra of crystal B 109 (pure surface) and B 108 (with surface layer) for two angles of incidence and polarized ($E_{\overline{II}}$) light. T = 4,2 K.

a) c-axis $\perp$ plane of incidence ($E_{\overline{II}} \perp c$)

b) c-axis $\parallel$ plane of incidence ($E_{\overline{II}} \parallel \rightarrow \perp c$)

thin crystals and for small deviation from normal incidence the reflectivity gives an excellent spike at ω_L for crystals of all thicknesses and for greater angles of incidence (fig. 3b). We have found almost no influence of surface layers and their variations by the above mentioned treatments. The longitudinal spike at ω_L is, however, sensitive to damping, it broadens for example with increasing temperature. Calculations have shown, that the half-width of this spike is related to the damping constant in a simple way, so that measurements of this kind can be used to analyze the life time of excitons. This new anomaly in the reflection spectra might therefore be used much better than the layer type spike to characterize the perfectness of an unisotropic crystal.

5. ACKNOWLEDGEMENT

We wish to thank D. Maier and E. Birkicht für assistance during the measurements, M. Richard for performing the calculations and P. Tussing for his advice on some theoretical considerations.

6. REFERENCES

[1] Pekar, S.I.: Sov. Phys. JETP 6, 785 (1958); Sov.Phys.Solid State 4, 953 (1962)
[2] Hopfield, J.J.; Thomas, D.G.: Phys.Rev. 132, 563 (1963)
[3] Permogorov, S.A.; Travnikov, V.V.; Sel'kin, A.V.: Fiz. Tverd. Tela 14, 3642 (1972)
[4] Biellmann, J.; Grosmann, M.; Nikitine, S.: Proceedings of Taormina Conference on Polaritions, E. Burstein ed., 1973
[5] Tosatti, E.; Harbeke, G.: preprint
[6] Evangelisti, F.; Fischbach, J.U.; Frova, A.: Phys. Rev. B9, 1516 (1974)
[7] Maradudin, A.A.; Mills, D.L.: Phys. Rev. B7, 2787 (1973)
[8] Skettrup, T.: Phys. stat. sol. (b) 60, 695 (1973)
[9] Lempicki, A.: Proc. Phys. Soc. 74, 138 (1959)
[10] Hopfield, J.J.; Thomas, D.G.: Phys. Chem. Sol. 12, 276 (1960)

ANOMALOUS WAVES AND FABRY-PEROT MODES OF PHOTOEXCITONS (POLARITONS) IN THIN SEMICONDUCTING CRYSTALS

V.A. KISELEV, B.S. RAZBIRIN, I.N. URAL'TSEV
A.F. Ioffe Physico-Technical Institute,
Leningrad, USSR

In crystals with spatial dispersion additional
waves can propagate through the medium besides
the classical wave of light. In the case of
sufficiently thin platelets that were studied
by us this manifests itself in a fine structure
of interference optical spectra. This structure
is due to a compound interference of the ad-
ditional and classical waves. Analysis of the
obtained spectra gives much information about
the properties of photoexcitons in crystals.

The ability of excitons in semiconductors to move through crys-
tal matter gives rise to one or more additional electromagnetic
waves propagating through the medium besides the ordinary clas-
sical wave of the refracted light [1-3]. These waves are called
anomalous or additional waves and may be characterized by the
same frequency and polarization but different wave-vectors.
Fig. 1(a) shows dispersion branches of photoexcitons (polaritons)
1 and 2 in the region of the exciton $A_n = 1$ in CdSe crystals. The
short-wave part of branch 1 coming from the dispersion of the
mechanical exciton with positive mass is the dispersion of the
anomalous wave, propagating through the crystal together with
the classical wave 2.
This report deals with the investigation of the anomalous photo-
exciton waves in CdS and CdSe semiconducting crystals.
Reflection, transmission and emission of more than 20 thin
single-crystals of high quality grown from vapour were photo-
graphed at 4.2 and 1.8 K using a grating spectrograph with a
linear dispersion of 1.9 Å/mm. The thickness of the platelets
ranged from 1.7 to 0.08 µm.
The obtained spectra show fine interference structure in the
excitonic region. See, for example, the reflectivity spectrum
of a CdSe crystal ~0.36 µm thick shown in Fig. 1(b). The
peculiarities of all spectra can be well understood in terms of
three types of interference of photoexciton waves 1 and 2 taking

Fig. 1(a): experimental curves for dispersion of photo-
excitons $A_n = 1$ in CdSe crystals - solid lines
1, 2; theoretical dispersion - broken lines;
(b): microphotometre trace of reflectivity
of CdSe at T = 4.2 K, E⊥c, 1 = 0.356 µm;
(c): computed spectrum (see text).

place in thin crystals. The structure at the longitudinal
exciton A_L (see Fig. 1) is due to the classical interference [4]
or Fabry-Perot modes of wave 1. In thin crystals it is localized
in the exciton resonance region and is spaced widely enough to
show a change in the refractive index of wave 1. In some spectra
we can see two or three lines of the interference of this kind
even above the A_L frequency (minima turning into maxima). This
indicates the identity of the light interference and the size
effect in the transverse motion of excitons through crystal when
the photoexciton interaction is taken into account.

Farther to the short-wave side the interference structure doubles
its period. This kind of structure is due to a compound inter-
ference of waves 1 and 2, that is, when reflected from the
boundary of a crystal wave 1 can give rise to wave 2 and vice
versa, and this changes the optical length in such a way that
the period is approximately doubled. In some CdSe crystals this
structure is seen in a wide frequency region of ~ 200 cm^{-1} up to
the exciton $B_{n=1}$.

The compound interference stands out against the background of
the classical Fabry-Perot interference of wave 2, the most dis-
tinct and wide minimum of which is seen in Fig. 1(b) in the
vicinity of the frequency 14754 cm^{-1}.

Pekar's formulas [1] were used to compute the reflectivity of
the thin crystal at normal incidence. The result is shown in
Fig. 1(c). The value 0.58 m_o [5] for the effective mass of the
exciton A was taken as the basic parameter. All other parameters,
including the thickness of the crystal $1 = 0.356$ μm, were chosen
to fit the experimental spectrum.

We obtained $\varepsilon_o = 10.8 \pm 0.1$ for the background dielectric constant
and $\hbar\omega_{LT} = 7.5 \pm 0.1$ cm^{-1} for the LT-splitting of the exciton.
The thickness was controlled using an interference microscope.
Damping of the mechanical exciton was taken constant, $\hbar\Gamma = 1$ cm^{-1}.
This treatment improves the results of our previous paper [6].
The computation of spectra allows us to find the exact corres-
pondence of peculiarities in the spectra and the frequencies de-
fined by the well-known interference condition

$$2nl = \lambda N, \quad (N = 1, 2, 3, \ldots)$$

where n is the refractive index and N is the order of inter-

ference. Knowing it, one can draw the dispersion curves using
the experimental spectra in Fig. 1 (a,b).
Solid lines 1 and 2 in Fig. 1(a) shows the branches of photo-
excitons constructed in this way. A slight nonparabolicity of
the e-citon band is seen in comparison with the theoretical
curve (broken line), the computation of the spectrum (c) being
made for the parabolic band.
Analogous spectra were taken for CdS crystals. All spectra in
reflection, transmission and emission correlate in accordance
with the above theoretical considerations.

We thank Dr. Permogorov, Dr. Sel'kin and Dr. Solov'ev for
valuable discussions and Dr. Flögel for supplying some of the
crystals investigated by us.

REFERENCES

[1] Pekar, S.I.: Zh. Teor. and Eksp. Fiz. $\underline{34}$, 1176 (1958).
[2] Agranovitch, V.M.; Ginzburg, V.L.: Kristallooptica s
 uchetom prostranstv. disp. i teor. eksitonov. Nauka,
 Moscow, 1965.
[3] Hopfield, J.J.; Thomas, D.G.: Phys. Rev. $\underline{132}$, 563 (1963).
[4] Fuchs, R.; Kliewer, K.L.; Pardee, W.J.: Phys. Rev. $\underline{150}$,
 589 (1966).
[5] Wheeler, R.G.; Dimmock, I.O.: Phys. Rev. $\underline{125}$, 1805 (1962).
[6] Kiselev, V.A.; Razbirin, B.S.; Ural'tsev, I.N.: Pis'ma v
 Zh. Teor. and Eksp. Fiz. $\underline{18}$, 504 (1973).

DISORDERED SEMICONDUCTORS

CONTENTS

STRUCTURES OF AMORPHOUS TETRAHEDRALLY-BONDED SEMICONDUCTORS[+]

G.A.N. CONNELL

Harvard University, Cambridge, Massachusetts, USA

The structural properties of amorphous tetra-
hedrally-bonded semiconductors are reviewed.
Topologically different random networks are
required to explain the behaviour of the ele-
mental and binary materials.

1. INTRODUCTION

The rapid development of interest in amorphous semiconductors
is manifest in the proceedings of these conferences. The first
contributed paper (on amorphous Ge) appeared at Kyoto in 1966
[1], but six years later at the Warsaw meeting, two whole
sessions and a plenary paper were devoted to them [2]. The time
scale for this review is therefore comparatively short.
Nevertheless, for the review to be complete, in the sense of
covering, in detail, some part of the structural, vibrational
and electronic properties, it is necessary to restrict it to a
small number of materials and to a selection of the most recent
developments. Thus I shall limit the discussion to the struc-
tural properties of the tetrahedrally co-ordinated semiconductors.
These materials have the advantage of relative structural sim-
plicity and yet exhibit the effects of topological and quanti-
tative disorder [3]. These are the structural properties that
distinguish the amorphous forms from their corresponding
crystalline phases and are the necessary underpinnings for the
development of reasonable theoretical models for the electronic
and vibrational states [4]. Therefore, in section 2, I discuss
the structures of these materials, concentrating on those
features which turn out to be independent of preparation. The

[+]Work supported by the National Science Foundation, the Joint
Services Electronics Program and the Division of Engineering
and Applied Physics.

smaller variations of the structure of a given material with
preparation conditions are ignored, although the origin of these
small structural changes, the defects within the network
('dangling bonds'), seriously affects the number and distrib-
ution of states across the pseudogap [5] and hinders the invest-
igation of the existence and extent of the band tails, the
existence and sharpness of any mobility gap between localized
and extended states [6] and the value of the energy gap itself
in the ideal fully co-ordinated material. Finally, in section 3,
I briefly summarize some of the outstanding experimental and
theoretical problems that need solution before a full understand-
ing of these materials can be obtained.

2. STRUCTURES

The experimental determination of the structure of an amorphous
solid is difficult since the scattering techniques employed to
determine crystalline structures lead only to a radial density
function, $J(r) = 4\pi r^2 \rho(r)$, where $\rho(r)$ is the average atomic
density at a distance r from any given atom. $J(r)$, therefore,
does not uniquely define a structure. This problem is demon-
strated by $J(r)$ of amorphous Ge [7], shown in Fig. 1(a). There
are 4 nearest neighbours centered at r_1 = 2.47 ± 0.01Å, the
spread in r arising almost entirely from the thermal vibrations
of the netowrk. The 12 second nearest neighbours are spread
around r_2 = 4.00 ± 0.04A ($\approx \sqrt{8/3} r_1$). The atoms are therefore
tetrahedrally co-ordinated, but with an average bond angle devi-
ation, determined from the thermally corrected width of the
second peak, of about $10°$. The structure beyond this tetrahedral
sub-unit is determined by the interconnection of neighbouring
tetrahedra. For example, two interpenetrating tetrahedra may be
rotated with respect to one another, without bond angle distortion,
about an axis connecting the centres of the tetrahedra, the angle
of rotation being called the dihedral angle, θ_d. θ_d is set to 0
for the eclipsed configurations of the tetrahedra and to 60 for
the staggered configurations. The distribution of dihedral angles,
$P(\theta_d)$, will thus be instrumental in determining the distribution
of three-bond neighbour distances, where it is usual to define a
pair of atoms as n-bond neighbours when they are separated by

n-bonds and no fewer. The number of such three-bond neighbours
will be determined by the closure of the chains of atoms into
rings. For example, 36 three-bond neighbours occur in the ring-
free Bethe lattice, but in diamond the formation of 6-membered
rings and the subsequent overlap of atomic positions limits the
three-bond neighbours to 24. The formation of 5-membered rings
can produce a further reduction, since within the ring only two-
bond neighbours exist. Thus $J(r)$ of amorphous Ge beyond the
second neighbour peak depends on both $P(\theta_d)$ and the ring stati-
stics and unfortunately these cannot be determined uniquely from
experiment. From a theory relating to $P(\theta_d)$ to $J(r)$, Temkin has
concluded that θ_d greater than 30 is more probable than θ_d less
than 30, but the large uncertainties in $P(\theta_d)$ are sufficient to
rule out any conclusions on the ring statistics [8].

The $J(r)$'s of Si [9], $Ge_x Si_{1-x}$ [10] and $Ge_x Sn_{1-x}$ [11] are very
similar, although the alloys exhibit additional effects associ-
ated with the random distribution of the atoms on the network
sites. Somewhat surprisingly, the $J(r)$'s of the 3-5's [12] (and
possibly the 2-6's) are not dissimilar beyond the tetrahedral
sub-unit. An example, amorphous GaAs [13] is shown in Fig. 1(a).
The major difference is that the peak near 6Å is shifted to
higher r in GaAs relative to Ge, a result which is also apparent
in the work of Shevchik and Paul on other 3-5's [12]. This can-
not be explained as resulting from the difference between nearest
neighbour distances, but
rather must be explained
from (possibly large) changes
in the topology of the net-
work. This view is an impor-
tant refinement of the origi-
nal view that the close simi-
larity of the $J(r)$'s indicated
a close similarity in the
structures [12]. Indeed, the
bond length and bond angle
distributions are similar,
but the proposed changes in
the topology have important

Fig. 1(a). $J(r)$'s of Ge [7] and
GaAs [13] and the 201-atom and 238-
atom CRN's of [17] and [18] respec-
tively. (b) The insert shows $\Delta G(r)$
for $\Delta J(r) = J_{Ge}(r) - J_{GaAs}(r)$: ——— and
$\Delta J(r) = J_{201}(r) - J_{238}(r)$: - - - -.

consequences for the interpretation of the vibrational and electronic properties.

Further insights into the experimental $J(r)$'s have been obtained from the analysis of tetrahedrally-coordinated models. Polk [14] and Shevchik and Paul [15] showed that a continuous random network (CRN) could be constructed that had bond length and bond angle distributions in agreement with experiment. In later work, Polk et al (PB) [16] and Steinhardt et al [17] demonstrated that a model, containing atoms relaxed to their positions of lowest elastic energy, had a $J(r)$ that agreed well with that of Ge for $r \leq 10\text{Å}$. In Fig. 1(a), the $J(r)$'s of the 201-atom model of Steinhardt et al and amorphous Ge are compared. The model $J(r)$ has been artificially broadened beyond $r = 3\text{Å}$ to ensure exact agreement for the second neighbour distribution. The feature at 6.05Å can now clearly be established as the three-bond neighbour distance arising from chains of atoms that occur when tetrahedra are arranged in the staggered configuration. The surprising result is that the distribution of these distances is the same as the distribution of second neighbour distances [7], which contradicts the intuitive idea that broadening increases rapidly with r in amorphous solids.

Why is it now suspected that the PB model is not appropriate for the 3-5's and 2-6's, since, intuitively, it would seem that small modifications to it might produce the required changes in its $J(r)$? The answer is to be found in its ring statistics. In diamond and wurtzite there are 2 six-membered rings per atom. In contrast, the PB model contains 0.38 five-membered, 0.91 six-membered and 1.04 seven-membered rings per atom [17]. Thus, whereas the diamond lattice can be divided into two interpenetrating sublattices of A and B atoms, the odd-membered rings in the PB model necessitate the existence of like or wrong (A-A and B-B) bonds. A wrong bond fraction of about 10% has been estimated [18]. Experimental evidence for the existence of wrong bonds has been given [15,19] but no estimate of their number. Recently, Shevchik et al pointed out that the core electrons should be sensitive to the chemical ordering, since their energies depend upon the Coulombic potential induced by neighbouring atoms

[20], i.e. the chemical shift at the j^{th} atom depends upon the local Madelung constant α_j. Thus in zinc blende crystals, in which the two different atoms have a single type of site, there should be only one peak in the core level spectra associated with each atomic species corresponding to $\alpha_j = \alpha_o = 1.64$. However, for the PB model of an amorphous material, there is no reason to expect a single peak to occur in the core level spectra of every atomic species [20], since α_j fluctuates from site to site by approximately $\pm\alpha_o$ about α_o [21] and a significant smearing of the core level spectra should result.

In Fig. 2, the electron energy distribution curves (EDC's) for the In 4d level in In metal and amorphous and crystalline InAs are compared [20]. In the pure metal the 4d level is split by spin orbit interaction into its $J = 5/2$ and $3/2$ components. This is maintained in the semiconducting alloys. The important point, however, is that no difference between the amorphous and crystalline phases can be detected. It was therefore suggested that these data and similar data on other 3-5's and 2-6's were generally inconsistent with the wrong bond fraction found in the PB model. Some materials which exhibited core level features and Raman spectra [22] of the cation were determined to be phase separated.

Fig.2. In 4d levels in amorphous and crystalline InAs + metallic In [20].

As a result of earlier studies on optical [19] and crystallization [23] phenomena of 3-5's and independent of the core level data, Connell and Temkin (CT) constructed an even-membered ring CRN of 238 atoms [18]. Its J(r), artificially broadened beyond $r = 3\text{Å}$ again to ensure exact agreement for the second neighbour distribution, is shown in Fig. 1(a). Small differences between the J(r)'s of the two models occur at the position of the peak near 6Å and in the sharpness of the feature near 8Å, although the latter is subject to considerable uncertainty, owing to the finite size of the constructions. In contrast, the difference in the topology of the models is enormous. This is evident from their $P(\theta_d)$'s [18] and their ring statistics. The 238-atom model has 2.3 six-

membered rings per atom, somewhat greater than in the diamond
and wurtzite structures. More important, it can be divided into
two sublattices and consequently has no wrong bonds when the
atoms of a binary compound are so assigned. The local Madelung
constant now fluctuates only by $\pm 0.05\ \alpha_o$ about α_o from site to
site [21]. Thus, a CRN model with even-membered rings does not
require a significant broadening of the core level spectra and
as such is in agreement with experiment.

The similar behaviour of the models and Ge and GaAs is demon-
strated more clearly in Fig. 1(b). Here $\Delta G(r) = \Delta J(r)/r$ (i.e. the
difference between the densities of the models is zero) is plotted
for $\Delta J(r) = J_{Ge}(r) - J_{GaAs}(r)$ and $\Delta J(r) = J_{201}(r) - J_{238}(r)$ and
$5 \leq r \leq 8\text{Å}$. More quantitative agreement can be obtained by in-
corporating fewer odd-membered rings in the model for amorphous
Ge [18]. These results are discussed in detail by Temkin [13],
but the conclusion here is that the differences in $J(r)$ between
Ge and GaAs are consistent with large changes in the topology or
connectivity of the network, and these differences must be con-
sidered in the analysis of the vibrational and electronic prop-
erties.

Howie et al [24] have proposed a variation of the microcrystal-
line model for the structure of amorphous Ge and Si, based on
their off-set bright field and dark field electron micrographs.
In the first case, the transmitted beam and a segment of the
first diffracted beam were allowed through the objective aper-
ture to produce fringes that were interpreted to correspond to
the 3.3Å layer spacing in crystalline Ge and to arise from phase
coherence over small domains approximately 5-14Å across. In the
second case, segments of the first and second diffracted rings
were allowed through the objective aperture to produce bright
spots that were interpreted as direct images of the coherent
domains. This view has been contested by several workers [25].
They showed that the dark field imagery could be produced by
CRN's, but the off-set bright field effects are still the subject
of much discussion. Although a CRN can produce fringes, Cochran
[26] demonstrated that the visibility of these fringes is in
general much too low to be observable. On the other hand,
Graczyk and Chauhari noted that the PB model is anisotropic over

limited volumes and has planes of high atomic density, which occur with a periodicity equal to the 3.3Å spacing of successive planes in the diamond cubic structure [27]. Perpendicular to these directions, the fringe visibility is considerably enhanced over the calculated value, i.e. over the average value. Furthermore, as Cochran has suggested [25], the observed fringe contrast is probably enhanced by more selective electron-optical filtering of Fourier components than has been envisaged in the calculations to date. It seems likely therefore that these electron micrographs alone do not distinguish between a CRN and microcrystalline model; the choice must be made from an over-all comparison of available model and experimental data.

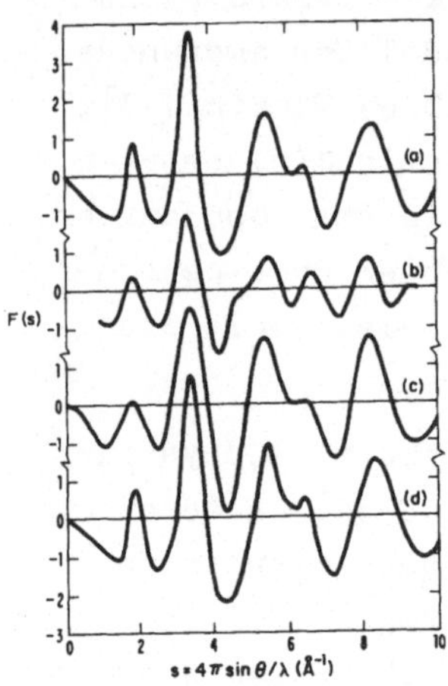

Fig. 3. F(s) for Ge:
(a) Experiment [7];
(b) the modified microcrystallite model of [24];
(c) the RTM of [7];
(d) the CRN of [17].

The diffraction interference function F(s), calculated by Howie et al [24] for their microcrystalline model which consists of wurtzite units, 14Å in diameter, inter-connected by a tissue of random network, is shown in Fig. 3(b) for amorphous Ge. The J(r) of the interconnecting tissue was represented by the first and second nearest neighbour peaks of the PB model but had no structure thereafter. For comparison, F(s) of amorphous Ge [7] is shown in Fig. 3(a). In Fig. 3(c), F(s) has been calculated from J(r) of the interconnecting tissue alone. This last approximation to the amorphous structure is called the random tetrahedral model (RTM) [7] and has been used to demonstrate what part of F(s) arises solely from the ordering of the first and second nearest neighbours. It is immediately apparent that there is more similarity in the experimental and RTM F(s)'s than in the experimental and microcrystalline F(s)'s. In fact, the former agree quantitatively at large s. Therefore, the criterion for the success of a sophisticated model, i.e. one including information about ring statistics, should be that its F(s) is in quantitative agreement with experiment at all s, since the semi-quantitative agreement stems solely from the tetrahedral units. In the case

of the microcrystalline model discussed by Howie et al, the
introduction of wurtzite-correlations beyond the tetrahedral
unit worsens the agreement with experiment! On the other hand,
Graczyk and Chaudhari [28] showed that F(s) of the 61 atom model
of Henderson and Herman [29] was in good agreement with experiment.
Similarly, Fig. 3(d) shows the excellent agreement between ex-
periment and the 201-atom model discussed earlier. Therefore, I
conclude that CRN's are the appropriate description of the
structures, although large variations in the connectivity of the
network are expected as material parameters are changed.

3. CONCLUSION

I hope that I have demonstrated that considerable progress has
been made and suggested some areas where new effort would be
valuable. The importance of structure in determining the
electronic and vibrational properties and in limiting the range
of theoretical models has been clearly established [3,4] but the
difficulty of its full determination and statistical description
has not yet been completely overcome. Further work is needed to
explain convincingly the origin of the bright-field electron
micrographs, which in turn may lead to a better understanding of
the structural correlations that exist beyond the tetrahedral
subunit.

4. ACKNOWLEDGEMENTS

I have greatly benefitted from the collaboration of Bill Paul,
Rick Temkin and, more recently, Adam Lewis. I also thank all
those people who so kindly sent preprints of their work and
regret that so much of it had to be omitted.

5. REFERENCES

[1] Tauc, J.; Grigorovici, R.; Vancu, A.: J. Phys. Soc. Jap 21
 (Suppl.), 123 (1966)
[2] Paul, W.: Proc. 11th Int. Conf. on Phys. Semicond.
 (Warsaw, 1972), p. 38
[3] Weaire, D.; Thorpe, M.F.: Phys. Rev. B4, 2508 (1971)
 Thorpe, M.F.; Weaire, D.: Phys. Rev. B4, 3518 (1971)
[4] Ziman, J.M.: Proc. Roy. Soc. Lond. A318, 401 (1970)
[5] Paul, W.; Connell, G.A.N.; Temkin, R.J.: Adv. Phys. 22,
 529 (1973)

[6] Mott, N.F.; Davis, E.A.: Electronic Processes in Non-
 Crystalline Materials (Oxford, 1971)
 Weiser, K.: Comm. Sol. State Phys. (to be published)
[7] Temkin, R.J.; Paul, W.; Connell, G.A.N.: Adv. Phys. $\underline{22}$,
 581 (1973)
[8] Temkin, R.J.: Proc. Int. Conf. on Tetrahedrally Bonded
 Semicond. (Yorktown Heights, 1974), to be published
[9] Moss, S.C.; Graczyk, J.F.: Proc. 10th Int. Conf. on Phys.
 Semicond. (Boston, 1970), p. 658
[10] Shevchik, N.J.; Lannin, J.S.; Tejeda, J.: Phys. Rev. $\underline{B7}$,
 3987 (1973)
[11] Temkin, R.J.; Connell, G.A.N.; Paul, W.: Sol. State
 Commun. $\underline{11}$, 1591 (1972)
[12] Shevchik, N.J.; Paul, W.:J.Non-Cryst. Sol. $\underline{13}$, 1
 (1973/74)
[13] Temkin, R.J.: to be published
[14] Polk, D.E.: J. Non-Cryst. Sol. $\underline{5}$, 365 (1971)
[15] Shevchik, N.J.; Paul, W.: J. Non-Cryst. Sol. $\underline{8-10}$, 381
 (1972)
[16] Polk, D.E.; Boudreaux, D.S.: Phys. Rev. Letters $\underline{31}$, 92
 (1973)
 Duffy, M.G.; Boudreaux, D.S.; Polk, D.E.: to be published
[17] Steinhardt, P.; Alben, R.; Weaire, D.: to be published
[18] Connell, G.A.N.; Temkin, R.J.: Phys. Rev. B (June 15)
 Connell, G.A.N.; Temkin, R.J.: Proc. Int. Conf. on Tetra-
 hedrally Bonded Semicond. (Yorktown Heights) to be
 published
[19a] Connell, G.A.N.: Phys. Stat. Sol. 53(b), 213 (1972)
[19b] Connell, G.A.N.; Paul, W.: J. Non-Cryst. Sol. $\underline{8-10}$, 215
 (1972)
[20] Shevchik, N.J.; Tejeda, J.; Cardona, M.: Phys. Rev. $\underline{B9}$,
 2627 (1974)
[21] Connell, G.A.N.: unpublished
[22] Lannin, J.S.: Proc. Int. Conf. on Tetrahedrally Bonded
 Semicond. (Yorktown Heights, 1974) to be published
[23] Temkin, R.J.; Paul, W.: Proc. 5th Int. Conf. on Amor. and
 Liq. Semicond. (Garmisch, 1973), to be published
[24] Howie, A.; Krivanek, O.L.; Rudee, M.L.: Phil. Mag. $\underline{27}$,
 235 (1973)
[25] See W. Cochran's review: Proc. Int. Conf. on Tetrahedrally
 Bonded Semicond. (Yorktown Heights 1974) to be published
[26] Cochran, W.: Phys. Rev. $\underline{B8}$, 623 (1973)
[27] Graczyk, J.F.; Chaudhari, P.: unpublished
[28] Graczyk, J.F.; Chaudhari, P.: Phys. Stat. Sol. $\underline{58(b)}$,
 501 (1973)
[29] See Reference 28 for a description

GAPS AND ELECTRONIC STATES IN TETRAHEDRALLY BOUNDED AMORPHOUS GeS

L. ŠTOURAČ, M. ZÁVĚTOVÁ, A. ABRAHÁM

Institute of Solid State Physics, Czechoslovak Academy
of Sciences, Prague, Czechoslovakia

Experimental results of the investigation
of the reflectivity, optical absorption,
photoconductivity and electrical conduc-
tivity on both crystalline and amorphous
GeS are reported and critically compared.

1. INTRODUCTION

Chalcogenides of germanium form an interesting group of semi-
conductors, where the short range order is different in crys-
talline and amorphous materials. GeS, GeSe and GeTe crystal-
lize in a distorted NaCl structure; it is difficult to pre-
pare stoichiometric samples in a glassy form by quenching of
the melt. In amorphous modification they were obtained in the
form of layers. Both amorphous and crystalline chalcogenides
of germanium are studied only recently; least information
exists on GeS (see for example [1-8]).

2. EXPERIMENTAL

2.1 SAMPLE PREPARATION

Amorphous GeS was prepared either in a form of layers up to
0.9 mm thick by sublimation on warm substrates [8] , or by
flash evaporation on glass, quarz, Mylar or ceramics. Stoi-
chiometry is less certain for the latter method. Thin crys-
talline platelets of GeS were prepared by sublimation on a
cold substrate in H_2-H_2S atmosphere. GeS is orthorhombic
(lattice parameters a=4.29 Å, b=10.42 Å, c=3.64 Å). It easily
cleavs along the layers perpendicular to the b-axis, but
not in any other direction. Our samples were imperfect, which
made the transport measurements difficult.

2.2 RESULTS

Reflectivity was measured on natural cleavage surfaces of
c-GeS, and on polished amorphous samples, in the 0.5-4.8 eV

FIG. 1. Reflectivity of a-and
c-GeS (the latter
measured in polarized
light)

FIG. 2. Absorption edge of
a-and c-GeS (γ [2];
a-GeS ••• bulk,
▲▲▲ thin layers;
ooo c-GeS

region (FIG. 1). The latter reflection curves were normalized
to the values of R, calculated from index of refraction,
measured on prisms from a-GeS in the 0.6-1.4 eV region. The
absorption coefficient (FIG. 2) of a-GeS was measured on
bulk samples prepared by sublimation for $\alpha < 10^3$ cm^{-1}, on eva-
porated thin layers above this value. The R-T method was
employed, multiple reflections and the substrate effect (for
thin films) were taken into account.
Spectral dependence of photoconductivity (FIG. 3) was mea-
sured at room temperature on bulk and thin layers of a-GeS
and on c-GeS.
Dark d.c. electrical conductivity was studied on c-GeS in
the 150-330 K temperature range along the a crystalographi-
cal axis and on a-GeS in the 250-330 K region, the latter
measurements being limited by the extremely high resistance
of the samples at lower temperatures. At room temperature,

the conductivity of a-GeS is 6 times lower than that of the
crystal (FIG. 4).

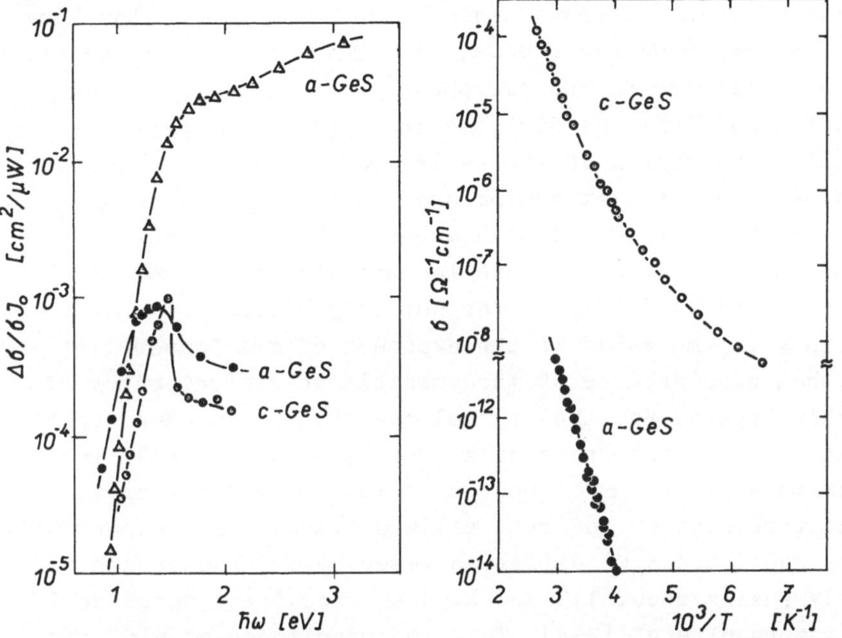

FIG. 3. Spectral dependence
of photoconductivity
of a- and c-GeS (nota-
tion see FIG. 2)

FIG. 4. Temperature dependence
of conductivity for
a-and c-GeS (notation
see FIG. 2)

3. DISCUSSION

As follows from the analysis of the radial distribution curves
[1], the short range order of a- and c-GeS differs. Conside-
rable differences in the physical properties of both modifi-
cations should be expected. On the other hand, position of
the $GeK\beta_2$ band in the X-ray emission spectra was found [7]
to differ but little for c- and a-GeS. This indicates that
the effective charge on the Ge atoms is the same during the
c-a transition.
As in other amorphous materials, the fine structure of the
reflectivity of crystal disappears, and only a broad peak
remains. The shift of the reflection maximum of a-GeS towards
lower energies is in agreement with previous VUV results [9] ;
the possible effect of the high frequency light scattering

was not estimated. The abrupt drop in the reflectivity of
c-GeS takes place in the energy range of the optical gap E_g^{opt}
determined from the transmission measurements. Similar beha-
viour was reported for c-GeSe [10] . This pronounced structure
detail disappears in the amorphous phase. The absorption
edge of a-GeS (FIG. 2) displays the typical three-part struc-
ture [11], the region of the tails ($\alpha < 5$ cm^{-1}) is followed by
an exponential part extending over about 3 orders of magni-
tude, and by an interval with a quadratic dependence of α.
Extrapolation of the last one to zero yields $E_g^{opt} \approx 1.5$ eV.
This value is subject to error not only because of the un-
certainty in the value of the exponent of the "quadratic"
part, but also because of the possible nonstoichiometry of
the thin layers. For the crystal the absorption edge can be,
due to unsufficient range of energies, almost equally well
approximated by the $\alpha^{1/3}$ and $\alpha^{1/2}$ law, corresponding to in-
direct forbidden or indirect allowed transitions respectively.
The extrapolated E_g^{opt} differ, however by few hunderedth of
eV only (being about 1.5 and 1.56 eV resp.) and agree well
with the results of [2-4]. Thus the magnitudes of E_g^{opt} for
both a- and c-GeS as well as the temperature dependence of
their absorption edges are rather close to each other.
Although the exponential part of the spectral dependence of
the photoconductivity of a-GeS follows that one of the ab-
sorption coefficient, the gap E_g^{PC}, evaluated by Moss rule
on bulk samples and by extrapolation of the quadratic part
for the layers, equals 1.1 eV. With c-GeS there exists a
dependence of E_g^{PC} on the crystal thickness [12], giving 1.35
and 1.5 eV for d=0.6 and 0.01 cm respectively.
The electrical conductivity of c-GeS possesses in the studied
temperature range a dependence typical for impurity conduc-
tivity in agreement with the measurements on polycrystalline
samples [4]. In the intrinsic conductivity range these authors
give the value 1.58 eV for ΔE. For amorphous GeS we found
an usual exponential temperature dependence of conductivity,
that can be expressed as $\sigma = \sigma_o \exp(-\Delta E/2kT)$. The values of
$\Delta E=1.1$ eV applies equally well for bulk material and the

layers. The evaporated layers display a larger spread of ΔE,
however, which is probably due to unsufficient stoichiometry
control during deposition.
The results of the transport and the optical measurements
agree well for c-GeS, but for the a-GeS the activation energy
of electrical conductivity ΔE, and photoconductive E_g^{PC} are
considerably lower than the optical gap E_g^{opt}. Such behaviour,
and the value of the gap difference about 0.4 eV is consis-
tent with the theoretical speculation about the compensated-
semiconductor model of glass [13], but the uncertainty with
impurity and nonstoichiometry effects should be eliminated.

We are grateful to Drs. A. Hrubý and T. Šimeček for the
preparation of the samples, to Mrs. M. Šilhavá for the re-
flectivity measurements and to Dr. B. Velický for his
valuable comments.

4. REFERENCES

[1] Červinka, L.; Hrubý, A.: Proc. 5th Int. Conf. on Amorph.
 and Liquid Semicond., Garmish-Partenkirchen 1973, in press
[2] Yabumoto, T.: Phys. Soc. Jap. 13, 599 (1958)
[3] Lidder, K. F.; Solovyev, L. E.: Solid State Phys. 4,
 1500 (1962) (in Russian)
[4] D'Amboise, M.; Handfield, G.; Bourgon, M.: Canad. J.
 Chem. 46, 3545 (1968)
[5] Kawamoto, Y.; Tsuchihashi, S.: J. Amer. Ceramic Soc.
 52, 626 (1969)
[6] Nielsen, S.: Infrared Phys. 2, 117 (1962)
[7] Drahokoupil, J.: Materials of the Conf. on Amorph.,
 Liquid and Glassy Semicond., Sofia, p. 123 (1972)
 (in Russian)
[8] Trousil, Z.; Pajasová, L.; Závětová, M.: Czech. J.
 Phys. B 21, 220 (1971)
[9] Pajasová, L.: Proc. 3rd Int. Conf. on VUV Rad. Phys.,
 p. 1pB2-8, Tokyo (1971)
[10] Lukeš, F.: Czech. J. Phys. B 18, 788 (1968)
[11] Tauc, J.: Mat. Res. Bull. 5, 721 (1970)
[12] Abrahám, A.; Hrubý, A.; Štourač, L.; Závětová, M.:
 Czech. J. Phys. B 22, 1168 (1972)
[13] Shklovskiy, B. I.; Efros, A. L.: Zh. Exp. Teor. Phys.
 62, 1156 (1972)

X-RAY AND FAR-UV PHOTOEMISSION STUDIES OF AMORPHOUS As_2S_3, As_2Se_3, AND As_2Te_3

S.G. BISHOP[+]

Naval Research Laboratory, Washington, D.C., USA

N.J. SHEVCHIK

Max-Planck-Institut für Festkörperforschung,
Stuttgart, Fed. Rep. Germany

Photoemission spectra of amorphous As_2S_3, As_2Se_3, and As_2Te_3 reveal that the valence bands in these materials are more than 15 eV wide and exhibit a pronounced minimum at about 7 eV. The spectra indicate that the valence bands below this minimum are predominantly s-like and those above are mostly p-like: s-p hybridization is not extensive.

1. INTRODUCTION

In semiconductors such as the arsenic chalcogenides which contain group-VI elements in two-fold coordination, unshared electron pairs (lone-pairs) are believed to make an important contribution to the valence band density of states [1]. For example the double peak structure exhibited by the interband optical spectra of crystalline and amorphous As_2S_3 and As_2Se_3 has been interpreted in terms of distinct p-like bonding and non-bonding (lone-pair) valence band states [2]. However recent molecular orbital calculations carried out for these materials [3,4] have brought into question this interpretation and indicate a large degree of s-p hybridization in the valence states. The X-ray photoemission spectra (XPS) and uv photoemission spectra (UPS) of the arsenic chalcogenides presented here provide the valence band density of states from which the extent of s-p hybridization, and total band width can be estimated.

[+]On sabbatical leave at Max-Planck-Institut für Festkörper-
 forschung

2. EXPERIMENTAL

A Hewlett-Packard 5950 ESCA spectrometer was used to obtain monochromatized XPS data for bulk samples of crystalline and amorphous As_2S_3, bulk and evaporated amorphous As_2Se_3, and amorphous sputtered films of As_2Te_3. Sputtered films of As_2Te_3 were also studied in a Vacuum Generators ESCA III system which is equipped with a differentially pumped capillary discharge lamp. The UPS data have a resolution of 0.1 eV compared with 0.6 eV for the monochromatized XPS data. Electrostatic charging precluded the possibility of obtaining reliable UPS data for the insulating As_2S_3 and As_2Se_3.

3. RESULTS AND DISCUSSION

Monochromatized XPS for amorphous As_2S_3, As_2Se_3, and As_2Te_3 are presented in Figs. 1-3. All of these spectra have been corrected for the effects of inelastically scattered secondary electrons by a straight line approximation, and the zero of binding energy has been defined by extrapolating the steeply rising onset to the background level. In each case the XPS exhibits a strong leading peak at a binding energy of 1.0-1.5 eV, followed by a pronounced shoulder at about 4.5 eV. All these spectra have a well-defined minimum at 7 eV, below which lies a broad band which for As_2S_3 and As_2Se_3 exhibits two peaks and extends to 18 eV. Comparison of the XPS data of Figs. 1-3 with previously reported XPS spectra of the constituent elements As [5], Se [6], and Te [6] reveals a remarkable similarity in the form of these spectra. This similarity is graphically illustrated by the composite XPS spectra for As_2Se_3 (Fig. 2) and As_2Te_3 (Fig. 3) which were constructed by simply adding the properly weighed XPS spectra for the amorphous forms of the constituent

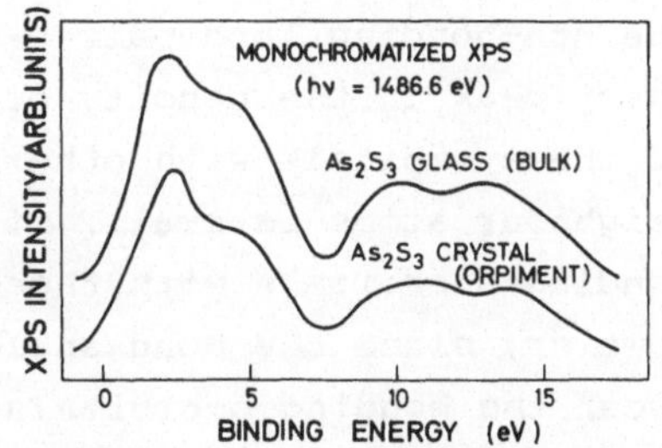

FIG. 1: Valence band XPS spectra for bulk glassy and cleaved crystalline As_2S_3.

FIGS. 2 and 3: Valence band XPS spectra for amorphous As_2Se_3
(left) and As_2Te_3 (right) compared with spectra
fabricated by adding properly weighted XPS
spectra for the amorphous elements.

elements. These fabricated spectra have been positioned in
energy so as to provide the best possible alignment of their
leading peaks and prominent minima with the experimental XPS
spectra. The electronic structures of the As chalcogenides are
related to those of the elemental chalcogens. In S, Se, and Te
the non-bonding lone-pair p-electron orbitals give rise to a
sharp peak in the density of valence states because the overlap
of these orbitals with others on the same atom and nearest
neighbour atoms is small. Of course the overlap of the two
bonding p-orbitals with those of neighboring atoms is large.
However, since the bond angle is nearly 90°, the coupling be-
tween the bonding p-orbitals belonging to the same atom is small.
As a consequence the bonding (and antibonding) p-bands are well
described by isolated bonds, which fact results in a sharp peak
on the density of states lying lower in energy than the lone
pair peak. In As_2S_3, As_2Se_3, and amorphous As_2Te_3 a similar
situation is expected because As and the chalcogens have the
same bonding configurations as their elemental forms. Hence the
lone-pair structure associated with the chalcogen should be the
same as that in the elements. In addition, because of the nearly
90° As-Chalcogen-As bond angle, the bonding p-states (in this
case derived from the As and Se orbitals) associated with each

As (S, Se, Te)$_{3/2}$ pyramid should couple only weakly with the
bonding p-states of neighboring pyramids. Thus the shape of the
upper p-like portion of the density of states is determined
primarily by the electronic structure of the isolated As (S, Se,
Te)$_{3/2}$ unit. That is, the gross features of the density of
valence states are determined primarily by the nearest neighbour
coordination numbers of the constituent atoms. This explains the
nearly identical XPS spectra observed for amorphous and crystall-
ine As$_2$S$_3$ (orpiment) (Fig. 1) since X-ray diffraction studies
indicate that the nearest neighbour coordinations of orpiment
are maintained in glassy As$_2$S$_3$. While amorphous As$_2$Te$_3$ has the
orpiment coordination numbers, crystalline As$_2$Te$_3$ is not iso-
morphic with orpiment; its Te atoms are <u>3-coordinated</u> and its
As atoms either tetrahedrally coordinated or octahedrally co-
ordinated [7]. These changes in the coordination numbers may
well manifest themselves in markedly different XPS and UPS
spectra for crystalline As$_2$Te$_3$.

On the basis of the foregoing dis-
cussion, it is possible to assign
the features of the XPS spectra of
Figs. 1-3 to contributions from s,
bonding p, and non-bonding lone
pair electrons. The sharpest upper-
most peak at 1.0-1.5 eV is as-
cribed to the lone pair p-electrons
of the chalcogen. Below this peak
lies the shoulder (or broader peak)
at 4.5 eV which is attributed to

FIG. 4: Composite density
of valence states for
amorphous As$_2$Te$_3$.

the bonding p-orbitals of As and the chalcogen. Note that these
two p-like features are much better resolved in the UPS
spectrum for As$_2$Te$_3$ shown in Fig. 4. The broad bands lying
below the minimum at 7 eV are identified with the As and
chalcogen s-levels. The fact that these bands are observed only
in the XPS data and not in the 21.2 eV UPS spectra for amorphous
As$_2$Se$_3$ and As$_2$Te$_3$ supports this identifications since the photo-
emission cross section for s-electrons, negligible at 21.2 eV,
becomes comparable to that of the p-electrons at the higher
photon energy. The total width of the observed XPS and UPS

spectra for these arsenic chalcogenides and the assignment of
the features above the 7 eV minimum to p-bands and those below
to s-bands implies that the s-p hybridization in these valence
states is much less than suggested by the earlier molecular
orbital calculations for As_2S_3 and As_2Se_3.

4. ACKNOWLEDGEMENTS

The authors wish to thank A. Barynin, G. Krutina, and W. Neu for
expert technical assistance and M. Cardona for useful dis-
cussions.

5. REFERENCES

[1] Kastner, M.: Phys. Rev. Lett. 28, 355 (1972)
[2] Drews, R.E.; Emerald, R.L.; Slade, M.L.; Zallen, R.: Solid
 State Comm. 10, 293 (1972)
[3] Chen, I.: Phys. Rev. B7, 3672 (1973)
[4] Chen, I.: Phys. Rev. B8, 1440 (1973)
[5] Ley, L.; Pollak, R.A.; Kowalczyk, S.P.; McFeely, R.;
 Shirley, D.A.: Phys. Rev. B8, 641 (1973)
[6] Shevchik, N.J.; Cardona, M.; Tejeda, J.: Phys. Rev.
 B8, 2833 (1973)
[7] Cornet, J.; Rossier, D.: J. Non-Cryst. Solids 12, 85 (1973)

RAMAN STUDIES OF THE CRYSTALLIZATION OF As_2Se_3

E. FINKMAN, A. P. DeFONZO AND J. TAUC,
Division of Engineering and Department of Physics
Brown University, Providence, Rhode Island 02912

Raman spectra of the photocatalyzed crystallization process of amorphous As_2Se_3 between the glass-transition temperature and the melting point were studied. The relationship between the crystalline and amorphous vibrational spectra is discussed, and conclusions are drawn about the phonon dispersion in crystalline As_2Se_3.

1. INTRODUCTION

An important group of amorphous semiconductors can be prepared in the bulk form by quenching the liquid. Recently, we have reported Raman studies on a typical representative of this group As_2S_3 in which we related the vibrational spectra of the amorphous solid with those of the liquid and suggested a model for the structural changes [1]. When we extended this work to As_2Se_3 we observed crystallization between the glass-transition temperature T_g and the melting point T_m. We have found that the rate of crystallization can be adjusted for easy observation by Raman spectroscopy by a proper choice of temperature, laser beam intensity and sample history. The method is very suitable for studying the influence of these factors on the crystallization since we can follow the process continuously [2]. In this communication we report on the Raman spectra of As_2Se_3 as a function of time during crystallization. From this study we can draw some conclusions about the crystallization process. The development of the crystalline spectra from the amorphous spectra shows in a particularly dramatic way the relationship between these two spectra.

2. EXPERIMENTAL METHOD AND RESULTS

The Raman spectra were measured by the reflection technique using the argon ion laser line 5145 A and a triple grating monochromator. The sample was sealed in an optical cell with crystalline sapphire windows cut perpendicular to the optical axis. The preparation of amorphous As_2Se_3 is described in Ref. [3]; the cooling rate from 450°C was about 20°C per minute.

In the range between T_g and T_m we could observe a gradual appearance of sharp peaks in the Raman spectrum of amorphous As_2Se_3 until eventually the whole spectrum became like that of crystalline As_2Se_3 (Fig. 1). This

Fig. 1 Changes of the Raman spectrum of As_2Se_3 during crystallization. The dotted line shows the reduced spectrum of amorphous As_2Se_3.

crystallization was not in general a spontaneous effect as we could hold the material in the amorphous form at a temperature between T_g and T_m for several days in the dark. The crystallization process was catalyzed by laser illumination which could be as low as 60 mW. The crystallized region was confined on the surface to the area illuminated by the laser beam. We could not induce the crystallization below T_g. The local rise of temperature induced by laser illumination was estimated to be less than 8°C at the highest power levels.

In order to emphasize the changes in structures in the low frequency region, the measured Raman intensities are plotted in Fig. 1 without reducing the spectra by the usual factors [4]. The low energy Raman intensity decreases during the crystallization.

In our experiments, the crystallization was completed within 1.5 to 4 hours. The rate of crystallization depended somewhat on the laser power but the dominant factors which influenced the rate of crystallization were the temperature T, the rate of the cooling of the material from the melt, and the time of annealing of the amorphous material between T_g and T_m. The crystallization was faster if T was higher, if the cooling of the melt was slower, and if the annealing period was shorter. The time dependence of the growth of the crystalline peaks could be fitted to a well known formula for crystallization kinetics [5].

3. DISCUSSION

The chemical bonding and the nearest neighbor configuration are the same in the crystalline, amorphous and liquid As_2S_3 and As_2Se_3 [1,6-11]. The difference in their structure are due either to differences beyond the first

coordination numbers (the short-range order), and to differences in the
correlations in molecular configurations extending over many units (the
long-range order). If the short-range order is the same for the amorphous
and crystalline material then the crystalline dispersion relations for the
vibrational spectra can be used for determining the shape of the Raman
spectra as a function of the correlation length Λ of vibrations [4] if we
can assume that the matrix elements for each vibrational band are approxima-
tely k-independent.

In a one dimensional model, the contribution of a mode with wave-vector k
($0 \leq k \leq \pi/a$) to the Raman intensity (corrected for the population factor)
is proportional to

$$I_k \sim \frac{L}{\pi} \frac{\Lambda^{-1}}{(k-K)^2 + \Lambda^{-2}} \tag{1}$$

where K is the light scattering vector ($K << \pi/a$), L is the length of the
sample. If $\Lambda \rightarrow \infty$ ($\Lambda/a >> 1$ in the crystalline case), the above expression
becomes a δ-function centered at $k = K \approx 0$, and one observes lines only at
$k \approx 0$. With a finite Λ one obtains contributions weighted by the above expres-
sion from the whole BZ. It is interesting to note that if one could follow
the changes of the Raman bands during the initial stages of crystallization,
when Λ is a small multiple of the unit cell dimension, then one could
possibly obtain information about the dispersion curves.

The changes of the shape of the Raman band (Fig. 1) suggests that this may be
the case. However, a closer inspection shows that the bands are described
in the first approximation as a superposition of the crystalline I_c and
amorphous bands I_a: $I_{total} = xI_c + (1-x)I_a$ ($0 \leq x \leq 1$). Since intermediate
states are not observed one can conclude that Λ changes very rapidly in the
initial stages of crystallization.

The integrated reduced intensities were checked and found to be consistently
lower for the liquid than for the crystalline material at the same temperature.
Typical differences were 14% at 190°C and 30% at 250°C.

Assuming the matrix elements are to a first approximation independent of the
wavenumber, one obtains from Eq. (1) the following expression for the inte-
grated intensity of a single band

$$I \sim (\arctan \pi\Lambda/a_i)^n \tag{2}$$

where a_i is the unit cell dimension in the i-th direction and n is the dimen-
sionality. The coherence length Λ may now be estimated by comparing the dif-
ference in the total integrated intensity between the pure liquid ($x = 0$)

and the pure crystalline state (x = 1). The one dimensional case (n = 1) yields $\Lambda \simeq 5a$ and $\Lambda \simeq 1a$ for the temperatures of 190°C and 250°C, respectively. The two or three dimensional cases yield values of Λ less than the unit cell dimension. If the assumptions used in obtaining (2) are correct, then one would expect that the correlations at a site in the liquid state are greater in one direction than in the other orthogonal directions.

An interesting process is observed in the spectrum during the crystallization. It is clearly seen in Fig. 1 that some crystalline modes appear to grow faster than the others. The most prominent of those are the lines at 89, 145, 202 and 215 cm^{-1}. As a result, for example, the center of the main amorphous band at 232 cm^{-1} shifts in the crystalline spectrum towards lower energies where the most intense line appears at 215 cm^{-1}. If one would try to construct the amorphous spectrum as the envelope of the crystalline lines (as suggested for As_2S_3 [11]), the results will give a band broader than that observed which peaks at a lower energy. The amorphous spectrum can be constructed by broadening the three upper lines at 230, 247 and 272 cm^{-1}. The two lines at 202 and 215 cm^{-1} contribute little to the intensity in the amorphous case. This observation leads to an important

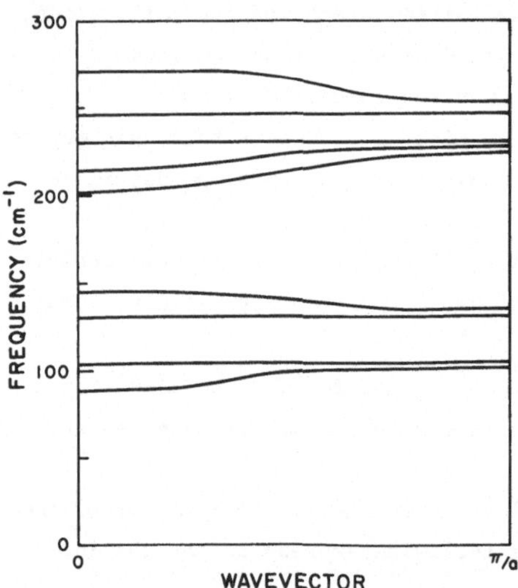

Fig. 2 Schematic representation of the dispersion relations of As_2Se_3. The bands which change little with k are schematically represented by horizontal lines.

prediction about the dispersion relation of some modes in As_2Se_3 which are probably applicable also to As_2S_3. With the loss of long-range order, the Raman intensity shifts from the energy of the phonons at k = 0 in the crystal to the energy of the phonon states of high density that usually occur at the zone edge. From the fact that the intensities of the broadened 230 and 247 cm^{-1} crystalline lines are almost identical to the spectrum of amorphous As_2Se_3, we conclude that these modes are almost flat throughout the Brillouin zone. The bands associated with the 202 and 215 cm^{-1} lines should be curved upward in order to give the

observed amorphous spectrum. In a similar way we predict schematically
the behavior of some other modes which are given in Fig. 2. The other lines
were too weak to account for the changes in the intensity. Our initial
results on the dispersion relation calculations for the high frequency pho-
non bands support the scheme outlined in Fig. 2. The phonon bands which have
a slight dependence on k are almost molecular in their nature. They give a
higher contribution to the density of states than the curved bands. The
latter will distribute their contribution over a broader energy range. In
this context we mention the reason for the success of the molecular model
of Lucovsky and Martin [7,8]. This is due to the fact that the high contri-
butions to the density of states in As_2Se_3 come from modes with a molecular
nature even in the crystalline form. From this point of view, the splitting
of some of these crystalline modes (at 247 and 272 cm^{-1} for example) can be
explained in a first approximation by the distortion of a pyramid and the
different sites for different As and Se atoms. We may conclude that the
success of the molecular model does not rule out the possibility that rem-
nants of the crystalline structure are maintained in As_2Se_3 and As_2S_3.

4. ACKNOWLEDGMENTS

We thank F. J. DiSalvo for the samples, T. R. Kirst for technical assistance
and G. B. Fisher for his comments on the manuscript.

This work was supported in part by a grant from the National Science Founda-
tion and a Cottrell grant. It has also benefited from the general support
of Materials Science at Brown University by the NSF.

5 . REFERENCES

[1] Finkman, E.; DeFonzo, A. P.; Tauc, J.: Proc. 5th Int. Conference on
 Amorphous and Liquid Semiconductors, Garmisch Partenkirchen, 1973,
 in press
[2] Exarhos, G. J.; Risen, W. M. Jr.: to be published
[3] DiSalvo, F. J.; Menth, A.; Waszczak, J. V.; Tauc, J.: Phys. Rev. B6,
 4574 (1972)
[4] Shuker, R.; Gammon, R. W.: Phys. Rev. Lett. 25 222 (1970)
[5] Turnbull, D.: Solid State Physics (ed. F. Seitz and D. Turnbull) 3,
 222 (1956)
[6] Renninger, A. L.; Averbach, B. L.: Phys. Rev. B4, 1507 (1973)
[7] Lucovsky, G.; Martin, R. M.: J. Non-Cryst. Solids 8-10, 185 (1972)
[8] Lucovsky, G.: Phys. Rev. B6 1480 (1972)
[9] Taylor, P. C.; Bishop, S. G.; Mitchell, D. L.; Treacy, D.: Proc. 5th
 Int. Conference on Amorphous and Liquid Semiconductors, Garmisch-Par-
 tenkirchen, 1973, in press
[10] Bishop, S. G.; Shevchik, N. J.: Solid State Communications, in press
[11] Kobliska, R. J.; Solin, S. A.: Phys. Rev. B8, 756 (1973)

VARIATION OF THE ABSORPTION COEFFICIENT AFTER OPTICAL EXCITATION IN As_2Se_3 AT 1.6°K

J. CERNOGORA, F. MOLLOT and C. BENOIT A LA GUILLAUME

Groupe de Physique des Solides de l'E.N.S.[+]

Paris, France

Absorption coefficient of amorphous As_2Se_3 increases after excitation at 1.6°K. This phenomenon is correlated with an "ageing effect" observed in radiative recombination experiments. An interpretation is given in term of a qualitative model involving relaxation of atoms in distorted regions.

1. INTRODUCTION

Recombination radiation on amorphous chalcogenide semiconductors is generally located at the midgap.

Furthermore, other important features occur [1].
a) The intensity of the radiative recombination decreases during excitation as if light excitation produces an "ageing effect".
b) This "ageing effect" can be partially released by thermal annealing at temperature lower than 300°K.
c) The radiative recombination presents an excitation spectrum whose maximum corresponds to an absorption coefficient of a few 10^2 cm^{-1}.

In order to explain these facts, the absorption tail of As_2Se_3 was investigated by means of optical absorption measurements before and after excitation by the same kind of light already used in radiative recombination experiments.

2. EXPERIMENTAL ARRANGEMENT

An excitation by a focused laser is performed upon a 1 mm^2 area of the 100 μm thick sample, immersed in pumped He ; absorption measurements are done before and after excitation through the same area with monochromatic light. Thus, the weak intensity of the measuring light beam has no ageing effect on the sample. On the other hand, this way of proceeding is not the best for

+ Laboratoire associé au C.N.R.S..

absolute absorption coefficient measurements.

3. EXPERIMENTAL RESULTS

Figure 1 shows the absorption spectrum before and after excitation by the 6764 Å line of a Krypton laser. This line was chosen because it is the most efficient one to give radiative recombination. The curves have been obtained from transmission results, supposing a constant reflection coefficient $R = \left(\dfrac{n-1}{n+1}\right)^2$ with $n = 3.5$ [2] in the energy scale studied.

FIG. 1: Absorption coefficient α
(1) before excitation
(2) after a 4 s excitation
(3) " 16 s "
(4) " 64 s "
Excitation λ = 6764 Å
 I = 60 mW/mm^2
 T = 1.6°K
(5) represents the corrected $\Delta\alpha_c$ corresponding to curve (4)

The curve (1) in Fig. 1 has qualitatively the same shape as that obtained by Edward [3].

One obtains a sensible increase of the absorption after excitation over all the energy scale studied which increases with the excitation time. Since the ageing effect occurs only in a region of width α_L^{-1} where α_L is the absorption coefficient of the exciting light $\Delta\alpha$ has to be corrected and the corrected value $\Delta\alpha_c = \Delta\alpha \, d \, \alpha_L$ is shown in Fig. 1 (curve 5). This ageing effect on the absorption is quite "metastable" since the transmission spectrum has been taken <u>after</u> the excitation. Furthermore this ageing effect is released at ambient temperature.

Figure 2 shows the variation $\Delta\alpha$ of the absorption coefficient versus the excitation time on a log log scale at a fixed energy, 1.7 eV, when absorption coefficient before the

excitation was of the order of 100 cm^{-1}.

FIG. 2: Variation of the absorption coefficient with time excitation.

$\Delta\alpha$ obeys a t^a law ; here a = 0.2. One has to recall that the recombination decay varies at t^{-b} (b depending on the excitation intensity). So it seems that the absorption increase is related to the radiative recombination decay. Besides at the same 1.7 eV fixed energy when the excitation line is at higher energy (6471 Å) the increase of the absorption coefficient is less efficient. We find a $t^{0.16}$ law. Nevertheless it seems too early to conclude that the ageing effect in absorption has the same excitation spectrum as the radiative recombination.

4. DISCUSSION

__4.1__ J.S. Berkes [4] observed photodecomposition of As_2Se_3 at T > 120°K leading to an increase of the absorption coefficient. This process is characterized by an induction time longer as the temperature decreases and an annealing temperature of the order of the melting temperature. Our phenomenon presents no induction time and its annealing occurs below 300°K. (Furthermore the shape of our curves showing the variation of the transparency versus excitation time is very different from the analogous curves of J.S. Berkes.)

We conclude that we are not concerned by such a photodecomposition.

__4.2 R.A. STREET'S MODEL [5] AND ABSORPTION VARIATION__

Let us now discuss Street's model of radiative recombination

predicting a change of absorption.

This model is a one electron model with fluctuations of band edges and neutral and charged centres in the middle of the gap. Radiative recombination occurs between an electron located at a potential minimum of the fluctuating band edge and a charged centre. Non radiative recombination consists in a tunnel effect to another neighbouring potential minimum of the band edge and trapping by a neutral centre. So charged centres density increases and according to R.A. Street optical absorption has to increase in the region of 10^2 cm^{-1}. We do not agree with this model which did not correlate the position of the charged centres with the position and amplitude of the potential fluctuations. Indeed when the electron is trapped as a neutral centre the new charged centre will push up the potential minimum. So, as the average, the potential fluctuation amplitude will decrease. So we argue that, except for strange model, the injection of carriers should decrease the potential fluctuations. Thus it seems necessary to take into account atomic displacement to explain the increase of the absorption coefficient.

4.3 THE PROPOSED QUALITATIVE MODEL

P.W. Anderson [6] shows that Franck Condon effect may exist in disordered structures and may be stronger when states are more localized. This is why we proposed a model represented by a configuration diagram shown in Fig. 3. This diagram looks like another configuration diagram proposed by B.H. Rose [7] in order to explain absorption by a F centre in BaO and creation of F^+ metastable centres.

Transition (1) corresponds to the optical absorption (α) in a distorted localized region. This region is now in an excited state and can lose energy by transition (2) which gives luminescence (RR) at a lower energy because there is a Franck Condon

FIG. 3: Configuration diagram.

effect. But there is another opportunity : an ionization (3) occurs (i.e. a carrier, an electron, for instance, moves out of the considered region) and there is a relaxation towards another metastable configuration as the one mentioned by Anderson [8]. This region can now present optical absorption at lower energy α' (4). It is the case for F centre in BaO ; absorption at lower energy explains the increase of absorption experimentally observed and gives arguments why there is no longer radiative recombination in the considered region. Indeed let us remark that absorption at lower energy involved more localized states. Thus we expect a stronger Franck Condon effect for transition (5), so strong that return to ground state is only done by relaxation as mentioned by transition (5) on figure 3. This model is coherent with the fact that the shape of luminescence spectrum does not change during excitation time, but only luminescence intensity decreases when these distorted regions go to other metastable configurations ; we are not surprised to observe that luminescence intensity and absorption coefficient follows similar laws of variation with excitation time.

5. REFERENCES

[1] Mollot, F.; Cernogora, J. and Benoit à la Guillaume, C.:
 Phys. Stat. Sol. (a) 21, 281 (1974)
[2] Kolomietz, B.T.: Phys. Stat. Sol. 7, 359 (1964)
[3] Edward, J.T.: Brit. J. Appl. Phys. 17, 979 (1966)
[4] Berkes, J.S.; Ing Jr., S.W. and Hillegas, W.J.: J. Appl.
 Phys. 42, 4908 (1971)
[5] Street, R.A.; Searle, T.M. and Austin, I.G.: J. Phys. C 6,
 1830 (1973)
[6] Anderson, P.W.: Nature 235, 163 (1973)
[7] Rose, B.H. and Hensley, E.B.: Phys. Rev. Lett. 29, 861
 (1972)
[8] Anderson, P.W.; Halperin, B.I.; Varma, C.M.: Phil. Mag. 25,
 1 (1972)

LOCALIZED STATES DISTRIBUTION IN TERNARY CHALCOGENIDE GLASSES

A. VANCU and R. GRIGOROVICI
Institute of Physics, Bucharest, Romania

The shape of the band tails in ternary chalcoge-
nide glasses was evaluated from absorption and
photoconductivity data. Optical transitions bet-
ween localized states must be supposed to occur
with about equal probability as between local-
ized and extended states.

1. INTRODUCTION

This paper represents an attempt to determine the density of
states distribution in the band tails of ternary glasses of
the systems As-Te-Ge and As-Te-Si with high Te content from
absorption and photoconductivity data.

2. THE SAMPLES

Samples of $As_{20}Te_{70}Ge_{10}$ and $As_{28}Te_{64}Si_8$ glasses were prepared
directly from the elements, special care being taken to avoid
oxidation and inhomogeneities. Slabs of 0.06 to 2.0 mm thick
were cut from the batch and their faces optically polished.

3. THE OPTICAL ABSORPTION EDGE

Transmittance and reflectivity measurements were performed at
room temperature in the 0.12 to 0.90 eV range with the aid of

FIG. 1: Separation of the
total light attenuation
coefficient $(\alpha + \beta)$ into
a true absorption coef-
ficient (α) and an at-
tenuation coefficient
by scattering (β).

a Grubb and Parsons IR spectrometer. The attenuation coeffi-
cient $(\alpha + \beta)$ due to both true absorption (α) and to light
scattering (β) was calculated. The results are displayed in
fig. 1. A relatively steep exponential absorption edge is fol-
lowed towards low energies by a flat tail which has been shown
previously by us [1] to be due to attenuation by scattering.
After subtracting β from $(\alpha + \beta)$ the true absorption edge

log α remains a straight line as far as it can be followed.
A calculation of the density of states distribution in the
band tails using these data can be performed under the assump-
tions that: a) the Fermi level always lies in the middle of
the mobility gap; b) the tails and bands are symmetrical;
c) optical transitions are subject only to the condition of
energy conservation. Two extreme cases will be considered:
Case 1: Optical transitions from both localized to extended
(ℓ-e) and from extended to extended (e-e) states are governed
by the same energy-independent matrix element; ℓ-ℓ transitions
are neglected.
Case 2: e-e, e-ℓ and ℓ-ℓ transitions are all equally probable.

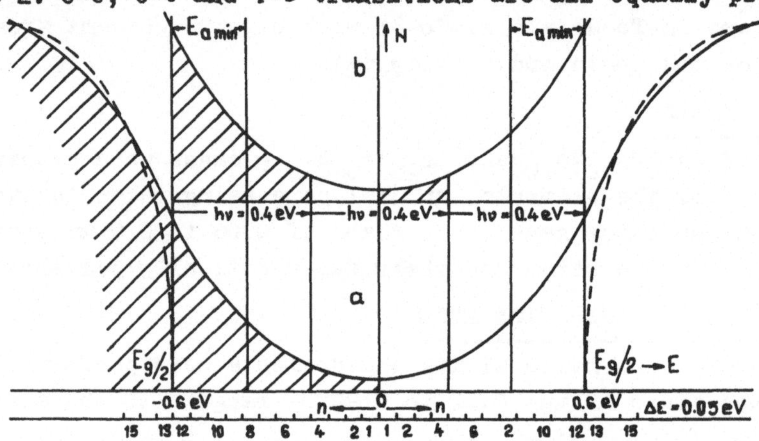

FIG. 2: Density of states distribution (schematically).—— Ac-
tual distribution; --- extended states. $\pm E_g/2$ - optical and
mobility edges in AsTeSi glass. a) Filled and empty states in
the dark. b) Filled and empty states and optical transitions
after prolonged illumination with 0.4 eV photons at low tem-
perature. $E_{a\,min}$ - lowest possible activation energy of dark
conductivity after illumination with 1 eV photons at low T.

With these assumptions the absorption coefficient α is given
by the convolution integral

$$h\nu \cdot \alpha \sim \int_{E_o}^{h\nu} N_1(E) \cdot N_2(E + h\nu)\, dE. \tag{1}$$

In case 1: $E_0 = E_g/2$; $N_1(E)$ and $N_2(E)$ are the density of sta-
tes distributions in the tails and in the bands beyond the mo-
bility edges, respectively ($N_{2c} = N_{2v} \sim |E - E_g/2|^{1/2}$).
In case 2: $E_0 = 0$; $N_1(-E) = N_2(E)$ are the densities of states
above and below the Fermi level ($E_F = 0$).

Knowing the dependence of α on $h\nu$, $N_1(E)$ can be obtained in case 2 by a Laplace transformation [2] or, in both cases, by replacing the integral (1) by the sum

$$\sum_{i=1}^{i=n} N_1\left[(i-\tfrac{1}{2})\Delta E\right].N_2\left[(n_0+n-i+\tfrac{1}{2})\Delta E\right]; \quad h\nu = n\,\Delta E; \quad E_0 = n_0\Delta E, \tag{2}$$

where ΔE is a finite but small enough energy interval and $\underline{n}$ is an integer increasing up and down from E_F (fig. 2). By using an iteration procedure, $N_1(n) = N_1(|n\,\Delta E - \Delta E/2|)$ is found to be given in case 1 by

$$N_1(n) \sim \sqrt{\tfrac{2}{\Delta E}}\sum_{i=1}^{i=n} i\,\Delta E\,\alpha_i\,A_{n-i+1}; \quad A_{n-i+1} = -\sum_{j=1}^{j=n-i}A_j\sqrt{2(n-i+1)-2(j-1)}; \quad A_1 = 1 \tag{3}$$

and in case 2 by

$$N_1(n) \sim \frac{1}{2\sqrt{\Delta E}\,\alpha_1}\left[n\,\Delta E\,\alpha_n - \sum_{i=2}^{i=n-1}N_1(i).N_1(n-i+1)\right]; \quad N_1(1) = \sqrt{\Delta E}\,\alpha_1; \quad N_1(2) = \frac{\alpha_2\sqrt{\Delta E}}{\alpha_1}. \tag{4}$$

The results of such a calculation performed on the AsTeSi glass are shown in fig. 3. As expected, the tails are practically exponential, but have different slopes in cases 1 and 2. So far no preference for one of the two cases emerged.

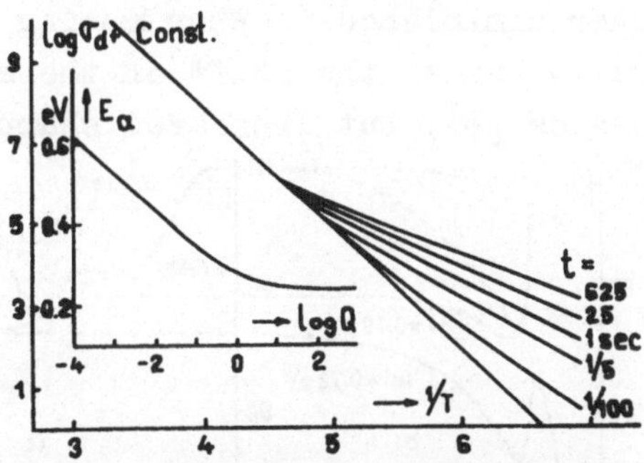

FIG. 3: Calculated density of states N in AsTeSi glass. 1 - from absorption; no l-l transitions; 2 - the same with l-l transitions; 3 - from photoconductivity.

FIG. 4: Temperature dependence of dark conductivity σ_d in AsTeSi glass after previous illumination of duration t. Insert: Activation energy E_a vs. quantity of illumination Q.

4. PHOTOCONDUCTIVITY AND LIGHT-INDUCED ABSORPTION

The samples were provided with two NiCr electrodes in planar arrangement about 2 mm apart and illuminated in vacuo at 130°K through a Zeiss SPM 1 monochromator.

The spectral response displayed two clear photoconduction edges, the first at 0.4 eV, the second at 0.7 eV. The low energy photoconductivity was enhanced up to 10 times by preliminary illumination in the 1 eV range; the decay of the high energy

photoconductivity was never complete (fig. 4). Log σ_d (σ_d - residual dark conductivity) and its activation energy E_a first decreased linearly with log Q (Q - illumination x time) and then reached saturation values ($E_{a\,min}$ = 0.23 eV) (fig. 5). This increased dark conductivity persisted for hours unchanged as long as the sample was not heated above ~230°K.

FIG. 5: Onset and decay of photoconductivity in glassy AsTeSi at 130°K after a) 1/5 sec light pulse; b) 650 sec illumination. xxx theoretical values after eq. (5).

A decrease in transmittance $\mathcal{T}$ could be both induced and detected at 130°K by using 1 sec light pulses (hν = 0.59 and 0.73 eV) (fig. 6a). Additional illumination with >1 eV photons further diminished $\mathcal{T}$. When heating the sample, $\mathcal{T}$ decreased gradually due to the shift of the absorption edge towards low energies [3], but increased sharply around 230°K (fig. 6b).

FIG. 6: Light-induced absorption in glassy AsTeSi. a) Reduction in transmittance $\mathcal{T}$ as a function of number n of 1 sec light pulses. b) Dependence of $\mathcal{T}$ on temperature during heating after preliminary illumination.

All these facts fit into the following model. The deep levels of the band tails act below 230°K as traps which, once filled, exchange carriers with their bands, but do not act as recombination centers. Therefore photons with h$\nu \geq E_g/3$ = 0.4 eV can rise the quasi-Fermi level by hν/2 by filling the traps to this height through ℓ-ℓ transitions and also promote the trapped carriers above the mobility edges. This explains both the lower photoconduction edge (fig. 2b) and the light-induced absorption. The traps empty thermally around 230°K.
Nearly all the carriers promoted into higher states by ℓ-e and e-e transitions will fall into the deep traps if they are emp-

ty. Therefore, for short enough illumination times, recombination being excluded, the decay of the photoconductivity may be shown to follow a law of the form

$$\sigma = \sigma_d \frac{1 + \frac{1-\sigma_d}{1+\sigma_d}\exp(-2\gamma\sigma_d t)}{1 - \frac{1-\sigma_d}{1+\sigma_d}\exp(-2\gamma\sigma_d t)} \tag{5}$$

which fits the experimental results very well (fig. 5a). From the experimental law: $E_a \sim \log Q$, the energy distribution of the density of states can easily be calculated under the same assumption of absent recombination and yields the results shown in fig. 3, curve 3. The better agreement with the slope of curve 2 shows that the contribution of ℓ-ℓ transitions to the optical absorption cannot be neglected. After long illumination times the decay of the photocurrent follows a different law due to the now intervening recombination (fig. 5b).

5. CONCLUSIONS

The band tails of our glasses extend exponentially from below the mobility edges to near the middle of the gap. Traps shallower than 0.23 eV empty even at 120°K. This agrees with earlier thermo-depolarization measurements on similar glases [4].

6. ACKNOWLEDGEMENTS

We gladly acknowledge the contribution of Dr N. Croitoru by valuable suggestions and useful discussions. We are also indebted to St. Slădaru for carefully preparing the samples.

7. REFERENCES

[1] Vancu, A.; Slădaru, St.; Grigorovici, R.: Proc. 5th Conf. Amorph. Liq. Semicond., Garmisch, 1973 (in press).
[2] Howard, W.E.; Tsu, R.: Phys. Rev. B 1, 4709 (1970).
[3] Tauc, J.; Menth, A.; Wood, D.L.: Phys. Rev. Letters 25, 479 (1970).
[4] Boţilă, T.; Slădaru, St.; Croitoru, N.: Proc. 5th Conf. Amorph. Liq. Semicond., Garmisch, 1973 (in press).

PHOTOLUMINESCENCE IN AMORPHOUS SELENIUM AND ITS ALLOYS

R. A. STREET, T. M. SEARLE AND I. G. AUSTIN,

Department of Physics, The University, Sheffield S3 7RH.

Photoluminescence properties of amorphous selenium
and preliminary data on some Se alloys are reported.
The alloying elements comprise the isoelectric ele-
ments O, S and Te, the univalent elements Cl and I,
and the branching elements As and Ge. The PL inten-
sity, the peak position, the shape of the excitation
spectrum and the temperature dependence of the inten-
sity are influenced in a variety of ways by the form-
ation of these alloys. The effects are discussed in
relation to the known transport properties of similar
compounds.

1. INTRODUCTION

The photoluminescence (PL) properties of chalcogenide glasses in the
system As_2X_3, (X = S, Se, Te) have recently been the subject of several
investigations [1, 2, 3, 4]. Common to all such compounds is a broad PL
band located near half the band gap energy (E_g), an excitation spectrum
that indicates a rapidly decreasing PL quantum efficiency (y_L) at energies
greater than E_g, and a temperature dependent intensity that follows the
relation

$$y_L = \text{const. exp } (-T/T_o) \tag{1}$$

In all the As chalcogenides y_L reaches a maximum value of order 0.1 at
T = 0 and T_o equals 20 K in As_2Se_3 and 26 K in As_2S_3.

In the present work the PL properties of Se and various Se alloys have
been investigated. The alloy materials are grouped into three categories:
the isoelectronic elements O, S and Te; the univalent elements Cl and I; and
the branching elements As and Ge. Measurements have been made on the peak
position, the PL intensity, the excitation spectrum, and the temperature
dependence of the intensity.

2. EXPERIMENTAL RESULTS

2.1 AMORPHOUS SELENIUM

As illustrated in figure 1, the PL properties of amorphous selenium
are similar to those of the As chalcogenides with a broad PL peak centred
near $E_g/2$ and a similar excitation spectrum (see figure 3b). The temper-
ature dependence is included in figure 2b and follows equation 1 with

$T_o = 19$ K. However the PL intensity is low, with the maximum y_L approxi-

FIGURE 1: PL spectrum, Excitation spectrum and Optical absorption of amorphous selenium.

mately 10^{-3}. Also the position of the band, at 0.8 eV, is lower than in As_2Se_3 by $\sim$ 0.1 eV despite a larger E_g.

2.2 ISOELECTRONIC ALLOYS

In figure 2a are shown the PL spectra of Se alloys with O, S and Te.

FIGURE 2: (a) PL spectra of Se alloys with O, S and Te.
(b) Temperature dependence of PL intensity.

The PL intensity shows a systematic change through the chalcogen series, increasing markedly with Te but reducing with O. The quenching effect of O has also been observed by Bishop (private communication). The shift in the peak position is probably associated with a change in E_g since this is increased by the addition of S but reduced by Te. Also there is a second PL peak in $Se_{95}S_5$ near 1.5 eV. The excitation spectra of all these alloys are similar in shape to pure Se but there are some differences in the temperature dependence as is shown in figure 2b. Thus in $Se_{95}Te_5$, $T_o = 22$ K, whilst in $Se_{95}S_5$, $T_o = 39$ K, which is larger than in Se by a factor of two.

2.3 UNIVALENT ALLOYS

The addition of Cl to Se produced no observable change in any of the

PL properties. There was some doubt as to the exact composition of this alloy since it was introduced as Se Cl_4 which is strongly deliquescent. For this reason alloys with I were also investigated. Here the peak position decreases slightly to 0.74 eV, the intensity is increased by a factor of $\sim$ 3 and T_o increases to 29 K

2.4 BRANCHING ALLOYS

The results of alloying with As are reported elsewhere [7] and it is found that the PL intensity increases proportionally with As concentration up to 40 at.% (i.e. As_2Se_3) as shown in figure 3a.

FIGURE 3: (a) Relation between PL intensity and As concentration in Se–As alloys.
(b) Excitation spectra of Se and alloys with As and Ge.

Throughout the As–Se system the shape of the excitation spectrum is similar to pure Se as is the value of T_o, and the PL peak is at .87 eV.

The behaviour of Ge alloys is rather different. The intensity of the 5% alloys is close to pure Se, as is the peak position. However T_o takes the value 34 K and the excitation spectrum is broader, as shown in figure 3b.

3. DISCUSSION

From the low energy of the PL peaks it is apparent that deep localized recombination centres are involved and the PL intensities indicate that the density of such states is much smaller in Se than in As_2Se_3. The same conclusion has been reached from transport measurements, particularly the observation of larger carrier lifetimes in selenium than in As_2Se_3 [5] and the sensitivity of the conductivity of Se to the presence of impurities [6].

The linear relation between intensity and As concentration (figure 3a) shows clearly that As atoms are responsible for the recombination centres in the Se–As alloys and elsewhere it is argued [7] that these centres are in fact As dangling bonds. In view of the close similarity of all the PL

properties of the As chalcogenides and Se, it seems very probable that the recombination centres are of similar type, and therefore that they are also dangling bonds in Se.

The present authors have shown that the characteristic temperature dependence of equation (1) can be explained by the assumption of long range potential fluctuations of the conduction and valence bands with magnitude 0.1 - 0.2 eV [8]. On this model $T_o = (4 \alpha^2 \gamma^2 k)^{-1}$ where α is the localization parameter and γ defines the parabolic potential minimum $(r/\gamma)^2$. Large values of T_o will correspond to larger fluctuations. Also there is evidence [9] that the PL spectral width may be due to small polaron formation. The lattice will deform around a carrier in a shallow level and lower its energy. Polaron formation is observed in orthorhombic sulphur [10] and crystalline As_4S_4 [11] and in each case it is associated with electron transport. We therefore suggest that the PL transition is between a shallow localized hole and an electron in a level associated with a positively charged dangling bond around which there is a strong electron–phonon interaction. The lattice deformation and hence the electron energy will be a characteristic of the selenium host and therefore similar for each alloy. This may explain why the PL peaks all occur at approximately the same energy in striking contrast to the changes in the conductivity activation energy of up to 0.5 eV in Cl and O alloys [6].

3.1 SE ALLOYS

Since this model suggests that the PL recombination centre is an electron level we expect a correlation between the PL properties and the electron mobility. Transport measurements have been reported by Schottmiller et al. [5] on most of the alloys used in the PL experiments. On the addition of both S and Te there is a reduction in the electron drift mobility, with the decrease being greater in the Te alloys. Since the mobility activation energy remains constant an increase in the density of traps is indicated. This correlates with the increase in PL intensity which rises faster in the Te alloys than the Se alloys. Schottmiller et al. further associate this behaviour with the appearance of features in the Raman spectra, also more pronounced in Te alloys, which they interpret as mixed rings Se_xS_{8-x}, Se_xTe_{8-x}. Recently, however, the identification of the ring and chain structure from Raman data has been questioned [12].

If the PL temperature dependence is due to long range potential fluctuations of the bands then we expect that a change in the value of T_o (see equation 1) such as is found in S and Ge alloys, will accompany marked changes in both the electron and hole mobilities. This appears not to be

the case, but it is perhaps significant that the mobility lifetime behaviour of the S alloys is totally different from the Te alloys. A possible explanation is that the fluctuations exist only in the vicinity of the recombination centre and do not influence the drift mobility of the bulk. An increased value of T_o corresponds to a lower escape probability of the electron from the centre and might therefore imply an increased probability of the capture of a hole. In fact Schottmiller et al. find the hole lifetime is reduced by a factor 100 in the S alloys.

Discussion of the other alloys is limited because, in the one case there is not much available transport data for the univalent alloys apart from the observation that the electron lifetime decreases very rapidly in Cl alloys, and in the other case the transport behaviour of the As alloys is very complex and the PL behaviour is dominated by the introduction of recombination centres due to As dangling bonds. With the present data it is not possible to separate the latter effect from the PL of pure Se. However, we point out that in the above discussion the electron lifetime is not associated with any PL property and so the absence of a change in the PL behaviour in the Cl alloys is perhaps not surprising.

4. REFERENCES

[1] KOLOMIETS, B.T., MANONTOVA, T.N. and BABAEV, A.A., J. Non Cryst. Solids, 4, 289 (1970).
[2] STREET, R.A., SEARLE, T.M. and AUSTIN, I.G., J. Phys. C. 6, 1830 (1973).
[3] BISHOP, S.G. and MITCHELL, D.L., Phys. Rev. B, 8, 5696 (1974).
[4] CERNOGORA, J., MOLLOT, F. and BENOIT A. LA GUILLAUME, C., Phys. Stat. Sol. (a) 15, 401 (1973).
[5] SCHOTTMILLER, J., TABAK, M., LUCOVSKY, G. and WARD, A., J. Non Cryst. Solids, 4, 80 (1970).
[6] TWADDELL, V.A., LACOURSE, W.C. and MACKENZIE, J.D., J. Non Cryst. Solids, 8-10, 831 (1972).
[7] STREET, R.A., SEARLE, T.M. and AUSTIN, I.G., to be published.
[8] STREET, R.A., SEARLE, T.M. and AUSTIN, I.G., Proceedings of Fifth International Conference on Amorphous and Liquid Semiconductors (1974).
[9] STREET, R.A., SEARLE, T.M. and AUSTIN, I.G., to be published.
[10] GIBBON, D.J. and SPEAR, W.E., J. Phys. Chem. Solids, 27, 1917 (1966).
[11] STREET, G.B. and GILL, W.D., Phys. Stat. Solids, 18, 601 (1966).
[12] LUCOVSKY, G., Proceedings of Fifth International Conference on Amorphous and Liquid Semiconductors (1974).

INFLUENCE OF PREPARATION CONDITIONS ON THE
RADIATIVE RECOMBINATION IN AMORPHOUS SILICON

D. ENGEMANN and R. FISCHER

Fachbereich Physik der Universität Marburg

The luminescence spectrum of amorphous silicon was investigated as a function of deposition temperature T_S between $40^{\circ}C$ and $250^{\circ}C$. Samples deposited at low T_S show a weak luminescence band at about 1 eV which moves to higher energies, changes its shape, and grows considerably when the substrate temperature increases. We explain this behaviour by a deposition-temperature dependent state distribution in the gap. With rising T_S the density of states decreases and maxima near the band edges disappear.

1. INTRODUCTION

The density of states in the forbidden gap of amorphous semiconductors determines the electrical and many optical properties of these materials, which can be varied by annealing or by different preparation conditions. It is remarkable that in the case of amorphous silicon prepared by the decomposition of silane, annealing has not much influence. Varying the deposition temperature, however, changes the behaviour of this material considerably. When the deposition temperature is enhanced, the photoconductivity rises [1] and the density of states in the gap decreases as field effect measurements show [2].

The luminescence spectrum gives additional information about the density of states distribution in the gap [3]. Thus we are able to investigate, by means of luminescence measurements, the influence of the deposition temperature on the density of states in silane-type a-Si more in detail. (For experimental details see ref. [3].)

2. RESULTS

The deposition temperature affects the luminescence spectrum in two ways: It influences the shape of the luminescence band and changes the luminescence efficiency. Let us first deal with the

FIG. 1: Luminescence spectrum of
a-Si prepared at different
substrate temperatures T_S.

effect on the spectral shape. Figure 1 shows the luminescence spectra, taken at 77 K, of six amorphous silicon samples which were deposited at the temperatures indicated on the right. The curves are normalized to their maximum value to demonstrate how the shape of the spectrum changes. For the sake of clarity successive curves are shifted on the vertical scale by 50 units.

All spectra consist of a broad band whose center of mass moves to higher energies when the substrate

temperature increases. At the same time additional maxima appear, which disappear again at high deposition temperatures. At 47°C the luminescence band has a maximum at 1.05 eV, which splits into two maxima at 0.93 eV and 1.13 eV, when the temperature is raised to 110°C. At 140°C another maximum at 1.23 eV appeares, so that this spectrum shows three peaks, labelled 1, 2 and 3. The sample prepared at 220°C only shows the maxima 1 and 2. Here the hump at 0.93 eV can no longer be detected. At 250°C only one maximum at 1.2 eV exists. When the substrate temperature increases, the spectrum as a whole becomes narrower; the halfwidth of the luminescence band decreases from 0.49 eV at 47°C to 0.28 eV at 250°C.

The effect of the deposition temperature on the luminescence efficiency is shown in Fig.2, where the total intensity is plotted logarithmically vs. substrate temperature. The effi-

FIG. 2: Luminescence intensity versus substrate
temperature.

ciency increases rather rapidly by about two orders of magni-
tude, when the deposition temperature is raised from 47^{o}C to
190^{o}C. Above 190^{o}C the luminescence intensity reaches a satura-
tion value. The quantum efficiency of the sample prepared at
250^{o}C was estimated to be of the order of unity.

The luminescence measurements were mostly performed at 77 K,
because the intensity of the luminescence does not depend on
temperature between 8 K and 77 K. This holds for all deposition
temperatures as can be seen in Fig.2, where the open triangles
are determined by measurements at 77 K and the filled triangles
result from measurements at 8 K.

3. DISCUSSION

The dependence of the height and the shape of the luminescence
band on the deposition temperature can be understood by a model
for the state distribution in the gap of amorphous silicon
which is shown in Fig.3. Here the density of states is schema-
tically plotted versus energy. We think, that the band edges of
crystalline silicon are more or less retained in the amorphous
phase as edges in the density of localized states. Therefore
the luminescence peak at about 1.2 eV is very likely to corres-

FIG. 3: Density-of-states model
for a-Si.

pond to the indirect gap of crystalline silicon. The other two maxima in the luminescence spectrum can be explained by transitions between the density-of-states edges and maxima in the state distribution as indicated by the arrows [3]. A very similar distribution of states follows from field effect measurements [2].

At low deposition temperatures the luminescence intensity is small and the maximum lies at 1.05 eV. We think, that here mainly transitions between the maxima and the opposite edges occur. As the maxima are relatively broad, this leads to one broad luminescence peak. In addition many localized states exist in the middle of the gap, which make it possible for a great portion of the excited electron-hole pairs to recombine without emitting photons. This case is described by curve a. When the substrate temperature rises, the luminescence intensity increases at first rapidly (Fig.2). The probability for non-radiative recombination thus decreases quickly, i.e. the density of recombination centers near the middle of the gap decreases. At the same time the maxima become narrower and above a substrate temperature of 110°C the transitions at 1.13 eV and 0.93 eV can be detected separately. When the deposition temperature rises further, the density of states in the gap becomes so small, that the transition at 1.23 eV can be detected. Above 140°C the intensity of the transitions at higher energies grows much faster with rising deposition temperature than that at 0.93 eV. In this model the maximum in the state distribution near the conduction band therefore disappears faster than that close to the valence band. This is demonstrated by curve b in Fig.3. On samples pre-

pared well above 200°C the transition at 0.93 eV can no longer
be detected. Here the radiative recombination between the tails
of valence and conduction bands predominates. The maxima in the
density of states have almost disappeared as is shown in curve
c. In addition the high quantum efficiency of the order of
unity at these samples shows, that recombination centers situ-
ated near the center of the gap do not play a major role at low
temperatures. -

In ideal amorphous silicon the absence of long-range order leads
to a smearing-out of the crystalline band edges without pro-
ducing other structure in the state distribution [4]. We con-
clude from our luminescence measurements, that this is approxi-
mately the case for the sample deposited at 250°C, because there
is only one luminescence peak. Defects, which depend on the
preparation conditions, create additional maxima in the distri-
bution of localized states. The most likely candidates for
these states are vacancies, divacancies, and surface states in
the bulk originating from voids [5]. Such defects may be res-
ponsible for the density-of-states maxima, though the energies
of these states, taken from the crystal [6,7,8], do not fit
very well to the energetic position of the maxima in the state
distribution found by our experiments. - The dependence of the
density of states maxima on the deposition temperature may be
connected to the surface mobility of the deposited Si atoms.
At higher substrate temperatures the surface mobility is higher
and therefore it is reasonable to assume that the probability
for creating defects becomes smaller.

4. REFERENCES

[1] Loveland, R.J. et al. : J.Non-Crystal.Sol. 13, 55 (1973/74)
[2] Spear, W.E. : Proc.5th Intern.Conf.on Am.and Liquid Sem.,
 Garmisch-Partenkirchen (1973)
[3] Engemann, D.; Fischer, R. : ibid.
[4] Cohen, M.H. : J.Non-Crystal.Sol. 4, 391 (1970)
[5] Mott, N.F.; Davis, E.A. : Electronic Processes in Non-
 Crystalline Materials, Clarendon Press, Oxford 1971
[6] Müller, W.; Mönch, W. : Phys.Rev.Letters 27, 250 (1971)
[7] Watkins, G.D.; Corbett, J.W. : Phys.Rev. 121, 1001 (1961)
[8] Kalma, A.H.; Corelli, J.C. : Phys.Rev. 173, 734 (1968)

CONTINUOUS RANDOM NETWORK MODEL FOR AMORPHOUS ARSENIC

G.N. GREAVES AND E.A. DAVIS

Physics and Chemistry of Solids, Cavendish Laboratory,
Cambridge, England.

A continuous random network has been constructed based on the
three-fold coordination and bond angle found in the double
layers of crystalline As. The model accounts naturally for
the lower density of the amorphous phase and suggests why it
is semiconducting. Structural data for the model is dis-
cussed in relation to the reduced spacial function, $G(r)$, and
the reduced intensity function, $F(k)$. Comparison with
experiment is good. Finally the contributions made by the
first, second and third nearest neighbour distributions are
examined.

Continuous random networks embody atomic short range order but lack any long
range periodicity. They can be generated by retaining, within limits, crystalline
nearest neighbour bond angles and lengths but allowing the dihedral angle to
float. This modelling procedure has been successful in predicting the
density, the bond angle distribution and the structure beyond next nearest
neighbours of many glassy materials. Previous models have concentrated on
structures that contain the tetrahedral unit: SiO_2 [1], Ge and Si [2],[3]
and III-V's [4]. In this paper we describe a hand-built model of 533 atoms
all of which are three-fold coordinated. The starting bond angle chosen was
97° - i.e. that associated with the A7 rhombohedral structure. The model
successfully reproduces for amorphous As, the density, the lack of inter-
action between atoms not directly bonded, and hence its semiconducting
properties [5], and the structural data obtained by X-ray diffraction [6].

In contrast to glasses with tetrahedral coordination, three-fold coordinated
glasses are usually 10 to 20% less dense than their crystalline counterparts.
The present model reproduces this density deficit in a natural way. Voids
occur without the breaking of bonds and are interconnected to make a cavern-
like structure. Some impression of the porous nature of the model can be
gained from the projection shown in Fig.1. Of particular interest also is
the way the random network minimises the possible electronic interaction
between atoms not directly bonded. Chemical bonding arguments link the
localization of the covalent bond with semiconducting behaviour [5]. Films
of amorphous As are indeed semiconducting with a band gap in the range 1.2
to 1.4 eV [6]. The conductivity of a typical film is reproduced in Fig.2.

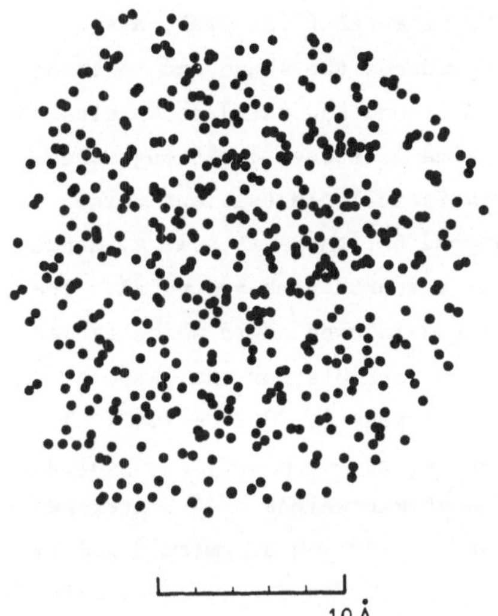

Fig. 1: Projection of spherical
model.

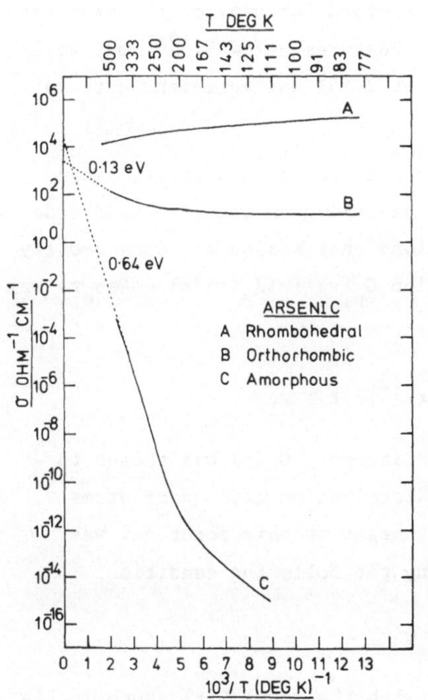

Fig. 2: Conductivities of allotropes
of As.

(Curve C). In orthorhomic (arseno-
lamprite) and rhombohedral (A7) As
the atoms are three-fold coordinated
within double layers, but atoms in
adjacent layers are so positioned
that resonant bonding between layers
is facilitated [5]. The total
coordination is effectively greater
than three. Compared to amorphous As
both forms of crystalline As are
vastly more conducting at room
temperature (Fig.2, curves A and B).
Orthorhombic As, which has an inter-
layer separation of 3.56Å, is a
narrow gap semiconductor with a band
gap in the range 0.2 to 0.3 eV. In
rhombohedral As the electronic
overlap between layers is greatest.
The interlayer separation is 3.15Å
and this allotrope is the familiar
semimetal.

Details of the construction of the
model and a discussion of the radial
distribution function, $J(r)$, and the
network statistics will be published
elsewhere [8]. For this presentation
we discuss the model in terms of the
reduced spacial function, $G(r)$
$(=(J(r)/r) - 4\pi\rho_o r)$, and its Fourier
transform, $F(k)$ $(\int_0^\infty G(r).\sin kr.dr)$.
$G(r)$ for the model plotted from 0 to
7.25Å is compared to experimental
$G(r)$'s for two forms of glassy As
deduced from the radial distribution
functions of Krebs and Steffen [6]
in the top half of Fig.3. The model,
$G(r)$, which was Gaussian broadened
to reproduce the height of the

Fig. 3: Model G(r) and F(k) com-
pared to experiment (6)

experimental first peak, also reproduces the shape and location of the second and third peaks plus the minima inbetween. Although the statistical fluctuations worsen beyond 6Å, there is some evidence for a fourth peak around 7Å. The spread in bond angle which is large-ly responsible for the shape of the second peak is $98.2^{O} \pm 7.4^{O}$. The density of the model, ρ_{o}, scaled to the atomic weight of As, averages out at 4.93 ± 0.31 gm/cm^{-3} and is quite close to reported densities of amorphous As. The densities of ortho-rhombic and rhombohedral As are 5.67 and 5.73 gm/cm^{-3} respectively. In the bottom half of Fig.3, F(k) for the model is plotted from 0 to 15Å^{-1} together with the corresponding experimental curves for amorphous As calculated from the diffraction data [6]. Beyond 1.5Å^{-1} the broad features of the experimen-tal F(k) are well reproduced by the model. Features below 1.5Å^{-1} may well be associated with the termination of G(r) at 7.25Å and agreement here is probably fortuitous.

The G(r)'s associated with the first, first and second, and first, second and third nearest neighbours for the model are shown on the left hand side of Fig.4. In order to obtain balanced functions that shared the same density as the total model, each partial distribution $G_{p}(r)$ $(=(J_{p}(r)/r) - 4\pi\rho_{o}r)$ was added to the following background function,

$$G_{b}(r) \quad = \quad \begin{cases} 0 & r < r_{1} \\ 4\pi\rho_{o}r(1-\exp(-\gamma^{2}(r-r_{1})^{2}) & r \geq r_{1} \end{cases}$$

where r_{1} is the average nearest neighbour distance. $G_{b}(r)$ was chosen to start at r_{1} because the actual background distribution made up of atoms beyond third nearest neighbours effectively began at this point. γ was obtained for each partial distribution using the following condition

$$\int_{o}^{\infty}(G_{p}(r) \quad + \quad G_{b}(r)) \quad r.dr = 0.$$

Comparison of the nearest neighbour G(r)'s with the total G(r) shown in Fig. 3 indicates that the third peak at r $\sim$ 5.7Å is largely the result of third nearest neighbours. (Connell and Temkin [4] have found the same to be true

Fig. 4: Model G(r) and F(k) for 1st, 1st and 2nd, and 1st, 2nd and 3rd nearest neighbours.

for their tetrahedrally coordinated model). Analysis of the present model has also shown, but to a lesser extent, that the structure around 7Å is mainly due to fourth nearest neighbours [8]. Breitling and Richter [9] have pointed out that the peaks in the radial distribution functions of many amorphous semiconductors correspond to successive distances along planar zig-zag chains. Our model demonstrates for a triply coordinated network how these chains might be cross linked to form a three dimensional structure.

On the right hand side of Fig.4, the corresponding F(k)'s are plotted. Because of its sharpness, the first nearest neighbour distribution centred about 2.5Å gives rise to a sinusoidal F(k) of period 2.5Å^{-1} ($\sim\frac{2\pi}{2.5}$) and the first proper maximum occurs at 2.9 Å^{-1}. The introduction of second nearest neighbours breaks up the structure below $\sim 7.5\text{Å}^{-1}$ and most of the features in F(k) for the whole model (Fig.3) are established. The first peak now occurs at 2.0Å^{-1}. When the third nearest neighbour distribution is finally included, the structure sharpens up; the peaks at 3.6 and 5.5Å^{-1} are readjusted in height and, in particular, the first peak shifts back out again - this time to 2.3Å^{-1}. It is clear from Fig.4 that the structure in F(k) up to $\sim7.5\text{Å}^{-1}$ is dominated by the second nearest neighbour distribution and beyond that by first nearest neighbours. The heights and detailed

positions of features in F(k), and especially the first major peak at $\sim 2.3 \mathring{A}^{-1}$, are also affected by structure beyond second nearest neighbours. This in turn is directly related to the network statistics and in a real glass would doubtless be a function of preparation conditions. It is perhaps significant that the position of the first major diffraction maximum, which is also one of the strongest and best defined, is different for β and γ As and also for precipitated and evaporated As [8].

ACKNOWLEDGMENTS

We would like to thank Dr. A.J. Leadbetter for helpful discussions about amorphous As and for supplying samples of arsenolamprite. We acknowledge the financial support of the SRC and one of us (G.N.G.) of Pilkington Bros. Ltd.

REFERENCES

[1] Bell, R.J.; Dean, P.: Phil. Mag. 25, 1381 (1972)

[2] Polk, D.E.; Boudreaux, D.S.: Phys. Rev. Letts. 31, 92 (1973)

[3] Shevchik, N.J.: Phys. Stat. Sol. (b)58, 111 (1973)

[4] Connell, G.A.N.; Temkin, R.J.: (to be published) (1974)

[5] Krebs, H.: J. Non-Cryst. Solids 1, 455 (1969)

[6] Krebs, H.; Steffen, R.: Z. anorg. allg. Chem. 327, 224 (1964)

[7] Greaves, G.N.; Knights, J.C.; Davis, E.A.: 5th International Conference on Amorphous and Liquid Semiconductors, Garmisch-Partenkirchen (Taylor & Francis) (1974)

[8] Greaves, G.N.; Davis, E.A.: Phil. Mag. (to be published) (1974)

[9] Breitling, G.; Richter, H.: Mater. Res. Bull. 4, 19 (1969)

DANGLING BONDS IN TETRAHEDRALLY BONDED AMORPHOUS SEMICONDUCTORS

TSUYOSHI UDA and EIZABRO YAMADA

Central Research Laboratory, Hitachi, Ltd.

Kokubunji, Tokyo, Japan

The electronic structure of dangling bonds in
tetrahedrally bonded amorphous semiconductors
is discussed using the Hall-Weaire Hamiltonian.
The allowed energy regions are derived for
general systems, and the density of
states is given for the Bethe lattice.

1. INTRODUCTION

Recently, Weaire has proposed a simple and interesting model for
topologically disordered systems[1]. However, it contains two
serious oversimplifications with respect to the actual semi-
conductors. One is that the interactions between atomic orbitals
are limited to only the short range ones, V_1 and V_2 . The other
is the omission of the existence of dangling bonds. This paper
deals with the effects of these bonds on the electronic
structure.

2. THEORY

The model Hamiltonian takes the same form as used by Weaire
et al.[2],

$$H = V_1 \sum_{i,j\neq j'} a^+_{ij} a^+_{ij'} + V_2 \sum_{\substack{i\neq i' \\ j}} a^+_{ij} a_{i'j} \qquad (1)$$

where, the subscript i refers to the atom and j to the bond, V_1
and V_2 are matrix elements between orbitals within the same atom
(and different bonds) and those associated with the same bond(
and different atoms), respectively. The effect of a dangling
bond is that one of the second terms in Hamiltonian(1) vanishes
when the atomic orbitals labelled ij is the dangling bond.
Substituting the wave function $\psi = \Sigma\, c_{ij} a^+_{ij}$ into the Schrodinger
equation $H\psi = E\psi$, the following equation is obtained between
the coefficients c_{ij} of the ij bonds and $\bar{c}_{ij} = c_{i'j}$ of the other
orbitals associated with these bonds,

$$Mc_i = -V_2\bar{c}_i \, , \qquad\qquad (2)$$

where,

$$M = \begin{bmatrix} -E & V_1 & V_1 & V_1 \\ V_1 & -E & V_1 & V_1 \\ V_1 & V_1 & -E & V_1 \\ V_1 & V_1 & V_1 & -E \end{bmatrix}$$

and c_i and $\bar{c}_i$ are four vectors composed of c_{ij} and $\bar{c}_{ij}$ ($j=1\sim 4$), respectively. Taking scalar products of both sides of Eq.(2), and summing up for all sites i, the following relations are obtained.

$$\max(\,\lambda_n^2\,) \geq v_2^2\,(\,1 - \Sigma\,|c_{ij}^d|^2\,)\,, \qquad\qquad (3)$$

$$\min(\,\lambda_n^2\,) \leq v_2^2\,(\,1 - \Sigma\,|c_{ij}^d|^2\,) \leq v_2^2\,, \qquad\qquad (4)$$

where, n=1 or 2, $\lambda_1 = -E+3V_1$, $\lambda_2 = -E-V_1$, and c_{ij}^d denotes the coefficient c_{ij} associated with a dangling bond.

For the case of $V_2/V_1 < 2$, the allowed energy regions are the same as those obtained for the case containing no dangling bonds. When $V_2/V_1 > 2$, there appear new states in the energy gap. To estimate the width and the location of this energy region, another relation is needed besides the inequalities (3) and (4), which depends on the dagling bond density and the correlation of their mutual positions.

In the situation that almost all of the atoms have one dangling bond, the quantity F defined by

$$F = \Sigma(\Sigma\,|c_{ij}|^2 + |c_{ij}^d|^2)$$

should satisfy the relation $F \leq 1$ from the normalization requirement. This gives the following relation,

$$E^2 + 3v_1^2\,(\,1 - 1/\alpha\,) \leq 0\,, \qquad\qquad (5)$$

where $\alpha = \Sigma\,|c_{ij}^d|^2$ is used. In the case of lower densities of dangling bonds in which two dangling bonds do not occur on adjacent atoms, the following inequality is derived.

$$(E + V_1)^2 (E - 3V_1)^2 + V_2^2 E^2 + 3V_1^2 V_2^2 (1 - 1/\alpha) \lesseqgtr 0. \qquad (6)$$

The possible energy regions satisfying (3),(4), and (5) are shown against α in Figs. 1:a), b), and c) for three values of the parameter V_2/V_1.

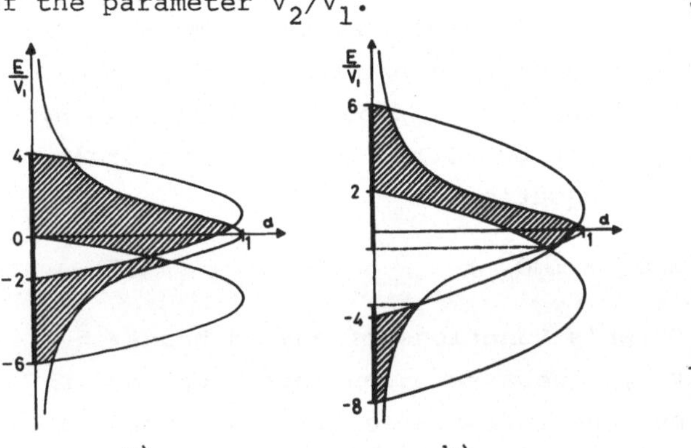

a) b)

Fig. 1: The possible energy vers. α.
The shaded regions are allowed
and unshaded are forbidden.
a) $V_2/V_1 = 3$, b) $V_2/V_1 = 5$, c) $V_2/V_1 = 7$.

The new states due to the dangling bonds appear near the center of the band gap. It should be noticed that the wave functions of these states have large probability amplitudes only on the dangling bonds($\alpha \approx 1$). The width of this energy region is large for small values of V_2/V_1 and decreases as the value of V_2/V_1 increases. In Fig.2, the allowed energy regions are re-plotted as a function of V_2/V_1. The solid and the dashed lines show the results for the present and the lower density case obtained by the same procedure from (3), (4), and (6).
It is also shown through simple considerations that the number of these states is equal to that of the dangling bonds.

Fig. 2: Bounds of the density of states as a function of V_2/v_1. The solid and the dashed lines correspond to the different densities of dangling bonds.

3. BETHE LATTICE

The density of states profiles are calculated for the Bethe
lattice for which the previous results should apply. The Bethe
lattice is a mathematical construct that is rather artificial
but it illuctrates a number of useful points[3]. We consider
the high density case that each atom has one dangling bond.
If the atomic function of the danglind bond is indicated by A
and that in the bonding state by B, the density of states $\rho(E)$
is given by the following equation.

$$\rho(E) = -\frac{1}{\pi}(\ 3\mathrm{Im}G_0^A + \mathrm{Im}G_0^B\) \tag{7}$$

where, G_0^A and G_0^B are Green's functions of any ij bonds in A
and B states respectively. These functions are developed in
terms of V_1 and V_2, then the infinite series are summed up in
closed forms as follow,

$$G_0^{A^{-1}} = \frac{E^2 - V_1^2}{E} - \frac{2V_1V_2}{(E-V_1)}(\lambda + \frac{V_2E}{2V_1(E+V_1)}) - \frac{2V_1V_2}{(E-2V_1)}(\lambda + \frac{V_1(E+V_1)}{V_2E}).$$

$$G_0^{B^{-1}} = E - \frac{3V_1^2}{E-2V_1 - \frac{2V_1V_2}{(E-V_1)}(\lambda + \frac{V_2E}{2V_1(E+V_1)})}. \tag{8}$$

where λ satisfies the following equation,

$$\lambda^2 - B\lambda + 1/2 = 0, \qquad B = \frac{(E+V_1)(E-3V_1)}{2V_1V_2} - \frac{V_2E}{2V_1(E+V_1)}. \tag{9}$$

The resulting densities of states are shown in Figs.3 and 4
for $V_2/V_1 = 3$, and 7, respectively. These figures clearly
reveal the general results of the previous section concerning
the allowed energy regions and the correlation between the
energy and the probability amplitudes on the dangling bonds.
Another interesting feature is that the states near the center
of the band gap due to the dangling bonds are made up from
states which otherwise form the δ function in the density of
states, namely those located on the tops of valence and
conduction bands.

a) b)

Fig. 3: Local densities of states. The solid line shows
the present results and the dash-dotted line
those of no dangling bond case[2],[3]. The
heavy lines indicate bounds on the density of
states by the general theorem. The δ functions
still remain, but their intensities become small.

a) b)

Fig.4: Same as Fig. 3, except $V_2/V_1 = 7$.

4. REFFERENCES

[1] D. Weaire; Phys. Rev. Letters <u>26</u>, 1541 (1971).

[2] D. Weaire and M.F. Thorpe; Phys. Rev. <u>B4</u>, 2508 (1971),
Phys. Rev. <u>B4</u>,3518 (1971).

[3] J.F. Nagle, J.C. Bonner, and M.F.Thorpe; Phys. Rev. <u>B5</u>,
2233 (1972).

ON SOME DIELECTRIC PROPERTIES OF DISORDERED SEMICONDUCTORS

V. L. BONCH-BRUEVICH

Faculty of Physics. Moscow University
Moscow. B-234. USSR

Nonlinear dielectric properties of a disordered semiconductor are considered. Hidden electronic ferroelectricity is predicted.

1. INTRODUCTION. HIDDEN FERROELECTRICITY

A disordered semiconductor (D.S.) often contains some kind of a random field acting upon the charge carriers. Two features of the electron energy spectrum of such materials seem now to be well established (see the review paper [1] containing references to the relevant original papers): a) there is an everywhere dense spectrum of discrete fluctuational levels in a mobility gap; b) there is a random gradual bending of the bands, the relevant characteristic length, ℓ , being small compared to the dimensions of a sample and big compared to the "electronic" length γ^{-1} . In the problem of interest the latter is the characteristic localisation radius of the carrier trapped at the fluctuation level of importance. In view of a) the energy spectrum of a D.S. should contain a number of discrete states possessing low enough ionization energies, W , and high enough polarizabilities.[+]

Hence the trapped carrier contribution, ε_t , to the total dielectric constant, ε , might become significant. Some consequences therefrom were studied in ref. [3] , including the temperature dependence of ε_t and the eventual influence of illumination ("optodielectric" effect). Another obvious consequence is the eventual dielectric nonlinearity

[+] Cf. the similar diamagnetic effect [2] .

showing up at relatively low field strengths. This might
be of interest in connection with the property b). In fact
the latter means that there is a random gradual field $\underline{E}$
acting upon the charge carriers. This is related to the
macroscopic Lorentz field $\underline{E}'$ and the polarisation vector
$\underline{P}$ by

$$\underline{E} = \underline{E}' + \xi \underline{P} \, , \tag{1}$$

ξ being some numerical coefficient (to make life simpler
we neglect both eventual tensor character of ξ and its
$\underline{P}$ dependence). The quantities $\underline{E}$ and $\underline{P}$ are to be un-
derstood here as local expectation values obtained by ave-
raging over the random field within the spatial region of
the linear dimensions ℓ . In absence of an external field
$\underline{E}' = 0$. Yet the vectors $\underline{E}$ and $\underline{P}$ may be locally non-
-zero provided the function $\underline{P}$ ($\underline{E}$) is nonlinear. Thus
we come to the idea of a "hidden ferroelectricity" in a
D.S.: such a material would resemble the multidomain ferro-
electric the spatial regions where $\underline{P}$ and $\underline{E}$ are approxi-
mately constant playing the role of randomly oriented do-
mains. To see if this is indeed possible we have to consi-
der the non-trivial solutions of eq.(1) at $\underline{E}' = 0$.

2. POLARISATION VECTOR AND THE FERROELECTRICITY CONDITION

Evidently, $\underline{P}$ is the sum of the three contributions, $\underline{P}_e$,
$\underline{P}_h$, $\underline{P}_l$, coming from the trapped electrons, trapped holes
and other mechanisms respectively. The latter we assume to
be linear: $\underline{P}_l = a_l \underline{E}$, $a_l \xi < 1$ (a "normal" dielectric).
On the other hand the functions $\underline{P}_e (\underline{E})$, $\underline{P}_h (\underline{E})$ have the
same form only if $E < E_\infty = e^{-1} W \gamma (W)$; the relevant values
of a_e and a_h were calculated in ref. $[3]$. At higher
field strengths the autoionization of the local levels comes
into play the carriers coming mainly to the region of the
quasicontinuous spectrum ($[1]$, § 9). Then the sum $\underline{P}_e + \underline{P}_h$
is easily seen to be bounded by a field independent quanti-
ty (until the semiconductor-metal transition takes place).

1059

Thus eq. (1) at $E'=0$, $E \neq 0$ is reduced to the form $\varphi(E)=1$ where

$$\varphi(E) = \xi \frac{\Phi(E)}{E} \rightarrow \begin{cases} a\xi, & E \rightarrow 0 \\ a_L \xi, & E \gg E_\infty \end{cases} \qquad (2)$$

and $a = a_L + a_e + a_h$.

Thus there are three cases to be considered: $\alpha)\ a\xi < 1$, $Max\ \varphi < 1$; $\beta)\ a\xi < 1$, $Max\ \varphi > 1$; $\gamma)\ a\xi > 1$. Note that, as contrasted to the conditions imposed on φ , the inequality $\gamma)$ does not contain too model-like assumptions, $a_e + a_h$ being expressible in terms of the density of states and the temperature.

In the case $\alpha)$ there is no non-trivial solution; in the case $\beta)$ both trivial and nontrivial solutions exist being either stable or metastable; in the case $\gamma)$ only the non-trivial solution is available, the trivial one, $E = 0$,being thermodynamically unstable. Since a_e and a_h are temperature- and illumination dependent the phase transitions between the three cases may take place. Since the electronic polarizability plays here a crucial role such materials might be called electronic ferro-electrica. According to [3] , a is very probably increased on raising the temperature, T, due to excitation of the localised carriers to the higher polarizability states (this lasts until the carriers become excited to the bands). The following possibilities arise here: 1) case $\gamma)$ is realised at $T=0$. The transition to the nonferroelectric case make take place only at high enough temperature when the thermal trap-to- -band excitation becomes of importance. Before that the "hidden ferroelectricity" might reveal itself, probably, via the residual polarisation only (and possibly via the dielectric losses). 2) case $\gamma)$ is realised at $T \geqslant T_o > 0$, the critical temperature T_o being easily obtainable from the results given in [3] . There is a phase transition at $T = T_o$ the ferroelectric phase corresponding to the higher temperatures. Jump of ε might be expected at $T = T_o$: at $T \lessgtr T_o$ $\varepsilon_t(T)$ should follow the "inverse" Curie law with the sign of $T - T_o$ opposite to the conventional one. 3)case

β) is realised. Hysteresis phenomena are to be expected
on polarising the sample by an external field.
Note that the phase transitions considered might be govern-
ed both by temperature and by illumination provided the
photon energy is less than the differences between the band
edges and the Fermi level.

ACKNOWLEDGEMENT

The author appreciates useful discussions with J.M.Lifshits,
M.J.Kaganov, A.G.Mironov and J.P.Zviagin.

REFERENCES

1 Bonch-Bruevich, V.L.; Mironov, A.G.; Zviagin, J.P.:
 La Rivista del Nuovo Cimento. $\underline{3}$,N Y, 391 (1973)

2 White, R.M.; Anderson, P.W.: Phil. Mag.,$\underline{25}$, 737 (1972)

3 Bonch-Bruevich, V.L.: Zurn. Eksp.Teor.Fiz., Pisma,
 $\underline{19}$, 198 (1974)

ELECTRONIC PROPERTIES OF TETRAHEDRALLY BONDED AMORPHOUS SEMICONDUCTORS

B. KRAMER and J. TREUSCH
Institut für Physik, Universität Dortmund
Fed. Rep. Germany

The Complex-Band Structure Method has been extended
to include changes of the pseudopotential form fac-
tors caused by short-range disorder. An effective
crystal model is introduced to calculate densities
of states and optical constants of amorphous tetra-
hedral semiconductors. Si and GaP are presented as
examples.

1. INTRODUCTION

The complex-band structure model (CBS)[1], which is based on the
pseudopotential scheme has been applied quite successfully to an
explanation of optical spectra of amorphous tetrahedral semicon-
ductors [2]. In this model the crystal structure is used as the
basis of a gaussian-broadened two-atom correlation-function.
Pseudopotential form factors are transferred **without change** from
the crystalline case. This assumption is obviously oversimpli-
fied, e.g. there is no reason for v_{2oo} to vanish in amorphous
Si, since inversion symmetry no longer holds microscopically. In
the following we will describe a first step of improvement and
demonstrate characteristic consequences that follow from a
change in effective form factors.

2. "EFFECTIVE CRYSTAL" APPROXIMATION AND ITS APPLICATION

We start from the Hamiltonian of an electron moving in the po-
tential of N identical atomic scatterers located at the posi-
tions r_j, in momentum representation

$$(1) \quad H = \sum_k |k\rangle\langle k|k^2 + \sum_{kk'} |k\rangle\langle k'| \sum_l \exp\{i(k-k')r_l\} v_{kk'}$$

where $v_{kk'}$ are the atomic form factors. Eq. (1) can be cast into
a "crystalline" form

$$(2) \quad H = \sum_k |k\rangle\langle k|k^2 + \sum_{kk'} |k\rangle\langle k'| \exp\{i(k-k')R_l\} v_{kk'}^l$$

where R_l are lattice points and v^l is given as

1062

(3) $v^l_{kk'} = \exp\{i(k'-k)(R_1-r_1)\} v_{kk'}$

Thus, we have transformed the structurally disordered system into a compositionally disordered one, because the potential v^l associated with the lattice points R_1 vary statistically. Now we introduce a new Hamiltonian describing an effective medium, namely

(4) $\hat{H}(z) = \sum_k |k\rangle\langle k|k^2 + \sum_{kk'} \sum_l \exp\{i(k-k')R_1\}\tilde{v}(z)|k\rangle\langle k'|$

where $\tilde{v}(z)$ is an effective, energy dependent atom potential. The usual single-site self-consistency condition for compositionally disordered systems reads [3]

(5) $\langle\hat{t}_1(z)\rangle = 0,$

where $\hat{t}_1$ is the scattering operator associated with lattice site R_1 relative to the effective medium:

(6) $\hat{t}_1(z) = (v^l-\tilde{v}(z)) + (v^l-\tilde{v}(z))\hat{G}(z)\hat{t}_1(z).$

Here v^l is defined by equ.(3) and the resolvent of the effective crystal is given as

(7) $\hat{G}(z) = (z-\hat{H}(z))^{-1}$

If now we restrict ourselves to the Born approximation

(8) $\hat{t}_1 \approx v^l - \tilde{v}(z),$

and assume the site potentials to be statistically independent (!) the self-consistency condition (5) implies, that

(9) $\tilde{v}_{kk'} = N^{-1}\sum_{l=1}^{N}\exp\{i(k-k')(R_1-r_1)\} v_{kk'}$

is independent of energy. Equ.(9) provides a reasonable estimate of the effect that structural changes have on the atomic pseudopotential form factors. The effects due to the lack of long range order are not expected to be treated satisfactorily. From earlier work [2], however, we know, that the lack of l.r.o. essentially yields a life-time broadening of the energy spectrum. Hence, the application of the method described above, in its simplest form following equ.(9) involves three steps:

(1) choose a suitable structural model and evaluate $\tilde{v}_{kk'}$,

(2) calculate the "band structure" and density of states of the effective crystal

(3) remove the fine structure in the density of states and op-

tical spectra by introducing an energy dependent life-time
broadening calculated along the lines of the CBS-method [3].

3. RESULTS AND COMPARISON WITH EXPERIMENTS

In a previous paper [4] the authors have described the simplest
structural models that may be used to calculate the effective
atom potential $\tilde{v}$, namely 17-atom clusters that contain a central
atom and all of its nearest and next-nearest neighbors. By rota-
ting two adjacent tetrahedra of this cluster by 60° around the
bond axis one goes from the "diamond"-to the "wurtzite"-struc-
ture. Rotating around more than one of the bond axes belonging
to the central atom one obtains atomic configurations that con-
tain five-fold rings. In the following we will confine ourselves
to a 30°-rotation around only one (111)-axis, which seems to be
a reasonable average angle of rotation if one looks at the Polk
model [5]. As in the case of Ge and GaAs [4] the form factors
v(111) belonging to the direction of rotation remain unchanged,
whereas the values of v(1-11) etc. decrease, thus leading to a
splitting of bands in the nonequivalent Λ-directions. v(2oo) is
no longer vanishing in Si. This leads to a splitting of the X-
point degeneracy in the valence band and in the lowest conduc-
tion band. The "band structures" calculated for crystalline and
30°-distorted Si and GaP are shown in figs. 1 and 2. Since
Cohen and Bergstresser's [6] form factors have been used as
starting point, quantitative agreement with experimental data is
not quite as good as possible [7] . The main point, however, is
clearly visible: the gap between the two lowest L-peaks in the
valence DOS of Si tends to close, whereas it remains quite pro-
nounced in GaP. This effect is still enhanced through a CBS-
broadening that is twice as large in the critical energy range
as compared to simple CBS-calculations. New van-Hove singulari-
ties along the X-W-axis in Si fill in additional states into the
lower gap of the DOS. So, the measured valence DOS with its
characteristic differences between homopolar and heteropolar
semiconductors [8] seems to be at least qualitatively explaina-
ble. The results for the conduction bands are also shown in
figs. 1 and 2. Besides the drastic life-time broadening of the

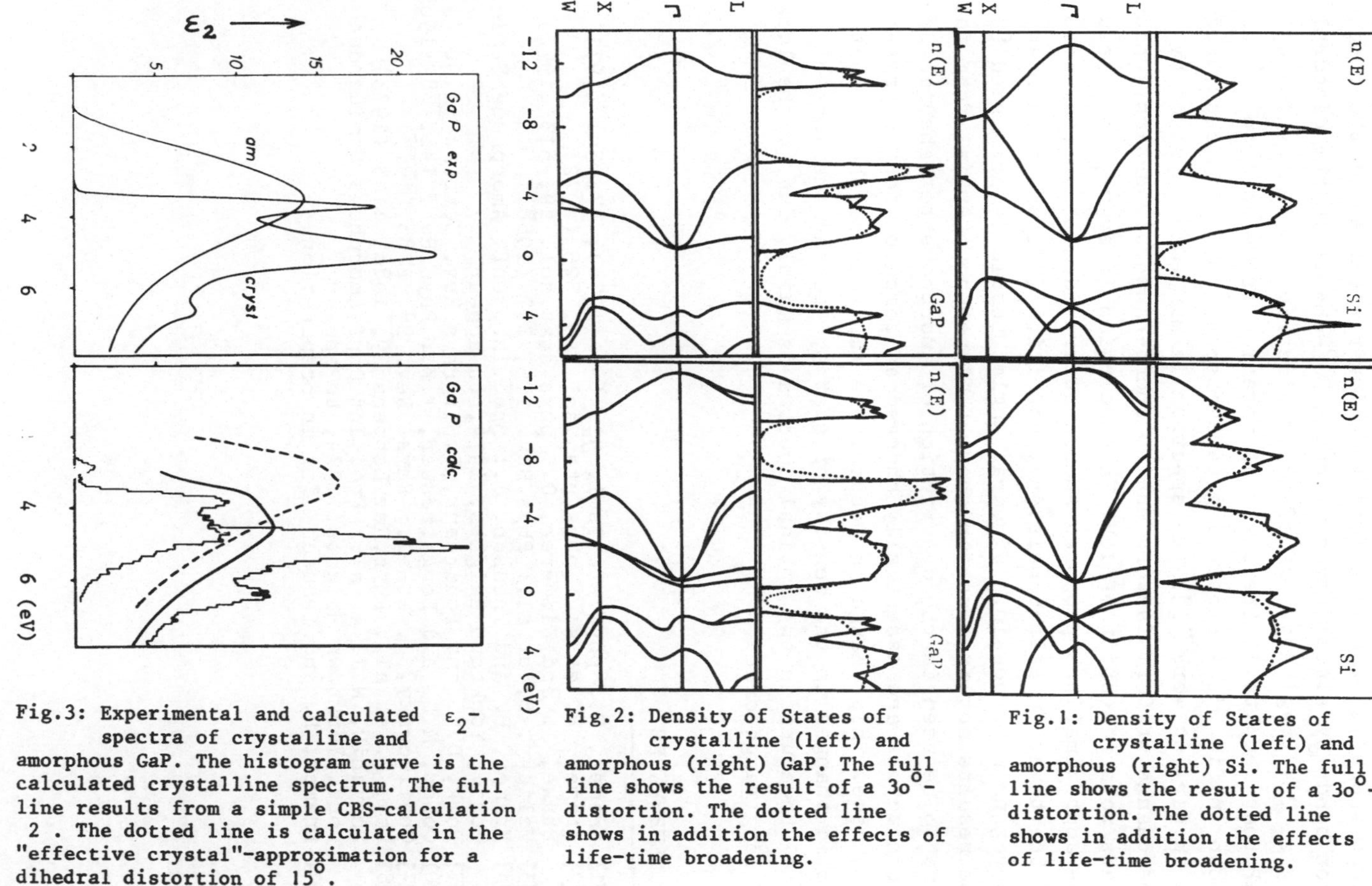

Fig.1: Density of States of crystalline (left) and amorphous (right) Si. The full line shows the result of a 30°-distortion. The dotted line shows in addition the effects of life-time broadening.

Fig.2: Density of States of crystalline (left) and amorphous (right) GaP. The full line shows the result of a 30°-distortion. The dotted line shows in addition the effects of life-time broadening.

Fig.3: Experimental and calculated ϵ_2-spectra of crystalline and amorphous GaP. The histogram curve is the calculated crystalline spectrum. The full line results from a simple CBS-calculation 2 . The dotted line is calculated in the "effective crystal"-approximation for a dihedral distortion of 15°.

bands there is an average shift to lower energies of the conduction bands by almost 2eV due to the 30°-change in the dihedral angle. The imaginary part of the dielectric constant $\varepsilon_2(\omega)$ of GaP that results if the angle of rotation is reduced to 15° is shown in fig. 3. A broadening parameter $\alpha = o.o7$ has been used as in earlier work [2]. The spectrum is markedly improved by the change of form factors due to short-range distortion, which leads to a red-shift of 1eV. In conclusion we argue, that though the approximations proposed should be extended to an evaluation of the t-matrix self-consistency condition (5), and though an averaging over possible dihedral angles should be performed [9], the results obtained so far show, that pseudopotential form factors to be used in the description of amorphous tetrahedral semiconductors are essentially changed as compared to the crystalline case. This change is responsible for some of the characteristic changes in the density of states, whereas the loss of line structure in the optical spectra is mainly due to the lack of long range order.

4. REFERENCES

[1] Kramer, B.: phys.stat.sol. 41, 649 (197o); 47, 5o1 (1971)
[2] Kramer, B. et al.: phys.stat.sol. 49, 525 (1972)
 Stuke, J. and Zimmerer, G.: phys.stat.sol. 49, 513 (1972)
[3] Velicky, B. et al.: Phys.Rev. 175, 747 (1968)
[4] Kramer, B. and Treusch, J.: Proc.Int.Conf. Amorphous Semicond., Yorktown Heights, 1974, to appear
[5] Polk, D.E. and Boudreaux, D.S.: Phys.Rev. Letters 31,92 (73)
[6] Cohen, M.L. and Bergstresser, T.K.: Phys.Rev. 141, 789(1966)
[7] Chelikowsky, J. et al.: Phys. Rev. B8, 2786 (1973)
[8] Ley, L. et al.: Phys.Rev.Letters 29, 1o88, 11o3 (1973)
 Shevchik, N.J. et al.: Proc.Int.Conf.Amorphous Semiconductors, Yorktown Heights, 1974, to appear
[9] Kramer, B. and Treusch, J.: in preparation

THE THERMAL CONDUCTIVITY OF LIQUID SEMICONDUCTORS

MELVIN CUTLER

Physics Department, Oregon State University, Corvallis, Ore.

Transport equations are presented in a form appropriate for considering ambipolar transport in liquid semiconductors. This is done by making use of Fermi-Dirac integrals to allow for near-degeneracy, and using a formulation which does not assume two separate bands. The theory is applied to published data for liquid Sb_2Se_3.

1. INTRODUCTION

The theory for the electronic thermal conductivity κ_e is well known, and it is usually considered in terms of two components, κ_b and κ_m. The single band contribution κ_b is related to the electrical conductivity σ by the Wiedemann Franz (WF) law $\kappa_b = W\sigma T$. The ambipolar contribution κ_m may be written in a two band situation in terms of electron (n) and hole (p) contributions to σ and the thermopower S:

$$\kappa_m = (S_p - S_n)^2 \, \sigma_p\sigma_n T/(\sigma_n + \sigma_p). \qquad (1)$$

The more familiar expression for κ_m incorporates the Maxwell Boltzmann (MB) approximation, according to which $(S_p - S_n) = (k/e)(const + E_g/kT)$ where E_g is the band gap [1].

In semiconducting liquids, the MB approximation for κ_m is generally inaccurate when κ_e is comparable to the atomic thermal conductivity κ_a because of statistical degeneracy. More complicated expressions should be used (but are frequently ignored) based on Fermi-Dirac integrals. A second complication occurs when the bands overlap, since there is now another contribution to the conductivity $\sigma(E)$ at each energy which cannot be associated with the individual bands [2]. The purpose of this paper is to discuss the appropriate theoretical treatment of κ_e in these circumstances. We shall derive the behavior of $\kappa_e/\sigma T$ as a function of E_g for the special case of symmetrical parabolic bands.

The theoretical curves for symmetrical bands are applicable to the thermal transport in liquid Sb_2Se_3. We find that ambipolar transport provides an adequate explanation of the behavior of κ_e. This conclusion differs from the one in an earlier study by Regel <u>et al.</u>[3]. The difference is probably due

to their use of the MB approximation in Eqn. 1. Our analysis
also leads to quantitative information about the behavior of
the pseudogap of liquid Sb_2Se_3.

2. THE ELECTRONIC THERMAL CONDUCTIVITY

The basic equations when the transport of charge and heat
in states at energy E is described by a conductivity $\sigma(E)$ and a
heat of transport $(E - E_f)$ are [4]

$$\sigma = - \int \sigma(E)(\partial f/\partial E)dE \qquad (2)$$

and

$$\kappa_e/\sigma T = \left(\frac{k}{e}\right)^2 \left\langle\left(\frac{E - E_f}{kT}\right)^2\right\rangle - S^2 , \qquad (3)$$

where $f = [1 + \exp\{(E - E_f)kT\}]^{-1}$, and

$$S = \frac{-k}{e} \left\langle\left(\frac{E - E_f}{kT}\right)\right\rangle. \qquad (4)$$

The expression $\langle F(E)\rangle$ refers to the special average

$$\langle F(E)\rangle = - \int F(E)[\sigma(E)/\sigma](\partial f/\partial E)dE. \qquad (5)$$

The first term on the right side of Eqn. 3 is $\kappa_e^*/\sigma T$, where κ_e^*
is the ratio of the heat flux $\vec{Q}$ to $-\vec{\nabla}T$ when the electric field
is zero. Ambipolar thermal conductivity arises when κ_e is in-
creased above the single band value because of a decrease in S
due to electronic transport in two bands. Therefore, the maxi-
mum value of κ_e for ambipolar transport is κ_e^*, and this occurs
when $S = 0$.

In liquid semiconductors with diffusive transport,
$\sigma(E) \propto [N(E)]^2$ [5]. If there is an overlapping valence band
with a density of states $N_v(E)$ and conduction band with density
$N_c(E)$, $\sigma(E)$ contains a term proportional to N_cN_v in addition to
the single band terms [2]. In such a situation, it is simpler
to evaluate $\kappa_e/\sigma T$, which differs from a single band value by an
amount determined mainly by the decrease in S. In the extreme
case where the bands are symmetrical with respect to E_f, $S = 0$,
and κ_e has a maximum value.

It is easy to present the results for symmetrical overlap-
ping bands with the variable parameter $\xi = (E_f - E_c)/kT =$
$-E_g/2kT$ if we know the form of $N(E)$. In another paper [2], we
found that $N(E) \propto (E - E_c)^{1/2}$ for molten Tl_2Te. This is the

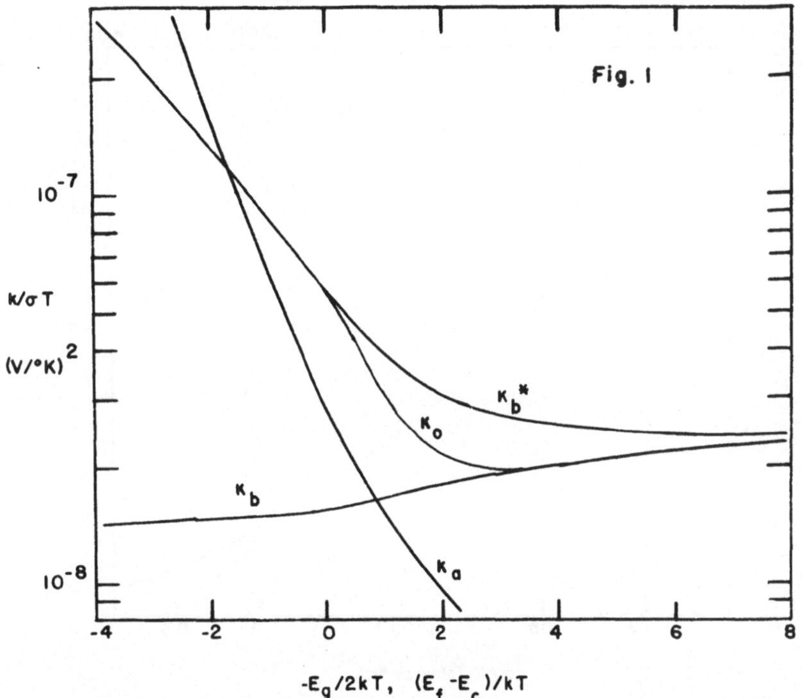

FIG. 1: Theoretical Contributions to $\kappa/\sigma T$ <u>vs.</u> ξ

standard form of $N(E)$, and using this in Eqns. 2, 3, and 4, we derive curve $\kappa_o/\sigma T$ shown in Fig. 1, where κ_o is κ_e for the overlapping bands. The calculation of the integrals for the overlapping band situation is discussed in Ref. 2. We also show $\kappa_b/\sigma T$ and $\kappa_b^*/\sigma T$ for a single band. The similarity of $\kappa_b^*/\sigma T$ and $\kappa_o/\sigma T$ shows that the overlap term $N_c N_v$ has a relatively small effect.

By comparing $\kappa_o/\sigma T$ with $\kappa_b/\sigma T$, one sees that the ambipolar contribution becomes appreciable when $\xi \overset{\sim}{<} 3$, i.e., while the band gap is still negative. σT decreases rapidly when ξ becomes negative, causing $\kappa_a/\sigma T$ to also increase. We put things into perspective by plotting also $\kappa_a/\sigma T$, using a typical value of the atomic thermal conductivity $\sim 4 \times 10^{-3}$ watt/cm deg. and b = $\sigma(E)/E \sim 2300(\text{ohm cm eV})^{-1}$ as observed in Tl_2Te [2].

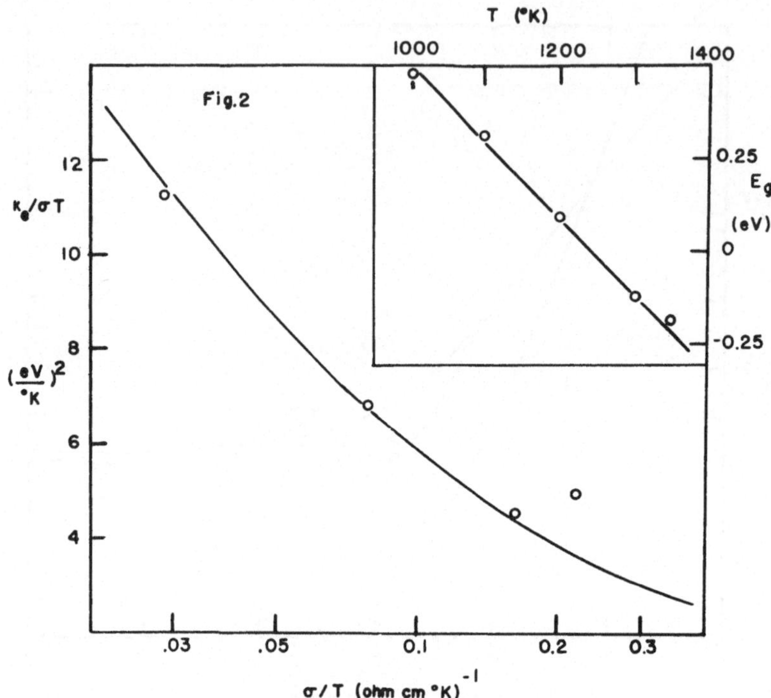

FIG. 2: Analysis of Sb_2Se_3 Data in Terms of the
Parabolic Symmetrical Band Model

3. THE BEHAVIOR OF ANTIMONY SELENIDE

The thermoelectric data for liquid Sb_2Se_3 reported by
Kazandzhan [6] shows that S is essentially zero over the temper-
ature range 1000 - 1350°K, whereas σ increases rapidly with T
from $\sim$ 10 to 300 ohm^{-1} cm^{-1}. This implies that the Fermi energy
is very close to the center of a rather symmetrical pseudogap
which goes to zero within the experimental range. In these cir-
cumstances, one would expect a large contribution to κ_e from
ambipolar transport, and our symmetric band calculations should
be applicable.

In addition to the curves shown in Fig. 1, the symmetric
overlapping parabolic band model yields σ in the form
$\sigma = bkTF_\sigma(\xi)$, where $F_\sigma(\xi)$ is a particular function of ξ. There-
fore we can compare the experimental curve for $\kappa_e/\sigma T$ <u>vs</u> σ/T
with the theoretical curve obtained by eliminating ξ from
$\kappa_e/\sigma T$ <u>vs</u> ξ. This is shown in Fig. 2, where we have fitted

theory and experiment with bk = 0.0775 (ohm cm deg)$^{-1}$. In doing
this, we have followed Regel _et al._ [3] in taking κ_a = 1.0
× 10^{-3} cal/cm deg sec. We have omitted the point at the lowest
T (1000°K) because $\kappa_e \sim 0.2\ \kappa_a$ in this case, causing a large
error. Using this value of bk, we infer from the experimental
σ(T) the values of ξ and hence E_g. E_g(T), shown on an inset in
Fig. 2, varies linearly with T (in °K) in accordance with the
equation E_g = (2.5 - 2.03 × 10^{-3} T)eV.

These results may be compared with corresponding parameters
determined for Tl$_2$Te [2], where bk = 0.228 (ohm cm deg)$^{-1}$ for
the conduction band and 0.072 (ohm cm deg)$^{-1}$ for the valence
band. The value of -dE_g/dT for Sb$_2$Se$_3$ is larger than Tl$_2$Te
(7.5 × 10^{-4} eV/deg), but comparable to As$_2$Se$_3$ (1.7 × 10^{-3}
eV/deg) [7].

We have neglected the effects of a mobility shoulder in
σ(E) because the data aren't sufficient for a meaningful analy-
sis which would take this additional factor into account.
Because of the diffusiveness of the Fermi Dirac factor, $\partial f/\partial E$,
the theoretical curves in Figs. 1 and 2 would not be altered
much by such a change in shape, but irregularities can be ex-
pected in the region of $\xi \sim 0$ where non-conducting states over-
lap in energy. The parabolic band edge is another arbitrary
assumption. (We have found that when the calculated curves are
based on N(E) α E instead of E$^{1/2}$, the results are not much dif-
ferent from those shown in Figs. 1 and 2.) Therefore our analy-
sis can only be expected to provide an approximate interpreta-
tion of the observed behavior within the constraints of these
simplifying assumptions.

4. REFERENCES

[1] Regel, A. R.; Smirnov, I. A.; Shadrichev, E. V.: Phys.
 Status Solidi (a)5 13, 1971.
[2] Cutler, M.: Phys. Rev. B9, 1762 (1974).
[3) Regel, A. R.; Smirnov, I. A.; Shadrichev, E. V.: J. Non-
 Crystalline Solids 8 - 10, 266 (1972).
[4] Cutler, M.: Phil. Mag. 25, 173 (1972).
[5] Mott, N. F.; Davis, E. A.: "Electronic Processes in Non-
 Crystalline Materials," Oxford (1971), p. 26.
[6] Kasandzhan, B. I.: Soviet Physics Semiconductors 2, 329
 (1968).
[7] Edmund, J. T.: Brit. J. Applied Phys. 17, 979 (1966).

LOW FREQUENCY MODES IN AMORPHOUS SEMICONDUCTORS

P.C.TAYLOR, U.STROM, J.R.HENDRICKSON and MARK RUBINSTEIN

Naval Research Laboratory
Washington, D. C. 20375 USA

Two techniques are employed to probe the
low frequency modes observed in specific
heat measurements of amorphous semiconduct-
tors. A disorder induced coupling of acoustic
modes to optical radiation in these materials
is demonstrated by the experimental observa-
tion of a far infrared and microwave absorp -
tion whose magnitude correlates with the
density of states inferred from the T^3 term
in the low temperature specific heat. Enhanced
densities of low frequency modes, which con-
tribute to the T term of specific heat in
glassy As_2S_3 and As_2Se_3, are confirmed from
As^{75} spin-lattice relaxation rates determined
by pulsed nuclear quadrupole resonance (NQR).

1. INTRODUCTION

At low temperatures the specific heat of amorphous semicon-
ductors contains two terms ($c_1T+c_3T^3$) [1]. The term linear
in temperature, which is not present in crystalline materials,
dominates C_V below ~1K ($C_V=c_1T$). The term proportional to
T^3 which also occurs in crystalline materials, is greater
than that predicted on the basis of the Debye model using
experimental sound velocities. Ultrasonic attenuation mea-
surements indicate the presence of modes below 1K which sat-
urate [2] and are not Debye thermal lattice excitations. It
has been suggested that these modes result from tunneling of
atoms or groups of atoms between two near equilibrium positions
in amorphous materials [3]. There is also evidence for Debye
excitations at microwave frequencies from Brillouin scattering
experiments at temperatures between 1.7 and 300K [4]. This
article describes two additional probes of the low frequency
vibrational properties of amorphous semiconductors-far infrared
and microwave spectroscopy and nuclear quadrupole resonance
(NQR) spectroscopy. Experimental details are available else-
where [5, 6]

2. FAR INFRARED AND MICROWAVE MEASUREMENTS

The frequency dependence of the room temperature conductivity
(or index of refraction times absorption, $n\alpha$) is presented
for 5 amorphous semiconductors in Fig. 1 on a log-log scale.

Fig. 1

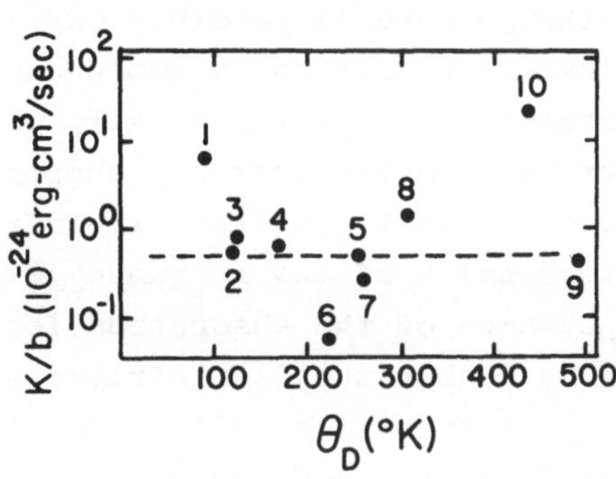

Fig. 2

Fig. 1: Room temperature, far infrared and microwave conduc-
tivity σ, (or index of refraction times absorption
coefficient, $n\alpha$) of selected amorphous semiconductors
as a function of wavenumber (or frequency) on a log-
log scale. (Data for $Tl_2SeAs_2Te_3$ were taken at 100K
to eliminate the large thermally-activated, band-type
conductivity observed at higher temperatures [6].
The α-Ge data show a thermally-activated conductivity
below 1 cm^{-1}).

Fig. 2: Contribution to the ν^2 conductivity of amorphous ma-
terials as a function of contribution to the T^3 term
in the specific heat where K and b are defined in
Eqs. 1 and 3, respectively. 1. $Tl_2SeAs_2Te_3$; 2. As_2Se_3;
3. Se; 4. As_2S_3; 5. PMMA; 6. PS; 7. B_2O_3; 8. GeO_2;
9. SiO_2; 10. $Na_2O-3SiO_2$. Values of b for As_2Se_3
and $Tl_2SeAs_2Te_3$ were estimated from θ_D.

The data for Fig. 1 can all be represented by an empirical
equation of the form

$$\sigma(\hbar\omega) = K(\hbar\omega)^\beta \tag{1}$$

where $\beta \lesssim 2$ and $\omega = 2\pi\nu$. For most materials β is 2 within
experimental error, but for one sample (Se), there is a devia-
tion from a strict ν^2 dependence ($\beta \approx 1.7$). In the five amor-
phous semiconductors of Fig. 1, the absorption is temperature
independent from 4.2 to 300K at 0.8 cm^{-1} and above. In all
amorphous materials, whether semiconducting or insulating, the

absorption is temperature independent [7] above 10 cm^{-1}. The
important features of the absorptions shown in Fig. 1 are the
nearly identical frequency dependences for a variety of struc-
turally diverse amorphous materials, the temperature indepen-
dence, and the magnitudes which are greatly increased over
those observed in corresponding crystalline materials.

Although several possible explanations for the observed far
infrared absorption in amorphous materials might be offered,
correlations with the density of states inferred from the T^3
term in specific heat [1] support a model based on a disorder-
induced coupling of the far infrared radiation to a Debye-
like density of low frequency modes. The lack of temperature
dependence of the absorption for all materials above 10 cm^{-1}
suggests that the far infrared absorption at a frequency ν is
due to interactions with a density of vibrational modes at
that frequency and not due to a relaxation broadening of modes
at higher energies. Density fluctuations, which are known to
exist in amorphous materials generally, could in principle
provide a mechanism for coupling to modes which are normally
acoustic in crystalline materials by inducing spatial fluc-
tuations in either static or dynamic charge or polarization.
Although several specific models might be proposed to explain
the data of Fig. 1, they all represent specific applications
of the general equation

$$\langle \sigma(\hbar\omega) \rangle = \langle M(E, \epsilon, e^*,...) \, g(E) \rangle \tag{2}$$

where g(E) is the density of phonon states and M is a genera-
lized coupling coefficient for the radiation which may depend
on energy, the dielectric constant of the medium, and a dis-
order-induced, dynamic or static effective charge or polari-
zation. The averaging procedure implied by Eq. 1 depends on
the details of the specific model and is affected by both the
extent and magnitude of the proposed fluctuations.

At low energies the density of vibrational states, as deter-
mined from the low temperature specific heat, can be written as

$$D(\hbar\omega) = a + b(\hbar\omega)^2 \tag{3}$$

where a is the magnitude of the anomalous linear term and
b is proportional to the magnitude of the T^3 contribution.
There is no particular correlation between the published val-

ues of a [1] and the magnitude of K for the materials studied. However, there is an apparent correlation between K and b as indicated in Fig. 2 for a set of 10 semiconducting and insulating amorphous materials. The ratio K/b is a measure of the coupling coefficient M of Eq. 2, while the ν^2 frequency dependence arises naturally from the phonon density of states. To the extent that the ratio K/b is nearly constant, the generalized coupling coefficients M of Eq. 2 are of the same order of magnitude for this set of structurally diverse materials. Furthermore, the greatest deviations in K/b occur in materials for which one would qualitatively expect large deviations. For example, in sample 10 (Na_2O-$3SiO_2$), the sodium is ionically bound in the glass, which enhances the absorption in glasses and even in crystalline materials [8]. The decreased value of K in sample 6 (PS) is due to partial crystallinity of the sample and is consistent with microwave absorption in partially crystalline polymers [9]. The enhanced absorption in $Tl_2SeAs_2Te_3$ could be due either to the partial ionic character of the thallium atoms or to local field effects enhanced by the large value of the dielectric constant (ϵ' = 30) as predicted in some specific models [10]. The remaining seven samples of Fig. 2 have ratios of K/b that vary by less than a factor of 3 about the average value indicated by the dashed line.

The consistency of these results for such structurally diverse materials supports the existence of Debye-like phonons in amorphous materials down to 0.1 cm^{-1} (0.14K) similar to those confirmed down to 1.7K in Brillouin scattering experiments [4].

3. NQR SPIN-LATTICE RELAXATION MEASUREMENTS

Since nuclear spin-lattice relaxation proceeds via vibrational modes, one might expect greater low temperature ^{75}As spin-lattice relaxation rates in glassy As_2S_3 over those in the crystalline material. The effective relaxation rate in glassy As_2S_3, T_1^{-1}, is proportional to T^2 from 77 to 1.1K and can be described fairly accurately by the empirical relaxation T_1^{-1} = 8 x $10^{-3}T^2$. The relaxation rate measured in the glass is much greater than the rate in the crystal at any given temperature,

and this difference is enhanced at lower temperatures where
the crystal relaxation rate decreases exponentially with tem-
perature, while the glass rate remains proportional to T^2.

In a crystalline material, a T^2 dependence of the relaxation
rate occurs whenever the dominant modes relaxing the nuclear
spins are at energies which correspond to temperatures lower
than the measurement temperatures [11]. The T^2 temperature
dependence in glassy As_2S_3 is strong evidence for the existence
of very low frequency vibrational modes in the glass which are
thermally excited even at ~1K and are strongly coupled to the
nuclear spins.

The T^2 dependence of the relaxation rates is consistent with
models invoking both atomic tunneling [3] and Debye thermal
lattice modes [1] if the interaction proceeds via a first order
Raman process [6]. Difficulties are encountered in attempts to
obtain numerical values for the relaxation rates given the den-
sity of low frequency modes from specific heat measurements [6].
Hence, the nature of these low frequency modes in glassy As_2S_3
is presently undetermined.

4. REFERENCES

[1] Zeller, R.C. and Pohl, R.O.: Phys. Rev. B4, 2029 (1971);
 Stephens, R.B.: Phys. Rev. B8, 2896 (1973).
[2] Hunklinger, S., Arnold, W., Stein, S., Nava, R. and Drans-
 feld, K.: Phys. Lett. A42, 253 (1972); Golding, B., Graeb-
 ner, J.E., Halperin, B.I. and Shutz, R.: Phys. Rev. Lett.
 30, 223 (1973).
[3] Anderson, P.W., Halperin, B.I. and Varma, C.M.: Phila. Mag.
 25, 1 (1972); Phillips, W.A.: J. Low Temp. Phys 7,351(1972).
[4] Love, W.F.: Phys. Rev. Lett. 31, 822 (1973).
[5] Strom, U. and Taylor, P.C.: Proc. V Int. Conf. on Liq. and
 Amorph. Semi. (Taylor and Francis, London, 1974) in press.
[6] Rubinstein, M. and Taylor, P.C.: Phys. Rev. B9, 4258(1974).
[7] Strom, U., Hendrickson, J.R., Wagner, R.J. and Taylor,P.C.:
 to be published.
[8] de Jong, C., Wegdam, G.H. and van der Elsken, J.: Phys. Rev.
 B8, 4868 (1973).
[9] Amrhein, E.M. and Schulze, H.W.: Kolloid Z. U. Z. Polymere
 250, 921 (1972).
[10] Bagdade, W. and Stolen, R.: J. Phys. Chem. Solids 29, 2001
 (1968); Whalley, E.: Trans. Faraday Soc. 68, 662, (1972);
 Schlömann, E.: Phys. Rev. 133, A413 (1964).
[11] Van Kronendonk, J.: Physica 20, 781 (1954).

HIGH RESOLUTION ELECTRON MICROSCOPY OF AMORPHOUS SEMICONDUCTORS AND OXIDE GLASSES

P. H. GASKELL[*] AND A. HOWIE

Cavendish Laboratory, Cambridge

Fringe structures in axial bright field high resolution micrographs have been examined in a number of amorphous materials. The tendency for these fringe structures to arise only in thin regions of the specimen could suggest some degree of local order. A structural model which may account for some of these effects is proposed for the tetrahedrally bonded cases.

High resolution transmission electron micrographs of a-Ge published by Rudee and Howie [1], showing fringe-like regions suggestive of locally ordered lattice planes, have excited controversy and speculation. Computed images of continuous random network (CRN) models lead to apparently contradictory conclusions even when the same coordinate data are used. A CRN model for silica [2] does not show fringe-like structure [3]. Opinion is divided on the nature of the images from the Henderson-Herman [4] and Polk [5] models for amorphous tetrahedral materials: one group claiming fringes for each model [6,7] the other failing to find them [8,9]. Berry and Doyle [10] using an essentially one-dimensional argument propose that the fringes will be produced by any model, random or otherwise if offset imaging conditions are employed. Spaepen and Meyer [11] obtained fringes in optical transforms of projections of the Polk model. More recently, Krivanek and Howie [9] compared images computed for a crystallite model and for the Polk model, finding fringes in the former case only and for specimens of thickness less than 50 Å. Current theory cannot account for the observation of fringes in thicker specimens of any structure. The experimental data of Rudee and Howie were obtained by tilting the incident beam of electrons so that the image is formed from the undiffracted beam and a portion ($\sim 1/3$) of the first diffraction ring. Although this technique has experimental advantages, including lower sensitivity to focus conditions than axial illumination, it is less clearly understood and gives rise to difficulties in correcting the astigmatism, which could result in spurious orientation - dependent features appearing in the image. In addition, the usefulness of the offset illumination method in reducing image overlap effects may have been overestimated in the case of small crystallites [3], since a given crystallite diffracts

[*] Permanent address: Pilkington Brothers Ltd., Ormskirk, England

simultaneously to diametrically opposite points on the ring. Accordingly,
micrographs of several amorphous materials have been recorded using axial
illumination. The image is formed from the undiffracted beam and one or
more diffraction rings which are accurately centred around the optic axis.
A JEM 100B was used with a pointed filament and precisely aligned optics.
Critical settings are current centre and minimum size of the projected fila-
ment image. For oxide glasses a low total beam current ($\leqslant 10\mu$Å) is
necessary to avoid degradation of the fringe structure. Large objective
apertures were used (120 μm) and in several instances the aperture was omit-
ted entirely. The importance of avoiding overlap effects in the image by
using very thin specimens is generally accepted and regions estimated to be
less than 100 Å thick were found to give optimum results. Specimens were
prepared either by vacuum deposition (C), R. F. sputtering (As,Ge) or by
collecting wedge-shaped fragments scored from a fresh fracture surface
(oxide glasses).

Figure 1 shows part of a through focal series for a-C. Fringes of spacing
~ 4 Å run in all directions (in contrast to the observations [1,3] with
tilted illumination) and parallel fringes frequently cover areas of 100 -
300 Å^2. Using Cochran's [8] criterion for fringe visibility, the proba-
bility of a fringe like region of area 100 Å^2 being generated by a random
structure is $\sim 10^{-18}$ or about one fringe system in $\sim 10^{11}$ micrographs.

Fig.1 Amorphous Carbon. Focussing steps ~ 200 Å

Figure 2 shows images of a-C foils photographed under identical conditions:
one (2a) prepared by slow vacuum evaporation (20 s) the second (2b) rapidly
deposited (2 s). Several fringes of area ~ 200 A^2 are visible in the slowly
evaporated film, none appear in the rapidly deposited film.

a b

Fig.2 Amorphous carbon films deposited slowly (a), rapidly (b)

Similar fringe-like structures are found in a-Ge, a-As and vitreous SiO_2
(Fig.3). A thick specimen of a partially crystallised lithium aluminosili-
cate glass is shown in Fig.4. A small ($\sim$50 Å) crystallite of β-eucryprite
is clearly visible against an 'amorphous' background indicating that imaging
conditions for lattice planes in crystalline and amorphous regions are
identical.

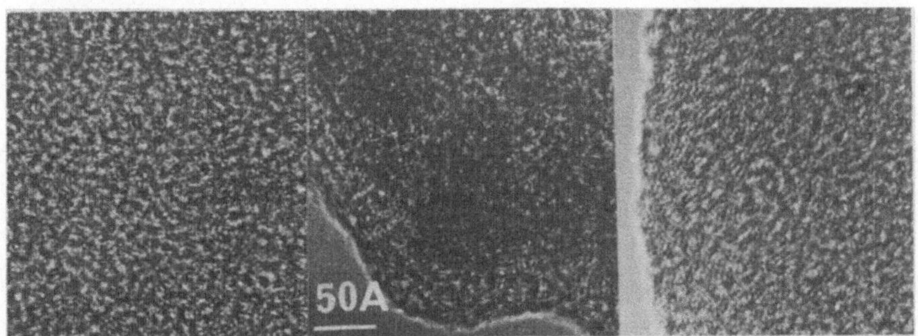

Fig.3 Amorphous Ge, As and vitreous SiO_2

Fig.4 Li-Aℓ-SiO_2 glass, partially crystallised

Our results indicate that large areas ($>$100 $\overset{\circ}{A}{}^2$) of fringe-like structure
tend to arise only in the thinner regions of wedged specimens (Fig.4). If

the typical structure, showing no such fringes, which occurs in thicker
regions (indicated with an asterisk in the figure) can be assumed to arise
because of the increased random superposition of atomic images (rather than
multiple or inelastic scattering), these observations would then suggest
some degree of order over distances comparable with the thickness in the
thinner regions. Although the micrographs are suggestive of ordered regions,
it is not clear what type of structure the ordered regions might have.
Single crystallite models give unsatisfactory fits either to scattering data
or optical measurements. The central difficulty lies in accounting for the
interface between crystallites - a point which was clearly recognised by
Warren in 1937 [12]. Apart from models in which a microcrystallite is sur-
rounded by a CRN matrix with modified skin [3] this problem seems to have
received little attention. If attempts are made to join microcrystals of
the diamond lattice, one non-trivial solution appears to have low bond dis-
tortion. Two (111) faces can be joined by a plane of eclipsed bonds. Two
planes meet in an edge containing 5-fold rings (Fig.5). Edges meet at a
corner where the 5-fold rings generate pentagonal dodecahedra. One model
which can be built in this way consists of 14-atom tetrahedra (diamond
cubic) bounded by (111) planes. Tetrahedra are stacked so that (111) planes
are contiguous. Lattices such as this are not space-filling; strain gener-
ated by pentagonal rings eventually leads to unacceptable distortion. The
effect of this strain is to broaden the first or second peak in the radial
distribution function (RDF) [the choice being determined essentially by the
near neighbour force-field and therefore model-independent]. The unbroad-
ened first and broadened second peak (R_2) in the experimental RDF for a–Ge
can thus be <u>contained</u> <u>explicitly</u> <u>within</u> <u>the</u> <u>model</u>. The breadth of R_2 depends
on the ultimate cluster size: a cluster of about 20-30 tetrahedra containing
several hundred atoms reproduces the observed width of R_2.

Fig.5 a) Two
14-atom units
(projection
along (110));
b) A 5-unit
model viewed
along the
5-fold axis.

Taking Tempkins data [13] for the shape of R_2, the positions and widths of
R_3 and R_4 distributions can be calculated <u>firstly</u> assuming that there is no

disorder except that contained in R_2. The model-dependent parameters intro-
duced at this stage are the number of 3-bond neighbours with ring geometry
specified by the model. The resulting RDF is shown in Fig.6(b) and is
already a good fit to the experimental data. The peak at 4.7 Å prominent
in the diamond lattice is diminished.

Fig.6 Experimental and calculated RDF for a-Ge

Introducing an additional broadening term into the calculations for R_3 and
R_4 due to the spread in Si bond angles (which is responsible for the spread
in R_2) agreement with experiment is further improved (Fig.6(c)). Additionally
this model seems likely to produce fringes at a spacing of d_{111} as the pro-
jection in Fig.5 indicates. The structure proposed here must be considered
as no more than an interesting speculation at this (very preliminary) stage
of its development. However, we are encouraged to note that nuclei with
complex but ordered structures are found experimentally in the early stages
of growth of gold and Ge particles [14,15]. We thank Pilkington Brothers
and the Science Research Council for financial support.

REFERENCES

[1] Rudee, M.; Howie, A.: Phil. Mag. 25, 1001 (1972)
[2] Bell, R. J.; Dean, P.: Phil. Mag. 25, 1381 (1972)
[3] Howie, A.; Krivanek, O. L.; Rudee, M. L.: Phil. Mag. 27, 235 (1973)
[4] Henderson, D; Herman, F.: J. Non-Cryst. Solids 8, 359 (1972)
[5] Polk, D. E.: J. Non-Cryst. Solids 8, 359 (1972)
[6] Graczyck, J. F.; Chaudhari, P.: To be published
[7] Chaudhari, P.; Graczyck, J. F.: Proc. 5th Int. Conf. on Amorphous and
 Liquid Semiconductors - to be published
[8] Cochran, W.: Phys. Rev. B8, 623 (1973)
[9] Howie, A.; Krivanek, O. L.: J. Appl. Cryst. (to be submitted)
[10] Berry, M. V.; Doyle, P. A.: J. Phys. C6, L6 (1973)
[11] Spaepen, F.; Meyer, R. B.: J. Non-Cryst. Solids 13, 440 (1973-4)
[12] Warren, B. E.: Appl. Phys. 8, 645 (1937)
[13] Tempkin, R. J.: Adv. in Physics 22, 581 (1973)
[14] Allpress, J. G.; Saunders, T. V.: Surface Sci. 7, 1 (1967)
[15] Mader, S.: J. Vac. Sci. Technol. 8, 247 (1969)

HEAVILY DOPED MATERIALS

CONTENTS

AN ANALOGUE OF MOTT TRANSITION IN COMPENSATED GaAs

B.M. VUL

Lebedev Physical Institute of the Academy of Sciences,
Moscow, USSR

The conductivity of compensated GaAs with $N_d \approx N_a \approx$
$\approx 2.5 \cdot 10^{17}$ cm^{-3} was investigated down to T = 0,5°K.
To explain the results of the measurements it was
suggested that the electrons at low temperature
are localized in a narrow continuum of states near
the bottom of the conduction band. At the appro-
priate concentration of the electrons a Mott tran-
sition occurs.

Impurities in semiconductor crystal form a system of "alien"
atoms distributed in a certain manner in the crystal. The in-
fluence of the medium into which impurity atoms are immersed is
taken into account by introducing its dielectric permeability
and the carriers' effective mass.

If a semiconductor contains only one type of impurity, for
instance, donors, and their concentration Nd is so low that
$N_d \cdot a_B^3 \ll 1$, where a_B is the Bohr orbit radius for impurity atoms,
the interaction between them can be neglected and we can regard
each donor as an independent entity. In another limiting case,
when $N_d \cdot a_B^3 \gg 1$, the impurity atoms are near to each other and
the strong interaction between them leads to the localization of
the electrons not bound by valency bonds.

The transition from semiconductor to metal conductivity as shown
by Mott [1] takes place at the condition

$$N_c^{1/3} \cdot a_B \approx 0.25 \tag{1}$$

where N_c is the concentration of donors, corresponding to the
transition.

The mentioned phenomena were extensively studied in non-compen-
sated semiconductors [2,3].

If both types of electrically active impurities-donors and accep-
tors are present, the interpretation of the transport phenomena
becomes more complicated [4,5,6,7]. Inversely charged impurity
atoms and the fluctuations in their concentration create randomly

distributed electric fields in the crystal. Gubanov [8] was the
first to suggest that near to the band edges, the electrons can
be localized. In 1958, Anderson [9] published his well-known
paper, showing that at the adequate conditions randomly distri-
buted potential wells can localize the electrons. Such tran-
sition of the semiconductor from conducting into non-conducting
state is known as the Anderson transition.

Going from an abstract model to the possible conditions of its
realization in some cases one can calculate the parameters of
potential wells taking into account the processes occurring
during the crystal growth from melt.

In a strongly compensated semiconductor, the Debye radius limit-
ing the fluctuation region is equal to

$$R_d = \left(\frac{\kappa kT_1}{4\pi Ne^2}\right)^{1/2} \tag{2}$$

where T_1 is a temperature near the melting point [10], κ - the
dielectirc permeability, k - the Boltzmann constant. The aver-
age magnitude of the fluctuation of the number of charged
impurity atoms in a sphere with radius R_d is equal to

$$\Delta N = \left(\frac{4\pi}{3} R_d \cdot N\right)^{1/2}$$

and the average value of potential near the surface of the
sphere
$$\overline{\varphi} = \frac{e}{\kappa R_d} \cdot \overline{\Delta N}$$

If with the ions, electrons and holes also play an appreciable
part in the screening, it can naturally influence the value of
R_d [11 + 14].

In studies of GaAs, important contributions were made by
D.N. Nasledov and his colleagues [15]. They investigated the
conductivity and Hall constants in a wide temperature range in
GaAs samples doped by different impurities [16].

Redfield and Crandall [17] have investigated high doped compen-
sated GaAs and interpret the results in terms of localization
effects of band tail.

In p-type GaAs, irradiated by varied fluences of neutrons Mora,
Loferski and Bermon observed at T=1, 42°K a kink in the slopes

of conductivity and Hall mobility temperature dependence from
the initial holes' concentration and came to the conclusion that
a Mott transition induced by neutron irradiation takes place [18].

It is known that compensated semiconductors can be regarded as
simplest disordered systems. The degree of disorder depends on
the impurity concentration and the method of impurity introduction.
To have small disorder effects we have chosen for our studies
impurity concentration when $N_d.a_B^3 \approx 1$. In this case the concept
of the "tails" of density of states is no longer valid. On the
other hand, the concept of isolated impurity atoms is not valid
either.

The studies were conducted on n-type gallium arsenide samples
with $N_d \approx N_a \approx 2,5.10^{17}$ cm^{-3}. The data for some of the samples
are presented in Table 1.

The results of resistivity measurements in the range between
room temperature and $T \approx 1°K$ for the samples with different
values κ are given in Fig. 1. From the fan-like spread of the
curves, a conclusion can be made that if electron concentrations
$n_o \geqslant 5.10^{16}$ cm^{-3}, they are in degenerated states, for the con-
centrations $n_o < 5.10^{16}$ cm^{-3} the conductivity is due to the
thermal activation.

Table 1

No	$n_o=N_d-N_a$, cm^{-3}	$N_i=N_d+N_a$, cm^{-3}	$K = \dfrac{N_a}{N_d}$, %
1	6.10^{15}	6.10^{17}	98
2	7.10^{15}	$5,5.10^{17}$	97
3	$1,3.10^{16}$	$4,8.10^{17}$	95
4	$1,5.10^{16}$	$4,6.10^{17}$	94
5	2.10^{16}	$4,8.10^{17}$	93
6	$2,2.10^{16}$	$4,7.10^{17}$	92
7	$2,6.10^{16}$	$4,5.10^{17}$	89
8	$4,0.10^{16}$	$4,5.10^{17}$	83
9	$4,8.10^{16}$	5.10^{17}	83

Fig. 1: Temperature dependence of the resistivity ρ.
(The index on the curve corresponds to that
of the samples in Table 1)

More precise data for the determination of electron activation
energy can be obtained from the results of measurements of Hall
effect temperature dependence. From these data it follows that
from room temperature down to T (5-7)°K, electron concentration

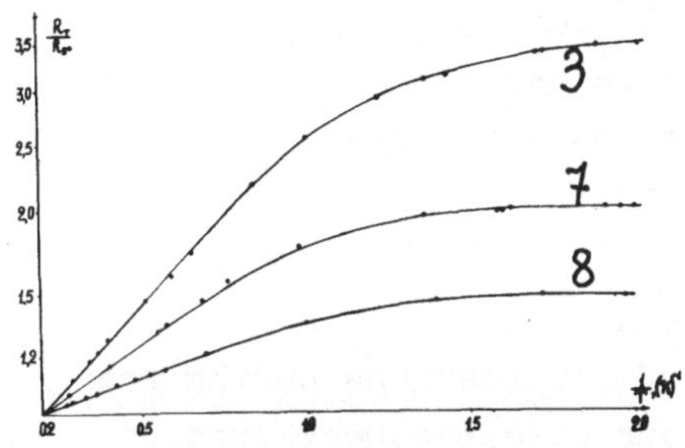

Fig. 2: The relative change of
the Hall constant.

remains practically constant,
from 5 to 1,5°K, it decreases
and at lower temperatures
does not change appreciably.

A relative change of Hall
constant R, dependent on I/T,
in the range $1,5 \leqslant T \leqslant 5°K$
for several samples is given
in Fig. 2. It may be noted
that in our samples the
transition from metallic to

non-metallic conductivity takes place at the condition
$n_c^{1/3} \cdot a_B = 0,37$. While the quantitative estimates for Mott's
criterion (1) are quite approximate [19], one can suggest that
the deviation, in our case, is due to the high compensation level.

The results of measurements of Hall mobility $\mu = R\sigma$ in the range
$1,5 < T < 300°K$ for some samples are presented in Fig. 3.

Fig. 3

Fig. 4

Fig. 3 and 4: Temperature dependences of Hall mobility

From these data it follows that in the temperature range between 20 and $100°K$, in a number of samples, a well pronounced dependence $\mu \sim T^{3/2}$ exists, which is typical for the case of carriers' scattering by charged centers.

For $T < 15°K$ as it is shown in Fig. 4, the mobility $\mu \sim T^{1/2}$ which is typical for scattering by dipoles or potential barriers. For these low temperatures, the electron wave-length is so large that the scattering occurs not by single centers, but by the impurity atoms' complexes.

The average sizes of potential wells as mentioned before were determined by the screening radius of fluctuations during the crystallization from the melt. In our case the Debye radius $R_d = 1,3.10^{-6}$ cm and $N.R_d^3 \sim 1$, therefore a concept of Debye screening is not quite correct. But the average distance between the impurities $R = N^{-1/3}$ is also equal to $1,3.10^{-6}$ cm therefore we can take this value as an average radius of potential wells.

It is evident that electron localization takes place in potential wells that can hold electrons. For a spherical well with radius R, the minimum depth is

$$U_{min} = \frac{\pi^2 \hbar^2}{8mR^2} \tag{3}$$

where m is the electron effective mass, and the depth of potential well

$$U_z = \frac{Ze^2}{\kappa R}$$

where Z is the number of elementary charges.

The localization of electrons occurs at the condition

$$U_z > U_{min} \quad \text{or} \quad U_z = \frac{Ze^2}{\kappa R} > \frac{\pi^2 \hbar^2}{8mR^2} \tag{4}$$

In our case we find that electrons localize if

$$Z > 1,2 \cdot \frac{a_B}{R} \approx 1 \tag{5}$$

and at $z = 1$, $U_1 = U_{min}$. In such potential wells the single energy level is at the edge of the well.

As deviations from average always exist in a system, this level is spread out into a band of states. The electrons at the levels inside the wells can only slightly influence the conductivity due

to random distribution of potential wells, the spread of the
levels inside them and the existence of barriers between them.
It is evident that electron mobilities on energy levels in the
wells are much lower than for electrons on the levels outside
the wells. These quasilocalized electrons can be regarded as
localized.

As it follows from the temperature dependence of conductivity
and Hall effect, the energies necessary for electron delocal-
ization are very small, and subsequently the potential "humps"
and wells only slightly warp the bottom of the conductance band.

Thus, as a model for calculations we have taken it that the
band edge is not warped and that near its bottom a narrow band
of localized states exists [19]. As even at $T \sim 5^\circ$K almost all
the wells are empty, the spread of the levels inside them is
near to $5 \cdot 10^{-4}$ eV, and, as the wells' concentration is near to
10^{16}cm^{-3}, the average density of states inside the energy band
is of the order of $\rho_1 \sim 10^{31}$cm^{-3}erg^{-1}. If, near the bottom of
band, the density of states is determined by the number of
wells, then, at higher neighbour energy levels the density of
states shall by determined mainly be donor concentration N_d.
The order of magnitude of the density of states approximately
[20] can be obtained from a relation

$$\rho_2 \sim \frac{2m}{\hbar^2} \cdot N_d^{1/3}$$

At $N_d = 3 \cdot 10^{17}$cm^{-3}and m $= 7 \cdot 10^{-29}$g, $\rho_2 \simeq 10^{32}$cm^{-3}erg^{-1}.

From our measurements, it follows that in the samples studied,
the total concentration of electron $n_o = N_d - N_a$ is larger than
the concentration of potential wells, $n_\ell(0)$ able to hold elec-
trons. At $T \to 0$, these levels are filled, and excess electrons
remain in the conductance band as a degenerate electron gas.

The physical speculation discussed above permit us to choose
as a simple model for calculations an energy structure of con-
ductance band drawn in Fig. 5.

At $T \to 0$, the electrons occupy an energy band $E_o = \dfrac{n_o}{\rho_1}$. A part
of the electrons is localized near the bottom in a band

$$\Delta = \frac{n_\ell(0)}{\rho_1}$$

Fig. 5: The energy model for density of states. The shaded range corresponds to the localized states, at T = 0.

Near the upper edge of this band the density of states changes abruptly from ρ_1 to ρ_2. In accordance with the accepted model one can calculate the parameters Δ and $\frac{\rho_2}{\rho_1}$ to fit a good agreement with the experimental data. In Fig. 6 the measured and calculated dependences $\frac{n_\ell}{n_0} = f(T)$ are shown.

Fig. 6: The temperature dependence of the concentration of localized electrons. The solid line corresponds to calculations, points - to the experimental data. Here

$$n_\ell = n_0 - n(T)$$

and n(T) is the concentration of the free electrons at the temperature T.

The results of the calculation are given in Table 2.

Table 2

No	$n_0, 10^{16} cm^{-3}$	$\Delta, {}^\circ K$	$E_0, {}^\circ K$	$\rho_1, 10^{31}$ $erg^{-1} cm^{-3}$	$\rho_2, 10^{32}$ $erg^{-1} cm^{-3}$	$n_\ell(o), 10^{16} cm^{-3}$
3	1,4	5,6	7,8	1,2	6,4	1,0
7	2,6	3,1	6,1	3,1	11	1,3
8	4,0	1,4	4,4	6,6	16	1,3

The concept of electron localization we tried to confirm by inducing Mott transition by laser excitation. The intensity of incident light with wavelength $\lambda = 0,69\ m\mu$ was changed from

5.10^{18} to 5.10^{20} photons/cm^2.sec. The measurements were made at $T = 1,6°K$ on a sample with total electrons' concentration $n_o \simeq 10^{16}cm^{-3}$. The results of these measurements are presented in Fig. 7, where the data are given for the dependence of the relation j/N_ϕ on the intensity N_ϕ, where N_ϕ is the number of photons penetrating the sample and j is the photocurrent. The dotted line corresponds to the values proportional to the quantum yield. As can be seen in Fig. 7, at a definite excitation intensity a jump of photocurrent takes place; it is natural to connect it with the delocalization of electrons from potential wells, due to screening by photoelectrons. The estimates show that delocalization takes place at electron concentration $n_c \sim 3.10^{16}cm^{-3}$ which is near to that calculated from the Mott relation.

Fig. 7:
The dependence of j/N_ϕ = d on N_ϕ. The dashed line corresponds to the case $N_\ell = 0$.

Thus it can be regarded that these measurements qualitatively confirm the basic suggestions made for the calculation and the physical picture of the discussed phenomenon. It is possible that the model used for our calculation is not a single one, to obtain an agreement with experimental data. However, the general picture of the phenomena correctly describes the reality.

The investigation was made together with E.I. Zavaritskaya, I.D. Voronova, N.V. Rojdestvenskaya and B.L. Voronov [21].

REFERENCES

1. Mott, N.F.: Proc. Phys. Soc. (London) 62, 416 (1949); Adv. Phys. 16, 49 (1967)

2. Alexander, M.N.; Holcomb, D.F.: Rev. Mod. Phys. 40, 815 (1968)

3. Fritzsche, H.: J. Phys. Chem. Sol. 6, 69, (1958) Fritzsche, H.; Cuevas, M.: Phys. Rev. 119, 1238 (1960)

4. Longo, T.A.; Ray, R.K.; Lark-Horovitz, K.: J. Phys. Chem. 8, 259 (1958)

5. Davis, E.A.; Compton, W.D.: Phys. Rev. 140, A2183 (1965)

6. Sasaki, W.; Yamanouchi, C.: J. Non-Cryst. Soc. 4, 183 (1970)

7. Shlimak, I.S.; Emtzev, V.V.: Phys. Stat. Sol.(b) 47, 325 (1971)

8. Gubanov, A.I.: Electr. Theory of Amor. Semicond., ed. by Academy of Science of USSR, 1963.

9. Anderson, P.W.: Phys. Rev. 109, 1492 (1958)

10. Keldysh, L.V.; Proshko, G.P.: Phys. Tverd, Tela (sov) 5, 3378 (1963)

11. Stern, F.: Phys. Rev. B3, 3559 (1971)

12. Bonch-Bruevich, V.L.: Optica i Spectrosk. (sov) 14, 495 (1963)

13. Volkov, B.A.; Matveev, V.V.: Phys. Tverd. Tela (sov) 8, 717 (1966)

14. Halpern, Yu.S.; Efros, A.L.: Phys. Tverd. Tela (sov) 6, 1081 (1972);
Efros, A.L.; Halpern, Yu.S.; Shklovsky, B.I.: Proc. 11th Int. Conf. Phys. Semicond., Warsaw, 126 (1972)

15. Emel'janenko, O.V.; Nasledov, D.N.: J.T.P. (sov) 28, 1177 (1958);
Nasledov, D.N.: J. Appl. Phys. 32, 2140 (1961)

16. Gassanly, I.Sh.; Emel'janenko, O.V.; Lagunova, T.S.; Nasledov, D.N.: Phys. i Tekhn. Polupr. (sov) 6, 2010 (1972)

17. Redfield, D.; Crandall, R.S.: Proc. 10th Int. Conf. Phys. Semicond., Boston, 574 (1970)

18. Mora, N.A.; Loferski, J.J.; Bermon, S.: Defects in Semicond.: Session 2, paper II, 103 (1972); Phys. Rev. Letts 27, 664 (1971)

19. Mott, N.F.: Phil. Mag. 6, 291 (1961)

20. Baltensperger, W.: Phil. Mag. 44, 1355 (1953);
Broody, T.B.: J. Appl. Phys. 33, 100 (1962)

21. Vul, B.M.; Voronov, B.L.; Voronova, I.D.; Zavaritskaya, E.I.; Rojdestvenskaya, N.V.: Phys. i. Tekhn. Polupr. (sov) 8 (1974).

OPTICAL RESPONSE OF HEAVILY-DOPED SEMICONDUCTORS

A. BRELOT, G. BOLBACH

G.P.S.E.N.S.[+] Université Paris VII - FRANCE

G. HOROWITZ, G. PELOUS

C.N.E.T., LANNION- FRANCE

Optical constants of semiconductors in the frequency range
below the energy gap are not well defined. The use of a set
of new methods will allow to improve the understanding of
optical absorption in the above mentioned wavelength range.
For instance we clearly confirm the $E^{3/2}$ energy dependence
of the relaxation time which means that ionized impurities
play the chief role in the scattering of free-carriers. We
also observed an extra-absorption at high frequencies in
p-type silicon which might have come from the valence inter-
band absorption. Finally we describe an interferometric
phase shift method that appears to be easy and fast to get
n and k.

1. INTRODUCTION

For frequencies lower than the gap energy, the optical constants n and k
may be written : $n^2 - k^2 = \varepsilon_1 - A <\tau f(\tau)>$ and $2\,nk = \varepsilon_2 + \frac{A}{\omega} <f(\tau)>$
where τ is the relaxation time, $\varepsilon_1 + i\varepsilon_2$ is the complex dielectric constant
excluding the free-carriers contribution, $A = \frac{e^2 C}{m^* \varepsilon_o}$, $f(\tau) = \frac{\tau}{1 + \omega^2 \tau^2}$, C is
the free carriers concentration.

We shall consider in this paper only the case of silicon. Thus ε_1 is inde-
pendent of ω and ε_2 is negligible for n-type silicon and $\varepsilon_2 = \frac{\lambda}{2\pi} n\,\alpha_{31}$
for p-type silicon (α_{31} is the valence interband absorption).

Schumann, P.A. et al [1] have determined n and k respectively by measuring
transmission and reflection up to 40μ for various doping levels (5×10^{16}
to 10^{19} carriers/cm^3). Till now this is the most complete set of experimen-
tal data on optical constants of silicon. The great uncertainty of experi-
mental data and the discrepency between theory and experiment above 12 μ
show clearly that much work has to be done in order to obtain a clear
understanding of absorption below the energy gap in heavily doped silicon.
The aim of this paper is to show how we shall obtain improved values by
using the results given by the set of the interconnected methods that we
describe in paragraph 2.

+ Laboratoire associé au C.N.R.S.

2. EXPERIMENTAL METHODS AND RESULTS

The precise determination of n and k can only be performed if samples of thickness a few times of α^{-1} can be prepared. For instance the measurement of 2×10^{19} Boron/cm^3 doped silicon requires a thickness less than 5 μ. It is difficult to prepare such samples with high and homogeneous doping levels by usual conventional method such as doping during growth. The technique that consits to diffuse impurities from the surface is not appropriated because the distribution of carriers is in general smooth and impossible to predict without systematic experiment. Low energy implantations were previously used in the same aim that ours [2]. But the proximity of the surface may modify the dopant atoms distribution while the annealing stage.

2.1 MULTIPLE HIGH ENERGY ION IMPLANTATION

Ion implantations in the 1-2 MeV energy range were carried out and annealed at 900° C. The thickness of the heavily doped region can be chosen by carrying out successive implantations whose the energy is increasing. A couple of high resistivity identical silicon samples were implanted. The characteristics of ions implantation for samples A and B were 0.95 MeV, 8.75×10^{14} cm^{-2} for A and 0.95 MeV, 8.75×10^{14} cm^{-2} plus 1.3 MeV, 9.25×10^{14} cm^{-2} plus 1.75 MeV, 1.10×10^{15} cm^{-2} for B. Figure 1 shows the corresponding free-carriers distributions that were calculated with L.S.S. theory [3] for samples A and B.

Fig.1

2.2 DISTRIBUTION THICKNESS CHECK BY INFRARED INTERFEROMETRY (I.I.R.) METHOD

The infrared wave reflected at the surface of a sample interferes with the wave reflected on the heavily doped region. We found [4] that the interference reflectivity spectrum contains all informations about the free-carriers distributions (we compared the I.I.R. method with capacitive and ion probe measurements [5]).

Fig. 2 Fig. 3

Reflectivity spectra which were calculated from rectangular distributions
(shown on figure 1 in dashed line) fit with the experimental spectra (fig 2
and 3). We deduced that the thickness of the heavily doped regions are res-
pectively d_A = 0.75 and d_B = 1.55μ for A and B samples.

2.3 ABSORPTION COEFFICIENT MEASUREMENT

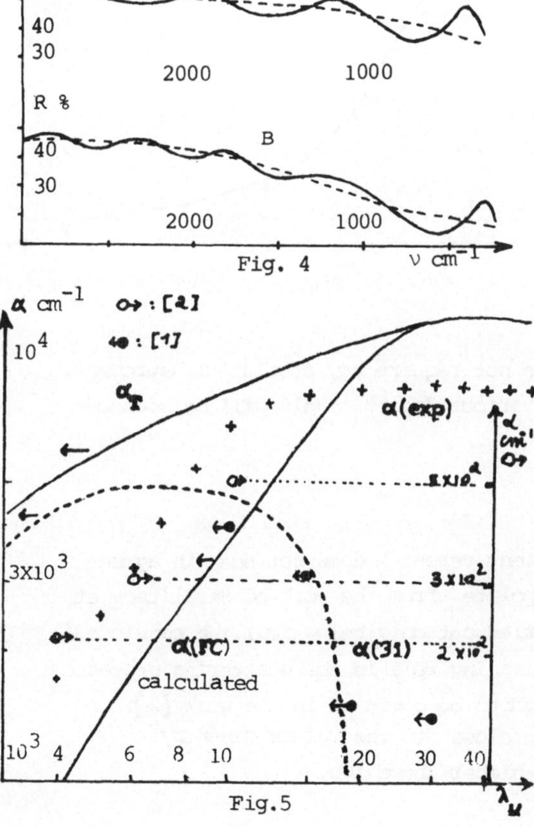

Fig. 4

Fig. 5

Figure 4 shows the transmis-
sion spectra of A and B samples.
Because $R^2 e^{-2\alpha d} \ll 1$ we may
write $\alpha = \dfrac{1}{d_B - d_A} \ln T_A/T_B$.
Absorption coefficient versus
the wavelength is plotted on
figure 5. The uncertainty cor-
responds to an error of 15 %
in the transmission measure-
ments (this is the maximum
error). The continuous curve
in fig. 5 is the free carrier
absorption α_{FC} that we compu-
ted by taking a $E^{3/2}$ dependen-
ce of relaxation time τ and a
non degenerate state density.
This calculation is different
from Lambert's one [6] who
took a relaxation time inde-
pendent of energy. The dashed
curve α_{31} is the interband
absorption as calculated by
Lambert. α_T curve is the total
absorption $\alpha(FC) + \alpha(31)$.

2.4 INTERFEROMETRIC PHASE SHIFT METHOD (I.P.S.)

The depth h of the buried heavily doped region is given by
$h = \dfrac{P - 1/2 + \Phi/2\pi}{2\,n_o\,\nu m}$ where P is the interference order, Φ is the phase shift
on the heavily doped region, νm is the wave number of the extremum.
The phase shift Φ is given by : $\tan\Phi = \dfrac{2n_o\,k}{n_o^2 - n^2 - k^2}$, where n_o is the refrac-
tion index of pure silicon.

In the high frequency region, Φ becomes negligible with respect to P and
we have $h = \dfrac{P - 1/2}{2\,n\nu}$. At low frequency we can determine n by :
$n^2 = n_o^2 - k^2 - \dfrac{2\,n_o k}{tg\Phi}$ (k is determined by absorption measurements)

The continuous curve of fi-
gure 6 is the calculated pha-
se shift Φ for 3.9 $\times 10^{19}$cm^{-3}
doped silicon substrates.
The crosses correspond to
the phase shifts that were
determined from experimen-
tal values of h and νm on
the reflectivity spectrum
of an undoped epitaxial layer
of thickness 4.3μ (see fig 1
of reference [7]). This
I.P.S. method allows to get
n by determining only the
wavelength of extrema in an
interference spectrum. Its

Fig. 6

interest is in the fact that it does not require any absolute measurement.
A complete set of curves n (λ) for various doping levels will be soon ob-
tained.

3. CONCLUSIONS

Our experimental absorption coefficient versus λ does not seem in agree-
ment with the one that we can interpolate from the work of Strel'tsov et
al [2] (their measurements were carried out only below 12μ). We concluded
that the ionised impurities play a leading rôle in the scattering processes
while acoustic vibrations were thought to be dominant in the work [2].
Our absorption measurements that are close to the values given by
Schumann et al [1] till 10μ, become higher above 10μ.

We believe the best precision that we obtained comes from the small thickness of our doped regions (here 1.55μ and 40μ in work[1]).

The form of the experimental $\alpha(\lambda)$ curve comes probably from the valence interband transitions. We found that the approximation τ = constant may introduce an error of a factor of about 2 below 10μ , in comparing with the calculation where $\tau = AE^{3/2}$ is taken. Thus we think that the Lambert's theory, where $\tau = C^t$ is taken, might be improved if this energy dependence was included. The I.P.S. method appears to be the easiest and fastest in order to get n for any doped semiconductors. It requires that a layer of a transparent material, must be created on the studied sample but this difficulty may be easily evercome for instance by evaporating amorphous undoped material or by using ion implantation.

We improved the determination of optical constants and consequently improved also the I.I.R. method that have a very large application field [8] [9]. Finally, this work is an example where ion implantation technique can be fruitful to get fundamental results outside of the field on this technique itself.

4. REFERENCES

1 Schumann, P.A. ; Keenan, W.A. ; Tong, A.H.; Gegenwarth, H.H. ; Schneider, C.P.. J. Elect. Soc $\underline{118}$, 145 (1971).
2 Strel'tsov, L.N. ; Titov, V.V. ; Sov. Phys. Semiconductors $\underline{4}$, 605(70).
3 See for instance : Johnson, W.S. ; Projected range statistics in semiconductors, Stanford (1969).
4 Brelot, A. ; Bolbach, G. ; Horowitz, G.. To be published in the proceedings of the International Conference on lattice defects. Frieburg.
5 Details may be found in D.R.M.E. reports n° 75/798 that is supporting the interferometry work.
6 Lambert, L.M.. Phys. stat. Sol $\underline{11}$, 461 (1972).
7 Brelot, A.; Bolbach G. ; Horowitz G.. Lagadec, Y. ; Pelous, G. Journal de physique suppl. au n° 11-12, 34, 35 (1973).
8 Brelot, A. ; Bolbach, G. ; Horowitz G. . To be published Proceedings of the conference on lattice defects. Frieburg, july 22^{th}-25^{th} (1974)
9 Brelot A. ; Bolbach G. ; Horowitz G.. To be published in the proceedings of the international conference on the applications of electrons and ions beams. Edited by " Société Française du Vide ". Toulouse October 8-11 th (1974)

MAGNETIC PROPERTIES OF CONDUCTION ELECTRONS IN METALLIC Si:P AND Si:B

G. C. Brown and D. F. Holcomb

Cornell University

Ithaca, N. Y., 14850, U.S.A.

New NMR data on Si:P(P^{31}) and Si:B(B^{11}) is presented for samples with impurity concentration from 5×10^{18} to 9×10^{19} cm^{-3}. These data give information about the magnetic properties of the impurity electrons. Comparison is made with recent NMR data for Si^{29} in similar samples by Sasaki, Ikehata, and Kobayashi. The most natural interpretation of the NMR data uses a model with a single, interacting, magnetic system of electrons. However, the data do not absolutely rule out the existence of localized magnetic moments. The NMR data is compared with recent measurements of electron spin susceptibility and specific heat.

1. INTRODUCTION

Heavily-doped silicon and germanium have attracted persistent attention as models of a disordered system in which an insulator-metal transition occurs as impurity concentration is varied. Mott [1] has given a recent synthesis of the state of interpretation of observations.

In this paper, we focus upon the magnetic properties of mobile carriers in Si:P and Si:B for samples with impurity concentration, n_I, from 4×10^{18} to 1.0×10^{20} cm^{-3}. Recent data and interpretation concerning the magnetoresistivity of Si:P [2] have supported and extended an early model [3] which postulated the existence of localized magnetic moments in the concentration range we consider. Recent spin susceptibility data [4] has been shown to be consistent with a model which separates the impurity electron system into two magnetic groups, one with a temperature-dependent Curie-Weiss susceptibility and a second with temperature-independent Pauli susceptibility. In this paper, we report new NMR measurements for the P^{31} system in Si:P and B^{11} system in Si:B which give further information about the magnetic character of the electron system. Our data complements new Si^{29} data of Sasaki, Ikehata, and Kobayashi [5], and is consistent with it.

After presenting our data and interpreting it in terms of a straightforward model, we first compare it with the Si^{29} NMR data. Then we match all NMR data against the ESR spin susceptibility data. We note that, although the two-component model of Quirt and Marko [4] has some attractive features, the data itself does not demand such a model. We then compare the existing specific heat data with implications of the direct magnetic data.

Although some minor discrepancies exist, there do not appear to be significant inconsistencies among types of experimental data where they can be directly compared. However, unambiguous evidence supporting any existing theoretical model is not yet at hand. In particular, the NMR, ESR, and specific heat data give little direct evidence concerning the nature of magnetic centers which may be responsible for the observed negative magnetoresistance.

2. DIRECT MAGNETIC DATA

In earlier studies of Si:P and Si:B, Sundfors and Holcomb [6] showed that, in the more heavily doped samples, the NMR properties are determined by interaction of the nuclei with the conduction electrons or holes. Figure 1 shows the P^{31} NMR absorption line at the extremes of the concentration range studied in the present work. Line shape and linewidth are determined by a

FIG. 1: NMR absorption lines for P^{31}. Signals were recorded by integrating the spin echo signal as the external field was swept through resonance. The reference zero for Knight shift measurements, the P^{31} reference in GaP, is marked.

distribution of values of K, the Knight shift, for the nuclei.

Table I gives experimental values of <K>, the value of K
at the center of gravity of the absorption line. Figure 2 plots

Table I

Values of <K> and P_F in Si:P

n_D	<K> x 10^4		χ_s(cgs/vol)* x 10^8		Wavefunctions at 4.2K	
	4.2K	1.3K	4.2K	1.1K	$\Omega_o P_F$ [†]	P_F [††]
8.0 x 10^{19} cm^{-3}	27.4 ± 2.4	32.6 ± 1.3	9.3	9.8	3400	0.28 ± 0.04 x 10^{24} cm^{-3}
3.2 x 10^{19}	37.0 ± 2.5	41.7 ± 1.6	7.0	7.5	6200	0.20 ± 0.02
1.9 x 10^{19}	41.5 ± 4.7	52.2 ± 3.0	5.8	6.3	8500	0.16 ± 0.02
0.75 x 10^{19}	(52.9 ± 9.7)**	71.3 ± 7.0	(6.9)**	7.2		

*Values of χ_s are taken from Fig. 1, Quirt and Marko, Ref. 4a.

**Value at 1.85K.

[†]Derived from the relationship $\langle K \rangle = \frac{8\pi}{3}(\chi_s)_{ESR}\Omega_o P_F$.

[††]Derived from previous column by setting $\Omega_o = 1/n_D$.

FIG. 2: Linewidth data for P^{31} in Si:P and B^{11} in Si:B, as a function of impurity concentration, n_I. ΔH is the full width at half maximum. For B^{11}, ΔH was obtained by measuring the free induction decay time, T_2^*, and using the relationship, $\Delta H = 2/\gamma T_2^*$.

values of the line width, ΔH.

In order to interpret the distribution of K values given by
the NMR data, we invoke the relationship

$$K = \frac{8\pi}{3}\, g\beta\, \frac{\langle S_z \rangle}{H_o}\, N\Omega_o P_F, \qquad (1)$$

where $\langle S_z \rangle$ is the mean value of the spin component, S_z, for
electrons at the Fermi energy, E_F; H_o is the applied magnetic
field; N is the number of electrons per unit volume; Ω_o is the
volume per impurity atom; and P_F is the electron wave function
density evaluated at the nuclear site. For a disordered system
such as Si:P or Si:B, there may be different local values for
the quantities, N, Ω_o, $\langle S_z \rangle$, and P_F.

The key to our analysis lies in the realization that any
temperature dependence in K must lie in $\langle S_z \rangle$, if we are at

sufficiently low temperatures.

Seeking the simplest interpretation of the NMR data, we further assume that all impurity electrons are in sufficiently strong magnetic communication with one another that there is a single value of $<S_z>$ which is appropriate to every electron. Two consequences follow.

1. For a sample at a given value of n_I, the line shape should be independent of field and temperature, and the line width should scale linearly with $<K>$. Data of Table I and Fig. 2, plus data not included herein [7], are consistent with this conclusion. To the level of detail included in their published paper, the Si^{29} data of Sasaki, et al., appears to us also to be consistent with this first consequence.

2. For a sample at a given value of n_I, the temperature dependence of $<K>$ should be exactly the temperature dependence of χ_s. For Si:P, comparison of the data of Table I with the data of Ref. (4) shows that the temperature dependence of $<K>$ is somewhat sharper than that of the χ_s data of Quirt and Marko and somewhat less sharp than the temperature dependence of χ_s found by Ue and Maekawa. We judge the data of Quirt and Marko to be somewhat more reliable.

The Si^{29} data of Sasaki, et al. shows a temperature dependence of $<K>$ which also appears to be consistent with the temperature dependence of χ_s.

Our conclusion is that the magnetic resonance data, both ESR and NMR, conform fairly well to the model suggested by Eq. (1), in which spin susceptibility and $<K>$ are directly connected by a single constant for a given sample. More particularly, the absence of any significant change in the shape of the NMR line as T or n_D is changed places significant constraints on models which require the presence of localized moments.

3. SPECIFIC HEAT

Two measurements of specific heat in Si:P have recently been performed. We focus on results for a sample with $n_D = 5.9 \times 10^{18}$ cm^{-3}. Marko, Harrison, and Quirt [8], working at zero magnetic field and at temperatures ranging down to 60 mK, see a small enhancement of the value of electronic specific heat, c_e,

below 1.0K which could be ascribed to electron correlations. Hedgcock, Heiniger, and Paoli [9], working at zero field and at 28.5 kG, from 1.5K upwards in temperature, searched for a magnetic contribution to specific heat.

In order to calculate the size of a magnetic term, we assume that all temperature dependence in the total susceptibility is contained in the spin term. (Total susceptibility data of Sasaki and Kinoshita [10] confirms this assumption.) From the Quirt and Marko data for $\chi_s(T)$, one can then calculate, directly from thermodynamics, a model-independent contribution to the specific heat at 3.5 kG, the field of the ESR measurements. From the data of Ref. 4a, we calculate this contribution to be 0.11 μJ/mole-K at 1.5K and 3.5 kG for the sample used by Hedgcock, et al.. Unfortunately, one must introduce a model in order to obtain a value appropriate to a field of 28.5 kG. We assume that the field dependence of specific heat follows a Curie-Weiss expression. We use the Curie-Weiss temperature of -3.7K determined by Quirt and Marko. The result of this extrapolation is to predict a magnetic contribution of 6.0 μJ/mole-K at 1.5K and 28.5 kG. The absence of such a term in the data of Hedgcock, et al. suggests some inconsistency between the experimental data for χ_s and c_H. Unfortunately, the necessity of extrapolating the ESR values of $\chi_s(T)$ to 28.5 kG eliminates the possibility of an unambiguous check.

4. REFERENCES

[1] Mott, N. F.: Adv. Phys. 21, 785 (1972).
[2] Khosla, R. P.; Fischer, J. R.: Phys. Rev. B 6, 4073 (1972).
[3] Toyozawa, Y.: J. Phys. Soc. Jap. 17, 986 (1962).
[4] Quirt, J. D.; Marko, J. R.: (a) Phys. Rev. Lett. 26, 318 1971; (b) Phys. Rev. B 7, 3842 (1973); (c) Ue, S.; Maekawa, S.: Phys. Rev. B 3, 4232 (1971).
[5] Sasaki, W.; Ikehata, S; Kobayashi, S.: To be published.
[6] Sundfors, R. K.; Holcomb, D. F.: Phys. Rev. 136, A810 (1964).
[7] Brown, G. C.; Holcomb, D. F.: Phys. Rev. B, To be published.
[8] Hedgcock, F. T.; Heiniger, F.; Paoli, A.: To be published.
[9] Marko, J. R.; Harrison, J. P.; Quirt, J. D.: To be published.
[10] Sasaki, W.; Kinoshita, J.: J. Phys. Soc. Jap. 25, 1622 (1968).

DYNAMICS OF ELECTRONS IN HEAVILY DOPED GaAs[*]

P.BROSSON, E.S.PINHEIRO, N.B.PATEL and J.E.RIPPER

Instituto de Física "Gleb Wataghin"
Universidade Estadual de Campinas
Campinas, S.P., Brasil

Spontaneous emission measurements in GaAs junction
lasers are interpreted assuming long thermalization
and recombination times of electrons injected into
the p region of the diode. Experimental results qua
litatively agree with calculations based on this model

1. INTRODUCTION

Up to date, thermal equilibrium has been assumed for electrons
injected into the p region of a semiconductor laser. We believe
that this assumption is not justified and we show through a
series of experiments that the thermalization time varies over
orders of magnitude, increasing rapidly as one goes deeper
into the gap. Long thermalization times lead to the possibility
of inhomogeneous broadening and hole burning which explains
the multimode [1] operation nearly always found in all semi-
conductor lasers. In addition, the long thermalization times
are confirmed by studies of time resolved spontaneous emission
intensity.

2. EXPERIMENTAL METHODS AND RESULTS

The homostructure lasers studied were formed by Zn diffusion
in a substrate of approximately 10^{18} donors/cm^3. The double
heterostructure lasers had a Si doped active region with a
net concentration of 10^{18} - 10^{19} acceptors/cm^3. Experimental
techniques used to measure spontaneous emission are
described in [2,3]. For time resolved spectral measurements,
the light was observed through a double monochromator with a

[*]Work supported by Telebrás S/A, Conselho Nacional de Pesqui-
sas and Fundação de Amparo à Pesquisa do Estado de São Paulo.

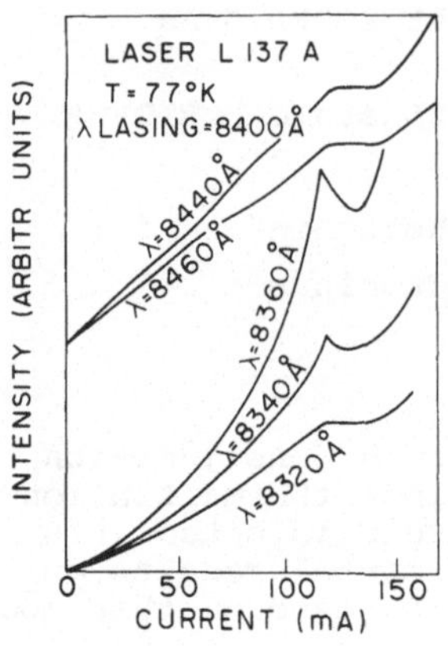

Fig. 1 Behaviour of the spontaneous emission intensity with current for a cw homostructure laser at 77 K. Upper part of the figure corresponds to the low-energy side of the laser line and the lower part to the high-energy side.

Fig. 2 Time behaviour of low energy (8560 Å) spontaneous emission of a DH laser below (132 mA) and above (400 mA) threshold. Beginning of the current pulse at zero time. $I_{th} \sim$ 300 mA.

cooled C 3104 RCA photomultiplier. Using a 50Ω as a load for the photomultiplier and a PM 3400 Philips sampling oscilloscope, a time resolution of about 5 ns was obtained.

In a first series of experiments, we measured the spontaneous emission intensity on each side of the laser line. Typical results are shown on Fig. 1 where we can observe a reduction of the spontaneous emission at threshold in a narrow energy (4-40 mev) range on the high energy side of the laser line and a partial saturation on the low energy side. This reduction is qualitatively different from the overall reduction of the spontaneous emission observed in single-heterostructure lasers which was caused by saturable absorbers [4]. Farther from the laser line, saturation of the spontaneous emission is observed as it was in Refs. [2,3]. Similar behaviour is observed for homostructure and double heterostructure lasers either pulsed or cw.

In a second series of experiments, we measured the emission intensity both during and after the excitation of the diode by fast current pulses. During the pulses, the high energy spontaneous emission reaches saturation within the response of the detection system (5ns), while on the low energy side, this time is quiet long (Fig. 2) even above threshold. The rise time of the

Fig. 3 Time resolved spectra of a DH laser below (a) and above (b) threshold. After the pulse the low energy peak shifts toward long wavelengths. Time is indicated near arrows. Vertical scale during the pulse is different from after the pulse.

spontaneous emission can increase by orders of magnitude as one moves toward longer wavelengths.

Time resolved spectra after the pulses are shown (Fig. 3) for current pulses below and above threshold. During the pulse, be low threshold we observe two peaks separated by about 50 meV, while above threshold we only observe the high energy peak corresponding to the laser line. The low energy peak seems to be hidden by the high energy peak. After the pulse, the low energy peak decreases slowly and shifts towards low energies while the high energy peak decreases faster without any noticeable shifting. We

Fig. 4(a) Decay of the low energy emission for a 400 mA current pulse.
(b) Time constant vs wavelengths for different currents. For 400 mA current pulse, two time constants τ1 and τ2 are observed.

measured the decay of the low energy emission in the range of
8400 Å - 8800 Å (down to 140 meV below the laser line). Ex-
perimental results (Fig. 4) show that this decay can be
approximately described by an exponential law with time
constants varying between 10 to 70 ns in this range of energy.
On Fig. 4(b) we observe increasing time decays as one goes
deeper into the gap. In addition a second time constant is
observed above threshold for low energy states.

3. INTERPRETATION AND DISCUSSION

We propose to explain the experiments assuming long radiative
and thermalization times increasing as one goes deeper into
the gap.
Above threshold nearly all the extra injected electrons
contribute to stimulated emission and if the thermalization
time of these electrons is much smaller than stimulated
recombination time, the electron spectral distribution should
remain constant and this would lead to a saturation [2,3,5]
of the spontaneous emission. Assuming long thermalization
times (at least comparable to the stimulated recombination
time) the states responsible for lasing action would not be
replenished as rapidly as they are being emptied and this
would lead to a reduction of the electron population for
states near and above the laser line, the effect beeing less
pronounced as one goes farther from this line. In other words
this would lead to a kind of hole burning in the electron
distribution near and above the laser line. Our measurements
of the high energy spontaneous emission very close to the
laser line show this behaviour and confirm the long
thermalization times contrary to presently held views [1].
Preliminary calculations based on this model confirm hole
burning and the observed behaviour.
The second series of experiments can again be explained by
long thermalization times of electrons in the band tail. As
one goes deeper into the tail, the thermalization time in-
creases more than the spontaneous recombination time τ_s. The
double time constant behaviour is explained by this model in

the following manner: deep in the tail, immediately after the end of the injection pulse, the states are nearly full and the spontaneous emission is dominated by τ_s. As the population gets depleted, the number of carriers and consequently the decrease in the spontaneous emission is controlled by how fast the carriers can thermalize from higher states. It is interesting to note that a very fast thermalization time would lead to the oposite behaviour with a very slow decrease of the spontaneous emission initially as the states are kept full by thermalization, followed by a very fast decrease when carriers from higher energy states get exausted. The experimental observations described above are in good agreement with results of computer calculations based on this model. Because of space limitations, details of these calculations are not included here and will be subject of another publication.

4. CONCLUSIONS

Thermalization times among states contributing to spontaneous and stimulated emission are much longer than previously thought. Even near the lasing line these times are long enough to allow for the burning of a hole on the injected population and the consequent multimode behaviour.

5. REFERENCES

[1] H. Statz, C.L. Tang, J.M. Lavine: J.Appl.Phys.35, 2581 (1964)

[2] J.E.Ripper, N.B.Patel, P.Brosson: Appl.Phys.Lett. 21, 98 (1972)

[3] P.Brosson, N.B.Patel, J.E.Ripper: IEEE J. Quantum Electron., QE9, 273 (1973)

[4] P.Brosson, N.B.Patel, J.E.Ripper, Appl.Phys. Lett., 23, 94 (1973)

[5] T.L.Paoli: Appl. Phys. Lett., 21, 101 (1972)

ELECTRONIC DENSITY OF STATES FOR A TIGHT-BINDING HAMILTONIAN WITH FLUCTUATING BOND INTERACTION

H.W. STREITWOLF and W. BRAUER

Zentralinstitut für Elektronenphysik der AdW der DDR

Berlin, German Democratic Republic

A generalization of Weaire's Hamiltonian shows un-
realistic δ-peaks in the density of states even if
one takes into account all interactions between
nearest neighbour bonds. Treating the bond inter-
action as a stochastic variable we calculated the
CPA-density of states. As expected the δ-peaks are
broadened according to the distribution function,
while the band gap increases with increasing
homogeneity of the solid.

1. INTRODUCTION

In order to investigate the electronic density of states for
tetrahedrally coordinated solids like silicon and germanium
the tight-binding Weaire-Hamiltonian was extensively studied
(see e.g. [1] to [3]). It takes into account the most impor-
tant nearest neighbour interactions leading to a gap in the
density of states between the fully occupied bonding and the
unoccupied anti-bonding bands which is responsible for the
semiconducting behaviour of the model. This result is indepen-
dent of the detailed positions of the atoms and the density
of states is determined only by the topology of the network.

2. GENERALIZED WEAIRE HAMILTONIAN

We define the Hamiltonian in a 4N-dimensional sp^3-orbital re-
presentation by means of two interaction operators

$$S|i\mathcal{R}\rangle = \sum_{j \neq i} |j\mathcal{R}\rangle \quad \text{(interaction at an atom)}$$

$$T|i\mathcal{R}\rangle = |\overline{i\mathcal{R}}\rangle \quad \text{(interaction at a bond)}$$

where $|i\mathcal{R}\rangle$ is an sp^3-orbital centered at atom $\mathcal{R}$ (i = 1,....,4)
and $|\overline{i\mathcal{R}}\rangle$ is the orbital covalently bound to the orbital $|i\mathcal{R}\rangle$.
Taking into account all interactions between nearest neighbour
bonds and denoting the interaction integrals according to fig. 1

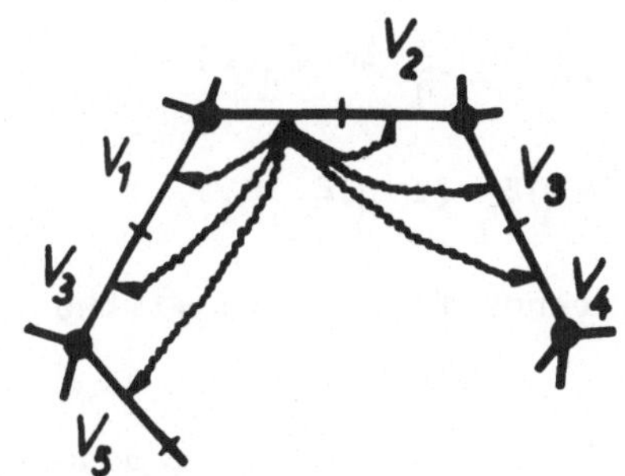

FIG. 1: Interactions in a tetra-
hedrally coordinated network

the Hamiltonian will be

$$H = V_1 S + V_2 T + V_3 (ST + TS) + V_4 TST \tag{1}$$

If the bond lengths and bond angles are the same throughout
the network, homologous interaction integrals between nearest
neighbour bonds will equally be the same and (1) will be a
realistic description of the network provided higher order
interactions (e.g. V_5) which depend on the dihedral angle be-
tween adjacent tetrahedra are unimportant.

Rewriting the Hamiltonian (1) as

$$H = 4(V_1 - V_4) P + H_2$$

(where $P = \frac{1}{4} (S + 1)$ is the s-projector [3]) one may easily
solve the eigenvalue problem of H_2 in terms of the eigenvalues
of the connectivity operator $T_s = PTP$:

$$H_2 \phi_\eta^\pm = E_\eta^\pm \phi_\eta^\pm \; ; \qquad E_\eta^\pm = \pm (V_2 + 2V_3 + 4V_4 \eta) - V_1 + 4V_3 \eta + 3V_4$$

$$H_2 \chi_\eta^\pm = E^\pm \chi_\eta^\pm \; ; \qquad E^\pm = \pm (V_2 - 2V_3) - V_1 - V_4$$

where $\phi_\eta^\pm = P^\pm \phi_\eta / \sqrt{(\phi_\eta , P^\pm \phi_\eta)}$, $P^\pm = \frac{1}{2}(1 \pm T)$ are the bonding
and anti-bonding projectors, ϕ_η are common eigenstates of P
and T_s ($P\phi_\eta = \phi_\eta$, $T_s \phi_\eta = \eta \phi_\eta$), and $\chi_\eta^\pm$ are arbitrary bonding
and anti-bonding states completing the $\phi_\eta^\pm$- set. Since the $\chi_\eta^\pm$
are pure p-states they are eigenstates of H with N-fold
degenerated eigenvalues at $\pm (V_2 - 2V_3) - V_1 - V_4$ yielding the un-
pleasant δ-functions in the density of states.

The integrated density of states may be calculated following
John [3] to give (E with a small positive imaginary part)

$$N(E) = -\frac{1}{\pi} \text{Im} \ln (E-E^+)(E-E^-) - \frac{1}{\pi N} \text{Im} \sum_\eta \ln Q_\eta(E)$$

where

$$Q_\eta(E) = E^2 - E(E_\eta^+ + E_\eta^- + E_1) + E_\eta^+ E_\eta^- + \frac{1}{2}\left[(1+\eta)E_\eta^- + (1-\eta)E_\eta^+\right]E_1$$

($E_1 = 4(V_1-V_4)$). If the η-spectrum extends from -1 to $+1$ the band edges of H are

$$\left.\begin{array}{l} V_2+3(V_1+2V_3+V_4) \\ V_2-(V_1+2V_3+V_4) \end{array}\right\} \begin{array}{l}\text{bonding}\\\text{band}\end{array} \qquad \left.\begin{array}{l} -V_2+3(V_1-2V_3+V_4) \\ -V_2-(V_1-2V_3+V_4) \end{array}\right\} \begin{array}{l}\text{anti-bonding}\\\text{band}\end{array}$$

provided the gap $-2(V_2-2V_1+2V_3-2V_4)$ remains positive. The δ-peaks of the p-states appear at the upper band edges as in the ordinary Weaire model.

3. FLUCTUATING BOND INTERACTION

In order to get rid of the unrealistic δ-peak at the upper band edges higher order interactions should be included, but even V_5 would be an interaction depending on the dihedral angle and therefore on the detailed structure of the network. A description in terms of the operators S and T would be impossible. On the other hand we could assume the bond interaction V_2 to fluctuate independently at each bond according to a distribution function $p(V_2)$. If we further write the Hamiltonian in a bonding – anti-bonding representation (made up of $P_\pm |i \mathcal{R}\rangle$) and neglect interactions between the two bands[1] we may easily apply the CPA to the problem. Using a semi-elliptical density of states for the connectivity matrix we have calculated the CPA-density of states for the two distribution functions inserted in fig. 2. The δ-peak is broadened according to the distribution function, the original band being only little affected. The band gaps are 2.61 eV and 2.30 eV for p_1 and p_2, respectively. The latter calculation

[1] The error made is expected to be negligible $\left[4\right]$

FIG. 2: CPA-density of states for the bonding band using
the inserted distribution functions. The band
edges of the undistorted semi-elliptical band
are denoted by arrows $V_2^0 + 3V_1$ and $V_2^0 - V_1$.

(V_1 = -2.5 eV, V_2^0 = -6.75 eV, Δ = 1 eV)

may be compared with a value of 2.35 eV given by Thorpe [6]
who assumed that each atom has two bonds with interaction
constants $V_2^0 + \Delta$ and $V_2^0 - \Delta$. The distribution function $p_1(V_2)$
means a further increase of the homogeneity of the solid which
may be connected with a decrease of the free energy and leads
to an increase of the gap supporting a conjecture made by
Phillips [7] that the free energy is minimized by a "self-
consistent arrangement of atoms which maximizes bond energies
by ejecting states that would otherwise fall in the "tail"
region to above the band edge".

4. REFERENCES

[1] Weaire, D.; Thorpe, M.F.: Phys. Rev. B 4, 2508 (1971)

[2] Thorpe, M.F.; Weaire, D.: Phys. Rev. B 4, 3518 (1971)

[3] John, W.: phys. stat. sol. (b) 55, K9 (1973)

[4] Streitwolf, H.W.: phys. stat. sol. (to be published)

[5] Velický, B.; Kirkpatrick, S.; Ehrenreich, H.: Phys. Rev.
175, 747 (1968)

[6] Thorpe, M.F.: J. Phys. C, Solid State Phys. 6, L75 (1973)

[7] Phillips, J.C.: Comments Sol. St. Phys. IV, 9 (1971)

FIELD-DEPENDENT CONDUCTIVITY IN BAND TAILS[+]

DAVID REDFIELD

RCA Laboratories, Princeton, N. J. 08540, U.S.A.

The electric field dependence of the conductivity
of controlled-occupancy band tails has been measured
for several positions of the Fermi energy and at
several (low) temperatures. At moderate field strengths
(~50 V/cm) we find $\sigma \simeq a(T) \exp (bF/T)$ which extends to
as low as a few V/cm. At "high" fields (200 V/cm) σ
becomes nearly independent of both F and T.

Detailed measurements of the temperature dependence of the conductivity σ
were reported last year [1] for electronic conduction in energy band tails
of heavily-doped, closely compensated GaAs. In these materials we not only
know the approximate width and shape of the band tail but we can also
adjust the Fermi energy E_F within it by precise control of the
compensation [2]. The unexpected result of the measurements described in
Ref. 1 was the accurate relation $\sigma = \sigma_o \exp [-(T_o/T)^{1/2}]$ at low
temperatures (<20 K) for nearly zero electric field F. Since this result
only appeared after a substantial refinement of earlier measurements [3],
it was decided to extend also the limited field dependence measurements
previously reported [3].

The serious experimental constraints on such measurements must be mentioned.
First, the known sensitivity of σ to temperature requires that very low
power dissipation and effective cooling be used. The latter need was
found to require immersion of the samples in liquid He, thus placing an
upper limit of T = 4.2 K. The former need was met by using pulsed currents
with low duty cycles for all but the lowest fields. This introduced
transient effects due to circuit capacitance and high sample resistance (in
some cases). As a result, the shortest pulses that could be used were
~1 μsec and these produced some sample heating during each pulse at the
higher power levels. To avoid effects of contact resistance, all data
were taken with 4-point measurements.

+ Supported in part by U. S. Office of Naval Research under Contract No.
 N00014-71-C-0371.

FIG. 1: Field-dependence of the conductivity at 4.2K of a single sample at three values of its Fermi energy E_F. Increasing circled numbers indicate the progressively lower E_F as the compensation approaches 1. D.C. measurements are shown by circles, pulsed measurements by triangles and squares.

Fig. 1 shows $\sigma(F)$ at 4.2 K for 3 stages of compensation; the lower curves correspond to E_F being deeper in the tail. It can be seen that: (i) Over a considerable range of σ, there is nearly an exponential dependence of σ on F; thus $\sigma = \sigma_1 \exp(F/F_o)$. In no case was Poole-Frenkel ($\sigma \sim \exp(F^{1/2})$) behavior followed. (ii) As E_F moves lower in the tail, σ_1 diminishes while F_o increases somewhat. It will be shown elsewhere that σ_o also diminishes and T_o increases as E_F gets lower. (iii) The exponential dependence on F extends to very low fields (~ 1 V/cm in some cases). (iv) σ tends to saturate at high fields; this occurs in spite of evidence that sample heating is present for the highest fields. Such heating would normally increase σ further.

In addition, the combined dependence $\sigma(T,F)$ has been measured for the most closely compensated case of Fig. 1 (labelled ③). The temperature range is limited by experimental factors to 2.25-4.2 K. The results for three of these temperatures are shown in Fig. 2. For the central portions of these curves (i.e. away from the low-field or high-field behavior) it is found that the data can be fitted approximately by the relation $\sigma = a(T) \exp(b F/T)$ with $a(T) \simeq 9 \times 10^{-5} T^2$ $(\Omega \, \text{cm})^{-1}$ and $b \simeq 0.07$ K/(V/cm). This behavior extends to still lower fields (< 1 V/cm) as the temperature is reduced. But another most unusual feature occurs in the high-field regime where σ becomes nearly independent of both T and F. Again, this is in spite of evidence that sample heating is present in this regime.

At present these results remain unexplained although physical models for them are now being sought. There are good reasons to believe the conduction takes place by hopping but none of the available theories account for the temperature and field dependences observed. Furthermore, the non-ohmic behavior at very low field strengths implies that some very large characteristic length ℓ must be involved. Taking $eF\ell \simeq kT$ as a criterion for observation of field-dependent effects with $T \simeq 4K$ and **using the slopes from Fig. 2, we find $\ell \simeq 10^{-5}$ cm.**

We remark finally that, apart from the saturation effects, these results resemble those recently reported for field-dependent conductivity in several amorphous semiconductors [4]. In those experiments, however, the temperatures were ~ 100 K - 300 K and the field strengths were much

FIG. 2: Combined temperature- and field-dependence of the conductivity of the sample condition shown as ③ in Fig. 1.

higher than ours ($\gtrsim 10^4$ V/cm). Even so, an explanation for those results is also not yet available [4].

REFERENCES

[1] Redfield, D.: Phys. Rev. Lett. <u>30</u>, 1319 (1973).

[2] Redfield, D,; Crandall, R. S.: <u>Proc. X Intl. Conf. on Phys. of Semi-conductors</u>, Edited by Keller, S. P.; Hensel, J. C.; Stern, F.; U.S. Natl. Tech. Information Serv.; Springfield, Va. (1970), p. 574.

[3] Redfield, D.: <u>Proc. XI Intl. Conf. on Phys. of Semiconductors</u>, PWN-Polish Scientific Publishers, Warsaw (1972), p. 218.

[4] Marshall, J. M. and Miller, G. R., Phil. Mag. <u>27</u>, 1151 (1973).

IMPURITY BREAKDOWN IN HEAVILY DOPED CLOSELY COMPENSATED SEMICONDUCTORS AS A MODEL OF SWITCHING IN AMORPHOUS SEMICONDUCTORS

S.M.RYVKIN, I.S.SHLIMAK AND A.G.ZABRODSKI

A.F.Ioffe Physico-Technical Institute,Academy
of Sciences of the USSR, Leningrad, USSR

Impurity breakdown in HDC-Ge was investigated.
It is shown that this process is similar to electronic
switching phenomenon in amorphous semiconductors.
Experimental results are explained on the base of
impact ionisation in manylevels system.

As is well known,charge transfer in amorphous and heavily
doped closely compensated(HDC) semiconductors occurs by hop-
ping conduction via localized states.The electronic mechanism
of switching to a high-conduction state can involve impact
ionization(II) transferring the carriers to a high-mobility
state.This process is similar to the impurity breakdown in
lightly doped semiconductors, the essential distinction being
the existence of a quasicontinuum of localized energy states
between the Fermi level and the percolation level defining
the boundary of high-mobility states in a disordered system.
In a first approximation,this quasicontinuum can be reduced
to an averaged level with a subsequent consideration of II in
a three-level system.We intued to show that such a model can
account for the principal experimental features of switching,
viz. the S-shape I-V characteristic and the existence of a
certain switching time delay.

The transition diagram is presented in Fig.I.The corres-
ponding kinetic equations are:

$$\begin{cases} \dfrac{dn}{dt} = C_1(N_d-N_a)+A_1 n(N_d-N_a)-B_1 nN_a +C_2 m +A_2 mn-B_2 n(M-m), \\[2mm] \dfrac{dm}{dt} = -C_2 m -A_2 mn +B_2 n(M-m)+C_{12}(N_d-N_a)+A_{12}n(N_d-N_a)-B_{21}mN_a. \end{cases} \quad (I)$$

Here N_d-N_a is the electron concentration at ground le-
vel I, M the density of states at intermediate level II, m
and n are the electron concentrations at levels II and III,
$n+m \ll N_d-N_a$. The coefficients A_1 , A_2 refer to the

process of II; B_1, B_2, to recombination capture; C_1, C_2, to thermal(possibly field-induced) transfer to level III. The coefficients A_{12}, B_{21} and C_{12} describe direct carrier exchange between levels I and II. It follows from egs. (I) that if these coefficients are zero, introducing intermediate level II does not change the steady state solution for the two-level model and, therefore [I], cannot account for the S-shaped I-V characteristic. Hence these transitions are of major significance for the switching.

From egs. (I) one can readily determine the field dependence of the steady-state carrier concentration $n_o(E)$ at level I:

$$n_o^{(1,2)} = \left(\delta \pm \sqrt{\delta^2 - S\rho} \right) S^{-1},\tag{2}$$

where

$$S = (A_2 + B_2)(A_1 + A_{12} - q^{-1}B_1),$$
$$2\delta = -C_2(A_1 + A_{12} - q^{-1}B_1) - (A_2 + B_2)(C_2 + C_{12}) - B_{21}N_a\left[A_1 - q^{-1}(B_1 + N_a^{-1}MB_2)\right],$$
$$\rho = C_1 B_{21} N_a + C_2(C_1 + C_{12}),$$

$q = (N_d - N_a)N_a^{-1}$ is the ground state occupancy.

Eg. (2) describes the forward and dropping branches of the I-V characteristic, i.e. essentially its S-shaped feature. The breakdown critical voltage E_c is found from the conditions $n_o^{(1)} = n_o^{(2)}$, the sustaining voltage E_s being determined from the condition $n_o^{(2)} \to \infty$. As the breakdown develops and level I approaches depletion, the condition $n + m \ll N_d - N_a$ naturally becomes violated. As a result, the II rate will decrease, and the trapping rate, increase, producing the second forward branch in the I-V characteristic. Taking this effect into consideration reduces to a substitution in the initial equations of $N_a + n + m$ for N_a. The corresponding $n_o(E)$ graphs are presented in Fig.2.

It may be shown that E_c decreases with increasing temperature thus making the S-shaped feature less pronounced which agrees with experimental data for HDC-Ge(Fig.3).

We turn now to discussing the mechanism defining the switching time delay. In the case of a small overvoltage the

first of egs.(I) can be rewritten as

$$\frac{dn}{dt} = C_1(N_d - N_a) + C_2 m + (A_2 + B_2) mn.$$

(3)

As seen from eg.(3), at low concentrations n increases exponentially.When $n > n_c = C_2 (A_2 + B_2)^{-1}$,the term quadratic in concentration becomes predominant after which the concentration begins to rise superexponentially by a "hyperbolic" law,so that any large value of n can be reached with the time interval tending to a finite limit.Thus the time in which the concentration n_c is reached determines the delay time τ_d .It can be evaluated from egs.(I) and (3) neglec - ting terms quadratic in concentration:

$$\tau_d \simeq 2v^{-1} \ln v [n_c - n(0)][C_1(N_d - N_a) + C_2 m(0)]^{-1},$$

(4)

where

$$v = -(C_2 + B_{21} N_a) + \sqrt{(C_2 + B_{21} N_a)^2 + 4 C_2 [A_{12}(N_d - N_a) + B_2 M]},$$
$$n(0) = n|_{t=0}, \quad m(0) = m|_{t=0}.$$

The dependence of τ_d on electric field arises from the field dependence of the coefficient A_{12} characterizing inelastic collisions of free and localized electrons.These collisions induce tunneling transitions of the latter to higher energy status.Therefore A_{12} involves the tunneling transition pro- bability W .It may be shown [2] that $W \sim \exp(-\mathrm{const} \times E^{-1/4})$ Therefore for the low temperature region,taking into conside- ration(4),we obtain

$$\ln [\tau_d(E)^{-1} \tau_d(E_c) - 1] = (2/3) \ell (E - E_c)(aE_c)^{-1},$$

(5)

where a is the radius of localization, ℓ the hopping length.

The dependences deduced from (5) using the experimental values for E_c , $\tau_d(E_c)$ and ℓa^{-1} in HDC-Ge [3] are shown with solid lines in Fig.4.

Note that eg.(5) agrees also with the empirical law found for the delay time in amorphous semiconductors [4] : $\tau_d \sim \exp(-V/V_0)$,where V is the voltage across the sample, V_0 a constant.

Thus in accordance with the present model the develop -

ment of switching in the following way.On application of the
switching voltage,inelastic collisions of hot and localized
carriers result in a buildup of electrons in high energy sta-
tes,in other words,there occurs redistribution of electrons in
the system of localized states.As the carrier concentration
at the high energy states reaches a certain value, II ava-
lanche develops.At the same time,the sharp increase in the
free electron concentration produces a drop in the breakdown
sustaining voltage due to intensification of the localized
electron"heating"processes,as well as to the increase of the
free carrier lifetime because of the filling of intermediate
states.

It should be noted in conclusion that the above results
are applicable also to lightly doped semiconductors if the
first excited state of the impurity atoms is considered as the
intermediate level [I,2].For lightly doped semiconductors the
probability of tunneling transitions between spatially sepa-
rated levels in naturally negligible.As the degree of doping
and compensation increases,the excited states disappear and
the tunneling transition probability increases.The states of
the major impurity which are shifted in energy because of a
random distribution of potential of charged donors and accep-
tors can act now as intermediate states.The ideas developed
here can be directly applied to HDC semiconductors which were
used for their experimental verification [2] .One can ex-
pect that the above switching model is valid also for amor-
phous semiconductors because of similar features in the ener-
gy spectra of states in these systems [5] .

REFERENCES
I. A.A.Kastalskii, Phys.Stat.Sol(a) 15, 599(1973)
2. A.G.Zabrodskii, I.S.Shlimak, Fiz.Tverd.Tela to be published
3. A.G.Zabrodskii, A.N.Ionov, I.S.Shlimak, Fiz.Tekhn.Polupr.
 8, 503(1974)
4. S.R.Ovshinsky, Phys.Rev.Lett, 21, I450(1968)
5. S.M.Ryvkin, I.S.Shlimak, Phys.Stat.Sol(a) 16, 5I5(1973)

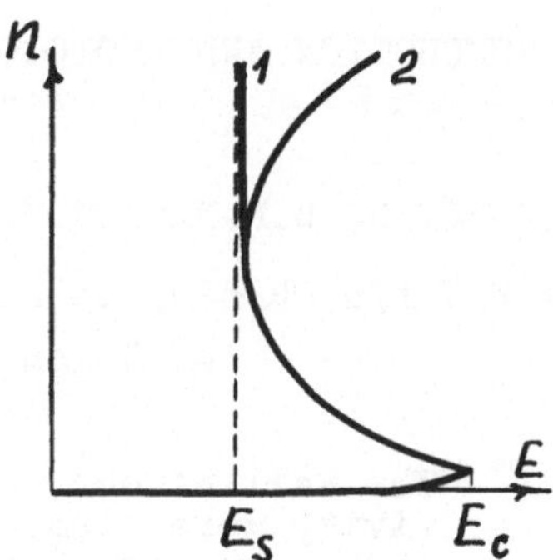

Fig. 1: Diagram of transitions. Fig. 2: n vs E dependence
without (1) and with (2)
depletion of ground level.

Fig. 3: Temperature dependence Fig. 4: τ_d^{-1} vs E dependence
of E_c and E_s for HDC-Ge sample. for the same sample.

CONDUCTION AND PHOTOCONDUCTION MECHANISMS IN COMPENSATED InSb CRYSTALS

V.G.KOROTIN, D.N.NASLEDOV, YU.S.SMETANNIKOVA,T.K.TASHKHODZHAEV

A.F.Ioffe Physico-Technical Institute of the USSR Academy of Sciences, Leningrad, USSR

The Hall effect, conductivity and photoconductivity were studied in compensated n- and p-type InSb crystals. Experimental results are interpreted on the basis of the "percolation" theory which takes into account large-scale fluctuations of charged impurities. Anomalous low Hall mobility at kT lesser than the amplitude of fluctuations is due to the specific Hall effect in an inhomogeneously conducting medium.

1. INTRODUCTION

As is known, a number of properties of strongly compensated semiconductors do not find an explanation from the point of view of the classical theory of carrier transport phenomena. Considerable interest in this problem aroused among theorists and experimentalists led to new concepts of the conduction mechanisms in such crystals.

In this paper we present some results on the galvanomagnetic and photoelectric properties of n- and p-type InSb crystals with different impurity compensation degrees over a wide temperature range.

2. EXPERIMENTAL RESULTS

Fig. 1a shows the temperature dependencies of the Hall coefficients (R_H) and conductivity (σ) for n-type InSb single crystals doped with Te and compensated with Ge. The compensation degree (K) increases with the curve number from K = 0.90 to K = 0.997. A drastic change in R_H in the temperature range from 50 to 100°K results, as has been already pointed out [1], from ionization of deep impurity centers. At T≤50°K straight line sections of $R_H(\frac{1}{T})$ dependence are observed whose slope increases with a rise in compensation degree, and the onset

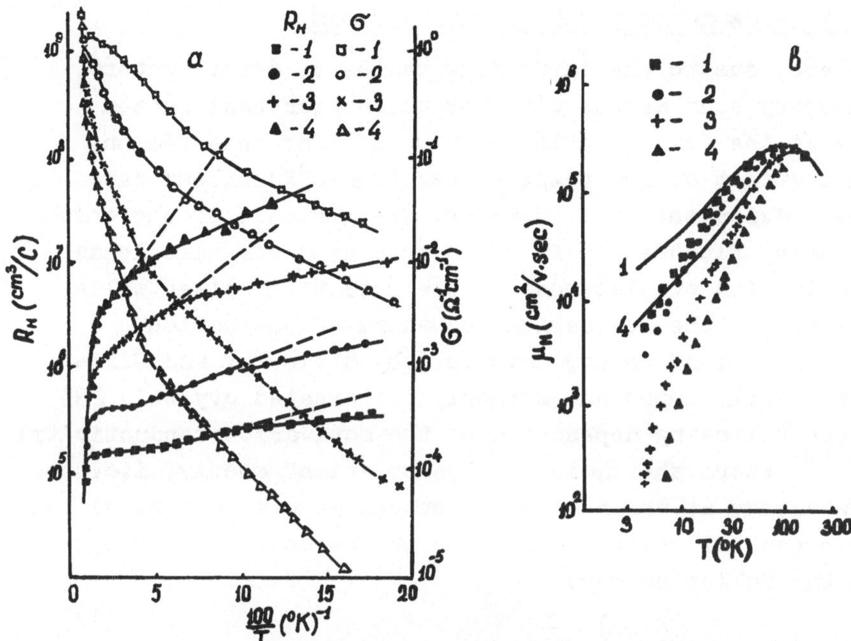

Fig. 1: Temperature dependence of the Hall coefficient R_H, conductivity σ (a) and the Hall mobility μ_H(b) for the n-type InSb crystals.
Compensation degree K: (1) 0.90; (2) 0.98; (3) 0.99; (4) 0.997. Solid line in (b) denotes calculated electron mobility: (1) $N_i = 7 \cdot 10^{14}$ cm^{-3}; (4) $N_i = 1.8 \cdot 10^{16}$ cm^{-3}.

shifts towards the high-temperature region. The Hall mobility $\mu_H = R_H \sigma$ for the same samples is represented in Fig.1b. As seen from this figure, μ_H drops with decreasing temperature, the drop being better pronounced at higher K.

Fig.2a and 2b show respectively the temperature dependence of R_H and the hole lifetime (τ_p) in p-type InSb single crystals doped with Mn, Te with the compensation degree from K = 0.4 to K = 0.998. For the temperature range of 50 to 100°K, the deep level of structural defects important for recombination processes is characteristic. Handling of the $R_H\left(\frac{1}{T}\right)$ curves shows that the activation energy of this level increases with rising K.

3. DISCUSSION

3.1 THE CONDUCTION MECHANISM IN InSb CRYSTALS IN THE PRE-

SENCE OF LARGE-SCALE IMPURITY FLUCTUATIONS

In n-InSb, due to the large Bohr radius of donor centers,
the impurity band merges with the conduction band bottom
already at the concentration $N_D \approx 10^{16} cm^{-3}$. For this reason,
the Hall effect at low temperatures $(T \leqslant 40^{\circ}K)$ in our case
must be independent of T. However, the presence of the char-
ged impurity fluctuation in the compensated crystals leads
to localization of electrons in the potential relief wells,
and, hence, to the activation mechanism of conduction.

The "percolation" theory developed by Shklovsky and Efros
for the heavily doped and strongly compensated crystals [3]
gives the following dependence of the activation conductivity:
$\sigma = \sigma_o e^{-\varepsilon_p/k_T}$, where the ε_p is the "percolation" energy, i.e. the
minimum energy at which an electron can move around hamps
of the potential relief. ε_p depends on the compensation deg-
ree in the following way:

$$\varepsilon_p = \gamma \frac{e^2 N_D^{1/3}}{\varkappa (1-K)^{1/3}} \tag{1}$$

where $\varkappa$ is the dielectric susceptibility, γ a factor of the
order of unit.

It seems natural to suppose that from the Hall effect one
can obtain the concentration of free electrons with the
energy $\varepsilon > \varepsilon_p$. Thus, from the activation energy estimations
$(\varepsilon_a \approx 7 \cdot 10^{-4} \div 7 \cdot 10^{-3} ev)$ we have obtained a good agreement with
(1). In case of lightly doped and strongly compensated semi-
conductors (p-type InSb), the theory [4] gives the following
result: with K increasing, the impurity center activation
energy rises by the value of ε_o, compared with the ionization
energy of an isolated atom ε: $\varepsilon_a \approx \varepsilon_o + \varepsilon$, where ε_o is obtained
from (1). The solution of neutrality equation has shown that
the difference between a real crystal activation energy and
an ideal one coincides with the calculated value from (1)
at $\gamma \approx 1.8$ [2].

As to the strong temperature dependence (τ_p) for compensated
p-type InSb crystals, its interpretation presented difficulties.

On the basis of the two-level recombination model we have
obtained the following expressions for τ_n and τ_p:

Fig. 2: Temperature dependence of the Hall coefficient R_H (a)
and the hole lifetime τ_p (b) for the p-InSb crystals.
Compensation degree K: (1) 0.40; (2) 0.15; (4) 0.90;
(6) 0.98; (7) 0.993; (8) 0.996; (9) 0.998; (10) K > 0.998
Recombination level parameters: $N_t = 3 \cdot 10^{13}\,\mathrm{cm}^{-3}$, $\gamma_{nt} = 5.5 \cdot 10^{-7}$
$\mathrm{cm}^3/\mathrm{sec}$, $\gamma_{pt} = 5.5 \cdot 10^{-9}\,\mathrm{cm}^3/\mathrm{sec}$, $\Delta E_t = 0.104 + \mathcal{E}_o$ ev; $N_s = 1.8 \cdot 10^{14}\,\mathrm{cm}^{-3}$,
$\gamma_{ns} = 10^{-4}\,\mathrm{cm}^3/\mathrm{sec}$, $\gamma_{ps} = 5 \cdot 10^{-10}\,\mathrm{cm}^3/\mathrm{sec}$, $\Delta E_s = 0.073 + \mathcal{E}_o$ ev.

$$\tau_p = \tau_n \left(1 + \frac{\dfrac{\gamma_{ns}}{\gamma_{ps}} \cdot \dfrac{N_s}{1 + P_{vs}/P_o}(P_o + P_{vt}) + \dfrac{\gamma_{nt}}{\gamma_{pt}} \cdot \dfrac{N_t}{1 + P_{vt}/P_o}(P_o + P_{vs})}{(P_o + P_{vt})(P_o + P_{vs}) + (P_o + P_{vs})n_t + (P_o + P_{vt})n_s} \right)$$

$$\tau_n = \left(\frac{\gamma_{ns} N_s}{1 + P_{vs}/P_o} + \frac{\gamma_{nt} N_t}{1 + P_{vt}/P_o} \right)^{-1}; \quad n_s = \frac{N_s}{1 + P_o/P_{vs}}; \quad n_t = \frac{N_t}{1 + P_o/P_{vt}}; \quad P_o = 1/R_H e$$

where $\gamma_{nt}, \gamma_{ns},\ \gamma_{pt},\ \gamma_{ps}$, are capture rates of electrons and
holes on T and S levels with concentrations N_t, N_s respec-
tively. In these calculations we had to consider the changes
in recombination level position with increasing K and a satis-
factory agreement between the calculated curves (solid lines
in Fig.2b) and experimental results has been obtained.

3.2 THE HALL EFFECT AND MOBILITY AT A LARGE AMPLITUDE OF POTENTIAL RELIEF ($\mathcal{E}_p \gg kT$)

One of the peculiarities of strongly compensated semiconduc-

tors which have found no interpretation as yet, is the Hall
mobility μ_H behaviour at low temperatures. μ_H decreases with
temperature more drastically than it follows from the classi-
cal theory. We believe that this peculiarity can be accounted
for as follows. With decrease in temperature there occurs the
freezing-out of electrons in the potential relief wells, and
the electron contribution varies: $n = n_0 e^{-\varepsilon_p/kT}$. However, at $kT < \varepsilon_p$
conduction will be mainly determined by the carriers with the
energy close to ε_p. In this case the crystal can be represen-
ted as having different conduction regions: those with metal-
lic conduction (electron drops) and weakly conducted ones
(the potential relief hamps). A.Ya.Shik considered transport
phenomena in a similar medium in crossed constant electric
and weak magnetic fields[5]. He has shown that at $kT \ll \varepsilon_p$ the
Hall voltage depends on the relation between sizes of these
regions:
$$V_H = \mu J \frac{H}{\sigma c} \cdot \frac{\Delta(T)}{L}$$
where L is the characteristic length of the potential relief,
$\Delta(T) \sim T^{\alpha(T)}$, $\alpha(T) \geqslant 2$. Thus, at the existence of large-scale fluc-
tuations R_H must deviate from linear dependence of $\ln R_H = f\left(\frac{1}{T}\right)$
with decreasing temperature, which is observed experimentally.
The Hall mobility $\mu_H = R_H \sigma$ will also have more sharp temperature
dependence: $\mu_H \sim T^{3/2 + \alpha(T)}$. Dashed lines in Fig.1a represent the tem-
perature dependence of $\sigma \mu$, where μ is the calculated value of
mobility. It is seen that the curve slope coincides with that
of the onset parts of experimental curves (when $\varepsilon_p < kT$). The
value of $\alpha(T)$ in our case is of the order of 2 and depends
weakly on temperature.

4. REFERENCES

[1]. I.M.Ismailov, D.N.Nasledov, Yu.S.Smetannikova, V.R.Feli-
tsiant, Phys. Stat. Sol. 36, 747 (1969)
[2]. D.N.Nasledov, Yu.S.Smetannikova, T.K.Tashkhodzhaev,
Phys. Stat. Sol. (a) 20, 101 (1973)
[3]. B.I.Shklovsky, A.L.Efros, Fiz. Tekh. Polupr.4, 305 (1970)
[4]. B.I.Shklovsky, Fiz. Tekh. Polupr. 6, 1197 (1972)
[5]. A.Ya.Shik, Zh. exper. teor. fiz., 66 (1974)(in print)

ABSORPTION OF SUB-MILLIMETER RADIATION IN HEAVILY DOPED n-TYPE GERMANIUM

KAZUO YOSHIHIRO, MADOKA TOKUMOTO and CHIKAKO YAMANOUCHI

Electrotechnical Laboratory

Mukodai-machi, Tanashi, Tokyo 188, Japan

The absorption of monochromatic radiations of 337 and 311 μm in wavelength has been investigated in Sb-doped Ge single crystals at temperatures below 4.2 K. The concentration ranges from 1.2×10^{16} to 3.6×10^{17} Sb atoms/cm^3. The absorption coefficient measured is substantially independent of temperature and rapidly decreases from 2×10^3 to 1 /cm as the concentration is lowered, showing a hump around $n = 3 \times 10^{16}$/cm^3. The observed hump suggests that the absorption is caused by a transition from the donor ground state to a delocalized conducting state which is supposed to lie between the ground state and the conduction band in the intermediate concentration region, and that the Coulomb interaction between an empty donor and a charge carrier in the delocalized state is appreciable.

1. INTRODUCTION

· The purpose of the present investigation is to study the far-infrared properties of heavily doped semiconductors. In germanium samples with antimony concentrations between 1×10^{16} and 1×10^{17} /cm^3, at low temperatures, there is a temperature region where the resistivity is characterized by an activation energy ε_2 [1]. The activation energy ε_2 depends strongly on the impurity concentration [1], magnetic field [2]-[4], deformation [5],[6] and compensation [7] in contrast to the donor ionization energy ε_1 and the activation energy ε_3 [1] for hopping type conduction. These behaviours of ε_2 have been understood in terms of the change in overlapping of donor wave functions, and ε_2 has been supposed to be an energy needed to excite electrons from the donor ground state into an unbound state in which an electric current flows [1],[5]. The present study aims at observing the optical transitions which would occur if this picture is real.

Most of previous works on the far-infrared properties at liquid helium temperatures have been made in the range of photon energies larger than 6 meV. The value of ε_2 is known to change from 5 meV to zero with increasing concentration [1]. For Sb-doped Ge, the first excited state of an isolated Sb donor lies 5 meV, about a half of ε_1, above the donor ground state. The use of a radiation whose photon energy is not enough to cause the transition between the ground and the first excited states is essential for the present study.

The absorption and reflectivity of monochromatic 337 and 311 µm radiations produced by a CW HCN laser [8] have been investigated in Sb-doped n-Ge single crystals. The excess donor concentration ranges from 1.2×10^{16} to 3.6×10^{17} Sb atoms/cm^3. No compensating impurity was intentionally doped and the compensation ratio was estimated at less than 5 %. The activation energy ε_2 was obtained from the temperature dependence of resistivity at temperatures below 20 K ; ε_2 decreases from 5.8 to 0.1 meV as the concentration changes from 1.2×10^{16} to 1.1×10^{17} Sb atoms/cm^3. The metallic type impurity conduction is observed at higher concentrations.

2. EXPERIMENTAL

The samples were bridge or brick shaped and about $10 \times 1 \times (0.05 - 1.0)$ mm^3 in size for the measurements of absorption, and $6 \times 2 \times 1$ mm^3 for the measurements of reflectivity. Both dc and optical measurements were made for each sample. The samples were immersed in liquid helium to avoid the possible temperature rise on illumination. The radiations of 337 or 311 µm from the HCN laser were chopped at 20 Hz. The transmitted or reflected radiations were detected by an n-InSb single crystal. The signal voltage of 20 Hz generated across the detector was fed into a lock-in amplifier. The incident power of radiation was monitored by a Golay cell with a silicon window continuously during the measurements. The photoresponses of the InSb detector were proportional to the power of incident radiation which was below mW level.

3. EXPERIMENTAL RESULTS AND DISCUSSIONS

Figure 1 shows the absorption coefficient $\alpha(337\mu m)$ and the resistivity ρ for various samples plotted against the excess donor concentration $N_D - N_A$, temperature being the parameter. The absorption coefficient rapidly increases with $N_D - N_A$ and substantially independent of temperature.

For the samples with $N_D - N_A > 1 \times 10^{17}/cm^3$, the resistivity is independent of temperature; the conduction is known to be of the metallic type. The absorption is roughly interpreted as due to free carriers in the conduction band.

For the samples with $N_D - N_A < 1 \times 10^{17}/cm^3$, the absorption

Fig. 1: Absorption coefficient at 337 μm and resistivity as functions of $N_D - N_A$.

coefficient rapidly decreases from 1×10^3 to 1 cm^{-1} as the concentration is lowered, showing a pronounced hump around $N_D - N_A = 3 \times 10^{16}/cm^3$. The resistivity exibits strong temperature and concentration dependences; the conduction is known to be of the nonmetallic type. These facts suggests that the rapid decrease in α seen in the range of $7 \times 10^{16} < N_D - N_A < 1 \times 10^{17}/cm^3$ is not associated with the free carrier absorption but is understood in terms of the transitions from the donor ground state into the conduction band whose bottom is lowered [9] due to the interaction between donors. The hump corresponds to the absorption associated with the activation energy ε_2 as will be discussed below.

The absorption of 311 μm radiation shows almost the same dependence on the concentration as that seen in Fig. 1, but the hump shifts to the lower concentration side.

In Fig. 2 is shown the reflectivity of 337 μm radiation

for various samples as a function of $N_D - N_A$. The reflectivity depends little on temperature. For the samples with $N_D - N_A < 1 \times 10^{17}/cm^3$, the reflectivity is constant; the value is approximated to that of pure crystalline germanium. For the samples with $N_D - N_A > 1 \times 10^{17}/cm^3$, the reflectivity increases up to about 0.75 with increasing $N_D - N_A$. The observed concentration dependence of reflectivity indicates that the absorption by free carriers is negligible for the samples with $N_D - N_A < 1 \times 10^{17}/cm^3$.

Figure 3 shows the absorption cross-section of 337 and 311 µm radiations plotted vs the activation energy ε_2 for respective samples. The absorption cross-section σ is obtained from the relation $\sigma = \alpha/N_D$, where the donor concentration N_D is replaced by $N_D - N_A$ in the present case because $N_A \ll N_D$. The vertical dashed lines indicate the photon energies of 337 and 311 µm radiations, $\hbar\omega = 3.7$ and 4.0 meV respectively.

Fig. 2: Reflectivity at 337 µm of various samples expressed in terms of that of an Al mirror.

Fig. 3: Absorption cross-section vs ε_2 for various samples.

For the samples with ε_2 larger than $\hbar\omega$, σ falles rapidly with increasing ε_2. This fact suggests that the absorption is associated with the optical transition from the donor ground state to a delocalized state which lies between the ground state and the conduction band, and that ε_2 is the energy needed to cause the transition.

For ε_2 smaller than $\hbar\omega$, σ passes through a maximum, which

corresponds to the hump in Fig. 1, and then decreases to about
60 % of maximum value showing a minimum as ε_2 is reduced, fol-
lowed by a rapid increase. The fact that σ is not a monotonic
increasing function of $N_D - N_A$ is difficult to be explained in
terms of the transitions associated with tail states extended
below the conduction band, since the absorption caused by the
tail states shows an exponential dependence on the photon
energy [10].

The width of the peak of σ vs ε_2 curves is about 1 meV.
If for the delocalized state a rigid band model is valid in the
narrow concentration range around the peak, and the Coulomb
interaction between an empty donor and a charge carrier in the
state is negligible, σ should reflect the density of states; the
width should correspond to the band width. But the observed
width appears too small to assure the one-electron interpreta-
tion. The narrow width indicates an appreciable effect of
Coulomb interaction in the delocalized conducting state [11].

4. ACKNOWLEDGEMENTS

The authors would like to acknowledge helpful discussions
with J. Kondo, K. Yamaji, H. Sumi and S. Ogawa. They wish to
thank Professor W. Sasaki and Professor T. Kurosawa for stimu-
lating discussions.

5. REFERENCES

[1] Fritzsche, H.: J. Phys. Chem. Solids 6, 69 (1959)
[2] Yamanouchi, C.: J. Phys. Soc. Japan 18, 1775 (1963)
 Yamanouchi, C.: ibid. 20, 1029 (1965). Yamanouchi, C.;
 Mizuguchi, K.; Sasaki, W.: ibid. 22, 859 (1967)
[3] Sadasiv, G.: Phys. Rev. 128, 1131 (1962)
[4] Tufte, O. N.; Stelger, E. L.: Phys. Rev. 139, A265 (1965)
[5] Fritzsche, H.: Phys. Rev. 125, 1552 (1962)
[6] Kinoshita, J.; Yamanouchi, C.; Yoshihiro, K.: J. Phys. Soc.
 Japan, 36, 1493 (1974)
[7] Davis, E. A.; Compton, W. D.: Phys. Rev. 140, A2183 (1965)
[8] Yoshihiro, K.; Yamanouchi, C.: Rev. Sci. Instrum. 45,
 (June 1974)
[9] Penin, N. A.; Zurkin, B. G.; Volkov, B. A.: Soviet Phys.
 Solid State 7, 2580 (1966)
[10] Wood, D. L.; Tauc, J.: Phys. Rev. B 5, 3144 (1972)
[11] Yoshihiro, K.; Tokumoto, M.; Yamanouchi, C.: J. Phys. Soc.
 Japan 36, 310 (1974)

SMALL-GAP SEMICONDUCTORS

CONTENTS

PHOTON-PLASMON PROCESSES

J. MYCIELSKI

Institute of Theoretical Physics, Warsaw University

Warsaw, Poland

Dissipative part of the dielectric function of free carriers is conditioned by their interaction with lattice imperfections and originates from scattering of individual carriers or from generation of collective plasma oscillations (plasmons). Photon – plasmon – ionized-impurity (defect) processes at frequencies close to plasma frequency are discussed and shown to be important in polar semiconductors with high static (lattice) dielectric constant and high carrier concentration. Experimental evidence for plasmon generation in n-PbSe and n-Pb$_{1-x}$Sn$_x$Se in infrared is presented.

1. INTRODUCTION

Let us consider a one-component semiconductor plasma consisting of free carriers (electrons or holes) of a scalar, energy independent effective mass. We will allow the existence of several equivalent band extrema for these carriers. Suppose the magnetic field is absent, the crystal is perfect, and charges of ionized donors or/and acceptors are smeared out to give a uniform charge density. In such conditions, a very small dissipative part of the carrier dielectric function (i.e., imaginary part, corresponding to conductivity) exists only because of finite radiation wavelength [1,2]. If the radiation electric field is assumed to be homogeneous, as it will be done in the following, the high-frequency conductivity vanishes and so does the free-carrier absorption [3-5]. The plasma response to the electric field is then purely reactive. In particular, there is no contribution to free-carrier absorption from carrier-carrier or carrier-plasmon interactions, in contrary to some published calculations [6,7] in which the influence of radiation electric field on the carrier-plasmon interaction was erroneously omitted [5].

Carrier-carrier interaction may give some free-carrier absorption in the case of energy-dependent effective mass (nonparabolic band) [4,8,9]. It is

well known, however, that the main contribution to free-carrier absorption is given by interactions with crystal imperfections. These interactions supply sink for carrier momentum and make the simultaneous conservation of energy and momentum possible. Free-carrier absorption consists then in individual--carrier excitations (individual-carrier scattering), as well as in collective--carrier excitations (plasmon generation). The former contribution was studied extensively, using both classical (Drude theory) or quantum approach (see, e.g., [7]). In this paper we will be concerned with the latter contribution which was discussed less frequently up to now.

It should be stressed that we are not interested here, at least in principle, in magnetoplasma. We will also omit all photon - long-wavelength-plasmon processes connected with the finite dimensions of the semiconductor sample, as volume-plasmon generation at oblique incidence of light on thin layer [10-12] or surface-plasmon generation on rough surface [13,14]. Only the contribution of short-wavelength-plasmon generation to high-frequency bulk conductivity (i.e. to energy dissipation) will be treated. Of course, this dissipation may, in turn, broaden the line of resonance consisting, e.g., in volume-plasmon generation in thin layer, mentioned before.

Photon-plasmon processes were studied, as a rule, in the more general framework of photon-plasma-imperfection interactions, in which also individual-carrier excitations were taken into account. Both nondegenerate and degenerate plasmas were considered. Usually, theory is formulated in the language of dielectric function for free carriers in perfect crystal. For the case of plasma-ion interactions in homopolar semiconductors, theory was developed by Ron and Tzoar [15,16] using the analogy with gaseous plasma [17]. The contribution of photon-plasmon-ion processes to free-carrier absorption was shown to be only a small fraction of the contribution given by photon - individual-carrier - ion processes [16]. The case of plasma - optical-phonon interaction in polar semiconductors was also investigated [9,18]. The most general formulation of the theory is due to McCumber who took into account simultaneously plasma-ion, plasma-acoustic-phonon, and plasma--optical-phonon interactions in polar semiconductors [19].

Here we are interested only in photon - plasmon - ionized-impurity (defect) processes in polar semiconductors, and not in the contribution given

by individual-carrier excitations. We also limit our considerations to the case of plasma frequency much higher than LO-phonon frequency to decouple the plasmon modes and screened LO-modes. Because of that we can use a much simpler model of plasma and assure the physical insight in the process of plasmon generation. In the following we derive and discuss the formula for the contribution to high-frequency conductivity given by photon - - plasmon - ionized-impurity (defect) processes [20]. We show that this contribution may be comparable with contribution from individual-carrier processes in polar semiconductors with high static (lattice) dielectric constant and high carrier concentration.

Only recently an experimental evidence was obtained for the contribution to free-carrier absorption given by photon - short-wavelength-plasmon processes in n-type PbSe [21] and n-type $Pb_{1-x}Sn_xSe$ [22]. These data will be also discussed in the following.

2. THEORY

Let us consider a dense, degenerated electron plasma, of the density N_e electrons per unit volume. The number of equivalent conduction band minima among which these electrons are distributed will be denoted by w. We assume that they are ions of S types in semiconductor, randomly distributed; $Z_s e$ and N_s (s=1,...,S) is the charge and concentration of the s-type ion, respectively (-e is the electron charge). Neutrality requirement yields

$$\sum_{s=1}^{S} Z_s N_s = N_e .\tag{1}$$

Of course, all results will hold also for hole plasmas.

We assume

$$\omega , \omega_p(0) \gg \omega_{LO} ,\tag{2}$$

where ω is the radiation frequency, $\omega_p(q)$ is the frequency of plasmon with wavevector $\vec{q}$, and ω_{LO} is the LO-phonon frequency. Because of (2), the ionic part of the lattice polarization (i.e. this connected with the difference $\varepsilon_o - \varepsilon_\infty$, where ε_o and ε_∞ denote the static and high-frequency lattice dielectric constants, respectively) and the corresponding polarization charge density $\varrho(\vec{r})$ are approximately constant in time. This is ob-

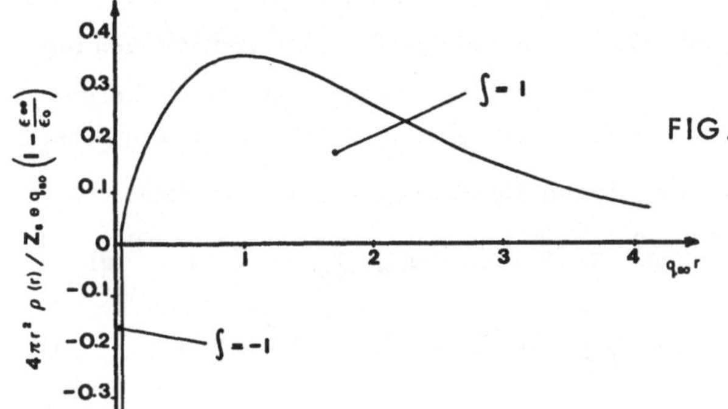

FIG. 1: Polarization charge density $\zeta(r)$ in the vicinity of an ion. The curve corresponds to the function
$$q_{so} \; r \; \exp (-q_{so} \; r) \; .$$

viously not true for the electronic part of lattice polarization (i.e. this connected with ε_∞). Therefore, we can treat the plasma as moving in radiation electric field

$$\vec{E} (t) = Re \left[\vec{E} \exp (-i \omega t) \right] \; , \tag{3}$$

in medium of dielectric constant ε_∞ and in presence of immobile ions and polarization charges.

Because of static free-electron screening of the ion field, the polarization charge density $\zeta(\vec{r})$ in the vicinity of an s-type ion has the form shown on Fig.1. The reciprocal of Thomas-Fermi screening radius (independent, of course, of ion type) is denoted by q_{so} . It involves the static dielectric constant ε_o . It is important to observe that at distances bigger than q_{so}^{-1} the electric field of the ion and polarization charge (plasma excluded) is $\simeq Z_s e / \varepsilon_\infty \; r^2$, while it is smaller, namely equal to $Z_s e / \varepsilon_o r^2$ in the absence of screening electrons.

We can introduce two artificial, mutualy compensating uniform charge densities $+eN_e$ and $-eN_e$. Our model consists now of an ideal plasma perturbed by the electric field and by the presence of charge density of ions, polarization charges and a uniform negative charge density. Let us use the noninertial reference system connected with the center of mass of electron plasma [16]

$$\vec{r}' = \vec{r} - (e/m^* \omega^2) \; Re \left[\vec{E} \exp (-i \omega t) \right] \; , \tag{4}$$

where m^* is the electron effective mass. In this noninertial reference system plasma is unaffected by the radiation field except for perturbation coming from oscillation of the electric potential of ions and polarization charges (which are moving in this reference system). It can be shown that the time--dependent perturbing part of the plasma Hamiltonian is of the form

$$H' = -(e\hbar^{1/2} N_e^{1/2}/2^{1/2} m^{*3/2} \omega^2)\ \mathrm{Re}[\vec{E}\exp(-i\omega t)] \sum_{\vec{q}}{}' q\,\vec{q}\,\omega_p^{-1/2}(q)$$

$$\times\left(\left\{\sum_{s=1}^{S}\left[\sum_{\vec{\delta}_s\in V_1}\exp(i\vec{q}\vec{\delta}_s)\right]\int\exp(i\vec{q}\vec{r})\,V_{is}(\vec{r})\,d^3r\right\}\hat{b}_{\vec{q}} + h.c.\right)\ .\quad (5)$$

Here we sum over all plasmon modes except for $\vec{q} = 0$, over all positions $\vec{\delta}_s$ of s-type ions in unit volume V_1, and over ion types. $\hat{b}_{\vec{q}}$ is the annihilation operator for the plasmon mode $\vec{q}$, and $V_{is}(\vec{r})$ is the electron potential energy in the field of an ion and corresponding polarization charge and part of uniform negative charge density.

Taking into account the randomness of ion positions and using the explicit form of $V_{is}(\vec{r})$, q_{so} and of plasmon dispertion (as given, up to the q^2-terms, by random phase approximation) we can calculate the net power absorption given by plasmon creation minus plasmon annihilation. Power absorption, in turn, is proportional to conductivity. The result can be stated in the perhaps simplest form when we use the high-frequency ($\omega\tau\gg 1$) Drude formula for conductivity

$$\sigma(\omega) = (\varepsilon_\infty /4\pi)\ [\omega_p^2(0)/\omega^2]\ \tau^{-1},\quad (6)$$

where τ is the momentum relaxation time. Of course, in our problem (6) is nothing but definition of τ (which may be frequency-dependent). It has nothing to do with some averages over energy-dependent relaxation times for individual carriers [23,24].

Our final result is that for $\omega < \omega_p(0)$ there is $\tau_{pl}^{-1} = 0$, and for $\omega > \omega_p(0)$

$$\tau_{pl}^{-1} = (5^{3/2}\pi/2^5 3^{3/2}\,w)\ D\ \omega_p(0)\ [\hbar\omega_p(0)/E_F]^3\ \{1 - [\omega_p(0)/\omega]^2\}^{1/2}$$

$$\times\{1 - [1 - (\varepsilon_\infty/\varepsilon_o)]\left(1 + (9\varepsilon_\infty/5\varepsilon_o)\{[\omega/\omega_p(0)]^2 - 1\}^{-1}\right)^{-1}\}^2.\quad (7)$$

Here E_F is the Fermi level and D is defined

$$D = \sum_{s=1}^{S} Z_s^2\ (N_s/N_e)\ .\quad (8)$$

In the particular case of one type of positive and one type of negative

ions with charges $\pm Ze$ we have from (1) and (8)

$$D = Z(1 + K)/(1 - K) , \qquad (9)$$

where K is the compensation ratio. It can be shown also that for $\hbar\omega_p(0) \lesssim E_F$ it is enough to assume

$$[\omega/\omega_p(0)] - 1 < 0.25 \qquad (10)$$

to avoid Landau damping of the generated plasmons.

It follows from (7) that τ_{pl}^{-1} is temperature-independent (if $\omega_p(0)$ and E_F are also independent). The magnitude of τ_{pl}^{-1} is also independent of electron concentration (if D is independent), the only effect of change of N_e being a change of ω-scale. Moreover, at ω close to $\omega_p(0)$ (i.e. wavelength of generated plasmon bigger than q_{so}^{-1}), τ_{pl}^{-1} is unaffected by ε_o because of the influence of electron screening on the polarization charge density discussed before.

On the other hand, high dielectric constant ε_o suppresses very seriously individual-carrier scattering by ions, collision parameter being smaller than q_{so}^{-1}. Therefore, we can expect that plasmon-generation processes may play an important role in free-carrier absorption of semiconductors with very high ε_o (and high concentration).

3. EXPERIMENT

A discrepancy of nearly two orders of magnitude between the values of relaxation time measured by optical and dc methods (τ_{opt} being shorter) was observed in far-infrared magnetoreflectivity measurements at normal incidence on n-PbSe [25] at low temperatures and was interpreted by inelastic scattering of carriers, in particular by optical phonons [26]. A similar effect was also observed in p-PbSe in transmission measurements [27].

In [25] the value of relaxation time was obtained by fitting the theory of Wallace et al. [28] with the experimental magnetoreflectivity curves. Recently, however, it was observed that the fit can be improved if $\tau = \tau(\omega)$ is allowed [21]. Also the magnetoreflectivity measurements on n-type $Pb_{1-x}Sn_xSe$ [29] can be reinterpreted in that way [22].

One must be careful in allowing $\tau = \tau(\omega)$ because Wallace's theory with frequency-dependent relaxation time violates the dispersion relations. Be-

cause of that two fitting methods were used. First, which seems to be more justified, consists in fitting only the minima of magnetoreflectivity curves. The values of τ^{-1} determined by this method for four n-PbSe samples are shown on Fig.2, and for one n-Pb$_{1-x}$Sn$_x$Se sample on Fig.3. The second

FIG. 2: Damping (τ^{-1}) in four samples of n-PbSe at 30 K, as determined by magnetoreflectivity-minima fitting method [21]. Sample characteristics are: N = 0.495, 1.33, 3.34, 5.16 $\times 10^{18}$ cm^{-3}, E$_F$ = 18, 33, 55, 69 meV, $\omega_p(0)$ = 3.36, 5.51, 7.43, 9.15 $\times 10^{13}$ s^{-1} for samples A, B, C, D, respectively. Magnetoplasma frequency $\omega_{pc} = [\omega_p^2(0) + \omega_c^2]^{1/2}$ at highest magnetic field is also indicated.

method consists in finding an overall optimal fit for magnetoreflectivity curves. The determined values of τ^{-1} are presented for two n-PbSe samples on Fig.4 and compared with the values obtained from the first method.

There is a bump on the $\tau^{-1}(\omega)$ curve, with an edge corresponding to $\omega_p(0)$ (or rather to magnetoplasmon frequency ω_{pc} at highest magnetic field) but not to, e.g., $2\omega_{LO}$ [30] or $\omega_p(0) + \omega_{LO}$. This suggests a photon – – short-wavelength-plasmon – ion process of free-carrier absorption.

To see if the orders of magnitude of the expected and observed effects are the same, the values calculated from (7) are presented in Fig.5 and compared with experiment for two n-PbSe samples. The parameters used were: w = 4, $\varepsilon_\infty / \varepsilon_o$ = 0.0805 (ε_∞ = 26, ε_o = 323), D = 7 (as, according to (9), at Z = 1, K = 0.75, or Z = 2, K = 0.56). It should be stressed that there exists evidence for two-fold charged defects, at least in PbTe, and for rather high compensations [31-33]. It should be also noted that the dependence of the maximal theoretical value of τ^{-1}_{pl} (as given by (7)) on concentration

FIG. 3: Damping (τ^{-1}) in a sample of n-Pb$_{1-x}$Sn$_x$Se, as determined by magneto-reflectivity-minima fitting method [22]. Plasmon and magnetoplasmon frequencies are indicated.

FIG. 4: Damping (τ^{-1}) in two samples of n-PbSe at 30 K, as determined by best-overall-fitting method (solid lines) [21]. On the upper pictures the limits of error of this method are indicated, and on the lower pictures there is a comparison of results of the two fitting methods (points being taken from Fig.2).

(in Fig.5) is caused by the nonparabolicity of the conduction band. It seems

FIG. 5: Damping (τ^{-1}) in two samples of n–PbSe at 30 K, as determined by two fitting methods (see lower part of Fig.4), compared with damping calculated from formula (7).

that the calculated contribution to free-carrier absorption of the photon – – plasmon – ion processes is roughly of the same magnitude than the bump on the experimental curve. However, further measurements are needed.

4. ACKNOWLEDGMENT

The author is grateful to Professor M. Balkanski, Professor M. Suffczyński, Dr. A. Aziza, Dr. J. Blinowski and Dr. A. Mycielski for helpful discussions.

5. REFERENCES

[1] Suffczynski, M.: Phys. Rev. 140, A147 (1965)
[2] DuBois, D.F.; Kivelson, M.G.: Phys. Rev. 186, 409 (1969)
[3] Hopfield, J.J.: Phys. Rev. 139, A419 (1965)
[4] Vlasov, G.K.; Mashkevich, V.S.; Timonina, E.A.: Fiz. Tverd. Tela 14, 3397 (1972) [Sov. Phys. – Solid State 14, 2870 (1973)]
[5] Blinowski, J.; Mycielski, J.: to be published
[6] von Baltz, R.: phys. stat. sol. (b) 43, K133 (1971)
[7] von Baltz, R.; Escher, W.: phys. stat. sol. (b) 51, 499 (1972)
[8] Wolff, P.A.: Phys. Rev. 132, 2017 (1963)
[9] Foo, E-Ni; Tzoar, N.: Phys. Rev. 187, 1000 (1969)
[10] Iwasa, S.; Sawada, Y.; Burstein, E.; Palik, E.D.: Proc. Intern. Conf. Phys. Semicond., Kyoto 1966, Jour. Phys. Soc. Japan 21, Suppl., 742 (1966)
[11] Melnyk, A.R.; Harrison, M.J.: Phys. Rev. Lett. 21, 85 (1968)

[12] Fuchs, R.; Kliewer, K.L.: Phys. Rev. 185, 905 (1969)
[13] Elson, J.M.; Ritchie, R.H.: Phys. Letters 33A, 255 (1970)
[14] Fischer, B.; Marschall, N.: 11 Intern. Conf. Phys. Semicond., Proceedings, Warszawa 1972 (PWN, Warszawa 1972), vol. 2, p. 1435
[15] Ron, A.; Tzoar, N.: Phys. Rev. 131, 12 (1963)
[16] Ron, A.; Tzoar, N.: Phys. Rev. 131, 1943 (1963)
[17] Dawson, J.; Oberman, C.: Phys. Fluids 5, 517 (1962)
[18] McCumber, D.E.: Phys. Rev. 154, 790 (1967)
[19] McCumber, D.E.: Rev. Mod. Phys. 38, 494 (1966)
[20] Mycielski, J.; Mycielski, A.: to be published
[21] Mycielski, A.; Aziza, A.; Mycielski, J.; Balkanski, M.: to be published
[22] Aziza, A.; Balkanski, M.: to be published
[23] Lyden, H.A.: Phys. Rev. 134, A1106 (1964)
[24] Kukharskii, A.A.; Subashiev, V.K.: Fiz. Tekh. Poluprov. 4, 287 (1970) [Sov. Phys. - Semicond. 4, 234 (1970)]
[25] Mycielski, A.; Aziza, A.; Balkanski, M.; Moulin, M.Y.; Mycielski, J.: phys. stat. sol. (b) 52, 187 (1972)
[26] Mycielski, A.; Mycielski, J.; Aziza, A.; Balkanski, M.: 11 Intern. Conf. Phys. Semicond., Proceedings, Warszawa 1972 (PWN, Warszawa 1972), vol. 2, p. 1214
[27] Maitre, A.; Le Toullec, R.; Balkanski, M.: 11 Intern. Conf. Phys. Semicond., Proceedings, Warszawa 1972 (PWN, Warszawa 1972), vol. 2, p. 826
[28] Wallace, P.R.; Chandler, K.K.; Harnad, J.: Can. J. Phys. 46, 243 (1968)
[29] Aziza, A.; Balkanski, M.; Faure, M.: Proc. Intern. Conf. "Transition Semiconducteur - Semimetal", Nice, Sept. 1973, in press
[30] Ganguly, A.K.; Ngai, K.L.: Phys. Rev. B 8, 5654 (1973)
[31] Parada, N.J.; Pratt, Jr., G.W.: Phys. Rev. Lett. 22, 180 (1969)
[32] Logothetis, E.M.; Holloway, H.: Solid St. Commun. 8, 1937 (1970)
[33] Parada, N.J.: Phys. Rev. B 3, 2042 (1971)

TEMPERATURE-DEPENDENT FAR INFRARED MAGNETO-TRANSMISSION IN HgTe

R. J. Wagner and L. H. Ngai

Naval Research Laboratory, Washington, D. C. 20375 USA

Intraband magneto-absorption has been studied in HgTe at wavelengths between 513. and 47. μm and at temperatures between 4.3 and 40.K. Only by raising the sample temperature above 4.3K could cyclotron and combined resonances be distinguished from interband and impurity level transitions. The energy vs magnetic field dependencies and selection rules for these lines have been experimentally studied and compared to previous results, both experimental and calculated.

1. INTRODUCTION

Tuchendler et al. [1] have reported extensive far infrared (FIR) magneto-absorption measurements on HgTe. They observed spectral features associated with interband, intraband and impurity transitions. Plasma-shifted cyclotron resonance was identified and analyzed in light of the "anomalous" dielectric constant of this zero gap material. All other inter- and intraband lines were fitted using a Luttinger model for the p-like Γ_8 conduction and valence bands. Earlier Groves et al. [2] studied the $\Gamma_8^v \rightarrow \Gamma_8^c$ magnetoabsorption and developed a model which included directly the interactions of the s-like Γ_6 and p-like Γ_7 valence bands with the Γ_8 bands. This present work concentrates on the intra (conduction) band transitions with comparisons to the model and fitting parameters of Tuchendler et al. [1] and Groves et al. [2].

2. EXPERIMENTAL DETAILS

The thin samples (d $\approx$ 10 - 25 μm) were prepared from an unoriented portion of a single ingot of HgTe. This ingot provided material for a transport study [3] with n (4.2K) = 1.5 x 10^{15} cm^{-3}. These samples were mounted in a light pipe system in the bore of superconducting or Bitter-type solenoidal magnets. Experiments in both Voigt ($\underset{\sim}{s}_\perp \underset{\sim}{B}$) and Faraday ($\underset{\sim}{s}_\parallel \underset{\sim}{B}$) geometries were possible. All results reported here were obtained with a fixed FIR laser wavelength incident on the sample while the magnetic field was varied. The pulsed laser was either an optically-pumped molecular gas laser or an electric-discharge H_2O laser.

3. CONDUCTION BAND LANDAU LEVELS AND INTRABAND SELECTION RULES

Two approaches have been used to describe the conduction (Γ_8^c) band Landau levels near the degeneracy point. Tuchendler et al. [1] use a Luttinger model [4] in which the p-like Γ_8 bands are assumed to be spherical, crystal momentum k_H along the field negligible and spin-orbit splitting large. Two sets of levels result associated with the p-like band-edge functions transforming as m_J = 3/2, -1/2 and + 1/2, -3/2 respectively. In this approximation analytic expressions for the energy levels can be obtained [4]. Tuchendler et al. have been able to fit both inter- and intraband results with Luttinger parameters γ_1^L = -12.8, $\overline{\gamma}^L = \gamma_2^L = \gamma_3^L$ = -8.3 and K^L = -10.5. The parameters also satisfy an associated work on the $\Gamma_6^v \to \Gamma_8^c$ magnetoabsorption by Guldner et al. [5]. Alternatively, Groves et al. [2] treat the interactions between s-like and p-like bands exactly in order to fit their interband magnetoabsorption results, $\Gamma_8^v \to \Gamma_8^c$ and $\Gamma_6^v \to \Gamma_8^c$. The Luttinger parameters from this procedure are: γ_1^L = -16.8, γ_2^L = -10.4, γ_3^L = -9.2 and K^L = -11.2. The transitions under discussion here are identified in Fig. 1 where the Landau level energies are those of Groves et al. The selection rules for the possible intraband transitions are as follows: Cyclotron resonance (CR), Oa → 1a, is strongly allowed; combined resonance (KR), Oa → 1b, while allowed in $\underline{E}_\parallel\underline{B}$, is not as strongly allowed as CR; spin resonance (SR), is not allowed [2,6].

FIG. 1: Field dependence of the lowest conduction and valence band Landau levels taken from the results of Groves et al. [2]. The vertical arrows indicate cyclotron (CR), combined (KR) and spin (SR) resonances. The symbol E_F indicates the estimated zero field Fermi energy.

4. EXPERIMENTAL RESULTS

The magneto-transmission of 393 μm radiation through HgTe is
shown in Fig. 2. In order to clearly identify CR, the sample
temperature had to be raised to about 12K. This became in-
creasingly important when the resonance was observed at higher

FIG. 2: Magneto-trans-
mission of HgTe
in the Faraday
geometry as a
function of
sample tempera-
ture.

frequencies. This indicates the necessity of populating the
initial state, 0a, from either impurity levels or the valence
band levels. The development of zero field absorption at T =
29.4K is not understood. At the field position of the antici-
pated SR transition, there is no evidence of resonant absorp-
tion. One concludes that SR is at least two orders of magnitude
weaker than CR.

FIG. 3: Magneto-
transmission of
HgTe in the Voigt
geometry with E∥B
as a function of
sample tempera-
ture.

Figure 3 shows the transmission results for the $\underset{\sim}{E}_\perp\underset{\sim}{B}$ geometry, λ = 96.5 μm and T = 4.3 - 35.9K. At 4.3K, the spectrum is dominated by interband and (or) impurity transitions. As was noted for the case of CR, heating the sample should increase the intensity of the KR transition. As is evident in the Figure, the interband/impurity transitions broaden and weaken as the sample temperature is increased. At the higher temperatures a line develops at the anticipated position of KR. Comparison of this result ($\underset{\sim}{E}_\parallel\underset{\sim}{B}$) with a similar experiment in $\underset{\sim}{E}_\perp\underset{\sim}{B}$ indicates that the observed transition is only present in $\underset{\sim}{E}_\parallel\underset{\sim}{B}$. While a numerical comparison between experimental and predicted line intensity has not been made, the line intensity is weaker than CR as expected. From these observations, this line is identified as KR. The other broad line down on Fig. 3 has not been identified.

The energy vs magnetic field dependencies of CR and KR have been determined using a number of laser frequencies. A compilation of this data is shown in Fig. 4 along with the data-fitted line positions of Tuchendler et al. [1] and the calculated line positions of Groves et al. [2]. The CR results agree

FIG. 4: Experimental and calculated resonant frequencies for CR, KR and SR in HgTe.

with the measurements of Tuchendler et al. within the limits of experimental accuracy. In contrast, the KR line position is significantly different from that observed by Tuchendler et al. On the other hand, it is in reasonable agreement with the calculated results of Groves et al. Attempts to identify

electron- (LO,TO) phonon interactions in the KR spectrum
between 30 and 40 kG have been unsuccessful. It was found that
increasingly high temperatures were required to identify KR as
the resonance frequency/magnetic field was increased. At these
temperatures (> 50K), broadening of the KR line became so severe
that the weak splittings would have been washed out.

5. CONCLUSION

The intra- (conduction) band magneto-optical resonances CR and
KR have been studied in HgTe. By heating of the samples, agree-
ment has been achieved between the anticipated and observed
line intensities and selection rules. The line position of KR
is in agreement with the calculated line position of Groves
et al.

ACKNOWLEDGMENTS

The authors wish to thank S. H. Groves for providing HgTe
eigenvalues and eigenfunctions, J. Furdyna for providing the
HgTe sample, R. Kaplan for suggesting sample heating, and B. D.
McCombe for helpful discussions regarding matrix elements.

6. REFERENCES

[1] Tuchendler, J.; Grynberg, M.; Couder, Y.; Thomé, H.;
 Le Toullec, R.: Phys. Rev. B8, 3884 (1973).
[2] Groves, S. H.; Brown, R. N.; Pidgeon, C. R.: Phys. Rev.
 161, 779 (1967).
[3] Galazka, R. R.: Phys. Lett. 32A, 101 (1970).
[4] Luttinger, J. M.: Phys. Rev. 102, 1030 (1956).
[5]Guldner, Y.; Rigaux, C.; Mycielski, A.; Grynberg, M.:
 Phys. Rev. B8, 3875 (1973).
[6] Baramidze, G. A.; Gurgenishvili, G. E.; Khutsishvili, G. R.:
 Sov. Phys.-Solid State 4, 2168 (1963).

ANOMALOUS MAGNETORESISTANCE EFFECT IN HgTe UNDER UNIAXIAL COM-
PRESSION AT LOW TEMPERATURES

K. TAKITA, N. TANIMURA and S. TANAKA

Department of Applied Physics, Faculty of Engineering
University of Tokyo, Bunkyo-ku, Tokyo

The degeneracy of the Γ_8 conduction and valence band of
HgTe was lifted and very small band-gap(E_g=2~3meV) was
introduced by the application of the uniaxial compression
up to 10^9dyne cm^2. The resistivity of the uniaxially com-
pressed HgTe decreased greately by the application of
strong magnetic field up to 46 kOe both in transverse
and longitudinal directions. These anomalous magneto-
resistance may indicate the anomalous Landau level spac-
ing of the conduction and the valence bands of uniaxially
compressed crystals.

1.INTRODUCTION

The conduction band and the valence band of HgTe are degenerate
at Γ-point and this degeneracy is induced by the cubic symmetry.
Therefore, the degeneracy is lifted by the application of the
uniaxial stress and very small band-gap appears[1][2]. The gal-
vanomagnetic properties under this situation was reported pre-
viously by the authors[3]. The temperature dependences of the
resistivity and the weak field Hall coefficient indicated the
formation of the band-gap of about 2meV in the case of the<100>
compression of about 5×10^8dyne cm^{-2}. An unusual behavior of the
negative magnetoresistance under a certain compression was also
reported there. Previously, however, the measurement of the
magnetoresistance was limited to the transverse magnetic field
up to 18 kOe. In the present experiment, we expanded the magnet-
ic field up to 46 kOe and investigated both of the transverse
and the longitudinal magnetoresistances. This extention of the
experimental condition gave new result as presented here.

The resistivity which was first raised by the application of
the uniaxial compression reduced greatly by further application
of the longitudinal magnetic field. This means that the thermal
energy-gap introduced by the uniaxial stress decreases again by
the application of the magnetic field. This comes from the fact
that the ground Landau levels both of the conduction and va-
lence bands of deformed crystal have a complicated character
due to the quantum effects[4].

Fig. 1 ; Transverse magnetoresistance of sample T1-B5 at 1.5K under various values of compression along <100> direction. The unit of the indicated values of the compression is 10^8dyne cm^{-2}. (The broken line shows the data obtained for the other pair of probes under the highest value of compression 7.14×10^8dyne cm^{-2})

2. EXPERIMENTAL METHODS AND RESULT

The samples were prepared by the same method described in the previous paper[3]. The use of the split-off-type superconductive magnet in this experiment made it possible to measure both of the transverse and the longitudinal magnetoresistance under compression.

The transverse magnetoresistance at 1.5 K is plotted against magnetic field in Fig.1 for some different values of the compression along <100> . Under the compression of 3 or 4×10^8dyne cm^{-2}, it showed a broad minimum of about -20% around 25 kOe. Under 5.3×10^8dyne cm^{-2}, the negative magnetoresistance reached 37% and the minimum point shifted to strong field of 42 kOe. It should be noted that, in the case of the compression along <100>, the magnetoresistance turned to positive in a wide range of magnetic field under stronger compression of 7.14×10^8dyne cm^{-2}.

Fig. 2 ; Transverse magnetoresistance of sample T2-C13 under various values of compression along <111> at 1.5 K. The values of the compression are indicated by the unit of 10^8dyne cm^{-2}.

The transverse magnetoresistance under the compression along <111> is shown in Fig.2. The negative magnetoresistance was also observed under the compression above 7.1×10^8 dyne cm^{-2} and it increased monotonically with increasing compression up to 9.5×10^8 dyne cm^{-2}. The magnetic field dependence of the highest compression along <111> was similar to that of the mediate compression of <100> as is seen in Fig.1 and Fig.2. As is discussed later, this means that the <100> compression induced larger band-gap than the <111> compression by factor of about 2.

On the other hand, in the longitudinal magnetic field, much pronounced negative magnetoresistance was observed as is shown in Fig.3. The longitudinal magnetoresistance at 1.5 K of the sample compressed along <111> reached to the value of -85% for 8.7×10^8 dyne cm^{-2}. It must be mentioned that the longitudinal magnetoresistance at 4.2 K were very similar to that at 1.5 K.

In Fig.4, the resistivity at 1.5 K under longitudinal magnetic field and uniaxial compression along <111> is plotted against magnetic field for various values of compression. Under zorostress, after showing a small minimum around 9 kOe, the resistivity increased rapidly with increasing magnetic field above 20 kOe. This seems to indicate the carrier freeze-out effect at strong field. The resistivity under no magnetic field also increased from $5.5 \times 10^{-2} \Omega$cm to $2.4 \times 10^{-1} \Omega$cm with increasing compression. This increase of the resistivity indicates that the population of the conduction band electrons reduced by the appearance of the band-gap. Contrary to these behaviors, the magnetic field

Fig. 3; Longitudinal magnetoresistance of sample T2-C13 at 1.5 K. The sample is compressed along <111> by the indicated value of compression in the unit of 10^8 dyne cm^2.

Fig.4 ; The resistivity at 1.5 K under longitudinal magnetic field and compression along <111> is plotted against magnetic field. The values of the compression are indicated near the lines by the same unit of 10^8dyne cm^{-2}. In this case, the current, the magnetic field and the stress axis are all parallel to <111> direction.

dependences of the resistivity of the uniaxially compressed HgTe is anomalous. They decreased drastically with magnetic field and crossed the zero-stress curve around 25 kOe as is seen in Fig.4. Under the compression above 6.2×10^8dyne cm^{-2} along <111>, the resistivity became smaller than that at zero-stress and no magnetic field. This suggests that the ground Landau level of the conduction band of deformed crystal anomalously approaches to the valence band by the application of the magnetic field and therefore the resistivity first raised by the uniaxial stress reduces again due to the electron population in this conduction band level.

3.DISCUSSION

The energy gap introduced by the application of the uniaxial compression is considered to be about 2 meV at 5×10^8dyne cm^{-2} in the case of the compression along <100> and at 10^9dyne cm^{-2} in the <111> compression, respectively, under no magnetic field. Under these compressions, the negative magnetoresistance were most pronounced in both transverse and longitudinal magnetoresistances. The Luttinger parameters of undeformed HgTe have been determined by the experiments of Γ_8-Γ_8 interband magnetoreflection[5] and Γ_6-Γ_8 interband magnetoabsorption[6]. The Landau level spacing between the ground Landau levels of conduction band and valence band of undeformed HgTe is 2.5 or 3 meV at 40 kOe because of the quantum effects[4]. On the other hand, the large negative magnetoresistance observed in the deformed crystal means that the ground Landau level of the electron approaches

energetically to that of hole by about 2 meV with increasing magnetic field contrary to the case of undeformed crystal.

The Landau levels of the valence bands of deformed Si and Ge were treated by many authors[7][8][9]. We apply their results to the case of HgTe compressed along <111>. The ground Landau states of conduction band and valence band are a(1)-state and b(2)-state, respectively, but b(0)-state exists near the b(2)-state in HgTe. The energy of these states under the stress χ along <111> are expressed as follows, if we neglect the k_H-dependence.

$$E_{a(1)} = \frac{3}{2}(\gamma_1 - \gamma_3) - \frac{1}{2}\kappa - \frac{1}{3} D_u' c_{44}^{-1} \chi, \tag{1}$$

$$E_{b(2)} = \frac{3}{2}\gamma_1 + \gamma_3 - \frac{1}{2}\kappa - [(\frac{1}{3} D_u' c_{44}^{-1} \chi + \frac{3}{2}\gamma_3 + \gamma_1 - \kappa)^2 + \frac{2}{3}(\gamma_2 + 2\gamma_3)^2]^{\frac{1}{2}} \tag{2}$$

$$E_{b(0)} = \frac{1}{2}(\gamma_1 + \gamma_3) - \frac{3}{2}\kappa + \frac{1}{3} D_u' c_{44}^{-1} \chi, \tag{3}$$

where $\gamma_1, \gamma_2, \gamma_3$ and κ are the Luttinger parameters ($\gamma_1 = 14.8, \gamma_2 = 9.0, \gamma_3 = 8.2$ and $\kappa = 10.2$ according to [5]) and D_u' and c_{44} are the deformation potential constant and the elastic stiffness constant, respectively. We neglect the effect of dilation because it does not contribute to the energy splitting. As is seen from the above equations, if D_u' is negative, the ordering of the levels of a(1) and b(0) is inverted by the application of a sufficient compression. Therefore, the spacing between $E_{a(1)}$ and $E_{b(0)}$ decreases with the application of the magnetic field. If this is the case in HgTe, we can expect the large negative magnetoresistance because the electron population of the ground Landau level increases with magnetic field. By comparison with the experimental result, we obtained the deformation potential constant, $D_u' = -1.5$ eV. Althogh the analysis of the transverse cases is much more complicated, the negative magnetoresistances in those cases also may be attributable to the same effect of the Landau level spacing.

4.REFERENCES

[1] Pikus,G.E.;Bir,G.L.:Soviet Physics-Solid State 1 1502(1960)
[2] Bir,G.L.;Pikus,G.E.:Soviet Physics-Solid State 3 2221(1962)
[3] Takita,K.;Tanimura,N.;Tanaka,S.:Proc.Internat.Conf.Physics of Semimetals and Narrow Gap Semiconductors,Nice (1973)
[4] Luttinger,J.M.: Phys. Rev. 102 1030 (1956)
[5] Uchida,S.;Yoshizaki,R.;Tanaka,S.:Proc.Internat.Conf.Semimetals and Narrow Gap Semiconductors,Nice(1973):private communication
[6] Guldner,Y.;Rigaux,C.;Grynberg,M.;Mycielski,A.:P.R.B8 3875(1973)
[7] Kleiner,W.H.;Roth,L.M.:Phys.Rev.Letters 2 334 (1959)
[8] Hensel,J.C.;Feher,G.: Phys.Rev. 129 1041 (1962)
[9] Gurgenishvili,G.E.:Soviet Physics-Solid state 5 1510 (1962)

LOW TEMPERATURE GALVANOMAGNETIC PROPERTIES OF n-TYPE $Hg_{.8}Cd_{.2}Te$

R.Dornhaus,H.Happ,K.-H.Müller,G.Nimtz,W.Schlabitz,P.Zaplinski
II.Physikalisches Institut der Universität zu Köln,5 Köln FRG
G.Bauer
I.Physikalisches Institut der RWTH Aachen,51 Aachen FRG

It is shown that neither thermal nor magnetic freeze-out of carriers occur, and that the mobility is not governed by ionized impurity scattering due to localized charged impurities. A Te vacancy without bound state is proposed to explain the experimental results. Hot carrier experiments were performed in magnetic fields and Landau level transitions observed.Radiative relaxation of Landau transitions was measured at a wavelength of 100 μm

1. INTRODUCTION

The narrow band gap semiconducting II-VI alloy $Hg_{.8}Cd_{.2}Te$ has some interesting properties for basic solid state physics as well as for infrared device [1,2]. In this semiconductor non-parabolicity effects of the conduction band are very pronounced and magnetic quantum effects easily observable due to the coupled small effective mass and the extremely large g-factor. In order to study those quantum effects, however, the interaction mechanisms between free carriers and the lattice and its imperfections have to be known. Up to now the low temperature galvanomagnetic properties have not been investigated completely to arrive at a definite conclusion about the relevant interaction mechanisms in this II-VI alloy.

2. HALL-EFFECT AND CARRIER MOBILITY

The experiments reported here were carried out with five samples of n-type high mobility material ($n = 9 \times 10^{14}$ cm^{-3}, $\mu =$ 220 000 cm^2/Vs at 77 K, Cominco). In this material the Hall coefficient R_H and the conductivity σ do not show an oscillatory behaviour as a function of temperature as it is usually observed in material with $\mu < 200\ 000$ cm^2/Vs [3]. In Fig. 1 Hall

coefficient and Hall mobility μ are presented dependent on temperature. At temperatures above 60 K the decrease of carrier

Fig. 1: Hall coefficient and Hall mobility versus temperature T. R_H was measured at a magnetic field of 0.5 kG. The broken line represents data of the calculated drift mobility [2]. Experimental values x of another high mobility crystal are from ref. 7.

Fig. 2: conductivity σ versus electric field E at various magnetic fields B. The broken line at B = 0 represents calculated values of the field dependence of the conductivity for ionized impurity scattering dominating momentum relaxation and polar optical scattering dominating the energy exchange.

mobility is due to polar optical phonon scattering [2] and R_H decreases since the intrinsic carrier excitation becomes dominant [1]. Below 20 K the data show that both R_H and μ are constant, thus far being independent on temperature. This result is in contradiction to earlier investigations [1]. Furthermore it is shown in Fig. 1 that R_H does not increase with magnetic fields up to 50 kG at 4.2 K [3].From these experimental results we deduce, that neither thermal nor magnetic freeze-out of carriers take place (in p-type $Hg_{.8}Cd_{.2}Te$ we observed a carrier freeze-out at 4.2 K with $R_H \propto B^{1/3}$; $p = 10^{17} cm^{-3}$). Usually the low temperature mobility in a semiconductor is governed by ionized impurity scattering. A calculated drift mobility for ionized impurity scattering is plotted in Fig.1 and shows that the expected increase of mobility with T between 10 and 40 K is not in agreement with the experimental data [2]. Obviously ionized impurity scattering is not the dominant scattering process.This conclusion is supported by hot carrier experiments carried out with the same samples at 4.2 K which are represented in Fig.2. The measured curve σ = f(E) for B = 0 does not agree with the

theoretical curve. The calculated increase of σ above 1 V/cm is characteristic for ionized impurity scattering.

Since we did not observe any evidence for bound states and ionized impurity scattering, we take the extrinsic electrons to be due to Te vacancies. Similarly to PbTe [4] the Te vacancy is assumed not to have a bound state. The vacancy model is supported by the fact that Hg-rich material is n-type and it is performed by annealing the crystal in Hg vapor.It was earlier suggested [1] that Hg atoms which exceed the number needed for stoichiometry might reside interstitially in the lattice. Such an interstitial imperfection is ruled out by our experimental results of the mobility. Such imperfection should act as a charged impurity, whereas vacancies do not have a net charge. Erginsoy [5] had shown that the momentum relaxation time τ of the carriers for interaction with neutral hydrogen-like impurities is independent of the electron velocity v. This follows from the proportionality of the total scattering cross-section $Q \propto 1/v$. Such a velocity independent scattering mechanism would explain the observed low temperature mobility independent of temperature and electric field. The calculation of τ for the investigated material at 4.2 K with the number of neutral impurities equal to the number of free electrons results in 1.6×10^{-13} sec which is much smaller than the measured 1.4×10^{-12} sec, which were calculated by the expression $\tau = \sigma \cdot m/(ne^2)$. This result is plausible as the potential of a vacancy is rather more located than that of a hydrogen-like impurity, especially in $Hg_{.8}Cd_{.2}Te$ with the extremely small effective mass.The vacancy potential is essentially repulsive [4] and might be compared with a spherical potential barrier. For the latter it is known [6] that in first order approximation also $Q \propto 1/v$ is valid, thus a velocity independent relaxation time is expected similarly to the hydrogen model. For a quantitative calculation of the scattering cross-section the vacancy potential has to be known.

3. NON-OHMIC CONDUCTIVITY AND LANDAU RADIATION

In Fig. 2 the dependence of σ on E is shown for B = 0 and transverse magnetic fields up to 15 kG. The curves do not give any evidence for ionized impurity scattering nor impurity break-

down at high magnetic and electric fields. An electron temperature model taking into account ionized impurity scattering (Brooks-Herring formalism, n = N_I) and polar optical scattering yielded the broken curve. For ionized impurity scattering dominating the momentum relaxation σ increases up to 20 V/cm with E in striking contrast to the experiments.

At T = 4.2 K and E = 0 all carriers are within the 0^+ Landau level for magnetic fields in excess of 3.5 kG.With the application of electric fields the carriers gain energy and finally are transferred into higher Landau subbands. In contrast to n-InSb [8], where in the analogous magnetic and electric field region also impact ionization of the electrons from donors has to be considered, our ohmic and non-ohmic experiments show that this process is absent in the investigated n-HgCdTe. Thus these data seem to be appropriate for an application of Yamada's and Kurosawa's theory [9]. These authors have considered for the first time quantized subband structure, change of electron scattering rates, and population of higher Landau levels in high electric fields. Since in other zinc blende materials hot carrier interactions are additionally influenced by impurity breakdown, a comparison with the above cited theory is not unambigous in these compounds.

Fig. 3: d^2E/dj^2 vs electric field for a transverse magnetic field of 10 kG and 7.5 kG.4.2 K

Fig. 4: Intensity of FIR radiation vs B for an applied electric field of 1.5 V/cm. 4.2 K.

As shown in Fig. 3 we have looked for fine structures in the σ(E) characteristics. In the second derivative d^2E/dj^2 there appear dips which are shifted to higher electric fields as the magnetic field is increased. These dips seem to be correlated with the transfer of electrons from the lower to higher Landau

subbands 0^- and 1^+ [10].

Electrons, which have gained energy enough from the electric field to populate higher Landau subbands, can relax to the lowest Landau level or to donor levels (e.g. in n-InSb, n-GaAs) either by emission of phonons or also by emission of photons [11,12]. In the range of magnetic fields applied in our experiments for $Hg_{.8}Cd_{.2}Te$ the radiative recombination is to be expected in the far infrared region. The emission process is not favoured by the relatively high carrier concentration in our material. Using a Ge/Sb detector, which had its maximum sensitivity near 9.8 meV, we observed photoconductivity signals originating from radiation from our samples. The experimental setup was similar to that described by Kobayashi et al. [12]. For an electric field of 1.5 V/cm the detector output vs magnetic field is shown in Fig. 4. The main peak corresponds to radiation originating from N = 1 to N = 0 transitions.

4. ACKNOWLEDGEMENT

We thank Dipl.Phys. Stahl (AEG-Telefunken) and Dr. H.Kölker (Consortium) for supplying us with samples.

5. REFERENCES

[1] Long, D.; Schmit, J.L.: in Semiconductors and Semimetals, ed. R.K.Willardson a. A.C.Beer (Academic,1970) Vol.5, p.175
[2] Nimtz, G.; Bauer, G.; Dornhaus, R.; Müller, K.-H.:Phys.Rev. B 15 July 1974, in press
[3] Dornhaus, R.; Nimtz, G.; Schlabitz, W.; Zaplinski, P.:Solid State Commun. June 1974, in press
[4] Parada, N.J.; Pratt Jr., G.W.: Phys.Rev.Letters 22,180 (69)
[5] Erginsoy, C.: Phys.Rev. 79, 1013 (1950)
[6] Mott, N.F.; Massey, H.S.W.: The Theory of Atomic Collisions (Clarendon Press, Oxford, 1965) third edition, Chapter II,5
[7] Kölker, H.: private communication
[8] Kotera, N.; Yamada, E.; Komatasubara, K.F.: J.Phys.Chem.Solids 33, 1311 (1972)
[9] Yamada, E.; Kurosawa, T.: J.Phys.Soc.Japan 34, 603 (1973)
[10] Kobayashi, K.L.I.; Otsuka, E.: J.Phys.Chem.Solids 35, 839 (1974)
[11] Gornik, E.: Phys.Rev.Letters 29, 595 (1972)
[12] Kobayashi, K.L.I.; Komatsubara, K.F.; Otsuka, E.: Phys.Rev. Letters 30, 702 (1974)

ACCEPTOR STATES IN HgTe

G. BASTARD, Y. GULDNER, A. MYCIELSKI[*], P. NOZIERES[**], C. RIGAUX

Groupe de Physique des Solides de l'Ecole Normale Supérieure,

24 rue Lhomond, 75231 Paris Cédex 05, France

Evidence for acceptor states in HgTe is obtained in
magnetooptical studies. Calculations performed within
the framework of scattering theory show the stability
of acceptor states which induce conductivity reson-
ances already reported.

1. EXPERIMENTS

We performed magnetooptical experiments at photon energy between 300-350meV,
in fields up to 60 kG, on HgTe thin plates (2μm thick) annealed under re-
duced pressure of Hg, the Hg vacancies acting as acceptor states. Cu-doped
samples were also investigated. Magnetotransmission experiments were carried
out at 4.2°K for σ^+ polarization and with linear $\vec{E}//\vec{H}$ polarization. Experi-
mental spectra, typically described on Fig.1, show a broad and intense

FIG. 1 Magnetoabsorption spectrum
for $\vec{E}//\vec{H}$.

FIG. 2

impurity absorption, for $\vec{E}//\vec{H}$ and σ^+ polarizations beside the sharp lines
due to $\Gamma_6 \rightarrow \Gamma_8$ magnetooptical transitions. For variously doped samples, the
minimum of the impurity absorption (Fig.2) depends linearly on the magnetic
field and extrapolates to 2.2 ± 0.3 meV above the interaction gap $E_{\Gamma_8} - E_{\Gamma_6} =$
302.5 meV.

[*] Permanent address : Institute of Physics, P.A.N., Warsaw, Poland.

[**]Permanent address : Institut Max von Laue-Paul Langevin, B.P. 156,Grenoble

From the linear field dependence of the energy, this absorption is unambiguously identified as the transition from 0^- level of Γ_6 to an acceptor state with 2.2 ± 0.3 meV binding energy. Previous magnetotransmission data [1,3] have also shown the presence of an acceptor state with 0.7 meV ionization energy. Both energies are in good agreement with the results of transport phenomena [2].

2. THEORY [4,5]

For free electrons in the Γ_8 bands, the effective mass Luttinger Hamiltonian, neglecting warping, can be written :

$$H_{\underline{k}} = \frac{9\varepsilon_{k_1} - \varepsilon_{k_2}}{8} + (\hat{k}.\vec{J})^2 \frac{\varepsilon_{k_2} - \varepsilon_{k_1}}{2} \tag{1}$$

in terms of electron $\varepsilon_{k_1} = \frac{\hbar^2 k^2}{2m_c}$ and hole $\varepsilon_{k_2} = \frac{-\hbar^2 k^2}{2m_v}$ energies, $\hat{k}$ is the unit vector in the direction of the wave vector $\vec{k}$ and $\vec{J}$ is the 4×4 angular momentum matrix for a state $J = 3/2$. The propagator $G^o_{\underline{k}}$ associated with $H_{\underline{k}}$ is the 4×4 matrix :

$$G^o_{\underline{k}} = \frac{9g_{k_1} - g_{k_2}}{8} + \frac{(\hat{k}.\vec{J})^2}{2}(g_{k_2} - g_{k_1}) \tag{2}$$

where $g_{k_1} = (E - \varepsilon_{k_1})^{-1}$ and $g_{k_2} = (E - \varepsilon_{k_2})^{-1}$ are the free electron and hole propagators.

The influence of impurity potential on the eigenstates can be easily determined under simplified assumptions :

- We assume that the Fourier transform of the potential $V(\vec{r})$ is a separable function, i.e. $V_{\underline{kk}'} = V u_{\underline{k}} u_{\underline{k}'}$, where the cut-off factor $u_{\underline{k}}$ vanishes at large k.

- We also postulate the isotropy in the k space of the $u_{\underline{k}}$ factor.

Under these assumptions the matrix propagator $G_{\underline{kk}'}$ is easily obtained from the Dyson equation :

$$G_{\underline{kk}'} = G^o_{\underline{k}} \delta_{\underline{kk}'} + \sum_{\underline{k}''} G^o_{\underline{k}} V_{\underline{kk}''} G_{\underline{k}''\underline{k}'} \tag{3}$$

If we define the $\tilde{G}$ and $\tilde{G}_o$ matrices as :

$$\tilde{G} = \sum_{\underline{kk}'} G_{\underline{kk}'} u_{\underline{k}} u_{\underline{k}'} \qquad \tilde{G}_o = \sum_{\underline{k}} u_{\underline{k}}^2 G^o_{\underline{k}}$$

$\tilde{G}_o$ and $\tilde{G}$ take a simple scalar expression in terms of $g_1 = \sum_{\underline{k}} u_{\underline{k}}^2 g_{k_1}$ and $g_2 = \sum_{\underline{k}} u_{\underline{k}}^2 g_{k_2}$: $G_o = \frac{g_1 + g_2}{2}$; $G = \frac{g_1 + g_2}{1 - V \frac{g_1 + g_2}{2}}$

Also, the density of states $\rho = \frac{1}{\pi} \mathrm{Tr}\, \mathrm{Im}\, G$ can be written in terms of g_1, g_2 as :

$$\rho = 2 \; \frac{\rho_1 + \rho_2}{\left| 1 - V \left(\dfrac{g_1 + g_2}{2} \right) \right|^2} \tag{4}$$

where ρ_1 and ρ_2 are the densities of free electrons and holes. $\rho(\varepsilon)$ takes the following expression :

$$\rho = 2 \; \frac{\rho_1 + \rho_2}{\left| 1 - V \{ \alpha(m_v - m_c) + \beta [m_c^{3/2} \sqrt{-\varepsilon} - m_v^{3/2} \sqrt{\varepsilon}] \} \right|^2} \tag{5}$$

where $\alpha = \dfrac{1}{2\pi^2} \int u_k^2 \, d^3k$; $\beta = \dfrac{\sqrt{2}}{2\pi \hbar^3}$

EXISTENCE AND LIFETIME OF THE BOUND STATES

a) Neglecting first the imaginary part of the binding energies, the localized states are given by the roots of the real part of the denominator $|D(\varepsilon)|^2$ in Eq.(4) :

$$\text{For } \varepsilon > 0 \quad \beta m_v^{3/2} \sqrt{\varepsilon} = \frac{1}{V_c} - \frac{1}{V}$$

$$\text{For } \varepsilon < 0 \quad \beta m_c^{3/2} \sqrt{-\varepsilon} = \frac{1}{V} - \frac{1}{V_c} \qquad \left(\frac{1}{V_c} = \alpha(m_v - m_c) \right)$$

For $V > V_c$ (potential sufficiently repulsive) degenerated acceptor states with the conduction band exist, whereas donor states located in the valence band only exist for $0 < V < V_c$ (as $m_v > m_c$ $V_c > 0$).

b) To discuss the lifetime of the levels, the width Γ of the states should be determined by accounting for the complex energies of the resonant states :

$$\Gamma = - \text{Im} D(\varepsilon) \left[\frac{d}{d\varepsilon} \text{Re } D(\varepsilon) \right]^{-1}_{\varepsilon = \varepsilon_0}$$

ε_0 is the real part of the energy.

As $\dfrac{\Gamma}{2|\varepsilon_0|} = - \left(\dfrac{m_c}{m_v} \right)^{3/2}$ for states lying in the conduction band,

and $\dfrac{\Gamma}{2|\varepsilon_0|} = - \left(\dfrac{m_v}{m_c} \right)^{3/2}$ for states lying in the valence band, only acceptor states are stable. The heavy hole mass $\left(\dfrac{m_v}{m_c} \approx 10 \right)$ precludes any stability of resonant donor states in the valence band.

The resolvent formalism enables us to deduce the conductivity $\sigma(T)$ in the limit of dilute impurities :

$$\sigma(T) \sim \frac{1}{N_A} \int_0^\infty \frac{\partial f}{\partial \varepsilon} \frac{\rho_1(\varepsilon) d\varepsilon}{\sin^2 \delta(\varepsilon)}$$

where $f(\varepsilon)$ is the Fermi-Dirac function and $\delta(\varepsilon)$ the phase shift :
$\delta(\varepsilon) = \text{Arg } D(\varepsilon)$.

FIG. 3 Calculated conductivity
(arbitrary units) versus
temperature.

Figure 3 shows $\sigma(T)$ versus T taking into account either the 0.7 meV acceptor state (circles) or the 2.2 meV level (crosses). In both cases a linear temperature dependence of the Fermi level is postulated for the sake of simplicity ($E_F(T) = \frac{2.2}{30} T$). Both curves clearly point out conductivity minima which occur when the Fermi level crosses the acceptor state. These minima have already received experimental evidence in transport phenomena [2].

We are greatly indebted to Dr. J. Blinowski for enlightening discussions on this subject.

3. REFERENCES

[1] Bastard, G.; Guldner, Y.; Rigaux, C.; Nguyen Hy Hau ; Vieren, J.P.;
 Menant, M.; Mycielski, A. : Physics Letters, 46A n°2, 99 (1973).
[2] Finck, C.; Otmezguine, S.; Weill, G.; Verié, C. : Proc. of the 11th
 Intern. Conf. on the Physics of Semiconductors, Warsaw (1972) and
 quoted references.
[3] Tuchendler, J.; Grynberg, M.; Couder, Y.; Thomé, H.; Le Toullec, R. :
 Phys. Rev. B.8, 3884 (1973).
 Liu M.; Brust D. : Phys.Rev.157, 627(1967).
 Gelmont B.L.; Dyakonov M.I. : Soviet Phys.-JETP, 21,713(1972).

RESONANCE CONDITIONS FOR ACCEPTOR LEVELS IN HgCdTe ALLOYS

A.MAUGER, S.OTMEZGUINE, C.VÉRIÉ

Laboratoire de Physique des Solides, CNRS, 92190, Meudon, France

J.FRIEDEL

Université de Paris-Sud, Lab. de Physique des Solides, 91405,Orsay,France

Using galvanomagnetic effects, the behaviour of substitutional impurity and Hg-vacancy acceptor states is investigated at low temperature on HgTe and HgCdTe within the semimetal-to-semiconductor transition induced by hydrostatic pressure. A qualitative understanding of the experimental results is provided by a theoretical study starting with the Koster-SLATER theory extended to the zero-gap configuration and using a potential highly localized on the scattering center site.

1. INTRODUCTION

The electronic structure of acceptor states in semiconductors (hereafter SC) is usually depicted by discrete levels lying in the band gap. In materials with symmetry of the diamond-zinc blende type, relativistic effects together with the degeneracy rules of the cubic symmetry may lead to an electronic structure configuration with no bandgap between the conduction and valence bands. In this situation, exemplified by α-Sn, HgTe and HgCdTe alloys, the resulting interference of impurity states with a quasi-continuum of extended band states has been studied theoretically [1,2] . Experimental evidences of resonant acceptor states were reported in HgTe and HgCdTe with alloy compositions corresponding to the semimetal configuration (hereafter SM) [3-10]. In this work, the extrinsic properties of p-type HgTe and $Hg_{1-x}Cd_xTe^*$ within the SM $\rightarrow$ SC transition induced by hydrostatic pressure are investigated.

2. EXPERIMENTAL RESULTS

Galvanomagnetic data (obtained in the ranges 1.7-4.2°K, 0-10KG and 0-8Kbar) are analyzed by fitting the magnetic field (B) dependence of the Hall coefficient R, at each pressure, according to a two-carriers expression of R, as

*Acceptors are due to Cu impurities, observed in the range 0.1 - 1 at. ppm and to mercury vacancies V_{Hg} produced by controlled non-stoechiometry.

FIG.1: Fit of non-oscillating part of the reciprocal relative variation of the Hall coefficient using a two-carriers expression: two sets of electrons, $n_1 = 8 \times 10^{14} cm^{-3}$, $\mu_1 = 4.3 \times 10^5 cm^2/Vs$; $n_R = 7 \times 10^{14} cm^{-3}$, $\mu_R = 9 \times 10^3 cm^2/Vs$

FIG.2: Galvanomagnetic data vs. pressure of $Hg_{1-x}Cd_xTe$; P_C: Transition SM→SC.

illustrated by FIG.1. In addition, useful informations are obtained from an observation of quantum oscillations on the curves R(B) and magnetoresistance vs.B: for example a good determination of the pressure coefficient of the negative $\Gamma_6 - \Gamma_8$ gap E_g in HgTe at 1.7°K, $dE_g/dP = +8 \pm 0.5 meV/Kbar$.

Whereas the pressure dependence of R at low B gives a direct evidence of the pressure-induced SM→SC transition as illustrated on an alloy HgCdTe with x = 0.14(FIG.2), a subsequent analysis (through the magnetic field variation studies) is needed to show an experimental evidence of the resonant configuration of acceptor states A_o and A_1 corresponding respectively to Cu [5,6,10] and to V_{Hg} [3,5,8,9]. The main results concerning heights E_A of virtual states (zero energy is at the band degeneracy point Γ_8) and widths of resonances ΔE, are the following [11] :

(i). From magnetic field variation studies of R(B), a two-carriers analysis shows that both electrons and holes are present for x =0.14 at 4.2°K. A plot of their concentrations, through p-n vs. P, with $p-n = N_{A1}^- - N_D$ and $N_{A1}^- = N_{A1} \times [1+4\exp\{(E_{A1} - E_F - \epsilon E_g)/kT\}]^{-1}$, (ϵ= o in the SM case and $\epsilon = 1$ in the SC one), indicates a strong upward shift of A_1 induced by pressure. This is illustrated by FIG.3 showing the relative position of E_{A1} and Fermi level E_F .

(ii). Using a theoretical study of ionized impurity scattering appropriate to the SM→SC transition [11], the experimental electron mobility data (FIG.2) can be fitted with a calculated curve (labelled 2 in FIG.4), taking into account of the de-ionization of charged centers (i): in FIG.4, where N_{cc} accounts for the concentration of ionized acceptors N_{A1}^- and donors N_D , the agreement is good with $N_D = 4 \times 10^{14} cm^{-3}$, except around P=3.7Kbar where $E_F = E_{A1}$ (Fig.3).Within this pressure range, the observed feature is attributed to a pressure-induced

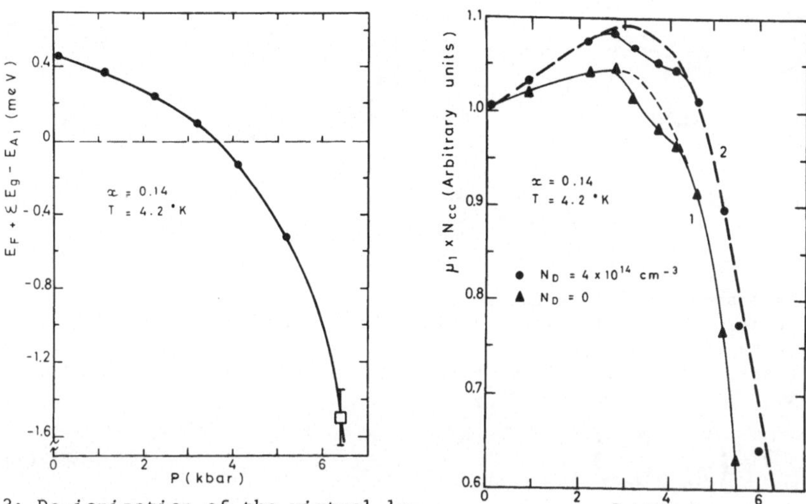

FIG.3: De-ionization of the virtual le-
vel A_1. The point at 6.4Kbar was obtained from an evidence of the crossover
of the O-Landau level with A_1.
FIG.4: Evidence of ionized impurity scattering of electrons and anomaly
around 3.7Kbar: curve 2, theory; ▲,● experimental data taking into account of
ionized donors with different concentrations.

resonance scattering of the electrons belonging to the extended states of the

conduction band.

(iii). A plot of the difference of experimental and calculated values of $\mu_1 \times N_{cc}$
versus P, together with the above value of dE_g/dP, yields for A_1 a resonance
width $\Delta E = 0.3$meV for x=0.14 at 4.2°K, i.e. essentially limited by the thermal
distribution spread-out. Indirect informations about ΔE are obtained by con-
sidering the observation of "slow electrons" through studies R(B) as reported
for HgTe in FIG.1. For this set of electrons, $\mu_R \simeq 9 \times 10^3 \mathrm{cm}^2/\mathrm{Vs}$ in HgTe whereas
$\mu_R \ll 2 \times 10^3 \mathrm{cm}^2/\mathrm{Vs}$ in $Hg_{0.86}Cd_{0.14}Te$. These slow electrons are attributed to
electrons quasi bounded to completely ionized resonant A_o states. These results
suggest that the resonance width of virtual acceptor levels is broader in
HgTe than in SM HgCdTe alloys.

3. THEORETICAL STUDY

In the framework of a two bands model, i.e. the conduction and valence bands
degenerated at Γ_8, the scattering problem is solved starting with a basis of
a set of doubly degenerate localized wannier functions (two orthogonal wannier
functions i_c and i_v per site i associated with each band). To examine the stri-
king features exhibited by A_1, the perturbing potential $\bar{V}$ induced by a scatte-
ring center is chosen highly localized on the defect site, noted O [12]. In
the representation $\{O_c, O_v\}$, $\bar{V}$ is a 2 x 2 matrix,

FIG.5: Resonance conditions for impurity states in zero-gap SC, illustrated by a virtual acceptor level E_A.

FIG.6: Pressure-induced shift of the virtual acceptor A_1 represented by the points ✚, using $dE_g/dP = 8\text{meV/Kbar}$.

$$\bar{V} = \begin{pmatrix} V_{cc} & V_{cv} \\ V_{vc} & V_{vv} \end{pmatrix} \tag{1}$$

where V_{cc} and V_{vv} are the intraband coupling terms corresponding to the conduction and valence bands respectively, V_{cv} and V_{vc} being the interband ones. The interband coupling is supposed to be strong and for simplification the matrix elements of (1) are taken equal,

$$V_{cc} = V_{vv} = V_{cv} = V_{vc} = V \tag{2}$$

Under such approximations, using the Green function formalism, it can be established that the resonance conditions are given by

$$\left. \begin{aligned} V\, F_o(E) &= 1 \\ F_o(E) &= \mathcal{P}\!\int \frac{g_o(E')}{E - E'}\, dE' \end{aligned} \right\} \tag{3}$$

where $\mathcal{P}$ means the "principal part of", $g_o(E)$ is the sum of the density of states of conduction and valence bands corresponding to the unperturbed Hamiltonian. The results show (FIG.5) that: (i) donors ($V<0$) do not lead to any resonance; (ii) acceptors ($V>0$) may produce a resonance in the conduction band if the potential is strong enough ($V>V_o$). In the latter case, the extra density of states induced by the defect in the vicinity of the resonance energy E_A is given by

$$\Delta n(E) = \frac{\Gamma}{(E-E_A)^2 + \Gamma^2} \tag{4}$$

where, the resonance width Γ is expressed as

$$\Gamma = \pi\, g_{oc}(E_A)\, V^2\, |\langle 0_v | \varphi_L | \rangle|^2 \tag{5}$$

with $g_{oc}(E_A)$ representing the conduction band density of states at $E = E_A$ and $|\varphi_L\rangle$ the localized state of the resonant acceptor.

4. DISCUSSION

As mentioned previously [3,5], the energy of A_1 is strongly dependent upon the density of the conduction band states $g_c(E)$ interfering with it. This is shown in FIG.6 where in addition to data vs.pressure, previous compositional studies are illustrated by dots [3,5]. Correlation of g_c (E) with E_g is obtained from the $\vec{k}.\vec{p}$ approach. This strong shift is qualitatively understood through our theoretical study showing that, for a given localized potential $V > V_o$, the location of E_{A1} (FIG.5) is dependent upon $g_c(E)$ through expression(3). Such a large shift of A_1 is mainly due to the virtual transitions induced by the potential $V_{cv}(1)$ which couples the extended states belonging to the two bands. The fact that its strength must be sufficiently large is probably a plausible feature of our highly localized potential since it is supposed to describe a virtual state corresponding to a mercury vacancy. On the other hand, we do not observe such large shift for A_o when $g_c(E)$ varies. This is understandable according to a theory for substitution impurity [2]. In that case, A_o is essentially attached to the valence band, in contrast with A_1 which remains in the conduction band even for an opened-gap alloy with $E_g = +50$meV, as illustrated by our experiments (FIG.6). Finally, considering the resonance width ΔE, the situation is different. Interband transitions do not significantly affect ΔE since the broadening of the acceptor levels is essentially due to virtual transitions between the acceptor states and the extended states of the conduction band. It is the magnitude of $g_c(E)$ in the vicinity of the resonance which is essential. This is illustrated by a comparison of the mobility values of the slow electrons in HgTe and HgCdTe obtained experimentally.

5. REFERENCES

[1] Liu M.; Brust D. : Phys.Rev.157, 627(1967).
[2] Gelmont B.L.; Dyakonov M.I. : Soviet Phys.-JETP, 21,713(1972).
[3] Vérié C. : New developments in semiconductors, ed.Wallace, Harris, Zuckermann,Noordhoff International Publish.,Leyden,p.513(1973).
[4] Elliot C.T.; Melngailis J.; Harman T.C.; Kafalas J.A.; Kernan W.C. : Phys.Rev. B5, 2985(1972).
[5] Finck C.; Otmezguine S.; Weill G.; Vérié C.: Proc.11th Int.Conf. Phys. Semicond, Warsaw, p.944. Vérié C. : Ibid.p.350.
[6] Tuchendler J.; Grynberg M.; Couder Y.; Thome H. : Ibid.p.346.
[7] Gelmont B.L.; Dyakonov M.I.; Ivanov-Omskii V.I.; Kolomiets B.T.; Ogorodnikov V.K.; Smekalova K.P. : Ibid.p.938.
[8] Mauger A.; Otmezguine S.; Weill G.; Vérié C. : Proc.Int.Conf."Transition Semiconducteur → Semimetal", Nice, Sept.1973, in press.
[9] Bastard G.; Guldner Y.; Rigaux C.; Nguyen Hy'Hau: Mycielski A.: Ibid.
[10] Uchida S.; Yoshizaki R.; Tanaka S.: Ibid.
[11] Otmezguine S.; Mauger A.; Vérié C. : submitted to Phys.Rev.
[12] Mauger A. : "Thèse 3e cycle", Orsay, April 1974; to be published.

NONLINEAR ELECTRIC AND OPTICAL EFFECTS IN GRADED MIXED SEMICONDUCTORS[+]

Jacek K. FURDYNA[++]

Institute of Physics, Polish Academy of Sciences

Ludwik LEIBLER and Jerzy MYCIELSKI

Institute of Theoretical Physics, Warsaw University

Warsaw, Poland

For semiconductor with position-dependent effective mass an additional force is acting on free carrier. Taking this force into account, the Boltzmann equation is solved in the case of two oscillating electric fields. The second order conductivity tensor is discussed in both $\omega\tau \gg 1$ and $\omega\tau \ll 1$ cases.

A lot of work was devoted in recent years to investigate graded mixed semiconductors (quasiheterojunctions) with slowly varying compositions. For such crystals, band parameters (energy gap, effective mass, etc.) are treated usually as position-dependent.

There are several specific physical effects which can be observed in graded mixed semiconductors (see e.g. [1-5]). In the present paper we are concerned with effects originating from position dependence of effective mass.

Classical Hamilton function of a carrier with position-dependent isotropic effective mass $m^*(\vec{r})$ is [6,7]

$$H = \vec{p}^2/2m^*(\vec{r}) + U(\vec{r},t), \tag{1}$$

where $\vec{p}$ is the carrier momentum canonically conjugated with $\vec{r}$ and $U(\vec{r},t)$ is the potential energy in external fields.

[+] Work sponsored by Institute of Physics, Polish Academy of Sciences.

[++] Now back at the Department of Physics, Purdue University, Lafayette, Indiana.

From the Hamilton function (1) we obtain equations of motion

$$\dot{\vec{p}} = \left[\vec{p}^2/2m^{*2}(\vec{r})\right] \vec{\nabla}_r m^*(\vec{r}) - \vec{\nabla}_r U(\vec{r},t), \qquad (2)$$

$$\dot{\vec{r}} = \vec{p}/m^*(\vec{r}). \qquad (3)$$

Thus, an additional force appears in equation of motion (2).

Equations (2) and (3) are nonlinear. Hence, a nonlinear response to external electric fields may be expected. Grinberg and Kastalskij have already suggested and estimated some nonlinear optical effects produced by effective-mass gradient [8] . However, they incorrectly omitted the first term in (2).

We have solved the Boltzmann equation for free carriers obeying (2) and (3) in the presence of two homogeneous oscillating electric fields of arbitrary polarization:

$$- \vec{\nabla}_r U(\vec{r},t) = q\left[\vec{E}_1 \exp(-i\omega_1 t) + \vec{E}_2 \exp(-i\omega_2 t)\right] . \qquad (4)$$

Here and in the following we use complex-valued quantities. q is the charge of carrier and $\vec{E}_1$, $\vec{E}_2$ are complex vectors. We assumed that $\vec{E}_1$ and $\vec{E}_2$ are perpendicular to $\vec{\nabla}_r m^*$. The collision term in Boltzmann equation was taken in the simplest form $-(f-f_0)/\tau$, where f and f_0 are the actual and equilibrium distribution functions, respectively, and τ is a constant relaxation time. We have neglected all terms of the second and higher orders in $\vec{\nabla}_r m^*$, terms involving the second or higher derivatives of m^*, and terms of the third and higher orders in $\vec{E}_1$ or/and $\vec{E}_2$.

The solution for f shows that the carrier density is constant in time. The current density is

$$\vec{J}(\vec{r},t) = \sigma^1(\omega_1,\vec{r})\vec{E}_1\exp(-i\omega_1 t) + \sigma^1(\omega_2,\vec{r})\vec{E}_2\exp(-i\omega_2 t)$$

$$+ \underline{\sigma}^2(\omega_1,-\omega_1,\vec{r})\vec{E}_1\vec{E}_1^* + \underline{\sigma}^2(\omega_2,-\omega_2,\vec{r})\vec{E}_2\vec{E}_2^*$$

$$+ \underline{\sigma}^2(\omega_1,\omega_1\vec{r})\vec{E}_1\vec{E}_1\exp(-2i\omega_1 t) + \underline{\sigma}^2(\omega_2,\omega_2,\vec{r})\vec{E}_2\vec{E}_2\exp(-2i\omega_2 t)$$

$$+ 2\underline{\sigma}^2(\omega_1,\omega_2,\vec{r})\vec{E}_1\vec{E}_2\exp\left[-i(\omega_1+\omega_2)t\right]$$

$$+ 2\underline{\sigma}^2(\omega_1,-\omega_2,\vec{r})\vec{E}_1\vec{E}_2^*\exp\left[-i(\omega_1-\omega_2)t\right] \; , \tag{5}$$

where

$$\sigma^1(\omega,\vec{r}) = q^2 n_0 \tau_\omega / m^* , \tag{6}$$

$$\underline{\sigma}^2(\omega',\omega'',\vec{r})\vec{E}'\vec{E}'' = \left[q^3 \tau^2_{\omega'+\omega''}(\tau_{\omega'}+\tau_{\omega''})n_0/4m^{*2}\right]$$

$$\times (\vec{E}'\vec{E}'')\frac{\vec{\nabla}_r m^*}{m^*} , \tag{7}$$

$$\tau_\omega = \tau/(1 - i\omega\tau). \tag{8}$$

$n_0(\vec{r})$ is the carrier density. Thus, we obtain doubled and combined frequencies in current density.

An important feature of the second-order conductivity tensor $\underline{\sigma}^2$ given by (7) is that it is independent of carrier statistics. The sign of $\underline{\sigma}^2$ depends on the sign of carrier's charge.

For a single electric field $\vec{E}\exp(-i\omega t)$ we have two limiting cases.

A. $\omega \ll 1/\tau$. For linearly polarized electric field

$$\mathrm{Re}\,\vec{J} = (q^2 \tau n_0/m^*)\vec{E}\cos\omega t$$

$$+ (q^3\tau^3 n_0/2m^{*2})(\vec{E}^2)\frac{\vec{\nabla}_r m^*}{m^*}(1 + \cos 2\omega t). \tag{9}$$

The electric field $\vec{E}_{comp} \cos 2\omega t$ compensating the oscillating component of current parallel to $\vec{\nabla}_r m^*$ is given by

$$\vec{E}_{comp}/(\vec{E}^2) = - (\mu^2/2q) \ \vec{\nabla}_r \ m^* \ , \qquad (10)$$

where $\mu(r)$ is the carrier mobility. Thus, measurement of $\vec{E}_{comp}$ may be used to determine effective-mass gradient.

Let us put $|\vec{\nabla}_r m^*| \approx 4 \ m_o/cm$ (m_o – free electron mass), and $\mu \approx 10^5 \ cm^2/V \cdot s$ which correspond roughly to n-type $Hg_{1-x}Cd_xTe$ – alloy with x changing from 0 to 1 on the distance of about one millimeter. Then, $E_{comp}/E^2 \approx 10^{-5} \ cm/V$. For $E \approx 1 \ V/cm$, $E_{comp} \approx 10 \ \mu V/cm$.

B. $\omega \gg 1/\tau$. The second-order electric susceptibility corresponding to the current density with frequency 2ω is

$$\chi_{ijk} = (q^3 n_o/16m^{*3}\omega^4) \ \delta_{jk} \frac{\partial}{\partial r_i} \ m^* \ . \qquad (11)$$

Let us put $n_o \approx 3 \times 10^{17} \ cm^{-3}$, $m^* \approx 0.02 \ m_o$ and $|\vec{\nabla}_r m^*|/m^* \approx 10^4 \ cm^{-1}$ which correspond roughly to n-type $Hg_{1-x}Cd_xTe$ – alloy with $x \approx 0.25$ at the given point but changing from 0 to 1 on the distance of about 20 μm. For the wavelength of CO_2 – laser (10.6 μm) $\chi \approx 6 \times 10^{-8}$ esu thus being of the order of nonlinear susceptibilities observed in homogeneous materials. For lower frequencies, the considered effect will dominate other mechanisms of second-harmonic generation.

ACKNOWLEDGMENT

We wish to express our gratitude to Dr J. Ginter and Dr W. Giriat for valuable discussions and remarks.

One of us, (J.K.F.) gratefully acknowledges the support of the Polish Academy of Sciences and the U.S. National Academy of

Sciences during his Scientific Exchange Visit at the Institute
of Physics in Warsaw (1972-1973).

REFERENCES

[1] Indradev, Van Ruyen, L.J.; Williams, F.: J. Appl. Phys.
 $\underline{39}$, 3344 (1968)

[2] Marfaing, Y.; Chevallier, J.: IEEE Trans. Electron
 Devices $\underline{ED18}$, 465 (1971)

[3] Cohen-Solal, G.; Marfaing, Y.: Solid State Electron $\underline{11}$,
 1131 (1968)

[4] Chattopadhyaya, J.K.; Mathur, V.K.: Phys. Rev. B $\underline{3}$, 3390
 (1971)

[5] Hill, R.; Williams, F.: Appl. Phys. Letters $\underline{11}$, 296 (1967)

[6] Gora, T.; Williams, F.: Phys. Rev. $\underline{177}$, 1179 (1969)

[7] Leibler, L.: to be published

[8] Grinberg, A.A.; Kastalskij, A.A.: Fizy. Tekh. Poluprov. $\underline{5}$,
 2030 (1971), in Russian. English translation Sov. Phys.
 Semiconductors (USA).

FAR INFRARED MAGNETO-TRANSMISSION MEASUREMENTS
OF $Bi_{1-x}Sb_x$ ALLOYS

B. D. McCOMBE and R. J. WAGNER

Naval Research Laboratory, Washington, D. C. 20375 USA

J. S. LANNIN

Max-Planck-Institut für Festkörperforschung

7000 Stuttgart 1, Fed. Rep. Germany

Far IR cyclotron resonance and electric dipole spin resonance have been observed in $Bi_{0.885}Sb_{0.115}$ (n-type). Results were fit with a two band model.

1. INTRODUCTION

The group V semimetals, Bi and Sb, form a continuous series of
solid solutions which exhibit a semimetal-semiconductor transi-
tion and a zero energy gap. In pure Bi the conduction band
minima are situated at the three equivalent L-points of the Bril-
louin zone and are separated from "mirror" valence bands by a
direct energy gap of approximately 12 meV. The semimetallic
behavior is due to an overlapping valence band at the T-point
of the zone. Between 4% and 8% Sb the L-point conduction and
valence bands cross, and at about 8% Sb the T-point valence band
moves below the conduction band giving rise to the semimetal-
semiconductor transition. At compositions between 8% and 22% Sb
in Bi the alloys are semiconducting.

Although there have been previous <u>inter</u>band magnetoreflec-
tion measurements [1] as well as microwave cyclotron resonance
studies [2] of this alloy system, no such studies have been car-
ried out in the far IR. This is a particularly interesting
region for this system since typical direct energy gaps occur
in the far IR.

2. EXPERIMENTAL

In this paper we present the results of far IR magneto-trans-
mission studies of a semiconducting (n-type) alloy ($x \approx 0.115$)
with a net electron concentration of approximately 2×10^{15}
cm^{-3}. Both electrical discharge and optically pumped far IR
lasers were used to span the photon energy region between $E_g/2$
and $1.5\ E_g$. The sample was a thin cleaved trigonal section.

FIG. 1: Transmission data for various orientations and polarizations.

Experiments were carried out with light propagation either parallel (trigonal axis) or perpendicular (binary and bisectrix axes) to the applied magnetic field with appropriate polarization in each case.

3. RESULTS AND DISCUSSION

3.1 $\vec{B}$ PARALLEL TO THE BINARY AND BISECTRIX AXES

With the magnetic field along the binary axis two of the anisotropic conduction band minima are equivalent and are characterized by extremely light cyclotron masses, while the third minimum is characterized by a much heavier mass. With the magnetic field along a bisectrix axis two distinct cyclotron masses are again expected. In this case the two equivalent minima have roughly twice the cyclotron mass of the third minimum, but both distinct masses are rather light.

Examples of the transmission data obtained at 4.3K are shown in Fig. 1. All electrons in the light mass minima are in the extreme quantum limit for the carrier density of this sample. In the upper trace of Fig. 1(a) ($\vec{B}_{\parallel}$binary) only the light mass cyclotron resonance transition is observed. Cyclotron resonance of the heavy mass electrons was not detected. In the lower trace of Fig. 1(a) ($\vec{B}_{\perp}$bisectrix) cyclotron resonance transitions due to both the "light" and "heavy" mass electrons are apparent.

The spectra for two different polarizations are shown in Fig. 1(b) with the magnetic field in each case directed along the binary axis. With $\vec{E}_{\parallel}\vec{B}$ a "tilted orbit" cyclotron

resonance [3] due to the two equivalent light mass minima is ob-
served. "Tilted orbit" cyclotron resonance can occur whenever
the magnetic field is not directed along a principal axis of the
effective mass tensor. The resonance frequency in this case is
determined by the usual cyclotron mass, i.e. $m_c^*(\mathcal{E}) = (\det \overset{\leftrightarrow}{m}^*(\mathcal{E})/$
$\hat{\eta} \cdot \overset{\leftrightarrow}{\underset{\approx}{m}}^*(\mathcal{E}) \cdot \hat{\eta})^{\frac{1}{2}}$, where $\overset{\leftrightarrow}{m}^*(\mathcal{E})$ is the effective mass tensor (a func-
tion of energy, $\mathcal{E}$, in the band), and $\hat{\eta}$ is a unit vector in the
direction of the magnetic field. On the other hand for $\vec{E}_\perp \vec{B}$,
magnetoplasma effects serve to shift the observed resonance
position away from the position determined from the above cyclo-
tron mass [3]. At the magnetic fields of interest the shifts
are small in this sample. Nonetheless, in the fitting discussed
below only the $\vec{E}_\parallel \vec{B}$ data were used.

It is well known that the simple two-band model [4,5] pro-
vides a reasonably good description of the magnetic energy level
structure in pure Bi for those magnetic field directions and
conduction band minima which are characterized by light cyclo-
tron masses. This is expected to hold true also for the small
gap $Bi_{1-x}Sb_x$ alloys.

The small effects of other bands have been treated in per-
turbation theory by Baraff [6], and his model has been simpli-
fied by comparison with interband magnetoreflection experiments
[7]. For this simplified Baraff model at low fields for elec-
trons with a positive g-factor the magnetic energy levels may
be written [8]

$$\mathcal{E}_{j,s}^{\pm}(0) = \pm \left(E_g^2/4 + E_g \hbar \omega_c\right)^{\frac{1}{2}} + 2s|G|\hbar\omega_c, \qquad j \neq 0$$

$$\mathcal{E}_{0,-\frac{1}{2}}^{\pm}(0) = \pm(E_g/2 - |G|\hbar\omega_c), \qquad j = 0 \tag{1}$$

where $\hbar\omega_c = |e|\hbar B/m_c^*(0)c$ with $m_c^*(0)$ the bottom of the band cy-
clotron mass, E_g is the energy gap, $\mathcal{E}(0)$ denotes energy evalu-
ated with the wavevector along the field set equal to zero,
$\pm$ denotes the conduction and valence bands, respectively, and
$j = n + \frac{1}{2} -s$ with the spin quantum number $s = \pm\frac{1}{2}$ and the Landau
quantum number $n = 0,1,2, \ldots$. The parameter G is a spin
splitting factor which is an explicit manifestation of correc-
tions to the simple two-band model g-factor due to other bands
($G \equiv 0$ in the simple two band model). Corrections to the orbital

energy levels are included implicitly in a redefinition of the effective mass tensor.

The cyclotron resonance transition energy in the extreme quantum limit is given by

$$\mathcal{E}^{+}_{1,-\frac{1}{2}}(0) - \mathcal{E}^{+}_{0,-\frac{1}{2}}(0) = (E_g^2/4 + E_g \hbar \omega_c)^{\frac{1}{2}} - E_g/2, \qquad (2)$$

i.e. the same result as the simple two-band model except for the implicit redefinition of the effective mass tensor.

The experimental cyclotron resonance results were fit to Eq. (2) with a least squares procedure, and a comparison of experiment and theory is displayed in Fig. 2. The best values of the band parameters are summarized in Table I.

FIG. 2: Fit of 2-band model to $E_\parallel B$ data (x, +→$E_\perp B$). Most points for $E_\perp B$ are omitted for clarity.

An interesting feature of the binary axis data is the rapid monotonic decrease in transmission at high fields for $\vec{E}_\parallel \vec{B}$ (lower trace in Fig. 1(b)). This behavior is to be constrasted with that for $\vec{E}_\perp \vec{B}$ (upper trace in Fig. 1(b)), where the transmission remains high at high fields. The magnetic field at which the onset of the high field absorption occurs was found to decrease as the photon energy increased (and vice-versa). We attribute this effect to a strong interband transition between the $j=0,-\frac{1}{2}$ states of the conduction and valence bands which is allowed only for $\vec{E}_\parallel \vec{B}$ [9]. This implies that the $\mathcal{E}^{\pm}_{0,-\frac{1}{2}}$ energy levels approach as the field is increased, in agreement with Eq. (1) and recent results on Bi [8], and is a direct manifestation of the corrections to the simple two-band model. The absence of similar behavior for $\vec{B}_\parallel$ bisectrix axis indicates $G \approx 0$ for this direction.

3.2 $\vec{B}$ PARALLEL TC THE TRIGONAL AXIS

In addition to a single strong cyclotron resonance from the three

equivalent minima with $\vec{B}_\parallel$ trigonal axis, a weaker, narrower line
was observed at slightly lower fields. Both combined and spin
resonance were considered as possible explanations; however, a
reasonable fit could be achieved only with the assumption that
the additional line was electric-dipole electron spin resonance
(e.g., $(0,-\frac{1}{2}) \rightarrow (1,+\frac{1}{2})$) with spin-splitting greater than orbital
splitting. For this case a fairly good fit to Eqs. (1) was ob-
tained with the parameters of Table I. This is at variance with
results for Bi [10] where the spin splitting is roughly half
the orbital splitting. In addition, the two-band model is not
expected to be adequate for this orientation. Further measure-
ments are required to completely substantiate the present inter-
pretation.

	$\vec{B}_\parallel$Binary	$\vec{B}_\parallel$Bisectrix	$\vec{B}_\parallel$Trigonal
m_c^*/m_o	0.0027 ± 0.0001 --------	0.00255 ± 0.0001 0.00470 ± 0.0001	0.022 ± 0.002
$\|G\|$	--------	--------	0.076 ± 0.003

TABLE I: Band parameters for $Bi_{0.885}Sb_{0.115}$ ($E_g = 17.3 \pm 1$ meV)

4. REFERENCES

[1] Tichovolsky, E. J.; Mavroides, J.G.: Solid State Commun.
 7, 927 (1969).
[2] Oelgart, G.; Herrmann, R.: Phys. Stat. Solidi (b) 58,
 181 (1973).
[3] Smith, G. E.; Hebel, L. C.; Buchsbaum, S. J.: Phys. Rev.
 129, 154 (1963).
[4] Cohen, M. H.; Blount, E. I.: Phil. Mag. 5, 115 (1960).
[5] Brown, R. N.; Mavroides, J. G.; Lax, B.: Phys. Rev.
 129, 2055 (1963).
[6] Baraff, G. A.: Phys. Rev. 137, A842 (1965).
[7] Maltz, M.; Dresselhaus, M. S.: Phys. Rev. B2, 2877 (1970).
[8] Vecchi, M. P.; Dresselhaus, M. S.: Phys. Rev. B9, 3257
 (1974).
[9] Wolff, P. A.: J. Phys. Chem. Solids 25, 1057 (1964).
[10] Smith, G. E.; Baraff, G. A.; Rowell, J. M.: Phys. Rev.
 135, A1118 (1964).

MAGNETIC FIELD-INDUCED SEMIMETAL-SEMICONDUCTOR TRANSITION IN Bi[*]

M. P. VECCHI[†¶], J. R. PEREIRA[§] and M. S. DRESSELHAUS[†]

Department of Electrical Engineering
and Center for Materials Science and Engineering
Massachusetts Institute of Technology, Cambridge, Mass. 02139

The unusual behavior of the lowest quantum-number
$j=0$ Landau levels of the electrons in Bi has impor-
tant consequences for the description of a magnetic
field-induced semimetal-semiconductor transition.
We discuss a quantitative treatment of the $j=0$
Landau levels, based upon the Baraff Hamiltonian,
and including the k_H-dependence. Magnetoreflection
experiments over a range of photon energies on Bi,
$Bi_{98}Sb_2$ and $Bi_{97}Sb_3$ for $\vec{H} \parallel$ binary axis provide
experimental confirmation for the model. For large
H, the k_H-dependence of the $j=0$ levels is anomalous,
showing a "camel-back" shape. As the $j=0$ conduction
band level at $k_H=0$ crosses E_F, magnetic field-
induced carrier pockets are formed away from $k_H=0$.
The implications of these magnetic field-induced
pockets on a semimetal-semiconductor transition
are discussed.

A magnetic field-induced semimetal-semiconductor transition
in Bi and BiSb alloys has been studied by Brandt et al. [1] and
experimental evidence supporting such a transition has been
reported. This transition has been described previously in
terms of an ellipsoidal parabolic model for the holes at the T
point, and for the electrons at the L points either an ellip-
soidal parabolic model [2] or an ellipsoidal non-parabolic two-
band model [1] were considered. Recent magnetoreflection exper-
iments in the low-quantum-limit [3] have shown that the lowest
quantum number $j=0$ Landau levels for the electrons in Bi are
significantly affected by corrections due to bands outside the
two-band model. These corrections can be treated satisfactorily

* Work supported by the National Science Foundation
† Visiting Scientist, Francis Bitter National Magnet Laboratory,
 M.I.T., supported by the National Science Foundation
¶ Instituto Venezolano de Investigaciones Científicas (IVIC)
 predoctoral Fellow
§ Instituto de Fisica, Universidade de São Paulo, Brazil,
 postdoctoral Fellow (supported by FAPESP, Brazil)

in terms of the Baraff Hamiltonian [4]. The resulting expression for the energies of the j=0 Landau levels at k_H=0 is given by [3]

$$E_{j=0}^{\pm}(k_H=0) = \pm \left[(E_g/2 - |G\beta^*|H)^2 + P(\beta^*H)^2 \right]^{1/2} , \qquad (1)$$

where H is the magnetic field, and the (+) and (−) signs refer to the conduction and valence bands separated by an energy gap E_g at H=0. The spin-splitting parameter $|G\beta^*|$ arises from the interaction with bands outside the two-band model, and the parameter $P(\beta^*)^2$ represents the interband coupling between the two j=0 levels. From Eq.(1) it follows that the two j=0 levels at k_H=0 approach each other at low H, reach a point of minimum separation at a magnetic field defined as H_{min}, and, as the interband coupling effect becomes significant, the two j=0 levels repel each other at high H.

The parameters in Eq.(1) were determined for Bi, $Bi_{98}Sb_2$ and $Bi_{97}Sb_3$ for $\vec{H} \parallel$ binary axis from magnetoreflection data over a range of photon energies. Fig.1 presents some typical magnetoreflection traces for Bi and $Bi_{98}Sb_2$.

FIG. 1: Magnetoreflection traces for $\vec{H} \parallel$ binary axis, T $\sim$ 20K.

The optical transitions involving the j=0 levels are indi-
cated in Fig.1 for both the interband transition (0→1) and the
cyclotron resonance transition (C.R.). The interpretation of
the lineshapes is related to the k_H-dependence of the Landau
levels and is discussed below. Fig.2 is a plot of $E_{j=0}^{\pm}(k_H=0)$
as given by Eq.(1). Notice that E_g is a decreasing function
of Sb concentration (up to 3%), in agreement with previous
results [5], and that H_{min} moves to lower values as E_g decreases.
This implies that at high magnetic fields $E_{j=0}^{\pm}(k_H=0)$ will rise
more rapidly for the smaller gap materials.

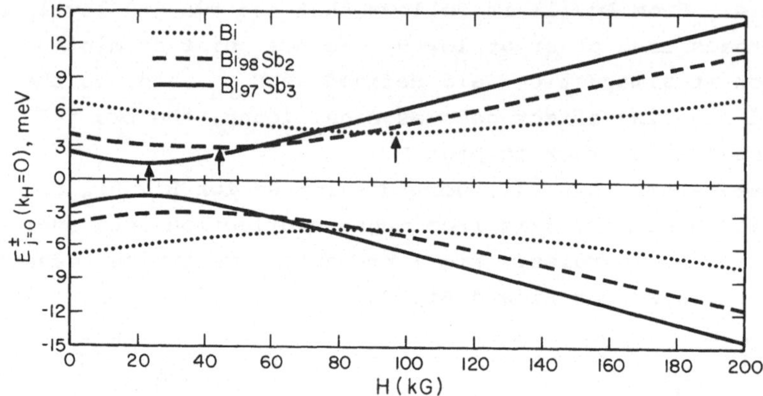

FIG. 2: Plot of $E_{j=0}^{\pm}(k_H=0)$ vs. H. The respective H_{min} for the
three materials are indicated by the arrows.

To calculate the density of states for the description of the
semimetal-semiconductor transition we need to know the k_H-
dependence of the Landau levels, and this dependence of the j=0
levels has been recently obtained [6] from the Baraff Hamilto-
nian. The physical picture that emerges is that for $H<H_{min}$ the
dispersion relation is normal, but that for $H>H_{min}$, when the
interband coupling effect becomes dominant, the dispersion
relation for the j=0 levels develops an anomalous "camel-back"
shape centered at $k_H=0$ (shown schematically by the inserts in
Fig.3). Returning to Fig.1, it is seen that the magnetore-
flection lineshapes for Bi at lower H (Fig.1A) differ substan-
tially from the lineshapes for Bi at higher H (Fig.1B) and from
the lineshapes for $Bi_{98}Sb_2$ (Figs.1C,D). Lineshape calculations
[6] have been successful in reproducing the observed magneto-

reflection results and have confirmed the validity of the model for the k_H-dependence of the $j=0$ levels derived from the Baraff Hamiltonian.

We have calculated the Fermi energy E_F and the carrier densities for the various pockets as a function of H for Bi, $Bi_{98}Sb_2$ and $Bi_{97}Sb_3$, using published values for the hole and the heavy electron parameters for Bi [7]. The hole parameters were taken to be independent of Sb concentration [8]. The parameters of the model used to describe the light electrons [6] were derived from our magnetoreflection experiments. The overlap energies were scaled linearly with Sb concentration [8]. Fig.3 shows the results of our calculations for $\vec{H} \parallel$ binary axis.

FIG. 3: Plot of the electron densities and E_F vs. H. The dashed lines mark the regime where the magnetic field-induced light electron pockets exist. E_F is measured with respect to $E^\pm_{j=0}(k_H=0)$ and $E^\pm_{min}$ is the minimum conduction band energy.

The density of light electrons (n_ℓ) for Bi increases with H
up to 400 kG, in agreement with published results [9]. The
heavy electrons (n_h) make a very small contribution and there-
fore particular attention must be given to the light electrons.
As H increases, the "camel-back" shape (see inserts Fig.3) be-
comes more pronounced and eventually $E_{j=0}^+ (k_H=0)$ crosses E_F, giv-
ing rise to magnetic field-induced carrier pockets away from the
L point (see lower insert in Fig.3). Since these magnetic
field-induced pockets are confined to a single Landau level,
they would not be observable by conventional de Haas-van Alphen
techniques. A semimetal-semiconductor transition will not oc-
cur, however, until the off-axis minimum E_{min}^+ crosses E_F. The
magnetic field dependence of the off-axis minimum E_{min}^+ is very
sensitive to the values of the parameters $|G\beta^*|$ and $P(\beta^*)^2$, and
for this reason it is difficult to estimate accurately the value
of H at which the semimetal-semiconductor transition will occur.
Our calculations are in agreement with the results of Brandt et
al. [1] insofar as they predict that the semimetal-semiconductor
transition will occur for H>>400 kG for Bi, and that this tran-
sition moves to lower H in the BiSb alloys. Our studies further
indicate that the detailed description of the transition is more
complex than previously discussed. In particular, attention
must be given to the magnetic field-induced carrier pockets
predicted in the present work.

Thanks are due to Dr. J. G. Mavroides for making the BiSb
alloy samples available.

REFERENCES

[1] N. B. Brandt; E. A. Svistova; Yu. G. Kashirskii: Soviet
 Physics JETP Letters (USA) 9, 136 (1969) -- N. B. Brandt;
 E. A. Svistova: J. Low Temp. Phys., 2, 1 (1970)
[2] H. T. Chu: Phys. Rev. B 5, 4714 (1972)
[3] M. P. Vecchi; M. S. Dresselhaus: Phys. Rev. B 9, 3257 (1974)
[4] G. A. Baraff: Phys. Rev., 137, A842 (1965)
[5] E. J. Tichovolsky; J. G. Mavroides: Solid State Commun. 7,
 927 (1969)
[6] M. P. Vecchi; J. R. Pereira; M. S. Dresselhaus: (to be
 published)
[7] G. E. Smith; G. A. Baraff; J. M. Rowell: Phys. Rev., 135,
 A1118 (1964)
[8] H. T. Chu; Yi-Han Kao: Phys. Rev. B 1, 2369 (1970)
[9] S. Takano; H. Kawamura: J. Phys. Soc. Japan 28, 348 (1970)

ELECTROREFLECTANCE OF PbTe – BAND SEPARATIONS AND PARITY AT L[+]

R. GLOSSER, J. KINOSHITA* and B. RENNEX
Department of Physics and Astronomy
University of Maryland
College Park, Maryland 20742 USA

The band separations at the L point of PbTe are de-
termined by electroreflectance. The results indicate
that the parity labelling of some of these states
may be incorrect.

1. INTRODUCTION

Band population effects in the electroreflectance spectrum of PbTe are used
to quantitatively determine the band separation at the L point of PbTe. The
results point to an incorrect parity labelling of some of the states at L.
Discussions of the experimental and theoretical background exist elsewhere
[1,2,3].

We make use of the knowledge that the band gap is at L and that it is possi-
ble to change the band filling so as to shift the Fermi level. This permits
the identification of corresponding shifting structure associated with the
Fermi level near L. This technique was used in making band structure identi-
fications in InSb [4] and PbSe [3].

2. EXPERIMENT

Bulk crystals, both n and p were studied at 80°K and 10°K over the spectral
range 0.3 to 6 eV. The samples were electropolished [5] and prepared as MIS
devices as described elsewhere [3,4].

3. RESULTS

The results shown were all taken at 10°K. Figure 1 depicts the electrore-
flectance (ER) spectrum of a p-sample of 3-5 x 10^{19} cm^{-3} carriers with varying
surface potential. Note the unshifting 1.20 eV negative peak.** Sample
p#1 was prepared twice and p#1,- is not part of the bias sequence of the
upper three traces. Trace p#1,- shows the extent of the shift of the 1.5

+ Research supported by NSF and ARPA through the Center of Materials Re-
search, University of Maryland

* On leave from the Electrotechnical Laboratory, Tokyo, Japan

** Polarity of ΔR is defined as $R(\varepsilon_2) - R(\varepsilon_1)$ where the applied field
$\varepsilon_2 > \varepsilon_1$. n#2, 1.0 and 3.0 V are excepted from this definition to show
inversion of 1.20 eV structure.

FIG. 1: $\Delta R/R$ (arb. units) of p-PbTe as a function of applied bias. p#1,- is not in the bias sequence.

continuous change of the Fermi level position from the valence to the conduction band within the space charge region. We justify this by our observation that n#1,3.0V and n#2,0V give essentially the same spectrum.[++] With this interpretation, as the Fermi level is moved from the valence towards the conduction band, the 1.20 eV peak remains unshifted and is inverted at n#2,1.0V indicating an accumulation layer [4]. The 1.5 and 2 eV structure disappear. New blue shifting structure appears just below 1.2 eV and also at 1.8 eV. For the trace of n#2,3.0 V two small

[++] n#1, broke down just above three volts.

and 2 eV structures. Note the positive rise at 1.54 eV. The sequence from −3 V to + 7.5 V shows the red shift with positive going bias of the 1.5 eV and 2 eV negative peaks and the uncovering of a positive peak at about 1.57 eV. At + 7.5 bias there is a clear indication of two structures. Figure 2 shows electroreflectance results for two n-samples n#1 of carrier concentration 1×10^{18} cm^{-3} and n#2 of 3×10^{17} cm^{-3} carriers. The results are displayed to show a

FIG. 2: $\Delta R/R$ (arb. units) of n-PbTe. The sequence of traces reading down to up is meant to follow the Fermi level from valence band to conduction band.

FIG. 3: ΔR/R (arb. units) of PbTe in the NIR. The sequence of traces reading down to up is meant to follow the Fermi level from the valence band to the conduction band.

negative peaks at 1.26 and 1.75 eV appear.

Near IR results are indicated in figure 3. The 1.2 eV region shows the same behavior described in figures 1 and 2 as the bulk carriers and surface potential are changed. A large structure is observed at 0.9 eV in p-material and an accumulation layer. In n-material with the bands bent up the peak is smaller and shifted to the blue. Biasing towards flat band, this structure disappears and does not reappear for increased bias. We note a correlation between the disappearance of this structure and the appearance of new structure just below 1.2 eV. Note that the separation is about the gap energy.

4. ANALYSIS AND CONCLUSIONS

The 1.20 eV structure which does not shift is associated with transitions somewhere in the Brillouin zone away from the Fermi level [4], probably

Σ [6]. The rest of the observed structure between 0.8 and 2.0 eV clearly results from band population effects and is associated with transitions involving the Fermi level near L. The results are summarized in figure 4. The bands are numbered in the same way as done by Herman et al [6] but no group labellings are attached. The assignments are made on the basis of appearance in n or p - surface and direction of shift with bias following the discussion in [4]. The 0.9 eV structure is associated with the 1.5 eV positive peak. The 1.5 and 2 eV negative peaks behave similarly. This is also true for the new structure just below 1.2 eV and at 1.8 eV when the sample has an n-surface. The 1.26 and 1.75 peaks are tentatively assigned as interconduction band transitions. These assignments are seen to be self consistent. Comparison with theory [1] show best agreement with the results of Herman et

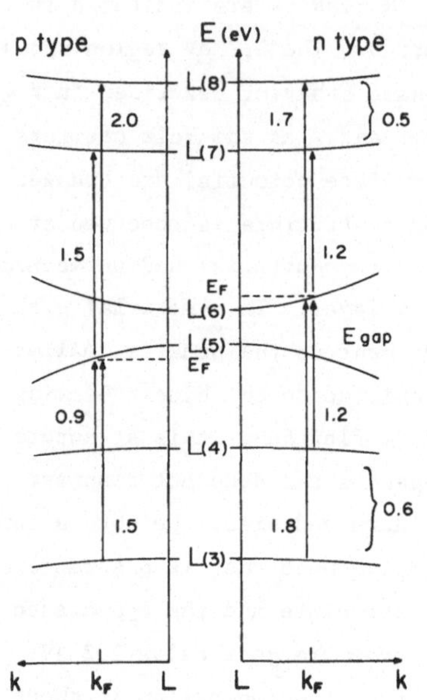

FIG. 4: Assignment of structures near L. (Not to scale)

al [6]. One question remains. Why is the Franz Keldysh effect apparently not operative at L? The 1.5 eV (negative peak) and 2 eV structures and the second 1.2 eV structure and 1.8 eV structure both should be present whether the surface is n or p type. If band population effects are not operative, then the Franz-Keldysh effect should be. We suggest that the reason may be that the accepted [1] parity labellings of some of the bands at L are incorrect and in fact some transitions at the L point are simply forbidden. If in fact the parity of the highest valence band at L is opposite to the next lower band and if the band shapes [1] are correct then the 0.9 eV structure may in part represent an M_3 transition.

5. ACKNOWLEDGEMENTS

We thank Drs. K. H. Chang, R. Guldi, K. Igaki, A. J. Strauss and D. C. Tsui for samples of PbTe.

6. REFERENCES

[1] Dalven, R.: in Solid State Physics, Ehrenreich, H., ed; Vol. 28 (Academic Press, New York, 1973).

[2] Kohn, S. E.; Yu, P. Y.; Petroff, Y; Shen, Y. R.; Tsang, Y and Cohen, M. L.: Phys. Rev. B8, 1477 (1973).

[3] Kinoshita, J. and Glosser, R.: Proceedings of the Int. Conf. on the Physics of Semimetals and Narrow Gap Semiconductors Nice-Cardiff, Dept. 1973 (to be published) Physics Letters A (in press).

[4] Glosser, R. and Seraphin, B. O.: Phys. Rev. 187, 1021 (1969. Glosser, R; Fischer, J. E. and Seraphin, B.O.: Phys. Rev. B1, 1607 (1970).

[5] Knorr, M. K.: J. Electrochem Soc. 109, 433 (1962).

[6] Herman, F.; Kortum, R. L.; Ortenburger, I. B. and VanDyke, J. P.: J. de Physique Suppl. 29, C4-62 (1968).

GRAPHITE LANDAU LEVELS IN THE PRESENCE OF TRIGONAL WARPING[+]

G. DRESSELHAUS

Lincoln Laboratory, Massachusetts Institute of Technology,
Lexington, Massachusetts, USA

M.S. DRESSELHAUS

Department of Electrical Engineering and
Center for Materials Science and Engineering
Massachusetts Institute of Technology, Cambridge, Mass., USA

The Bohr-Sommerfeld quantization condition is applied
to calculate the Landau levels in graphite in the
presence of trigonal warping. The off-axis extrema
arising from the trigonal warping result in unequally
spaced leg and central electron and hole Landau
levels which can be supported between two cut-off
energies. Application to the de Haas-van Alphen
effect and to interband Landau level transitions is
discussed.

There are several reasons why this study of the Landau level
structure of graphite in the presence of trigonal warping may be
of general interest. We describe here a technique, using the
Bohr-Sommerfeld quantization condition, which can be used to
solve the Landau level problem when an exact solution is diffi-
cult to find, or when perturbation theory is not rapidly con-
vergent or is difficult to apply. In the case of graphite, we
find that this technique gives rise to new Landau level ladders,
not easily obtained by use of perturbation theory. These new
Landau levels are annihilated when they cross upper and lower
cut-off energies, and the question arises as to whether the
annihilation of these levels can result in a de Haas-van Alphen-
type effect.

From the point of view of the electronic properties of graphite,
this technique provides a method for calculating Landau levels
for arbitrary values of the Slonczewski-Weiss-McClure (S-W-McC)
γ_3, which determines the trigonal warping of the Fermi surface

[+]This work was sponsored by the Department of the Air Force and
by the National Science Foundation.

[1]. The energy range in which the new Landau level ladders can be supported depends directly on the magnitude of γ_3, and in this way may provide a more direct determination of γ_3 than the magnetoreflection [2] or cyclotron resonance [3] lineshape studies, which presently are in disagreement over the magnitude of γ_3. If the creation and annihilation of these new Landau levels at the cut-off energies should give rise to a de Haas-van Alphen effect, then it may be also possible to find an identification for the minority de Haas-van Alphen period of $1.35 \times 10^{-1} G^{-1}$, which is one of the major unsolved problems in the electronic properties of graphite [1].

FIG. 1:
Splitting of the doubly degenerate E_3 level along the κ-Γ axis. For no trigonal warping (dashed curve) the κ-point is the only critical point. With trigonal warping (solid curve), there are two points of degeneracy κ_e and κ_{leg} and two off-axis saddle point extrema at κ_{e-sp} and κ_{h-sp}.

There is, of course, a direct method for calculating the Landau levels in graphite. The Slonczewski-Weiss-McClure (S-W-McC) (4x4) effective Hamiltonian in a magnetic field [1,2] can be written down formally, but because of the trigonal warping effect no exact solution can be found. In the past, the case for $\gamma_3 = 0$ has been solved exactly and the effects of $\gamma_3 \neq 0$ have been treated as a perturbation. The difficulty with this approach can be understood from Fig. 1 where we show the lifting of the degeneracy of the γ_3 levels in graphite as we move away from a point on the HKH axis of the Brillouin zone in the basal plane. In the absence of trigonal warping ($\gamma_3 = 0$, dashed curves) the only energy extremum is at the HKH axis ($\kappa = 0$) and we would expect to find a Landau level structure appropriate to a zero gap semiconductor. However, when trigonal warping is included (solid curves), off-axis extrema are formed. Special Landau levels can be supported between these energy extrema: for $\epsilon_c < \epsilon < \epsilon_{e-sp}$ we can have central election levels, for $\epsilon_{leg} < \epsilon < \epsilon_{e-sp}$ we can have leg electron levels, for

$\varepsilon_{h-sp} < \varepsilon < \varepsilon_e$ we can have central hole levels, for $\varepsilon_{h-sp} < \varepsilon < \varepsilon_{leg}$ we can have leg hole levels. For $\varepsilon > \varepsilon_{e-sp}$, majority electron Landau levels are found and for $\varepsilon < \varepsilon_{h-sp}$, majority hole Landau levels occur. The four energies ε_{e-sp}, ε_e, ε_{leg}, ε_{h-sp} therefore serve to define the regions in which particular Landau levels can be found. In Fig. 2 we show the k_z dependence of these critical energy points using the usual reduced wave vector notation [1]. The cross hatched regions in Fig. 2 correspond to the magnetic energy levels which arise from the trigonal warping of the Fermi surface and do not follow directly from the application of perturbation theory to the magnetic Hamiltonian.

The magnetic energy levels which correspond to each regime in Fig. 2 are calculated using the Bohr-Sommerfeld quantization condition [4]

$$S(E,k_z) = \int_{E=const} k_x dk_y = \frac{2\pi eH}{hc}(n+\delta)$$

FIG. 2:
Plot of k_z dependence of the critical points in Fig. 1. The majority carrier Landau levels occupy the regions above and below the cross hatched region.

in which the cross-sectional areas in momentum space are calculated from the S-W-McC band model and the phase factor for the majority levels δ is taken to be 1/2 (from the de Haas-van Alphen measurements [5]), appropriate to the range $0 < \xi < 0.4$ for which this calculation is of interest. The phase factor for the leg and central levels is taken = 0 for convenience. The resulting Landau levels obtained between the two cut-off energies ε_{e-sp} and ε_{h-sp} are shown in Fig. 3 at 3kG. The leg and central Landau levels exhibit very unequal spacings,

FIG. 3: The k_z dependence for some of the leg and central Landau levels is shown for H=3kG parallel to the c axis.

FIG. 4:
Magnetic Field Dependence of the
Landau levels at the κ-point. Leg
and central levels can be suppor-
ted between $\varepsilon_{h-sp} < \varepsilon < \varepsilon_{e-sp}$ and
majority levels elsewhere.
All levels are terminated discon-
tinuously on crossing ε_{h-sp} and
ε_{e-sp}.

characteristic of large non-
parabolic effects. However, the
Bohr-Sommerfeld quantization
condition guarantees that as the
magnetic field is increased, the levels for a given Landau ladder
pop through the cut-off energy with a periodicity in 1/H. The
question arises as to whether the annihilation of these levels as
they pass through the cut-off energy can result in a de Haas-van
Alphen period. Such a period could provide an attractive
mechanism for the identification of the observed minority period
in graphite of $1.35 \times 10^{-4} G^{-1}$ [1].

The magnetic field dependence of the Landau levels at the k-point
are shown in Fig. 4. Discontinuities at ε_{e-sp} and ε_{h-sp} corres-
pond to the creation and annihilation of leg, central and majority
Landau levels. From this figure, we can also consider interband
Landau level transitions which must satisfy conservation of energy,
selection rules and the requirements that the transition be made
from an occupied to an unoccupied state. Analysis of magneto-
reflection data in the far infrared provides tentative evidence
in support of transitions originating from the n=0 central level.

REFERENCES

[1] McClure, J.W.: Proceedings of the International Conference
 on the Physics of Semiconductors and Narrow Gap Semiconductors,
 Dallas, Texas (1970) (ed. D.L.Carter and R.T.Bate) Pergamon
 Press (New York) 1971, page 127
[2] Schroeder, P.R.; Dresselhaus, M.S.; Javan, A.: ibid. p. 139
[3] Ushio, H.; Uda, T.; Uemura, Y.: J. Phys. Soc. Jap. 33, 1551
 (1972)
[4] Onsager, L.: Phil. Mag. 43, 1006 (1952)
[5] Williamson, S.J.; Foner, S.; Dresselhaus, M.S.: Phys. Rev.
 140, A 1429 (1965).

OPTICAL PROPERTIES, RECOMBINATION

CONTENTS

COMPARISON OF HIGH-RESOLUTION ELLIPSOMETRIC AND ELECTROREFLECTANCE SPECTRA OF Ge, GaAs, AND Si

D. E. ASPNES

Bell Laboratories
Murray Hill, New Jersey 07974 U.S.A.

Joint-density-of-states third-derivative spectra
calculated numerically from high-precision
ellipsometric data, are compared to low-field
electroreflectance spectra for Ge, GaAs, and Si
over the energy range 2.0 - 5.4 eV. Relative
magnitudes of interband reduced masses are
obtained. The signs of these masses are
positive for each critical point observed.

1. INTRODUCTION

In the one-electron approximation, the relationship between the
field-induced change, $\Delta\varepsilon$, of the complex dielectric function,
ε, and its joint-density-of-states third energy derivative
(JDOS(3)), is given in the limit of a small electric field, $\vec{\mathcal{E}}$
by [1,2]

$$\Delta\varepsilon = \varepsilon(\vec{\mathcal{E}},E) - \varepsilon(0,E)$$

$$\cong \frac{(\hbar\Omega)^3}{3[1+id\Gamma/dE]^3} \frac{1}{E^2} \frac{d^3}{dE^3} [E^2\varepsilon(0,E)] \tag{1}$$

where $(\hbar\Omega)^3 = e^2\mathcal{E}^2\hbar^2/(8\mu_{\parallel})$ and Γ is the intrinsic lifetime
broadening energy. If $\varepsilon(0,E)$ is measured to sufficient
precision so that $JDOS(3) = E^{-2}d^3/dE^3[E^2\varepsilon(0,E)]$ can be cal-
culated numerically, and if $\Delta\varepsilon$ is determined by a low-field
electroreflectance (ER) experiment, then Eq. (1) shows that a
comparison between the two spectra should yield direct infor-
mation on $\mu_{\parallel}$ and $d\Gamma/dE$ [2]. This relationship is necessarily
approximate, since it does not include the effects of final-
state interactions. These generate interparticle forces between
the optically excited electron and hole and consequently modify
the simple uniform-acceleration (Franz-Keldysh) model leading
to Eq. (1), as numerical calculations have shown [3,4]. A
second complication occurs if $\Delta\varepsilon$ arises from a set of equivalent
critical points, in which case $\mu_{\parallel}$ represents an average value.

Nevertheless, the comparison described by Eq. (1) is useful
since relative mass values can be estimated and gross features,
such as the sign of $\mu_{||}$, can be determined. Here, we discuss
the results of this comparison for Si, Ge, and GaAs.

2. EXPERIMENTAL

High-precision dielectric function data were obtained by means
of a rotating-analyzer-photometric scanning ellipsometer. This
instrument [5] digitally processes the sinusoidally varying
transmitted intensity by accumulating 36 data points per cycle
for a number of 150 Hz optical cycles (typically 50-2000) at a
fixed wavelength. These data are then Fourier transformed to
obtain the azimuth and minor/major axis ratio of the polari-
zation state of the reflected light, from which the dielectric
function is calculated. The relative precision of the order of
10^{-5}, which is comparable to that attainable with high-precision
measurements of the reflectance [6]. High-resolution electro-
reflectance spectra were obtained using the Schottky barrier
technique [7]. All measurements were performed on n-type
samples of carrier concentration of the order of 10^{16} cm^{-3}.

3. RESULTS AND DISCUSSION

The ellipsometrically measured complex dielectric function, cal-
culated JDOS(3) lineshapes and the low-field ER spectrum for Si
are shown in Fig. 1a. The detailed comparisons of JDOS(3) line-
shapes and low-field ER spectra for Si, Ge, and GaAs are shown
in Figs. 1b - 1d. The JDOS(3) lineshapes were calculated by the
least-squares method [8]. The noise level in the JDOS(3) curves
for Si is typical. For direct comparison to ER spectra, we have
used the expression

$$\Delta R/R = -\mathrm{Re}\left[\frac{C\ \exp(i\theta)}{2n_a nD}\ \Delta\varepsilon\right] + A + B\cdot E \qquad (3)$$

where $\Delta\varepsilon$ is calculated from Eq. (1) with $d\Gamma/dE = 0$ and $\mu_{||}$
assumed constant, $[2n_a nD]$ gives the generalized Seraphin co-
efficients [9] for the three-phase substrate — 3.5 nm Ni film —
air Schottky barrier ER configuration, and $C \propto \mathcal{E}^2/\mu_{||}$ and

FIG. 1a: (above left) measured complex dielectric function, ε, numerically-calculated JDOS third derivative, and measured low-field ER spectrum for Si. (b), (c), (d) (top, center, and bottom right, respectively) comparison between numerically-calculated JDOS third derivatives (-----) and measured low-field ER spectra (———) for Si, Ge, and GaAs. The best-fit procedure is described in the text.

$\theta \propto -3\tan^{-1}[d\Gamma/dE]$ are the fitting parameters. The constants A and B were used to provide approximate baselines for the E_2 structures in Si, as indicated in Fig. 1b. ER spectra from different energy regions for the same material were scaled according to the modulation potential $V \propto \mathcal{E}^2$ to compensate the field dependence of C for different structures. Both JDOS(3) and ER curves use the same energy scale.

In general, the agreement between the best-fit JDOS third-derivatives and the low-field ER spectra is good. Structure in the former spectra are in a one-to-one correspondence with those in the ER spectra, except for the E_0' structures which are too weak to be resolved in the JDOS(3) curves for either Ge or Si. In addition, the relatively low interband reduced mass of these critical points makes them more easily resolved in ER. The absence of the E_0' structure in the JDOS(3) lineshapes of Si is consistent with wavelength-derivative measurements, which show the E_0' structure in Si at 80 K but not at room temperature [10]. Ellipsometric measurements at low temperatures should also resolve this transition in Si. The E_0' features are observed in both data for GaAs, as are all parts of the E_2 structures in all three materials measured. The parameters obtained by detailed comparison are given in Table 1, which lists $(1/C) \propto \mu_\parallel$ relative to the value for the E_1 transition for each material, and $-\theta \propto 3\tan(d\Gamma/dE)$ in the simple theory. A value of $(1/C) < 1$ means that $\mu_\parallel$ for the critical point in question is less than that for the reference E_1 critical point. The results for Ge are in qualitative agreement with Franz-Keldysh oscillation measurements [11], which show values of $\mu_\parallel$ slightly lower and substantially higher for the $E_1 + \Delta_1$ and E_2 transitions, respectively, relative to E_1. By contrast, $\mu_\parallel$ for the E_2 structures in GaAs and Si appears to be smaller than that of the reference E_1 transition, suggesting that the nature of the E_2 critical point in these materials is qualitatively different from that of Ge. This is supported by the fact that the threshold energies of the E_2 structures appear in different locations of the ε_2 spectrum for Ge than for GaAs and Si.

The phase angles, $-\theta$, in Table 1 all lie in a 60° range, indicating the same sign of $\mu_{||}$ for each of the transitions observed. The absolute values are in approximate agreement with those previously obtained for the E_1 and $E_1+\Delta_1$ transitions of Ge [2], and correspond to $d\Gamma/dE \sim 0.5$ in the simple model. This is clearly too large and probably reflects a breakdown of the one-electron approximation. The final-state interaction may be the cause but the model is too simple to permit a unique interpretation.

Transition	Si		Ge		GaAs	
	1/C (rel)	$-\theta$	1/C (rel)	$-\theta$	1/C (rel)	$-\theta$
E_1	1.00	60°	1.00	50°	1.00	60°
$E_1+\Delta_1$	–	–	0.7	40°	0.7	50°
$E_0'(1)$	–	–	–	–	1.4	50°
$E_0'(2)$	–	–	–	–	1.7	70°
$E_2(1)$	0.8	100°	1.5	70°	0.7	80°
$E_2(2)$	0.4	80°	–	–	–	–

Table 1. Best-fit parameter values for observed critical point structures.

4. REFERENCES

[1] Aspnes, D. E.; Rowe, J. E.: Solid State Commun. <u>8</u>, 1145 (1970); Phys. Rev. <u>B5</u>, 4022 (1972).
[2] Aspnes, D. E.: Phys. Rev. Letters <u>28</u>, 168 (1972).
[3] Blossey, D. F.: Phys. Rev. <u>B2</u>, 3976 (1970); <u>B3</u>, 1382 (1971).
[4] Dow, J. D.; Lao, B. Y.; Newman, S. A.: Phys. Rev. <u>B3</u>, 2571 (1971); Weinstein, F. C.; Dow, J. D.; Lao, V. <u>Y</u>.: Phys. Rev. <u>B4</u>, 3502 (1971).
[5] Aspnes, D. <u>E</u>.: Optics Commun. 8, 222 (1973).
[6] Sell, D. D.: Appl. Opt. <u>9</u>, 1926 (1970).
[7] Aspnes, D. E.; Studna, A. A.: Phys. Rev. <u>B7</u>, 4605 (1973).
[8] Savitzky, A; Golay, M. J. E.: Analytical Chem. <u>36</u>, 1627 (1964).
[9] Aspnes, D. E.: J. Opt. Soc. Am. <u>63</u>, 1380 (1973).
[10] Welkowsky, M.; Braunstein, R.: Phys. Rev. B5, 497 (1972).
[11] Aspnes, D. E.: Phys. Rev. Letters <u>31</u>, 230 (1973).

SATURATED ABSORPTION IN TERNARY ALLOYS USING MODE LOCKED LASER PULSES

J. F. REINTJES*, J. C. McGRODDY, A. E. BLAKESLEE
IBM Thomas J. Watson Research Center
Yorktown Heights, New York 10598

ABSTRACT: Saturation of the optical absorption near the band-
edge in the ternary semiconductor alloy systems $InAs_xP_{1-x}$ and
$Ga_{1-y}In_yAs$ been studied using mode locked pulses from a
Nd:glass laser. Transmission levels exceeding the linear value
by more than 4 orders of magnitude have been observed. Time
dependent measurements show that the absorption can recover in
times of the order of 30 psec, indicating that these materials
may be useful as mode locking saturable absorbers for infrared
lasers.

INTRODUCTION

Saturation of the optical absorption has been observed previously in a
variety of materials using many sources of excitation. The optical absorp-
tion of organic dye molecules has been observed to saturate and such dyes
are widely used for Q-switching and modelocking of laser oscillators. Sat-
uration of the direct band gap optical absorption in semiconductors on a
steady state basis has been reported using GaAs lasers[1], and Q-switching of
GaAs injection lasers has been related to saturation of the absorption of
unpumped[2] as well as damaged[3] regions. Saturation of acceptor level to con-
duction band transitions in semiconductors has also been observed[4].

In this paper we report observations of bleaching of the direct band-to-
band optical obsorption in the ternary alloys $In\,As_xP_{1-x}$ and $Ga_{1-y}In_yAs$
using mode-locked pulses from a Nd:glass laser. The alloy in each case
(the values of x and y) was chosen so that the band gap of the material
was slightly less than the photon energy of about 1.17 eV. Thus the equi-
librium optical absorption coefficient, α_0 , was of the order of 10^4.
The sample thicknesses were such that no linear transmission could be ob-
served. However, transmission of as much as 15% of the intense laser pulse

Work supported by the U. S. Office of Naval Research,
Contract No. N00014-70-C-0187.

*Present address: Naval Research Laboratory, Washington,
D. C. 20375

has been observed. At the highest intensities used the transmission (ratio of transmitted to incident pulse energy,) becomes independent of incident pulse energy, indicating that the saturable absorption has been fully bleached.

Using a weaker probing beam, we have also measured the time required for the absorption to recover after bleaching. This time is as short as 30 picoseconds, indicating that these materials are potentially useful as mode-locking saturable absorbers. Since the bandgap of these and other similar pseudobinary alloys can take on values between 0 and 2 eV, the material can be engineered to fit the photon energy of a particular lasing transition.

<u>EXPERIMENTS</u>

The experiments consisted in measurement of the transmission as a function of incident pulse energy, as well as a measurement of the dynamics of recovery of the absorption after saturation. The latter measurement consisted in a measurement of the transmission of a weak probing pulse as a function of its time position with respect to a strong pulse of fixed intensity. All the measurements were performed using pulses from a mode-locked N Glass laser operating at 1.06μ . The experimental arrangement is described in reference 5. The In $As_x P_{1-x}$ samples were prepared by standard liquid phase epitaxy on In P substrates and were kindly supplied by Jerry Woodall. The $Ga_{1-y} In_y As$ samples were grown from the vapor phase on GaAs substrates. In both cases defects near the heterointerface introduced undesired and un-bleachable optical absorption, limiting the maximum transmission to 15% in the case of In $As_x P_{1-x}$ and 5% in the case of $Ga_{1-y} In_y As$.

<u>RESULTS</u>

Typical transmission curves are shown in Fig. 1a and 1b for samples of the two alloys. Sample parameters are given in the caption. The behaviors indicated are typical of samples with energy gap slightly less than the photon energy. In samples with slightly larger energy gap only multiphoton absorption, leading to a decrease of transmission with increasing intensity, was observed. The lowest level of transmission reliably measurable with the pulsed setup was about 1%. In fact, the transmission of the In $As_{.19} P_{.81}$ sample at low levels was probably less than 10^{-6}.

Figure 1 Transmitted vs. Incident pulse energy for (a) 2μ thick epitaxial Ga_8 $In_{.2}As$ or GaAs and (b) 25μ thick $In As_{.19}P_{.81}$ on In P. The pulsewidth was 20 psec.

Fig. 2 shows typical bleaching recovery data. Details are given in the caption. The maximum transmission of the probing pulse was about twice the corresponding transmission of the strong main pulse which produced the bleaching. The width of the bleaching recovery curve in Fig. 2 is about 20 psec, limited by the 20 psec pulsewidth of the existing source. The recovery time of the absorption coefficient itself is probably longer - by at least a factor of the absolute value of the logarithm of the linear transmission.

<u>DISCUSSION</u>

One expects to be able to account for the bleaching effects in terms of a dynamic Burstein-Moss shift of the optical absorption edge. Then

$$\alpha = \alpha_0 (1-f_e-f_h)$$

α_0 being the equilibrium absorption coefficient, f_e and f_h the fractional populations of the relevant conduction and valence band states with electrons and holes, respectively.

Figure 2 Transmission of a weak probing pulse as a function of its arrival time relative to a strong exciting pulse of fixed amplitude. The sample is $Ga_{.8}In_{.2}As$, about $2\,\mu$ thick.

In the extreme transient limit in which the pulse width is so short that no scattering out of these states occurs during the bleaching, the energy required for bleaching is readily calculable by merely counting states and including the laser linewidth. (This is the regime in which self-induced transparency may be observable [6,7,8]). In the actual case we expect that the distributions will thermalize, increasing the required bleaching energy. In this limit (no recombination) the calculated energy is an order of magnitude lower than the experimentally observed energy if the electron and hole temperatures are taken to be the lattice temperature, 300K. There are two possible explanations of this discrepancy. The first attributes it to a higher electron and hole temperature - of the order of 900K. An alternative explanation, is that stimulated recombination occurs at a sufficiently rapid rate to deplete the non-equilibrium populations even during these short pulses. Full bleaching at energies substantially greater than the bandgap obviously implies large gains at lower energies[9]. A search for such super-radiant emission proved negative but our setup was not sufficiently sensitive as to be conclusive. In fact, any explanation of the large pulse energies required must require either super-radiant emission during or after the exciting pulse or rapid non-radiative recombination.

Conclusion

In this paper we report the direct observation of dynamic bleaching of the direct bandgap absorption in semiconductors and the recovery of the absorption. While the observed recovery times are fast enough to indicate some promise of the usefulness of these materials as passive mode-locking devices, present limitations on the quality of the material prevent a direct demonstration of such an effect. By the use of quaternary alloys it should be

possible to minimize the non-bleachable losses associated with misfit dis-
locations at the heteroepitaxial interface, providing samples suitable for
mode locking. In another direction, as intense mode-locked lasers with
shorter pulses and with wavelength tunability become available, measure-
ments such as these can be extended to measuring directly the relaxation
of initial hot electron distribution functions, relaxation of non-zero
net spin distributions and similar phenomena.

Acknowledgements

We gratefully acknowledge stimulating discussions with John
Armstrong, Marshall Nathan, Frank Stern, Peter Price, and Jerry
Woodall, and the skilled technical assistance of John Anderson,
Frederick Feigel and Bernhard Bischoff.

References

(1) P. D. Dapkus, N. Holonyak Jr., R. D. Burnham, D. L. Keune
 Applied Physics Letters $\underline{16}$ 93 (1970)

(2) G. L. Lasher
 Solid State Electronics $\underline{7}$ 707 (1964)

(3) E. S. Yang, P. G. McMullin, A. W. Smith, J. M. Blum and K. K. Shih
 Applied Physics Letters $\underline{24}$ 324 (1974)

(4) A. E. Michel and M. I. Nathan
 Applied Physics Letters $\underline{6}$ 101 (1966)

(5) J. F. Reintjes and J. C. McGroddy
 Physical Review Letters $\underline{30}$ 901 (1973)

(6) I. A. Poluektov and Yu. M. Popov
 Zh. E.T.F. Pis. Red. $\underline{9}$ 542 (1969)
 (English Trans. JETP Letters $\underline{9}$ 330 (1969)

(7) N. Tzoar and J. I. Gersten
 Physical Review Letters $\underline{28}$ 1203 (1972)

(8) A. Schenzle and H. Haken
 Optics Comm. $\underline{6}$ 96 (1972)

(9) G. Lasher and F. Stern
 Physical Review $\underline{133}$ A553 (1964)

FAR-INFRARED ABSORPTION IN p-TYPE GERMANIUM AND SILICON[+]

E. KACZMAREK

Institute of Physics, Polish Academy of Sciences,
Warsaw, Poland

Z. W. GORTEL

Institute of Theoretical Physics, Warsaw University
Warsaw, Poland

The absorption of electromagnetic radiation at
helium temperatures in the wavelength region from
450 to 3000 μm and from 100 to 1000 μm in p-type
Ge and Si, respectively, has been calculated in
the hopping model. Concentration of acceptors up
to 9×10^{16} cm^{-3} in Ge and 2×10^{17} cm^{-3} in Si, and
compensations up to 0.2 were considered.

The purpose of the work is to study the photon-induced hop-
ping transitions between acceptor states and their effect
on the absorption coefficient of the electromagnetic radia-
tion in p-type Ge and Si at liquid helium temperatures.
The experiments of Smith et al.[1] and Zwerdling et al.[2]
showed that the value of the absorption coefficient for
p-type Ge is about ten times higher than for n-type Ge, and
the dependence of the coefficient on the wave-length for
p-type Ge is not as strong as for n-type Ge. This fact was
difficult to understand till now in view of the similari-
ties of the ac hopping conductivity in both materials.
As it was previously shown [3], the description of the iso-
lated acceptor ground state by the set of envelope functions
introduced by Schechter [4] gives a picture of two center
states qualitatively different from the picture obtained

[+]Work supported in part by the Institute of Physics of
Polish Academy of Sciences.

when one describes the acceptor ground state by hydrogen-like
envelope function. A basic difference between these two pictu-
res is that in the n-type material as in the hydrogen-like ap-
proximation there are only two, while in p-type we have four
lowest two-center states of two impurities. All the four sta-
tes, denoted in the order of decreasing energy by ϵ_1, ϵ_2,
ϵ_3 and ϵ_4, have to be taken into account in the calculation
of absorption. A detailed analysis of the formulas for the ap-
propriate radiative transition rates [5] indicates that only
the transitions from ϵ_4 to ϵ_1 and from ϵ_3 to ϵ_2 states con-
tribute significantly to the total absorption rate. The appro-
priate energy differencies can be expressed as follows

$$\Delta\epsilon_{41} = \sqrt{4\,W_{R1}^2 + \Delta^2} \quad , \quad \Delta\epsilon_{32} = \sqrt{4\,W_{R2}^2 + \Delta^2} \quad . \quad (1)$$

Here Δ is the difference of the Coulomb energies at the posi-
tions of the acceptors in the field of the nearest donor ion.
The angular averages of the squares of the so called resonan-
ce energies W_{R1} and W_{R2} vs. normalized pair separations $2R/r_2$
are plotted in Fig. 1. The nonmonotonicity of $\langle W_{R1}^2 \rangle$ is due
to the presence of the d-like contribution in the Schechter
envelope functions.

FIG. 1: Plots of resonance
energies

The absorption cofficient, α,
was obtained by statistical
averaging, over the entire
crystal, contributions to the
absorption due to the close
environment of the donor ion.
The procedure is identical with
that developed by Blinowski
and Mycielski [6] . The formu-
la for α can be written as
follow

$$\alpha = (32\,\pi^4\,\kappa^{1/2} K N_A^3 / 3\hbar c)$$

$$\times (A_1 + A_2) \quad , \quad (2)$$

where

$$A_1 = \int_{\Omega_1(\omega)} dR \left[R^3 \left\langle W_{R1}^2 \right\rangle L(R,E_1)/E_1 \right]$$

$$\times \left[1 + \exp\left\{ \left[-\hbar\omega/2 + (E_1^2/4 + \left\langle W_{R2}^2 \right\rangle) \right]^{1/2}/kT \right\} \right]^{-1},$$

$$A_2 = \int_{\Omega_2(\omega)} dR \left[\exp\left\{ -(E_2^2/4 + \left\langle W_{R2}^2 \right\rangle)^{1/2}/kT \right\} \right.$$

$$\times R^3 \left\langle W_{R2}^2 \right\rangle L(R,E_2)/E_2$$

$$\times \left[1 + \exp\left\{ -\left[(E_2^2/4 + \left\langle W_{R1}^2 \right\rangle)^{1/2} - \hbar\omega/2 \right]/kT \right\} \right]^{-1},$$

$$E_{1,2} = (\hbar^2\omega^2 - 4\left\langle W_{R1,2}^2 \right\rangle)^{1/2}$$

The κ, K, N_A and ω are the dielectric constant, compensation, acceptor concentration and radiation frequency, respectively. The function L(R,E) is typical for the hopping theories and is defined in Ref. 4. The integration ranges $\Omega_1(\omega)$ and $\Omega_2(\omega)$ are so chosen that the E_1 and E_2 are real, respectively.

The numerical calculations were performed for the following values of variational parameters of the Schechter functions[4]:

| | r_1 | r_2 | c_1 | c_2 | c_3 |
	Å	Å	$10^9 cm^{-3/2}$	$10^{21} cm^{-7/2}$	$10^{21} cm^{-7/2}$
p – Ge	44.0	34.4	1.67	−2.15	4.68
p – Si	17.71	13.21	7.03	−28.6	10.23

The values of κ are 16 and 12 for Ge and Si, respectively. The ratio of the absorption coefficient to the compensation value, α/K, as a function of the wave-length is shown by the solid lines in Figs. 2a and 2b for Ge and Si, respectively. The experimental results of Smith et al.[1] and Zwerdling et al. [2] for Ge are presented by the dashed lines denoted by 1 and 2, respectively.

2a

2b

FIG. 2: Plots of absorption coefficient per unit compensation vs. wavelength of radiation for p-Ge,(2a), and p-Si,(2b), for various values of acceptor concentrations.

It seems that the presented theory improves the theories based on the hydrogen-like approximation because it gives the value of the absorption coefficient of the same order as experimentally observed. At present one should not pay very much attention to the absolute values of absorption coefficient. The discrepancy between the theory and experiment is of the same order as the discrepancy between the results of the two different experiments, and the theoretical values are very sensitive to the choice of the variational parameters. Moreover our theory explains the fact that the dependence of the absorption coefficient on the wavelength of incident radiation for p-type Ge is not so strong as for n-type materials and that the maximum of the absorption coefficient plotted vs wavelength is less pronounced and occurs at shorter wavelengths than that for n-type. These facts cannot be explained by the theory based on the hydrogen-like approximation.

The theory predicts a narrow peak which has no experimental confirmation. Its existance is related to the nonmomotonical dependence of $\left\langle W_{R1}^2 \right\rangle$ on the pair separation. It is possible that due to some factors not taken into account in the theory, e.g. angular dependence of W_{R1}^2, the peak is much broader than predicted and does not contribute to the predicted structure of the absorption curve. At present, one cannot exclude the possibility that the broad maximum reported by Smith et al. [1] has the same origin as the theoretical peak. However to have a definite answer to this question one should perform

experiments for at least two different acceptor concentrations.
There exists also a maximum on the absorption curves whose origin is the same as of the only maximum presented in n-type materials [6]. This maximum moves towards shorter wavelengths as the concentration increases and therefore overlaps the peak at higher concentrations. The maximum is better separated from the peak for Silicon than for Germanium.
The authors wish to thank Professor M. Suffczyński, Dr. J. Mycielski and Dr. J. Blinowski for many valuable remarks and discussions throughout all the phases of this work.

REFERENCES

[1] Smith, R.A.; Zwerdling, S.; Dermatis, S.N.; Theriault, J.P.:
 Proc. IXth Int. Conf. Phys. Semiconductors, Moscow 1968,
 Publ. House "Nauka", Leningrad 1968, p.149.
[2] Zwerdling, S.; Theriault, J.P.: Infrared Phys. 12, 165
 (1972).
[3] Kaczmarek, E.; Myszkowski, A.; Trylski, J.: Acta Phys.
 Polon. 30, 283 (1966).
[4] Schechter, D.: J. Phys. Chem. Solids, 23, 237 (1962).
[5] Gortel, Z.W.; Kaczmarek, E.: Acta Physica Polonica, A41,
 641 (1972).
[6] Blinowski, J.; Mycielski, J.: Phys. Rev. 136, A266 (1964).

THE TEMPERATURE SHIFT OF THE ABSORPTION EDGE IN CUBIC ZnS

Y. BRADA and B.G. YACOBI

The Racah Institute of Physics, The Hebrew University

Jerusalem, Israel

The temperature shift of the exponential absorption
edge of cubic ZnS was measured in the range from
10°K to 300°K. A Franz-Keldysh mechanism involving
phonon-created microfields, as postulated by Dow and
Redfield, was used to explain the results. Below
50°K only a very slight change in the edge energy E_g
was observed. Between 55°K and 100°K the energy of
the edge could be described by $E_g(T) = 3.798eV -
0.101$ $[1-\tanh(\hbar\omega_{LA}/2kT)]eV$, where $\hbar\omega_{LA} = 13.6meV$.
From 100°K to 300°K the edge changed as $E_g = 3.762eV-
0.400$ $[1-\tanh(\hbar\omega_{LO}/2kT)]eV$, where $\hbar\omega_{LO} = 43.6meV$.
The slope parameter σ of the Urbach rule behaved as
$\sigma = \sigma_o(2kT/\hbar\omega_o)$ $\tanh(\hbar\omega_o/2kT)$. Below 100°K: $\sigma_o=1.20$
and $\hbar\omega_o = \hbar\omega_{LA}$; above 100°K: $\sigma_o=2.50$ and $\hbar\omega_o=\hbar\omega_{LO}$.

The temperature shift of the exponential absorption edge [1]
of cubic ZnS was studied in detail. From known data on the
pressure dependence of the edge [2] and the temperature depen-
dence of the unit cell volume [3] it was found that the dila-
tion effect accounts for less than 10% of the absorption edge
shift. Several phonon dependent mechanisms were proposed pre-
viously. Recently Redfield [4] and Dow and Redfield [5] pro-
posed a Franz-Keldysh mechanism as being responsible for the
exponential edge of both ionic solids and doped covalent semi-
conductors. This theory proposes that charged imperfections,
such as ionic impurities, optical phonons or piezoelectric
phonons create electrical microfields which facilitate the
tunneling of the electron away from the hole. Therefore opti-
cal transitions are possible even at energies below the gap
energy. We have then a photon assisted tunneling process which
is also responsible for the exponential form of the Urbach
tail [7]. The mean square fields of the LO phonon was com-
puted by Dow and Redfield as

$$< F^2 > = \frac{\hbar\omega_{LO}q_c^3 \, (\varepsilon_o - \varepsilon_\infty)}{3\pi\varepsilon_o\varepsilon_\infty} \, \coth \, (\hbar\omega_{LO}/2kT) \qquad (1)$$

where $\hbar\omega_{LO}$ is the LO phonon energy, ε_0 and ε_∞ the static and high frequency dielectric constants respectively, k and T have the usual meaning and q_c is the polaron cut-off wave vector. q_c is in the range $\pi/a > q_c > \pi/2a$, where "a" is the exciton radius.

The Franz-Keldysh [6] effect connects the absorption edge shift with the action of electric fields. For an exponential edge

$$\Delta E_g = -(\hbar e^2/24\mu^*) \, (\sigma/kT)^2 \, F^2 = -\gamma F^2 \qquad (2)$$

where σ is the slope parameter of the Urbachs rule [7] $\alpha = \alpha_0 \exp[\sigma/kT(h\nu - h\nu_0)]$, where h, ν, $\hbar$, e, k, T have the usual meaning, μ^* is the reduced excitonic effective mass, F is the electric field, α is the absorption coefficient and α_0 is a constant. σ is known to be temperature dependent. We will assume for the time being that[+]:

$$\sigma = \sigma_0 \, (2kT/\hbar\omega_{LO}) \, \tanh \, (\hbar\omega_{LO}/2kT) \qquad (3)$$

where we use the same designations as before, and σ_0 is a constant.

If the Dow-Redfield theory is applicable in our case, we can combine (1), (2) and (3) into:

$$E_g(T_1) - E_g(T_2) = \Delta E_g = S[\tanh(\hbar\omega_{LO}/2kT_1) -$$
$$- \tanh(\hbar\omega_{LO}/2kT_2)] \qquad (4)$$

where $S = [e^2\hbar^2\sigma_0^2 q_c^3 (\varepsilon_0 - \varepsilon_\infty)]/[18\pi\omega_{LO}\varepsilon_0\varepsilon_\infty]$ and $T_2 > T_1$.

In the LO case, at $T_1 < 100°K$ the first term in the brackets will be almost unity. The experimental results for σ are shown in Fig. 1. We see there that at temperatures above 100°K the LO phonon is responsible for the changes in σ, below this temperature the interaction is due to the LA phonon. Fig. 2 shows the shift of the edge energy at $\alpha \cong 10^3 \mathrm{cm}^{-1}$. Figs. 3 and 4 correlate this change with $\tanh(\hbar\omega_{LA}/2kT)$ up to 100°K and with $\tanh(\hbar\omega_{LO}/2kT)$ up to 300°K, respectively. An excellent fit is obtained in these temperature regions. In this particular case

+ Generally $h\omega_0$ is written in eq.(3), where it is considered as the energy of an effective phonon [8].

Fig. 1
σ vs. T. Points - experimental values. Curves drawn according to eq. (3); the solid curve with $\sigma_o=1.20$ and $\hbar\omega_o=\hbar\omega_{LA}$; the dotted curve with $\sigma_o=2.50$ and $\hbar\omega_o=\hbar\omega_{LO}$. The values of the phonon energies are from Ref. [13].

Fig. 2
The temperature dependence of the absorption edge energy E_g in cubic ZnS at $\alpha = 10^3\,cm^{-1}$.

Fig. 3
Absorption edge E_g vs. $\tanh\,(\hbar\omega_{LA}/2kT)$.

Fig.4
Absorption edge E_g
vs. $\tanh(\hbar\omega_{LO}/2kT)$

the piezoelectric LA phonon produces a $< F^2 >$ which is larger than the $< F^2 >$ due to the LO phonon at temperature below 100°K.

It has to be pointed out however, that these relations cannot hold for higher temperatures, because then the large phonon-created fields cannot be introduced in conjunction with the weak field approximation [6] of eq. (2). In addition, the parameters entering S also show a certain temperature dependence. Using the results of Van Doorn [11] for 300°K - 900°K we observe that there the edge energy (at $\alpha \approx 10^2 cm^{-1}$) changes linearly with temperature.

From previously measured values of γ, both at room temperature and below [9,10], and the observed shifts ΔE_g, we can find the experimental values of the root mean square fields F_{rms}. For the LA phonon at 100°K, $F_{rms}=6\times10^4 V/cm$; for the LO phonon at 300°K, $F_{rms}=3.9\times10^5 V/cm$. The LO case agrees quite well with the computed fields of the Dow-Redfield theory. The uncertainty in choosing q_c [5] does not allow an accurate comparison at the moment.

Recently a band theoretical computation, using the pseudopotential method, was carried out by Tsay et.al.[12] for cubic ZnS. There the Debye-Waller factor was introduced to obtain the temperature dependence of the direct optical gap at the Γ

point. Their results have the general form obtained in our experiments, but as they pertain to the first exciton peak and not to the absorption edge, the agreement cannot be complete.

REFERENCES

[1] Piper, W.W.: Phys. Rev. 92, 23 (1953).

[2] Edwards, A.L.; Slykhouse, T.E.; Drickamer, H.G.:
 J. Phys. Chem. Solids 11, 140 (1959).

[3] Von Adenstedt, H.: Ann. der Physik (5) 26, 69 (1936).

[4] Redfield, D.: Phys. Rev. 130, 916 (1963).
 Trans. N.Y. Acad. Sci. 26, 590 (1964).

[5] Dow, J.D.; Redfield, D.: Phys. Rev. B1, 3358 (1970).
 ibid 5, 594 (1972).

[6] Franz, W.: Z. Naturforsch 13a, 484 (1958).

[7] Urbach, F.: Phys. Rev. 92, 1324 (1953).

[8] Mahr, H.: Phys. Rev. 125, 1510 (1962).

[9] Yacobi, B.G.; Brada, Y.: to appear in Phys. Rev. B15.

[10] Yacobi, B.G.; Brada, Y.: to be published.

[11] Van Doorn, C.Z.: Physica 20, 1155 (1954).

[12] Tsay, Y.F.; Mitra, S.S.; Vetelino, J.F.: J. Phys. Chem.
 Solids 34, 2167 (1973).

[13] Nilsen, W.G.: Phys. Rev. 182, 838 (1969).

ERRATA

1. Eq.(2) should read:

$$\Delta E_g = -(\hbar^2 e^2/24\mu^*) \, (\sigma/kT)^2 \, F^2 = -\gamma F^2$$

2. The expression for S (eq.(4)) should read:

$$S = [e^2 \hbar \sigma_0^2 q_C^3 (\epsilon_0 - \epsilon_\infty)] \, / \, [18\pi\mu^* \omega_{L0} \epsilon_\bullet \epsilon_\infty]$$

INFLUENCE OF INTERNAL FIELDS ON THE ABSORPTION IN ZnO

G. HVEDSTRUP JENSEN

Physics Laboratory III, Technical University of Denmark
2800 Lyngby, Denmark

Optical absorption measurements in the band gap
region of pure and Cu doped ZnO single crystals
have been performed at temperatures from 77 K to
500 K. The broadening of the absorption is consider-
ed to be caused by electric microfields from ionized
impurities, LO-phonon and piezoelectric LA-phonons.
The microfields are calculated as function of
temperature and doping and it is shown that acoustic
phonons are of importance in forming the Urbach
tail.

1. INTRODUCTION

Optical absorption measurements in the band gap region of un-
doped ZnO have been reported by several authors [1,2,3,4] and it
is concluded that LO-phonon broadened impurity and exciton
states are dominating in the band tail region. Also optical pro-
perties of Cu doped ZnO have been investigated both with regard
to infrared absorption (see [5] and references herein) and lu-
minescence [1] but no detailed measurements of the band gap ab-
sorption of Cu doped single crystals have been published. The
present work attempts to discuss the influence of internal
fields, microfields, on the absorption. The microfields discussed
are caused by ionized impurities, LO-phonons and piezoelectric
LA-phonons. By calculating the mean squared amplitude of these
fields as a function of temperature and doping their relative
importance is found.

2. EXPERIMENTAL

The crystals were grown from vapour phase at this laboratory.
They were mainly doped by diffusion of Cu from a Cu/ZnO cheramic
at temperatures from 900°C to 1300°C. Homogenization was per-
formed in O_2 at temperatures in the range 1000°C - 1300°C. Dif-
fusion times and temperatures were estimated by using the data
in [5]. Conductivity measurements showed acceptable agreement

FIG. 1a, b: Absorption coefficient α of highly (a) and critically (b) Cu-doped ZnO crystals. The impurity dominated crystals show downward curvature, while the critically doped crystals at higher temperatures show upward curvature. The low slope part of the curves at fig.(b) are due to the impurities. Polarization perpendicular to c-axis ($E \perp c$).

with the expected values. Also crystals doped by adding Cu to the charge before growing were used. No difference between the two kinds of samples were found. The spectral resolution was about 2 Å. A number of representative absorption curves at different temperatures and doping levels are plotted versus phonon energy, $\hbar\omega$, in fig. 1a and 1b. The samples in fig. 1a are heavily doped ($N \simeq 7\cdot10^{19}\,\mathrm{cm}^{-3}$, $N \simeq 10^{20}\,\mathrm{cm}^{-3}$), while the samples in 1b are doped to about $2\cdot10^{18}\,\mathrm{cm}^{-3}$, i.e. a doping level where the intrinsic and

(a)

(b)

FIG. 2: Isoabsorption curves of highly (a) and less doped (b) samples. The high temperature dependence of fig. (a) may be due to the LO-phonon broadening, while the low temperature dependence is in accordance with the band gap shift.

doped properties are of the same magnitude. When comparing with
[3] it is seen that the impurities tend to dominate at lower
temperatures. Fig. 2a and 2b show isoabsorption curves. In the
heavily doped sample (2a) the absorption follows the band gap
shift with temperature at least to 270 K yielding a slope in the
quasi linear part of -0.36 meV in agreement with the value in [8]
from reflection data. The anomaly at higher temperatures are seen
to be most important for low absorption coefficients. In fig. 2b
the slope is somewhat larger, showing a contribution from phonons
which are not at this lower doping level being dominated by the
impurities. It is seen, that the band gap shift may be obtained
from isoabsorption curves of heavily doped semiconductors, but
not from lightly doped ones because of the temperature dependent
absorption broadening in them.

FIG. 3: Mean squared fields $<F^2>$ from impurities, LA-phonons and LO-phonons. The LO-phonon and impurity contribution are temperature independent for T < 300 K.

3. DISCUSSION

The electric field distribution from LO- and piezoelectric LA-phonons is [6]:

$$P(F) = (\tfrac{2}{3}\pi <F^2>)^{-3/2} \; 4\pi \; F^2 \; \exp \{- \frac{3F^2}{2<F^2>}\} \qquad (1)$$

In [6] it is also shown that for LO-phonons of energy $\hbar\omega_{LO}$

$$<F^2_{LO}> = \frac{\hbar\omega_{LO} \; q_c^3}{3\pi\varepsilon^*} \; \coth \frac{\hbar\omega_{LO}}{2KT} \qquad (2)$$

where ε^* is the effective phonon permittivity and q_c a cutoff
wave vector. When affecting exciton states $q_c \simeq \frac{\pi}{2a_{exc}}$ [6], a_{exc}
being the exciton radius.

The electric field from a dilatation Δ in a piezoelectric crystal may be written

$$F_i = - \frac{e_{ijk}\,\Delta_{jk}}{\varepsilon} \tag{3}$$

e_{ijk} being the piezoelastic e-tensor, ε the dielectric constant. From (3) one obtains in the approximation of linear dispersion

$$\langle F^2_{LA}\rangle = \frac{e^2}{\varepsilon^2}\langle\Delta_,{}^2\rangle = \frac{e^2}{\varepsilon^2}\frac{(2kT)^4}{4\pi^2\rho s^5\hbar^3}\int_0^{\frac{\hbar q_c\cdot s}{2kT}} x^3 \coth x\; dx \tag{4}$$

Here ρ is the density and s the sound velocity.
A nearest neighbour distribution may describe the fields from point charges in a solid [7]:

$$P(F) = \frac{3}{2F}\left(\frac{F_o}{F}\right)^{3/2}\exp\left\{-\left(\frac{F_o}{F}\right)^{3/2}\right\} \tag{5}$$

with

$$F_o = \frac{e_o}{4\pi\varepsilon r_o^2}\;,\quad \frac{4\pi}{3}r_o^3 = N^{-1} \tag{6}$$

where N is the concentration of charges and e_o the charge pr. impurity. It may be obtained that

$$\langle F^2\rangle \simeq \frac{e^2}{16\pi^2\varepsilon^2\langle r^4\rangle} = 4\pi\frac{F_o^2}{\Gamma(7/3)} \simeq 10.5\,F_o^2 \tag{7}$$

The three contributions (2), (4) and (8) are plotted in fig. 3. It is seen that the phonons will dominate at impurity concentrations below $10^{18}\,cm^{-3}$ at higher temperatures, the LO-phonons yielding a nearly constant contribution. This is in agreement with the results in fig. 1b, 2a and 2b. Fig. 4 shows isoabsorption curves versus concentration and it is seen that the main features of fig. (3) are confirmed, but that the correspondence with the LO-phonon temperature dependence is poor. One has to include

FIG. 4: Isoabsorption curves versus concentration of charged impurities (and impurity field) at different temperatures. The impurity dependence at lower temperatures shows that acoustic phonons are of importance. The doping necessary for an impurity dominance is in accordance with fig. 3.

acoustic phonons to explain the temperature dependence. The high temperature behaviour in fig. 2a may be due to the increasing LO-contribution to the band gap shift. In conclusion it may be stated that the microfield model [6 and references herein] at least qualitatively describes the behaviour of the absorption tail in ZnO, but that both acoustic and optical phonons must be included to explain the temperature and doping dependence. Further it may be shown that a pure deformation i.e. not piezo-electric also will give qualitative agreement with the results. To check for quantitative agreement more exact measurements of the actual absorption tail's shape will be performed.

4. REFERENCES

[1] Andress, Bernhard: Z.f. Physik 170, 1 (1962)
[2] Dietz, R.E.; Hopfield, J.J.; Thomas, D.G.: J.Appl.Phys. suppl. 32, 2282 (1961)
[3] Jensen, G. Hvedstrup; Skettrup, T.: Phys.Stat.Sol. (b) 60, 169 (1973)
[4] Liang, W.Y.; Yoffe, A.D.: Phys.Rev.Lett. 20, 59 (1968)
[5] Müller, G.; Helbig, R.: J.Phys.Chem.Sol. 32, 1971 (1971)
[6] Dow, John D.; Redfield, David: Phys.Rev. B 5, 594 (1972)
[7] Redfield, David: Phys.Rev. 130, 914, 916 (1963)
[8] Jensen, G. Hvedstrup: to be published in Phys.Stat.Sol.
[9] Hutson, A.R.: J.Appl.Phys. suppl. 32, 2287 (1961)

DEPOLARIZATION OF LIGHT REFLECTED FROM CdS AND ZnO
IN THE EXCITON REGION

I. FILINSKI

Chemical Laboratory B, Technical University of Denmark
DK-2800 Lyngby, Denmark

T. SKETTRUP

Physics Laboratory III, Technical University of Denmark
DK-2800 Lyngby, Denmark

The method of photoelasticity is extended to the case
of reflection in the region of optical resonances.
Random strain fields in the crystals give rise to a
large amount of depolarization of the reflected light
near the resonances. Information about the crystal
perfection is thus obtained, and magnitudes of the
internal strains are derived.

1. INTRODUCTION

The method of photoelasticity has been widely used for checking
the perfection of transparent optical materials and crystals.
Here, we have extended this method to the case of absorbing cry-
stals by investigating the ellipticity and degree of depolariza-
tion of reflected light. Our method has, furthermore, the ad-
vantage that it points out the resonance energies of the ma-
terial since the depolarization is largest near the structure
in the optical spectrum.

2. EXPERIMENTAL METHODS AND RESULTS

The experimental set-up is shown in Fig. 1. The incident mono-
chromatic polarized light is normally reflected from the sample
and analyzed by the polarizing beam splitter. Modulators are
inserted so that the polarizer and compensator automatically are
adjusted for minimum signal. With this set-up the ratio of re-
flectivity ($\frac{R_\perp}{R_\parallel}$) and the phase difference ($\Theta_\parallel - \Theta_\perp$) for anisotropic
crystals can be directly measured [1]. The minimum signal de-
tected yields the amount of depolarized reflected light. The de-
gree of depolarization is defined as the ratio I_D/I_0 between the
intensities of the two main components of reflected light leav-

FIG. 1: Block diagram of ellipsometer. Polarizer and Compensator are automatically set to the positions of minimum signal I_D. The depolarization is defined as I_D/I_0. Both modulators are switched off when I_D is measured.

ing the polarizing beam splitter as shown in Fig. 1. The incident light was focused to a spot size of about 0.2 mm^2. The intensity of the depolarized light was dependent on the actual position of the spot on the surface. In order to get spectra from as perfect a crystal as possible, we have selected the positions with lowest intensities of depolarization. The spectral resulution was better than 0.1Å, and the depolarization level of the apparature itself

was about $3 \cdot 10^{-5}$. We have investigated the anisotropic crystals ZnO and CdS at 77K in the exciton region by this method. Furthermore, we have measured ZnO in the isotropic geometry where the incident light vector $\bar{k}$ is parallel to the optic axis $\bar{c}$, $(\bar{k}||\bar{c})$.

FIG. 2: a) Reflection and Phase of reflectivity from the A,B and C-excitons in ZnO for the 2 different polarizations, measured by means of the set-up in Fig. 1.
b) Depolarization spectra for $\bar{k}\perp\bar{c}$. The solid line is the experimental curve, while (1) and (2) are computed from equations (1) and (2) respectively.
c) Depolarization spectra for $\bar{k}||\bar{c}$. The solid line is the experimental curve, while (3) is computed from equation (3).

The reflection coefficient and phase of reflectivity for ZnO in the exciton region as measured by the ellipsometer in Fig. 1 are shown in Fig. 2a. From these curves the complex dielectric constant in this region may be computed. The depolarization spectra for $\bar{k}\perp\bar{c}$ and $\bar{k}||\bar{c}$ are shown in Fig. 2b and c in logarith-

mic scale. It is seen that very pronounced peaks (increase of several orders of magnitude) appear near the longitudinal resonance frequencies, where both R and Θ change much. The depolarization is very small at higher energies than the resonances, but relatively large at the low energy side, where the crystal is transparent, indicating that the whole volume of the crystal may contribute to the depolarization.

Similar depolarization was also observed in CdS as shown in Fig. 3. In this figure the readings of the polarizer and compensator from Fig. 1 are also shown.

FIG. 3: The solid line is the depolarization spectrum of CdS in the exciton region. Also shown are the ratio of reflection coefficients and the difference in the phases for reflected light polarized parallel and perpendicular to the optic axis.

3. DISCUSSION

The depolarization of the reflected light must be due to imperfections in the crystals. The strain around a dislocation may for example give rise to large depolarization [2]. We shall here only consider one important contribution to crystal imperfection, namely a distribution of local microstrains in the surface region where the reflection occurs and neglect volume effects. In the anisotropic case $\bar{k}\perp\bar{c}$ the presence of microstrains may cause depolarization because the resonance energies are fluctuating and the direction of the optic axis fluctuates. In the isotropic case $\bar{k}||\bar{c}$ only the fluctuating anisotropy (splitting of energy bands) yields depolarization.

We have computed the depolarization I_D/I_0 in the above mentioned 3 cases, assuming that the fluctuations involved are so small that expansions up to second order in the fluctuating parameter are valid. Then we find:

a) Anisotropic case, fluctuating resonance frequency ω_r with standard deviation σ_{ω_r}, and mean value ω_{r_0}:

$$I_D/I_o = \frac{\sigma_{\omega_r}^2}{4} \left| \frac{1}{\tilde{r}_1} \left(\frac{\partial \tilde{r}_1}{\partial \omega_r}\right)_{\omega_{ro}} - \frac{1}{\tilde{r}_3}\left(\frac{\partial \tilde{r}_3}{\partial \omega_r}\right)_{\omega_{ro}} \right|^2 \tag{1}$$

b) Anisotropic case, fluctuating shear stain e_{13} with standard deviation $\sigma_{e_{13}}$. The effect is to make the direction of the optic axis to fluctuate.

$$I_D/I_o = \frac{\sigma_{e_{13}}^2}{4} \left| \frac{\tilde{\varepsilon}_1 \tilde{\varepsilon}_3 \tilde{p}_{44}}{(\tilde{\varepsilon}_1 - \tilde{\varepsilon}_3)} \right|^2 \frac{|\tilde{r}_1 - \tilde{r}_3|^4}{|\tilde{r}_1 \tilde{r}_3|^2} \tag{2}$$

c) Isotropic case $(\bar{k} || \bar{c})$, fluctuating uniaxial stress X averaged over all directions, with standard deviation σ_X.

$$I_D/I_o = \frac{\sigma_x^2}{8} \left| \frac{1}{\tilde{r}_1} \left(\frac{\partial \tilde{r}_1}{\partial X}\right)_0 - \frac{1}{\tilde{r}_2}\left(\frac{\partial \tilde{r}_2}{\partial X}\right)_0 \right|^2 \tag{3}$$

Here $\tilde{r}$ is the complex amplitude reflectivity (usual reflection coefficient is $|\tilde{r}|^2$). Subscripts 1 and 2 refer to the isotropic plane while 3 refers to the optic axis. $\tilde{\varepsilon}$ is the complex dielectric constant, and $\tilde{p}_{44}$ a complex component of the photoelastic tensor. In deriving (1), (2) and (3) the fluctuating reflected intensities (not the fields) were added since it turned out that this inchorent summation gave the largest depolarization. It may be expected that such an incoherent summation method should be applied if the strained surface regions are much larger than the wavelengths involved. In Fig. 2b, (1) and (2) has been computed from the measured values of $\tilde{r}$ from Fig. 2a and compared with the experimental depolarization curve. It is seen that both (1) and (2) exhibit qualitative agreement with the experimental curve. In Fig. 2c, (3) is computed by means of an oscillator model for the A- and B-excitons and a splitting with a shear deformation potential $C_5 = 1.2$ eV [3]. Also here a qualitative agreement is obtained. It should be noted that the introduction of fluctuating parameters implies that the values of reflection and transmission coefficients measured in an experiment are average

values. The relations between these and the ideal values of the
perfect crystal (where $\sigma = o$) are not simple, but of great impor-
tance in the proper interpretation of optical spectra. Actual
magnitudes of the standard deviations may be obtained by fitting
the expressions (1), (2) or (3) to the experimental curves. We
obtain: $\sigma_{\omega_r} = 0.55$ meV and $\sigma_{e_{13}} \sim 10^{-3}$ (uncertain since p_{44} was
estimated in the resonance region, instead the fluctuation σ_α of
the direction of the optic axis may be found to be $\sigma_\alpha = 6.0^{\circ}$).
From Fig. 2c, we find $\sigma_x = 0.32$ kbar (corresponding to $\sigma_{e_{12}}$
$= 0.72 \cdot 10^{-3}$). The B- and C-excitons exhibit a shift of 2.7 meV/
kbar with hydrostatic pressure [4]. Then σ_{ω_r} may be transformed
to a $\sigma_x = 0.20$ kbar. Similarly, the compliance constant S_{44} [5]
may be used for transforming $\sigma_{e_{13}}$ to a $\sigma_x \sim 0.4$ kbar. All three
values of σ_x agree well. Similar values for internal stresses in
ZnO crystals have also recently been found by Berkowicz [6] in
transmission measurements.

4. CONCLUSION

We have presented a relatively simple method of checking the per-
fection of crystals. This can be performed in the convenient
spectral ranges where resonances occur. Like the usual spectro-
scopic modulation methods it has the advantage that it involves
the derivative of the reflection coefficient so that structure
in the spectra is more easily detected. It is thus a kind of
"internal" modulation method. From the depolarization curves one
may estimate approximate magnitudes of internal stresses.

5. REFERENCES

[1] Filinski, I.; Skettrup, T.: Sol.State Comm. 11, 1651 (1972)
[2] Nikitenko, V.I; Dedukh, L.M.: Phys.Stat.Sol. (a) 3, 383
 (1970)
[3] Skettrup, T.; Balslev, I.: Phys.Stat.Sol. 40, 93 (1970) and
 Langer, D.W.; Euwema, R.N.; Era, K.; Koda, T.: Phys.Rev.
 B2, 4005 (1970)
[4] Knell, R.L.; Langer, D.W.: Phys.Lett. 21, 370 (1966)
[5] Bateman, T.B.: J.Appl.Phys. 33, 3309 (1962)
[6] Berkowicz, R.: Private communication

THEORY OF MULTIPHONON ABSORPTION IN SEMICONDUCTING CRYSTALS

SEE-CHEN YING[*†]

Department of Physics, Brown University, Providence, R. I. 02912, USA

BERNARD BENDOW

Solid State Sciences Laboratory, Air Force Cambridge Research
Laboratories (AFSC), Bedford, MA 01730, USA

STANFORD P. YUKON[††]

Parke Mathematical Laboratories, Carlisle, MA 01741, USA

A correlation function approach is employed to cal-
culate the multiphonon absorption in semiconducting
crystals due to higher order electric moments, in
the presence of anharmonicity. We find an exponen-
tial-like behavior for $\omega/\omega_0 >> 1$, with broader absorp-
tion for increasing anharmonicity. Results of nu-
merical computations within a simplified single-
particle model are presented for representative crystals.

1. INTRODUCTION

Recently, the theory of multiphonon infrared absorption at frequencies ω far above the Rehstrahl ω_0 has been treated in a variety of papers [1,2]. The principal mechanism considered in the latter was anharmonic damping of the TO phonon. However, it is well known that multiphonon absorption may proceed as well via higher order, or nonlinear, electric moments [3]. The effect should be especially significant for the more covalent materials, in which linear moments arising from static-charge separation will be rela-tively small. We here first sketch a general correlation function treat-ment [1] of higher order moment absorption, and then turn attention to a simplified single-particle model which is especially amenable to "exact" numerical computations for representative crystals.

2. ABSORPTION DUE TO NONLINEAR MOMENTS

Adopting the bond concept of Flytzanis [4], the crystalline moment is a sum of bond moments, $\underset{\sim}{M} = \Sigma_{ij}\underset{\sim}{m}_{ij}$. The absorption coefficient is proportional to the imaginary part of the dielectric susceptibility χ, which can be ex-pressed as

$$\mathrm{Im}\chi_\omega = \frac{1}{2} <\underset{\sim}{M}(t)\underset{\sim}{M}(0)>_\omega \; [n(\omega) + 1]^{-1} \tag{1}$$

[*]Supported in part by NSF. [†]A.P. Sloan Fellow.
[††]Supported by AFCRL (AFSC) under Contract No. F19628-71-C-0142.

with n the Bose-Einstein function, and the subscript "ω" indicating evalua-
tion in ω-space; $< >$ indicates thermal average. Following the methods
described in [1], we express

$$\langle \underset{\sim}{M}(t)\underset{\sim}{M}(0)\rangle = \sum_{\underset{\sim}{q}_i \underset{\sim}{R}_i s_i} \exp[i\underset{\sim}{q}_1 \cdot (\underset{\sim}{R}_1 - \underset{\sim}{R}_2) + i\underset{\sim}{q}_2 \cdot (\underset{\sim}{R}_3 - \underset{\sim}{R}_4)]$$

$$\times m_{\underset{\sim}{q}_1}^{s_1 s_2} \, m_{\underset{\sim}{q}_2}^{s_3 s_4} Q \qquad (2)$$

$$Q \equiv < e^{\, i\underset{\sim}{q}_1 \cdot [\underset{\sim}{u}(\underset{\sim}{R}_1 s_1 t) - \underset{\sim}{u}(\underset{\sim}{R}_2 s_2 t)]} \, e^{\, i\underset{\sim}{q}_2 \cdot [\underset{\sim}{u}(\underset{\sim}{R}_3 s_3 0) - \underset{\sim}{u}(\underset{\sim}{R}_4 s_4 0)]} >$$

where m_q's are related to the Fourier transform of $m(\underset{\sim}{R})$ by phase factors in-
volving the basis vectors of the atoms s_i; here $\underset{\sim}{u}$'s are displacement vectors
and $\underset{\sim}{R}$'s equilibrium position vectors. One next expresses Q in terms of a
cumulant series defined in schematic form by $\log Q = \Sigma_i c_i <u_1 u_2 \cdots u_n>$, which
can be done in standard fashion [5]. Even when approximations are intro-
duced in the cumulant series, one still obtains contributions to α of all
orders in phonons. An example of a conclusion which follows directly from
Eq. (2) is: In the harmonic approximation the electric moment absorption,
when divided by ω^4, has the same frequency dependence as the anharmonicity
absorption from just linear moments, if only the interatomic potential v is
replaced by the moment m. Various other conclusions, as well as some ex-
plicit expressions for various approximations and choices of models, will be
presented elsewhere [6]. We here just briefly mention an approximation in
which the sums in Eq. (2) are restricted to a single cell, and the cumulant
series is truncated at the quadratic terms. Then suppressing the tensor
notation, and ignoring Debye-Waller corrections, one finds

$$\langle M(t)M(0)\rangle \sim \sum_{Rn} [m^{(n)}(\underset{\sim}{R}_0)]^2 \, D^n/n! \qquad (3)$$

where "(n)" indicates nth derivative and $D \equiv C_{11}(\underset{\sim}{R}t) + C_{22}(\underset{\sim}{R}t) - C_{12}(\underset{\sim}{R}t) -$
$C_{21}(\underset{\sim}{R}t)$, where $C_{ss'}(\underset{\sim}{R}t) \equiv <u(\underset{\sim}{R}st)u(0s'0)>$ is the displacement correlator of
the lattice. It can be shown that for reasonably smooth C_ω, $\alpha(\omega)$ will vary
smoothly at $\omega \gg \omega_0$. In this spirit, it is useful to consider a simple model
correlator, chosen to satisfy appropriate sum rules, such as those given in
Ref. [7]. If the R dependence is ignored, Eq. (3) may be evaluated ex-
plicitly for various choices of C, among them $C_\omega \sim \omega[n(\omega) + 1] \, b^{-3} e^{-\omega^2/b^2}$,
where b is an anharmonicity parameter (larger b implies greater anharmoni-
city). We here present just graphical results for the latter case in the

limit T → ∞, as displayed in Fig. 1.

FIG. 1:

$\mathrm{Log}_{10}\,\alpha$ vs. ω/ω_0 for Gaussian model correlator; $b^2 = 2.0$, 3.0, 5.0 for curves 1, 2, 3 respectively.

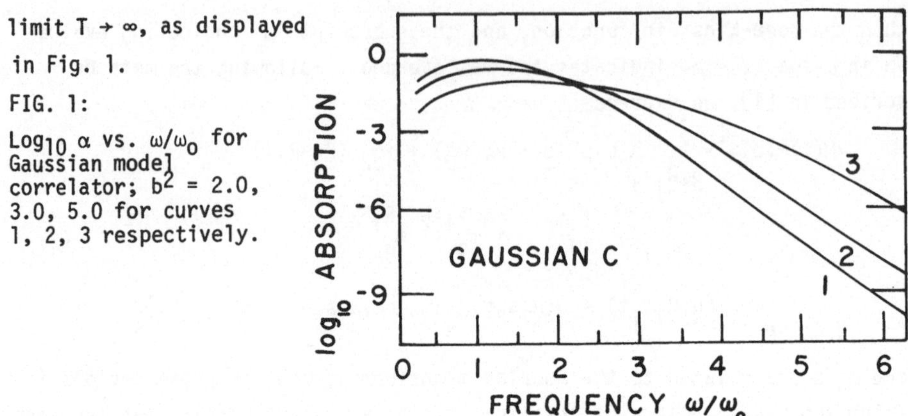

A characteristic exponential-like decrease in α for $\omega \gg \omega_0$ may be noted, as may a broadening of α associated with increasing anharmonicity. These features may be shown to hold for more general models as well.

3. SINGLE PARTICLE MODEL

This model has, in various forms, been utilized extensively [2,8] because of its simplicity and amenability to exact analyses. Although artificial, it has enjoyed reasonable success in the interpretation of experimental data. Briefly stated, the crystal is viewed as a collection of non-interacting cells, each containing a single particle (oscillator) moving in a potential $v(r)$, and possessing electric moment $m(r)$. For a diatomic crystal, e.g., r is the interatomic separation and v the interatomic potential. The spectrum is very simple for $T=0^{\circ}K$, where [8]

$$\mathrm{Im}\chi_\omega \sim \sum_n |<1|m|n>|^2 \delta(\omega-\omega_{n1}) \tag{4}$$

$$(\frac{p^2}{2\mu} + v(r))\,|n> = \omega_n|n>; \quad \omega_{n1} \equiv \omega_n-\omega_1$$

where "1" denotes the ground state. For v's of interest here $\mathrm{Im}\chi_\omega$ is a line spectrum. To obtain a continuous spectrum one may average over a distribution of harmonic frequencies [9]. However, nearly similar results are obtained more simply by just interpolating α between the ω_m's, a procedure which we utilize here. To apply Eq. (4) one requires a choice for m and v. If one takes a Morse v, and the form $m = e_s r + m_0 e^{-ar}$ for the moment, with e_s the static charge, then all parameters may be determined by combining the standard thermodynamic relations [10] with relations between experimental phonon pressure data and m and v, deduced from Ref. [11]. Details

of the method are described elsewhere [12]; here only the final results for α at T=0^0K will be given[†], as illustrated in Fig. 2. Also, exponential fits to calculated α's are listed in Tb. 1.

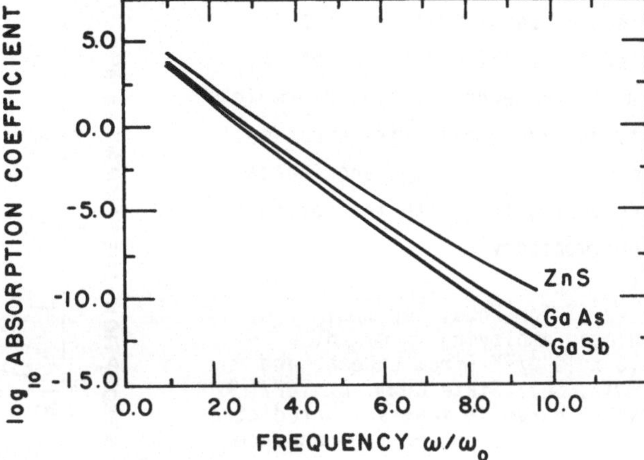

FIG. 2:

Log$_{10}$ α vs. ω/ω_{TO} for typical semiconductors.

TABLE 1: Exponential fits to α for Semiconductors

$$\alpha(cm^{-1}) = (A \times 10^4)\, exp(-\beta\omega/\omega_{TO})$$

Crystal	A	β
GaP	6.6	4.01
GaAs	6.2	4.02
GaSb	4.2	4.20
AlSb	5.8	3.95
ZnS	19	3.70
ZnSe	15	4.01
ZnTe	22	4.30

4. TEMPERATURE DEPENDENCE

As far as explicit temperature dependence, α in the n-phonon regime ($\omega/\omega_0 \sim n$) is approximately proportional to $A(T) \equiv [n(\omega_0) + 1]^n/(n(\omega) + 1)$, so that $\alpha \sim T^{n-1}$ for large T. However, implicit variations occur through the T-dependence of the phonon spectrum and other parameters in α. The major effect may be accounted for by taking [14]

$$\alpha \sim A(T)\, exp[-\beta\omega/\omega_0(T)] \tag{5}$$

where ω_0 is an appropriate average phonon, and β is only weakly dependent on T. Since $\omega_0(T)$ usually decreases with increasing T, the overall T dependence of α will be suppressed. This effect is illustrated for ZnSe in

[†] A controversial problem for the present case is the choice of e_s. For simplicity, we follow Ref. [11] by taking $e_s = 0$, although we note that Refs. [4,13] conclude $e_s \neq 0$.

Fig. 3.

5. CONCLUSIONS

The correlation function method provides a useful framework for investigating multiphonon absorption. Among the general features predicted are an exponential-like dependence for α vs. ω at $\omega \gg \omega_0$, and increasing broadening of α with increasing anharmonicity.

FIG. 3:

α vs. T for ZnSe, for ω/ω_{TO} = 6, calculated employing $\omega_0^{-1} d\omega_0/dT$ = 1.3 x 10^{-4}/°K [from LaCombe and Irwin, Sol. State Comm. **8**, 1427(1970)]; dashed curve indicates $\bar\alpha$ predicted in absence of phonon T-dependence.

6. REFERENCES

[1] Bendow, B.; Ying, S.C.; Yukon, S.P.: Phys. Rev. B**8**, 1679 (1973).

[2] Sparks, M.; Sham, L.J.: Phys. Rev. B**8**, 3037 (1973); Mills,D.L.; Maradudin, A.A.: Phys. Rev. B**8**, 1617 (1973); McGill, T.C.; Hellwarth, R.W.; Mangir, M.; Winston, H.V.: J. Phys. Chem. Sol. **34**, 2105 (1973); Namjoshi, K.V.; Mitra, S.S.: Phys. Rev. B**9**, 815 (1974); Rosenstock, H.B.: Phys. Rev. B**9**, 1973 (1974).

[3] Lax, M.; Burstein, E.: Phys. Rev. **97**, 36 (1955).

[4] Flytzanis, C.: Phys. Rev. B**6**, 1264 (1972).

[5] Maradudin, A.A.: in "Solid State Physics," Seitz, F., and Turnbull, D., eds., Vols. 18 and 19 (Academic Press, N.Y. 1966).

[6] Bendow, B.; Yukon, S.P.; Ying, S.C.: to be published.

[7] Kwok, P.C.: in "Solid State Physics", Seitz, F., and Turnbull, D., eds., Vol. 20 (Academic Press, N.Y. 1967).

[8] Yukon, S.P.; Bendow, B.: Optics Comm. **10**, 53 (1974).

[9] Rosenstock, H.B.: Opt. Cit., Ref. [2].

[10] Born, M.; Huang, K.: "Dynamical Theory of Crystal Lattices" (Oxford U.P., U.K., 1954).

[11] Humphreys, L.B.; Maradudin, A.A.: Phys. Rev. B**6**, 3868 (1972).

[12] Bendow, B.; Gianino, P.D.; Tsay, Y.F.; Mitra, S.S.: Applied Optics (in press).

[13] Lannoo, M.; Decarpigny, J.N.: Phys. Rev. B**8**, 5704 (1973).

[14] Bendow, B.: Appl. Phys. Lett. **23**, 133 (1973).

DETERMINATION OF THE OPTICAL CONSTANTS OF TELLURIUM IN THE FUNDAMENTAL ABSORPTION REGION BY THE EXCITATION OF ELECTRO-MAGNETIC SURFACE WAVES.

J. HUBER, A. OTTO, W. SOHLER

Sektion Physik der Universität München, 8 München 40
Fed. Rep. Germany

The optical constants parallel and perpendicular
to the cristallographic c-axis of trigonal Tellu-
rium have been evaluated from the surface wave
(surface polariton) resonance in attenuated total
reflection spectroscopy (ATR) at $300\,^{\circ}$K. The data
are compared to earlier results from Kramers Kro-
nig analysis of reflectance spectra and to theore-
tical pseudopotential calculations.

1. INTRODUCTION

Optical constants of semiconductors contain information of
the electronic band structure. The pseudopotential method of
band structure calculation allows the theoretical evaluation
of the imaginary part $\varepsilon_2(\omega)$ of the dielectric constant (see
e. g. [1]). The comparison of experimental to theoretical
ε_2-spectra is a check of the consistency of the pseudopoten-
tial method and is useful for the assignement of observed
structures.
For trigonal Tellurium, the most recent experimental data of
Tutihasi et al. [2] and Bammes et al. [3] may be compared to
the theoretical ε_2-spectra of Maschke [4] . In references [2]
and [3] $\varepsilon_2(\omega)$ is evaluated from near normal incidence reflec-
tivity by a Kramers Kronig analysis (KKA). Though both groups
used identical reflectivity spectra in the range of fundamen-
tal absorption between 1 and 3 eV, namely the values of Tuti-
hasi et al. [2] , their ε-values are quite different in this
energy range (see Fig. 3). This is due to different reflecti-
vities at higher energies. Hence the direct evaluation of
$\varepsilon(\omega)$ in the fundamental absorption range is a worthwhile ef-
fort as a check of the KKA data, to allow a definite compari-
son to theory.

In this work $\varepsilon(\omega)$ of Tellurium between 1.2 and 2.6 eV is
evaluated from the surface wave (surface polariton) resonance
in an attenuated total reflection (ATR) experiment. This me-
thod and relevant work has been reviewed recently [5] .

2. EXPERIMENT

Single crystals of trigonal Tellurium were cleaved along
($10\bar{1}0$) planes at 77°K. Only crystals with mirrorlike surfaces
were investigated. In an ATR experiment the sample is separa-
ted from the prism by a dielectric spacer layer of a thickness
of about one wavelength. A cleaved crystal is not optically
flat. Therefore it is not possible to have a uniform spacer
between the sample and a solid prism, which is necessary for
the evaluation of ε by fitting theoretical to experimental
ATR spectra (see Sect. 3). For this reason we replaced the
solid prism by a high index oil (DC 705), which matches
accurately the surface of the crystal at a distance given by
an evaporated cryolite coating. Cryolite has been chosen as
spacer material for two reasons: It has a low index of refrac-
tion and a high packing density, which prevents a soaking of
the oil. The oil was kept in the core of a fused quartz hemi-
cylinder. The thickness of the cryolite film was measured by
multiple beam interferometry. Within the fluid the incident
beam was parallel, as the focusing of the hemicylinder was
cancelled by a cylindric lens. The measured angle of incidence
α is accurate within $\pm3'$. The reflectivity ratio R_{TH}/R_{TE} for
p- (parallel to the plane of incidence) to s-polarized light
was taken versus α for two orientations of the sample: the
optical axis c is parallel respectively perpendicular to the
plane of incidence, but always in the surface of the crystal.
All spectra were taken at room temperature.

3. RESULTS

Fig. 1 displays as an example experimental spectra (bars) of
the reflectivity ratio R_{TH}/R_{TE} versus the angle of incidence
at a wavelength of 5900 Å. The c-axis lies in the plane of in-
cidence (x $\parallel$ c, upper spectrum), respectively perpendicular to

FIG. 1:

Fitting of theoretical ATR-spectra (solid line) to the experimental values (bars). Upper part: Experimental points for x ∥ c, i.e. c in the plane of incidence; $n_\parallel$ and $k_\parallel$ varied in the calculations. Lower part: Experimental points for x ⊥ c, i. e. c perpendicular to the plane of incidence; $n_\perp$ and $k_\perp$ varied in the calculations.

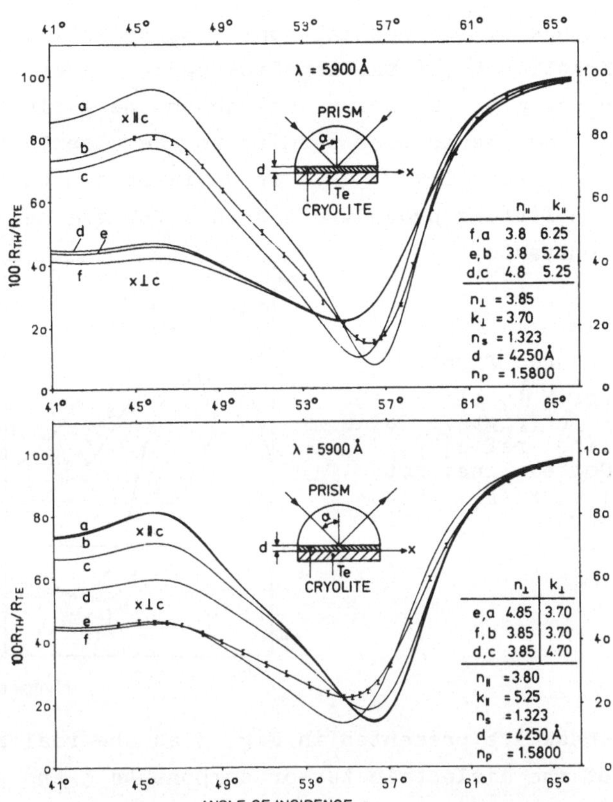

the plane of incidence (x ⊥ c, lower spectrum). Theoretical ATR-spectra, calculated with the reflectivity formulae for un-iaxial crystals presented in [6], are fitted to the experimental values. The x ∥ c (x ⊥ c) spectra are mainly sensitive to the optical constants $n_\parallel + ik_\parallel = \varepsilon_\parallel^{1/2}$ ($n_\perp + ik_\perp = \varepsilon_\perp^{1/2}$). $\varepsilon_\parallel$ and $\varepsilon_\perp$ are the dielectric tensor components parallel and perpendicular to the c-axis. The fitting procedure yields unambigous

FIG. 2:

The index of refraction n_s of the cryolite spacer layer of one sample with a thickness of 4250 Å.

values of the optical constants as well as the index of re-
fraction n_s of the cryolite spacer layer.
n_s is plotted versus the photon energy in Fig. 2. The accuracy
of the measurement permits to see clearly the dispersion of
the spacer layer. This is an important test for the quality of
the fitting procedure and thus for the accuracy of the optical
constants.

FIG. 3:

ε_1 and ε_2 of Te
for E ‖ c (left) and
E ⊥ c (rigth). Solid
line: ref. [2] ;
Dotted line: ref.
[3] ; x: ε_1, this
work; o: ε_2, this
work.

These are presented in Fig. 3 as the real and imaginary part
of the dielectric tensor components ε and ε . The errors de-
pend on the wavelength. This is due to the different strength
of the surface wave resonance in the experimental spectra,
which in turn depends on the thickness of the spacer layer.

FIG. 4:

Normal incidence reflec-
tivity of Te, measured
by Tutihasi et al. [2] (so-
lid line) and Stuke et al.
[7] (dashed-dotted line),
compared to the calcula-
ted values of this work(x).

In Fig. 4 the normal incidence reflectivity, calculated from
our ε_1 and ε_2-values of Fig. 2 is compared to the experimental

reflectivity spectra of Tutihasi et al. [2] and Stuke et al. [7]. The agreement with the result of [2] is good in view of the uncertainty of the reflectance of ±5%, as quoted in [2].

4. DISCUSSION

The agreement of $\varepsilon_{\parallel}$ and $\varepsilon_{\perp}$ with the KKA results of Bammes et al. [3] is very good with the exception of $\varepsilon_{2\parallel}$ (see Fig. 3). Here the present data are somewhat larger. Though the reflection spectra of Tutihasi et al. [2] are well reproduced by the present data (see Fig. 4), their KKA results are not only of different magnitude in ε_2, but also displaced in peak positions (see Fig. 3). The comparison of the results of the two KKA shows the influence of different reflectivities at higher energies on the optical constants in the visible, evaluated by a KKA and therefore the usefulness of an independent measurement.

The structures in ε_2 are caused by transitions between the highest valence band triplett and the lowest conduction band triplett [4]. Because of rather flat bands and strong variations of the oscillator strength within the Brillouin zone, they cannot be assigned to critical points or lines [4]. The peaks in the ε_2-spectra of Maschke [4] are displaced by 0.3-0.55 eV to lower energies compared to the data of this work.

5. REFERENCES

[1] Sandrock, R.; Festkörperprobleme X, p.283 (1970), O. Madelung (editor), Vieweg-Verlag

[2] Tutihasi, S.; Roberts, G.G.; Keezer, R.C.; Drews, R.E.: Phys. Rev. 177, 1143 (1969)

[3] Bammes, P.; Klucker, R.; Koch, E.E.; Tuomi, T.: phys.stat. sol. (b) 49, 561 (1972)

[4] Maschke, K.: phys.stat.sol. (b) 47, 511 (1971)

[5] Otto, A.: Festkörperprobleme XIV, (1974), H.J. Queisser (editor), Vieweg-Verlag

[6] Sohler, W.: Opt. Comm. 10, 203 (1974)

[7] Stuke, J.; Keller, H.: phys. stat. sol. (b) 7, 189 (1974)

OPTICAL ABSORPTION EDGE AND LUMINESCENCE OF PbI_2

F. LEVY, C. DEPEURSINGE, LE CHI THANH, A. MERCIER

E. MOOSER, J.P. VOITCHOVSKY

Laboratoire de Physique Appliquée, EPF - Lausanne

Switzerland

In the reflection spectrum of PbI_2, a new line
which can be attributed to the $n = 3$ exciton
state has been observed, indicating that the
exciton series in PbI_2 is normal and obeys a
Wannier Law with a groundstate ionization energy
of 30 meV.
A fluorescence peak of 4 meV above the exciton
ground state has been detected polarized with
E ‖ c which is probably due to the extraordinary
polariton mode.
Many spectra revealed the polytypic nature of
most PbI_2 samples.

1. INTRODUCTION

The exciton near the fundamental absorption edge of the layered semi-
conductor PbI_2 has been intensively investigated, [1-3]. Many inter-
pretations of the exciton series have been given [4,5] and new
features appearing in the absorption, reflectivity [6] and fluorescence [7]
spectra have been considered as characteristic of anisotropic crystals.
Nikitine [1] and Harbeke [2] described the exciton optical transitions
as a Wannier exciton series with an ionization energy of about 130 meV but
in which the ground state $n = 1$ does not fit the series. The ground state
anomaly was explained by a repulsive central cell correction originating
from the cationic nature of the valence and conduction band wave functions
[5]. In contrast with this, Baldini and Franchi [8] have attributed
the absorption peaks to overlapping normal exciton series without shift
of the $n = 1$ line. The corresponding ground state ionization energy of
about 60 meV is in better agreement with those of similar semiconductors.
Kleim and Raga [3] determined the exciton luminescence spectrum at
Helium temperature which is characterized by a strong emission line
attributed to bound exciton recombination. Phonon assisted transitions
were also observed. The rich impurity fluorescence spectrum has also been

intensively investigated [9] .

We present in this communication new absorption, reflectivity and photo-
luminescence measurements which have been carried out on high quality
single crystals grown in our laboratory. The results include:
- a series of new lines in the reflectivity spectrum as well as spectral
evidence for the complex polytypic structure of PbI_2,
- a series of new polarized emission lines and their interpretation in
terms of the model of the extraordinary polariton mode of uniaxial
crystals [6] .

2. ABSORPTION AND REFLECTION SPECTRA

Absorption measurements have been carried out with single crystal flakes,
a few microns thick, which were grown in temperature gradients of 400/500 -
300 $^{\circ}$C under an Argon pressure of a few torrs during three weeks. As
observed before [10] , the sublimated crystals are mostly of rhombohedral
symmetry (6R or 12R?). The absorption and reflection spectra at 1.5 $^{\circ}$K
typical of sublimated crystals are reported in Fig. 1

Fig. 1

Absorption and reflection spectra of
a sublimated single crystal, observed
at oblique incidence with E parallel
with the plane of incidence. These
spectra are typical for crystals of
rhombohedral symmetry.

A sharp absorption peak at 4954 Å corresponding to a dip in the
reflectivity is related to the bound exciton transition and disappears in
both spectra above about 20°K. The large value (20-25Å) of the halfwidth
of the absorption peak at 4940 Å is attributed to stacking fault broadening
[11] . This peak corresponds to the exciton ground state. The structure

at 4930 Å in the reflection spectrum appears at oblique incidence with a component of polarization parallel to the c axis of the crystal (R_p geometry). It corresponds to a window in the absorption spectrum and was attributed by Harbeke et al [6] to the extraordinary polariton mode.

Besides the already known exciton lines, a new reflectivity peak has been observed at about 4878 - 4885 Å. The same peak has also been detected in crystals of the other polytypes. The wavelength of all peaks are listed in Table 1, where the ionization energy R is given as well. Reflection measurements revealed the polytypic nature of many samples.

Table 1 : Exciton series in PbI_2

polytype	n=1 ⊥	n=1 ‖	n=2	n=3	∞	R
6R , 12R	4942Å	4929	4895	4885	4858	30 meV
2H	4962	4954	4918	4907	4886	30 meV

3. FLUORESCENCE SPECTRA

Fluorescence spectra have been investigated with large samples cleaved from Stockbarger crystals of different orientations. The emission spectrum in the bandgap region consists of:
- sharp luminescence lines due to exciton recombination,
- broad emission bands about 120 Å on the long wavelength side, which are related to crystal defects, probably to lead impurity centers.

The band edge fluorescence spectra are represented in Fig. 2, observed with oblique incidence and two different polarizations. The strongest emission peak at 4973 Å corresponds to the bound exciton recombination in PbI_2 -2H. The position of the peak is characteristic of this polytype and corresponds to the 4954 Å peak of the rhombohedral PbI_2 crystal [3,7] .

The peak at 4963 Å coincides exactly with the free exciton absorption and reflection lines. This coincidence as well as the line shape leads us to attribute this emission peak to free exciton recombination. The weak peak at 4953 Å is polarized with E ‖ c. The emitted photon originates from the intermediate branch of the mixed extraordinary polariton mode typical of this uniaxial crystal.

Fig. 2.

Fluorescence spectra of $2H-PbI_2$ observed at oblique incidence. The line at 4953 Å polarized with E ‖ c originates from the intermediate branch of the mixed extraordinary polariton mode.

On the long wavelength side of the strongest line, two phonon assisted transitions are observed at 4986 Å and 4997 Å. The intensity of the broad line at 5015 Å depends on the sample history, suggesting that the line originates from transitions into impurity states. A similar spectrum shifted by about 20 Å towards shorter wavelengths has been observed in crystals of rhombohedral symmetry.

Sensitive measurements of a rhombohedral crystal have revealed two further weak broad fluorescence lines at about 4897 and 4879 Å. These values correspond to the n = 2 and n = 3 exciton states determined in the reflectivity spectrum.

4. CONCLUSION

The measurement of a new n = 3 line in the reflectivity spectra of various PbI_2 single crystals and the occurrence of different polytypes in many samples, especially in crystals of rhombohedral symmetry, indicate that the exciton series in PbI_2 is <u>normal</u>. It obeys the Wannier law $E_n = E_g - R/n^2$, with a ground state ionization energy R = 30 meV.

In the sharp band edge emission spectrum, a new line polarized with E ‖ c has been observed. This line appears 4 meV above the intrinsic exciton line and is only observed at oblique incidence. By comparison with reflectivity measurements this peak can be interpreted in terms of the model of the extraordinary polariton mode for anisotropic crystals.

5. REFERENCES

[1] Nikitine, S.; Schmitt-Burckel, J.; Biellmann, J.; Ringeissen, J.:
 J. Phys. Chem. Solids 25, 951 (1964).
[2] Gähwiller, Ch.; Harbeke, G.: Phys. Rev. 185, 1141 (1969).
[3] Kleim, R.; Raga, F.: J. Phys. Chem Solids 30, 2213 (1969).
[4] Tubbs, M. R.: Phys. Stat. Sol. (b) 49, 11 (1972).
[5] Harbeke, G.; Tosatti, E.: Phys. Rev. Letters 28, 1567 (1972).
[6] Harbeke, G.; Bassani, F.; Tosatti, E.: Proc. 11th. Int. Conf. Phys.
 Semicond. Warsaw (1972), p. 163.
[7] Lévy, F.; Mercier, A.; Voitchovsky, J. P.: Solid State Commun.(1974).
[8] Baldini, G.; Franchi, S.: Phys. Rev. Letters 26, 503 (1971).
[9] Dugan, A. E.; Henisch, H. K.: Phys. Rev. 171, 1047 (1968).
[10] Hanoka, J. I.; Vand, V.: J. App, Phys. 39, 5288 (1968).
[11] Mooser, E.; Schlüter, M.: Nuovo Cimento 18 B, 164 (1973).

LINEAR AND NONLINEAR OPTICAL RESPONSE OF FREE HOLES IN TELLURIUM

Wlodek ZAWADZKI[*]

Groupe de Physique des Solides de l'Ecole Normale Supérieure,
24 rue Lhomond, 75231 Paris 05

Nonlinear and linear optical response of mobile holes
in the "camel-back" shaped valence band of tellurium
are calculated without expansion in powers of electric
field of the wave. We predict unusual nonlinear proper-
ties depending on intensity, frequency and polarization
of the driving field, as well as on the free hole con-
centration and hydrostatic pressure.

Optical nonlinearities due to mobile carriers in nonparabolic energy bands
of III-V semiconducting compounds are known to be among the strongest
ever observed. All theories of the effect[1-7] use an expansion of non-
linear current in powers of the driving electric field. Here we consider
nonlinear and linear optical response of free holes in the "camel-back"
shaped upper valence band of tellurium and demonstrate that, in this case,
one has to go beyond the usual expansion. This results in a number of
new possibilities. Our treatment is in the spirit of that used by
Keldysh[8] to calculate multi-photon absorption and tunneling in the
field of a strong electromagnetic wave. The upper valence band of tellur-
ium is described by a dispersion relation proposed by Betbeder-Matibet
and Hulin[9]

$$\varepsilon(\vec{k}) = -\alpha k_z^2 + (A^2 + B^2 k_z^2)^{1/2} - \hbar^2 k_\perp^2/2m - C \qquad (1)$$

where $\vec{k}$ is counted from the corner H of the Brillouin zone. The band para-
meters have been determined from the cyclotron resonance experiments by
Couder et al.[10]. Eq.(1) describes a saddle point at k = 0 and a "camel-
back" shape along k_z direction, with two maxima at $\pm k_z^m = 1.98 \times 10^6 \mathrm{cm}^{-1}$.
The longitudinal mass at the maxima is $m_{//}(\pm k_z^m) = 0.22\ m_o$. The zero of
energy is chosen at the band maxima and $\Delta = A-C = -2.18$ meV represents
the energy of the saddle point. Our purpose is to find optical response
of free holes induced by the electric component of a monochromatic in-
cident beam polarized along the z direction : $E_z(t) = E_o \sin \omega t$. In the
region of low losses $\omega\tau \gg 1$, and long wavelengths $\hbar\omega \ll \varepsilon_g$, the equa-
tion of motion, $d(\hbar k_z)/dt = eE_o \sin \omega t$, has the solution $k_z(t) = -K \cos \omega t +$
$+ k_{zo}$ where $K = eE_o/\hbar\omega$ represents the amplitude of oscillations and k_{zo}

[*] On leave from the Institute of Physics, Polish Academy of Sciences,
Warsaw, Poland.

the initial value. The usual expansion of nonlinear current in powers of electric field corresponds to an expansion of velocity in powers of K, which is in turn equivalent to the perturbation approach in quantum mechanics. In the case of tellurium, however, one can reach the values of K comparable to A/B and k_z^m, for which the above expansion is no longer valid, and one must calculate the current to all orders of K. The current induced in one band by a low frequency electric field of any magnitude is given by : $\vec{j}(t) = e \, \Sigma \, \vec{v}(t,\vec{k}_o) f(\vec{k}_o)$, where $f(\vec{k}_o)$ is the unperturbed distribution function. A small term due to distorsion of the Bloch bands by the electromagnetic field has been neglected[11,12]. The velocity is : $v_z(\vec{k}) = (1/\hbar)(\partial\varepsilon/\partial k_z)$, and upon using Eq.(1) one obtains $\vec{v}_z(t,K,\vec{k}_o)$. In III-V compounds the initial $\vec{k}_o$ values occur around k = 0, whereas in tellurium the holes are initially located in two pockets around $\vec{k} = (\pm k_z^m \, ; \, k_\perp = 0)$. When integrating over the initial values, the velocity $v(t,K,\vec{k}_o)$ is expanded around the points $\pm k_z^m$ to the quadratic terms in $\delta k_z = k_{zo} - k_z^m$. For hole concentrations $p < 2\times10^{16} cm^{-3}$ the Fermi energy is small compared to Δ and one can use ellipsoidal and parabolic $\varepsilon(\vec{\delta k})$ dependence in the vicinity of the maxima. Finally, in order to find exactly the linear and nonlinear response, the current is expanded in the Fourier series :

$$j_z(t,K) = j_1(K)\cos \omega t + j_3(K)\cos 3\omega t + \ldots \qquad (2)$$

which is not equivalent to the expansion in powers of K. For strongly degenerate hole gas one obtains :

$$j_1(K) = (4B/\hbar) \, pR_1 + (4\alpha/\hbar) \, pK \qquad (3)$$

$$j_n(K) = (4B/\hbar) \, pR_n \qquad (n = 3,5, \ldots.) \qquad (4)$$

where $R_n(K) = - \left[I_n - \dfrac{3}{10} \left(\dfrac{3\pi^2}{2} \right)^2 \dfrac{m_{//}}{m_{ds}} \dfrac{B^2}{A^2} p^{2/3} J_n \right] \qquad (5)$

for n = 1,3,5,.... with $m_{ds} = (m_{//} m_\perp^2)^{1/3}$, while I_n and J_n denote the Fourier coefficients :

$$I_n(K) = \frac{1}{\pi} \int_o^\pi \frac{(k_z'^{\,m}+K'\cos x)\cos nx \, dx}{\left[1+(k_z'^{\,m}+K'\cos x)^2\right]^{1/2}} \, , \quad J_n(K) = \frac{1}{\pi} \int_o^\pi \frac{(k_z^{\,m}+K'\cos x)\cos nx \, dx}{\left[1+(k_z'^{\,m}+K'\cos x)^2\right]^{5/2}} \qquad (6)$$

where $k_z'^{\,m} = (B/A)k_z^m$ and $K' = (B/A)K$.

Fig.1 shows the dimensionless nonlinear response R_3 and R_5 calculated for different free-hole densities. The standard expansion in powers of K breaks down completely for $K > 1 \times 10^6 cm^{-1}$, the third harmonics current

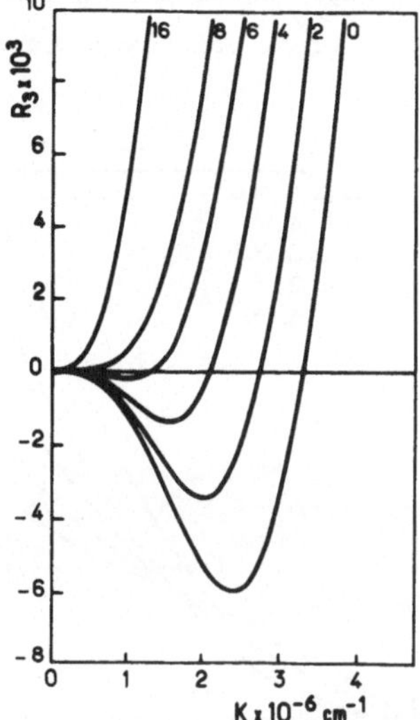

FIG. 1 : Nonlinear optical response of free holes in Te at third and fifth harmonics frequency _versus_ K = eE$_0$/ℏω, for different free-hole densities.

FIG. 2 : Nonlinear optical response of free holes (p=10^{15}cm^{-3}) in Te at third harmonics frequency _versus_ K=eE$_0$/ℏω, for different hydrostatic pressures (in kilobars).

having a maximum and changing sign. The negative anomaly decreases with increasing hole concentration. One can predict that for concentrations $p > 6\times10^{16}$cm^{-3} the response R_3 will be positive at all K values (this region has not been described since the parabolic $\varepsilon(\vec{\delta k})$ dependence ceases to be valid). Thus, at small K values the sign of j_3 should change with increasing hole concentration. It has been demonstrated by Anzin et al.[13] that hydrostatic pressure strongly affects the form of the upper valence band of tellurium. The local energy maxima become smaller, they occur at lower k_z^m values and finally, at P ≈ 14 kbar, the camel-back shape disappears. On the other hand, the hole density is almost pressure independent. Using the pressure coefficients determined in ref.14, the nonlinear optical response at 3ω has been calculated for the low hole density of $p = 10^{15}$cm^{-3} (which allows us to neglect the second term in Eq.(5)). The results for R_3 at different pressures are shown in Fig.2. The standard K^3 dependence breaks down at smaller and smaller K values as the pressure increases. At higher pressures R_3 is positive for all K values, so that at

small K the nonlinear current j_3 changes sign as a function of pressure. Finally, Fig.3 shows the optical response at the driving frequency ω as a function of K, calculated on the basis of Eq.(3) for different pressures.

FIG.3 : Linear optical response of free holes ($p=10^{15}cm^{-3}$) in Te versus $K = eE_0/\hbar\omega$, for different hydrostatic pressures.

The linear K-dependence is observed only at low K values. At the pressure for which the saddle point disappears, the quadratic k_z terms in Eq.(1) cancel out and one deals with a $\varepsilon(k_z)$ relation beginning with k_z^4. The linear current is then close to zero for small K and one deals with a medium in which the nonlinear response is almost comparable to the linear one. At higher K values, the linear response is highly nonlinear in K, which should result in strong self-focusing properties.

The HCN and water vapor lasers working in the pulse regime should provide adequate sources to test the above predictions.

REFERENCES

[1] Butcher, P.N. and McLean T.P.: Proc. Phys. Soc., 81, 219 (1963).

[2] Lax, B.; McWhorter, A.L.; Mavroides, J.G. : in Quantum Electronics, Proc. of the Third Intern. Congress, Paris 1964, ed. P. Grivet and N. Bloembergen (Columbia University Press, New York, 1964),p. 1521.

[3] Patel, C.K.N.; Slusher, R.E.; Fleury,P.A.: Phys. Rev. Letters, 17, 1011 (1966).

[4] Wolff, P.A.; Pearson, G.A.: Phys. Rev. Letters, 17, 1015 (1966).

[5] Lax, B; Zawadzki, W.; Weiler, M.H.: Phys. Rev. Letters, 18, 462 (1967).

[6] Jha, S.; Bloembergen, N.: Phys. Rev., 171, 891 (1968).

1246

[7] Wang, C.C.; Ressler, N.W. : Phys. Rev., <u>188</u>, 1291 (1969).

[8] Keldysh, L.V. : J. Exp. Teor. Fiziki, <u>47</u>, 1945 (1964); Engl. transl. Sov. Phys. JETP <u>20</u>, 1307 (1965).

[9] Betbeder-Matibet, O.; Hulin, M. : Phys. Stat. Solidi, <u>36</u>,573 (1969).

[10] Couder, Y.; Hulin, M.; Thomé, H.: Phys. Rev., <u>B7</u>, 4373 (1973).

[11] Butcher, P.N.; McLean, T.P. : in Quantum Electronics, Proc. of the Third Intern. Congress, Paris 1964, ed. P. Grivet and N.Bloembergen (Columbia University Press, New York, 1964), p. 1619.

[12] Genkin, V.N.; Mednis, P.M. : Soviet Physics JETP, <u>27</u>, 609 (1968).

[13] Anzin, V.B.; Bresler, M.S.; Farbstein, I.I.; Itskevich, E.S.; Kosichkin, Yu.V. ; Sukhoparov, V.A.; Telepnev, A.S.; Veselago, V.G.: Phys. Stat. Solidi (b), <u>48</u>, 531 (1971).

THEORY OF PHOTOELASTICITY OF SEMICONDUCTING CRYSTALS

BERNARD BENDOW and PETER D. GIANINO

Solid State Sciences Laboratory, Air Force Cambridge
Research Laboratories (AFSC), Bedford, MA 01730, USA

YET-FULL TSAY[†] and SHASHANKA S. MITRA[†]

Department of Electrical Engineering, University of Rhode
Island, Kingston, R. I. 02881, USA

A simplified extension of the empirical pseudo-
potential method is developed for calculation of
the electronic contribution to the photoelasticity
of semiconductors. Lattice contributions are cal-
culated from the theory of Humphreys and Maradudin,
by employing experimental phonon pressure dependence
data to determine the microscopic interatomic poten-
tial and dipole moment of the crystal.

1. INTRODUCTION

Photoelasticity, the change in refractive index induced by pressure or stress,
has long been a phenomenon of interest in the physics of solids [1,2]. How-
ever, microscopic, first-principles theories of photoelasticity have for the
most part been lacking, with the notable exception of the recent work of
Humphreys and Maradudin (HM) [3], who calculated the lattice contribution for
cubic diatomic crystals. We here investigate the electronic contribution for
semiconductors; also, we apply the HM theory to predict the infrared fre-
quency dependence of the elastooptic constants . We restrict attention to
zincblende semiconductors, for which we will be concerned with dn/dP, the
hydrostatic pressure derivative of the refractive index, and the three non-
vanishing elastooptic constants p_{11}, p_{12} and p_{44}. For these crystals one
also has the relation [2]

$$n^3 K(p_{11} + 2p_{12}) = 6 \ dn/dP, \tag{1}$$

where K is the isothermal compressibility.

2. ELECTRONIC CONTRIBUTION

Below the band gap, the electronic contribution to n^2 is well represented by
$n^2 = 1 + \omega_p^2/\omega_g^2$, where ω_p and ω_g are effective plasma and band gap frequencies,
respectively[4]. If ω_p is assumed to depend on P only through the density,
then

$$n^{-1}dn/dP = 1/2(1 - n^{-2})(K - 2\omega_g^{-1}d\omega_g/dP), \tag{2}$$

so determination of dn/dP reduces to determination of $d\omega_g/dP$.

[†]Supported in part by AFCRL (AFSC) under Contract No. F19628-72-C-0286

A variety of methods have been applied to the calculation of the pressure dependence of band gaps, among them variations of the OPW[5,6] and pseudopotential [7,8] approaches. In general, existing calculations are found to be either unnecessarily cumbersome, or to require fitting of pressure coefficients to experiments, vitiating their usefulness as a predictive technique. We here introduce a simpler scheme in which just atmospheric values of empirical pseudopotential form factors (PFF) and K are required as input. Namely, we assume that the principal effect of pressure on the band structure is accounted for by allowing the lattice constant a to vary with pressure, $a = a(P)$; the latter is deduced employing atmospheric values of K. PFF's appropriate for small changes in a are calculated within the rigid-ion approximation [6], starting from atmospheric values of Cohen and Bergstresser[9]. Application of the procedure requires knowledge of the pressure dependence of the dielectric function $\varepsilon(q)$, and determination of PFF's for reciprocal lattice vectors G at arbitrary P. The former, which involves $d\omega_g/dP$ itself, may be carried out self-consistently; the latter may be inferred from a polynominal interpolation scheme. These and other considerations will be discussed in detail elsewhere [10]. We here restrict ourselves to a presentation and discussion of just the final calculated results for $d\omega_g/dP$ and dn/dP.

Table 1 lists the pressure coefficients of several important band gaps of Si, Ge and the II-VI and the III-V semiconductors. The agreement of calculated coefficients with experiments is uniformly good for most of the gaps, with the possible exception of the indirect gap, $E_{\Gamma X}$. The calculated pressure coefficients are generally within 25% or so of experimental values. The discrepancy between calculated and experimental values of $dE_{\Gamma X}/dP$ may be due to its extreme sensitivity to small changes in PFF's. Since all the calculated $dE_{\Gamma X}/dP$ are rather small in magnitude, slight modification of the PFF's might effectively reverse the sign of the calculated $dE_{\Gamma X}/dP$. This is manifested in the case of GaP, where a negative $dE_{\Gamma X}/dP$ is predicted, in agreement with experiment. Also in Si, we found the indirect gap $E_{\Gamma X}$ first increases with P, then decreases, so that $E_{\Gamma X}$ at 10kbar is larger than at 20kbar, showing non-linearity of $E_{\Gamma X}$ as a function of pressure.

Previous high-pressure band structure calculations, except those by Melz [8], were concerned with IV-IV's such as Si and Ge. Except for the indirect gap $E_{\Gamma X}$, our calculated pressure coefficients for other important band gaps are at least in as good agreement with experimental data as those obtained from more complicated models. For III-V's our results are definitely better than Melz's. Our results are particularly gratifying in that only the

compressibility and normal pressure PFF's are used as input data.

If one carries out calculations for II-VI's in the same manner as for the IV-IV's and III-V's, one obtains results in poor agreement with experiment. For example, the calculated dE_0/dP are much too small. This seems to stem from overly strong pressure dependences of the anti-symmetric PFF's.

TABLE 1: Pressure Coefficients of Important Band Gaps (in 10^{-6}eV bar^{-1})

	dE_0/dP		dE_1/dP		dE_X/dP		$dE_{\Gamma L}/dP$		$dE_{\Gamma X}/dP$	
	Calc.	Expt.†	Calc.	Expt.†	Calc.	Expt.†	Calc.	Expt.†	Calc.	Expt.†
Si	1.3	*1+1	6.6	*6.2+0.4	3.6	3	5.5	-	0.5~-0.1	-1.5
Ge	16.2	1̄4.2	8.8	*7.8∓0.4	5.4	5.5	6.6	5.	2.7	-1.5
GaAs	13.3	10.7-11.7	7.4	5.0̄	4.6	-	6.2	-	1.5	-
GaSb	16.1	14.7	10.	7.5	6.5	6	8.5	5	3.1	-
GaP	8.6	10.5+1.6	3.5	5.8	1.8	-	2.1	-	-1.7	-1.1
InAs	11.1	9.6-10.8	6.3	7.0	3.5	-	4.8	-	-0.02	-
InP	13.4	8.5	7.5	-	4.6	-	6.8	-	1.8	-
InSb	15.2	15.5-17.6	9.5	8.5	5.9	6	8.3	-	2.7	-
A𝓁Sb	14.7	10+2	7.5	-	4.0	-	6.4	-	0.5	-1.5
ZnTe	8.1	7	4.0	++6.0	1.3	-	3.0	-	-1.6	-
ZnSe	5.8	7	2.5	-	-0.3	-	1.4	-	-3.7	-2
ZnS	3.6	6.5+1	0.86	-	-1.9	-	-0.1	-	-5.2	-
CdTe	2.8	8	1.5	++6.0	-1.0	-	0.5	-	-4.3	-

For this reason, results in Tb. 1 are based on calculations in which the anti-symmetric PFF's are assumed totally independent of pressure. Except for CdTe, the E_0 gap pressure coefficients are in fair agreement with available experimental data. Although not rigorously justified, the present procedure is motivated by Van Vechten's [11] conclusion that the heteropolar (ionic) gap C (which he takes as an appropriate average of the anti-symmetric PFF's) obeys dC/dP→0 in the perfectly ionic limit. Because the II-VI's possess relatively high ionicities, we have chosen to neglect entirely the P-dependence of their anti-symmetric PFF's.

For the effective gap in the calculation of dn/dP, we choose the point $k = 2\pi/a(\frac{1}{2},\frac{1}{2},0)$, as suggested by various recent works [12]. The results for $d\omega_g/dP$ and dn/dP are listed in Tb. 2, along with the predictions of Camphausen et al (CCP) [13] (deduced from an extension of Van Vecthen's theory [11]) and various experimental values. Our results are similar to CCP for diamond and III-V crystals, except for GaP, but differ substantially for II-VI's. Agreement with experiment is good for Si and Ge, fair for GaAs and poor for ZnS. The trend to increasingly positive values of dn/dP for the II-VI's, which are more highly ionic, appears reasonable. Discrepancy of our

†All values cited by CCP (see Text) except for those indicated by (*)[Schmidt and Vedam, Sol. St. Comm. 9, 1187 (1971)] and (††)[Langer, in "Proc. 7th Conf. Physics of Semiconductors" (Dunod, Paris, 1964)].

results with experiment may be a reflection of uncertainties in the calcu-
lated $d\omega_g/dP$'s, although adequate experimental data is lacking for a more
careful comparison.

TABLE 2: dn/dP of Semiconductors

Crystal	$\dfrac{d\omega_g}{dP}$ (10^{-6}eV/bar)	$\dfrac{1}{n}\dfrac{dn}{dP}$ $(10^{-6}/\text{bar})$		
		This work	CCP	Experiment (values cited by CCP)
Si	4.1	-0.31	-0.3+0.05	-0.3
Ge	7.0	-0.89	-1.0$\mp$0.2	-0.7 to -1.0
GaSb	9.1	-1.23	-0.8+0.2	-
GaAs	5.9	-0.44	-0.5$\mp$0.2	-0.7+0.1
InSb	9.2	-1.30	-1.1$\mp$0.2	-
InAs	5.9	-0.39	-0.7$\mp$0.2	-
GaP	2.0	+0.20	-0.3$\mp$0.2	-
AℓSb	5.7	-0.33	-0.5$\mp$0.2	-
InP	6.2	-0.45	-0.4$\mp$0.2	-
ZnS	1.9	+0.36	0.05+0.1	-0.1
ZnTe	4.9	+0.14	0.01$\mp$0.1	-
ZnSe	3.5	+0.32	0.07$\mp$0.1	-
CdTe	3.8	+0.53	0.1$\pm$0.1	-

The application of the above method to the calculation of band structure
under stress and the corresponding elastooptic constants will be given else-
where. We here indicate briefly some of the considerations involved: A
stressed crystal may be regarded structurally as a new, independent one, with
a lower symmetry. The new configuration depends on the applied stress, as
well as on internal strain parameters, for which a variety of information is
available. The PFF's may again be obtained by extrapolation from the un-
strained crystal values, within the rigid-ion approximation. When calcula-
ting the elastooptic constants it will, in general, be necessary to include
transitions other than that represented by the isotropic oscillator at
$[\tfrac{1}{2},\tfrac{1}{2},0]$ employed for the hydrostatic case. One expects that under uniaxial
strain regions of the zone of highest symmetry will be perturbed most. A
reasonable choice for the oscillators to be summed over in calculating the
p_{ij}'s will be those corresponding to Γ, Δ and X points, in addition to that at
$[\tfrac{1}{2}, \tfrac{1}{2}, 0]$.

3. LATTICE CONTRIBUTION

The lattice contribution to photoelasticity will be significant only for IR-
active crystals. If the susceptibility arises from a single TO mode ω_0, then

$$\frac{1}{n}\frac{d}{dx}\frac{1}{n^2} = -\frac{1}{n^4}[\alpha(1-\omega^2/\omega_0^2)^{-1} + \beta(1-\omega^2/\omega_0^2)^{-2} +\gamma], \quad (3)$$

where β and α are proportional to the derivative with respect to x of ω_0, and

of its associated oscillator strength, respectively; $\gamma \equiv d\varepsilon_e/dx$ is the contribution from electronic processes. If x is chosen as stress, then Eq. 3 in fact defines the elastooptic constants p_{ij}. The theory of HM provides expressions for α and β for zincblende semiconductors, which require the interatomic potential $v(r)$ and the dipole moment $m(r)$. Unfortunately, these are not well known for semiconductors; we choose $v(r) = v_0$ $(\exp[-2\S(r-r_0)] -2\exp[-\S(r-r_0)])$, and $m(r) = e_s r + m_0 e r_0 \exp[-2\S_1(r-r_0)]$. Here e_s is the static charge and r_0 the equilibrium interatomic separation. If e_s is known, then just $\S$, $\S_1$, v_0 and m_0 remain to be determined. Two conditions are: the standard thermodynamic relation between $v''(r_0)$, ω_0 and K; and the relation between the transverse charge and the moment [3], which for our case implies $e_T^* = e_s + 8m_0 e(1-\S_1 r_0)/3$. We have obtained additional conditions by relating phonon pressure dependence data to m and v via Eq. 1. These take the form $de_T^*/dP = -8m_0 eK(2\S_1^2 r_0 -2\S_1 r_0 -1)/9$, and $\gamma_{TO} = 8\sqrt{3}r_0(3\S r_0 -2)/9\mu\omega_0^2 K$, where γ is the mode Gruneisen parameter. We follow HM in taking $e_s = 0$; then application of the above described procedure yields the calculated values of α and β listed in Tb. 3.

TABLE 3: Elastooptic Parameters of Semiconductors

Crystal	α_{11}	β_{11}	γ_{11}	α_{12}	β_{12}	γ_{12}	α_{44}	β_{44}	γ_{44}
ZnS	71.9	5.9	-2.3	-33.5	16.9	0.2	107.6	-20.3	-1.8
ZnSe	67.3	4.2	-	-31.4	13.3	-	85.6	-12.9	-
GaP	15.0	1.0	10.8	-11.9	4.9	5.9	16.5	- 2.8	5.3
GaAs	27.9	2.3	19.6	-16.3	6.7	16.6	25.3	- 3.0	8.6

4. REFERENCES

[1] Born, M; Wolf , E: "Principles of Optics" (Macmillan, N.Y., 1964).

[2] Nye, J.F.: "Physical Properties of Crystals" (Oxford U.P.,U.K.,1964).

[3] Humphreys, L.B., Maradudin,A.A.: Phys. Rev. B6, 3868 (1972).

[4] Wemple, S.H; DiDomenico,J.: Phys. Rev. B3, 1338 (1971).

[5] Herman, F.; Kortum, R.; Kuglin, C.D.; Short, R.A.: in "Quantum Theory of Atoms, Molecules and Solids," Lowdin, P., ed. (Academic, N.Y.,1966).

[6] Goroff, I.; Kleinman,L.: Phys. Rev. 132, 1524 (1963).

[7] Brust, D.; Liu, L.: Phys. Rev. 154, 647 (1967).

[8] Melz, P.J.: J. Phys. Chem. Sol. 28, 1441 (1967).

[9] Cohen, M.L.; Bergstresser, T.K.: Phys. Rev. 141, 789 (1966).

[10] Tsay, Y.F.; Mitra,S.S.; Bendow,B.: Phys. Rev. B, to be published.

[11] Van Vechten, J.A.: Phys. Rev. 182, 891 (1969).

[12] Cohen, M.L.; Heine, V.: in "Solid State Physics," Vol. 24, Seitz,F.,and Turnbull, D., eds. (Academic, N.Y., 1970); Welkowsky, M.,; Braunstein, R.: Phys. Rev. B5, 497 (1971).

[13] Camphausen, D.L.; Connell, G.; Paul, W.: Phys. Rev. Lett. 26, 184 (1971).

ANISOTROPIC PHOTON DRAG IN p-TYPE TELLURIUM

J. AUTH, D. GENZOW, K. H. HERRMANN, M. WENDT

Sektion Physik der Humboldt-Universität zu Berlin
German Democratic Republic

The dragging of free holes in tellurium by 10.6
μm photons has been observed. The photon-drag
effect in Te is highly anisotropic due to an-
isotropic absorption and anisotropic band
structure. The drag effect connected with in-
tervalence band transitions of free holes is
dominating in the longitudinal effect as pre-
dicted by a simple theory.

1. INTRODUCTION

In thermal equilibrium the distribution of free carriers in
$\vec{k}$-space has the symmetry of the $E(\vec{k})$-function. If photons with
momentum $\hbar\vec{k}$ are being absorbed this symmetry is destroied. That
results in the photon drag current. At first it was observed
in p-type germanium [1] (recent results [2]). It is a matter of
interest to investigate this effect in p-type tellurium at
least for two reasons: (i) the relatively low symmetry of the
tellurium crystals; and (ii) the special case of free carrier
absorption due to allowed subvalence band transitions between
bands with nearly the same effective masses in $k_\perp$-direction
($\vec{k}\perp c$, c - trigonal axis). Both these facts make remarkable dif-
ferences to other semiconductors. Although the quantum energy
of CO_2 laser photons (117 meV) is lower than the energetic dif-
ference between H_4 and H_5 subvalence bands (122 meV at 77 K
[3]), the absorption cross section S_p of holes for CO_2 laser
radiation with polarization $\vec{e}\|c$ is the highest of any semicon-
ductor known ($S_{p\|} = (1.5 \pm 0.5)\cdot10^{-15}$ cm^2 at 77 K); absorption
is possible only due to collision broadening [3]. For $\vec{e}\perp c$ the
absorption cross section is very much smaller: $S_{p\perp} = 3\cdot10^{-17}$ cm^2.
This value is governed by Drude-absorption and by intersubband
transitions $H_4 \rightarrow H_6$ [4]. Concerning photon drag in tellurium,
up to now only a short note was published [5].

2. THEORY

The photon drag can be described by the tensor relation

$$j_i = \sigma_{iklm} I \varkappa_k e_l e_m, \tag{1}$$

with $\vec{j}$ — photon drag current density, $\hbar\vec{\varkappa}$ — photon momentum, $\vec{e}$ — unit vector in the direction of the electric field of the electromagnetic wave, I — light intensity. The σ-tensor σ_{iklm} has trigonal symmetry according to point group symmetry D_3. Furthermore it is symmetric with respect to last two indices. In the crystal axes system this σ-tensor has 37 non-vanishing components, 10 of them being linearly independent.
With respect to simplified calculations it is worthy to note, that in the case of cylindrical symmetry further components are zero and only 21 non-vanishing components are remaining, 7 of them being linearly independent. If the σ-tensor is the product of two second rank tensors s_{ik} and α_{lm} with the same symmetry

$$\sigma_{iklm} = s_{ik}\alpha_{lm}, \tag{2}$$

there are only 9 non-zero components, 4 being linearly independent. In this case there is no difference between trigonal and cylindrical symmetry. If we take for a first approximation $E_2(\vec{k}) = E_1(\vec{k}) - \Delta$ with E_1 referring to H_4 and E_2 to H_5 and write

$$E_1(\vec{k}) = Ak_\perp^2 + Bk_z^2 = E_2(\vec{k}) + \Delta, \tag{3}$$

we find from energy and momentum conservation, that transitions take place into final states in the H_4 band lying on a plane in $\vec{k}$-space described by the following in $\vec{k}$ linear form

$$A\varkappa_\perp k_\perp + B\varkappa_z k_z = \frac{1}{2}\left(\Delta + A\varkappa_\perp^2 + B\varkappa_z^2 - h\nu\right) = D, \tag{4}$$

with $k_{\perp,z}$ and $\varkappa_{\perp,z}$ being the components of $\vec{k}$ and $\vec{\varkappa}$ perpendicular and parallel to the c-axis, resp. It is essential to note that in the case of (3) in the energy conservation quadratic terms cancel each other and therefore we have no quadratic expression, leading to two possible quantum-states in each direction of $\vec{k}$. The main contribution will come from those $\vec{k}$-vectors $\vec{k}_e$ fulfilling eq. (4), that correspond to maximum energy by eq. (3). That leads to unreasonable high k_e-values of about $(2...3)\cdot10^8$ cm^{-1} taking into account the known values of

$D \approx 5$ meV; $A = 3.57 \cdot 10^{-15}$ eVcm2, $B = 3.77 \cdot 10^{-15}$ eVcm2, and $\kappa = 5.91 \cdot 10^3$ cm^{-1}. The difficulties are overcome regarding that the energy values in the subbands, assuming a scattering time of about 10^{-12} s, are unsharp at least to the amount of 4 meV. Under this condition transitions are possible to all the $\vec{k}$-values around the extremum of $E_2(\vec{k})$ and we can only use momentum conservation for calculating the photon drag current. This leads to a very simple theory, if we assume, that the electrons excited by radiation from $E_2(\vec{k} - \vec{\kappa})$ to $E_1(\vec{k})$ give a contribution to current density

$$\vec{j} = - e\left[\vec{v}_1(\vec{k})\tau_1 - \vec{v}_2(\vec{k} - \vec{\kappa})\tau_2\right] \frac{\alpha I}{h\nu},\qquad (5)$$

where α is the polarization dependent absorption coefficient, τ_1 and τ_2 are the momentum relaxation times in the two subbands and $\vec{v}(\vec{k}) = \hbar^{-1}\mathrm{grad}_{\vec{k}}E(\vec{k})$. In our case it may be justifiable to assume $\tau_1 = \tau_2 = \tau$. In this case we get with $\hbar^2/2m_{\perp,\parallel}$ for A and B, resp.

$$j_{\perp,\parallel} = \frac{e}{m_{\perp,\parallel}} \frac{\tau}{c} \alpha I \approx \mu_{\perp,\parallel}\frac{\alpha}{c} I,\qquad (6)$$

where μ is the hole mobility.

3. EXPERIMENTAL RESULTS

Radiation pulse with a peak power of 2.5 kW of a Q-switched CO_2 laser (half-width 200 ns, degree of polarization > 99.9 %) have been used. The samples have been prepared from single crystals with acceptor concentrations between 1.5 and 3.5 $\cdot 10^{14}$ cm^{-3} by a special treatment resulting in high mobility ($\mu_H = 9000$ cm^2/Vs at 77 K) and low carrier concentration simultaneously [6]. Investigations were carried out between 77 and 300 K. To avoid the competition processes reported earlier [7] the laser beam was focused weakly ($I < 100$ kW/cm^2). Signals were interpreted as photon drag only, if they were inversing their polarity with the direction of the laser beam. Further important arguments are constant pulse height, if the central region of the sample's front side is being scanned (FIG. 1), and linear dependence on laser power (FIG. 2). For $\vec{e}\parallel c$ the highest

FIG. 1: Responsivity profile. FIG. 2: Intensity dependence
 Scanning spot size of photon drag
 diameter 0.4 mm voltage

signals were observed in accordance with the estimation in
eq. (6). The sign of δ_{1133} was positive in the whole temper-
ature range (FIG. 3). At 77 K we found $\delta_{1133} = 5\cdot10^{-12}$ cm/V.
For $\vec{e}\perp c$ and $\vec{\kappa}\parallel c$ a photon-drag effect was found in the temper-
ature range between 77 and 180 K, the sign was always nega-
tive ($\delta_{3311} = -1\cdot10^{-12}$ cm/V at 77 K) (FIG. 4).

FIG. 3: Temperature dependence FIG. 4: Temperature dependence
 of photon drag voltage of photon drag voltage
 for $\vec{e}\parallel c$, $\vec{\kappa}\parallel x$ for $\vec{e}\perp c$, $\vec{\kappa}\parallel c$

4. DISCUSSION

Our results for $\vec{e}\|c$ are in agreement with eq. (6) in sign and
order of magnitude. Reasonable correlation between the values
of σ_{1133} and mobility u can be found (TABLE 1). In the case
of $\vec{e}\bot c$ there is no agreement in sign with eq. (6), this can be
understood in our simple model only if $\tau_2/m_2 > \tau_1/m_1$ in eq. (5),
as may be expected for $\vec{\varkappa}\|c$. In this case our simple model does
not apply, but it makes understandable as well as earlier
models proposed for germanium the possibility of a reversal in
sign of the photon drag. Further theoretical and experimental
investigations are necessary to get a clear picture of the
anisotropic photon drag in tellurium.

TABLE 1. Experimental results concerning σ_{1133} at T = 77 K

sample	p_0 cm^{-3}	u cm^2/Vs	responsivity mV/kW	$(\sigma_{1133}/\alpha_\|)$ /10^{-11} cm^2/V experiment	eq. (6)
66-1	$1.6\cdot10^{14}$	2700	1	2.2	1.5
66-4	$1.6\cdot10^{14}$	2700	1	2.3	1.5
66-2	$3.3\cdot10^{14}$	9000	0.55	3.9	5.1

5. REFERENCES

[1] Danishevskii, A. M.; Kastalskii, A. A.; Ryvkin, S. M.;
 Yaroshetskii, I. D.: Zh eksper. teor. Fiz. 58 (1970) 544
[2] Valov, P. M.; Ryvkin, B. S.; Ryvkin, S. M.;
 Yaroshetskii, I. D.: phys. stat. sol. (b) 53 (1972) 65
[3] Hardy, D.; Rigaux, C.: phys. stat. sol. 38 (1970) 799
[4] Bangert, E.; Fischer, D.; Grosse, P.: phys. stat. sol. (b)
 59 (1973) 419
[5] Panyakeow, S.; Shirafuji, I.; Inuishi, Y.: Appl. Phys.
 Letters 21 (1972) 314
[6] Kusnick, D.; Spitzer, M.:Thesis, Humboldt-Universität
 Berlin 1973 (unpublished)
[7] Herrmann, K. H.; Vogel, R.: Proc. Intern. Conf. Phys.
 of Semicond., Warsaw 1972, p. 870

OPTICAL RECTIFICATION AND PHOTON DRAG IN P-TYPE CaAs
AT 10.6μm AND 1.06μm

J.M. DOVIAK and S. KOTHARI

Department of Physics, University of Essex,
Wivenhoe Park, Colchester,
Essex, England.

Photoelectric effects occurring in p-type GaAs under
illumination by laser radiation are discussed. These
effects fall into two distinct types, optical recti-
fication and photon drag, which can be distinguished
by detailed examination of their tensor properties.

1. INTRODUCTION

Neglecting terms linear in electric field, higher order photoelectric
effects can be written as a series of terms which are combinations of powers
of the light electric field, $\vec{E}$, and the unit propagation vector, $\vec{q}$. Taking
this expansion to first order in $\vec{q}$ and second order in $\vec{E}$, we can write the
non-linear part of the photoelectric fields as:

$$E_i^{NL} = I\, R_{ijk}\, p_j\, p_k + I\, T_{ijkl}\, q_j\, p_k\, p_l \qquad (1)$$

where p_j is the component of the unit polarisation vector in the j direction
and I the light intensity. The first term on the right hand side of the
expression is optical rectification; the second term, linear in photon
momentum, is the photon-drag effect.

Optical rectification has been reported in Te [1] and KDP and K*DP [2].
It arises when electronic high frequency nonlinearities result in a net d.c.
shift. Photon drag, which appears in Ge [3,4], is the result of the moment-
um of the photon being distributed unequally in k-space amongst the excited
carriers, with a collective shift in the equilibrium distribution function
leading to the appearance of a net current. Since the classification of this
effect, detectors have appeared for use with high power CO_2 lasers.

An extention of the photon-drag effect to the visible region of the
spectrum should be possible; however, Ge would not be very suitable because
interband transitions lead to a very high absorption coefficient. For this
reason, GaAs has been investigated, with particular attention paid to the
region around 1μm. In this paper we report the observation of optical
rectification and photon-drag at 10.6μm and 1.06μm.

2. BACKGROUND

2.1. OPTICAL RECTIFICATION

The electric field generated by optical rectification is given by the
first term in Eq.1. GaAs is a III-V semiconductor possessing $\bar{4}3m$ symmetry;

this symmetry reduces the number of independent terms in the tensor R_{ijk} to one, the only non-zero terms being $R_{123} = R_{132} = R_{231} = R_{213} = R_{321} = R_{312}$. Carrying out the implied summation in Eq.1. and setting $D = R_{123}$, the following expression for the photoelectric field is obtained.

$$E_i = 2 \, I \, D \, P_{i+1} \, P_{i+2}$$

Representing by $E(\hat{r},\hat{q})$ the field measured in the $\hat{r}$ direction when the light propagates in the $\hat{q}$ direction, we can summarise the electric field polarisation direction dependence for a number of orientations. In Fig.1. is shown the convention used; $\hat{q}$, $\hat{a}$, and $\hat{b}$ form an orthogonal set of unit vectors with the light polarisation lying in the $\hat{a}\hat{b}$ plane.

Fig. 1. Convention used to describe polarisation unit vector $\hat{p}$ in terms of crystal axes a and b; the propagation direction unit vector, $\hat{q}$, completes the orthogonal set.

$E(100,100) = I \, D \, \sin 2\theta$	$E(001,100) = 0$	$a = [010] \quad b = [00\bar{1}]$
$E(110,110) = 0$	$E(1\bar{1}0,110) = -I \, D \, \sin 2\theta$	$a = [001] \quad b = [1\bar{1}0]$
$E(111,111) = -\dfrac{ID}{\sqrt{3}}$	$E(11\bar{2},111) = -\dfrac{2ID}{\sqrt{6}} \cos 2\theta$	$a = [1\bar{1}0] \quad b = [11\bar{2}]$

The optical rectification tensor R_{ijk} can be transformed into a non-linear susceptibility tensor d_{ijk}; using the frequency and subscript permutation rules in [5], this d_{ijk} is directly related to the linear electro-optic tensor r_{jki}. Combining these results gives;

$$R_{ijk} = \frac{2n^4 \varepsilon_o}{(\varepsilon - \varepsilon_o)} \, \frac{(\mu_o)^{\frac{1}{2}}}{(\varepsilon)^{\frac{1}{2}}} \, r_{jki} \tag{2}$$

We will use this relationship later in comparing the value of R_{ijk} measured with that obtained using Eq.2. and the accepted value for r_{jki}.

2.2. PHOTON DRAG

In [6] is set out the tensor nature of photon drag in a discussion of the results for Ge. The tensor for $\bar{4}3m$ materials will differ from that for Ge because of the lack of inversion symmetry. This fact leads to the appearance of a single new term Q associated with the anti-symmetric part of T_{ijkl}. With the definitions of S and P the same as given in [6], the electric field arising from the photon-drag effect can be written as:

$$E_i = I(Sq_i + (P-S) \, q_i P_i^2 + 2Qp_i \, (P_{i+1}q_{i+1} - P_{i+2}q_{i+2}))$$

Using the same convention given in Fig.1., we can summarise a few photon-drag fields.

$$E(100,100) = IS \qquad\qquad E(001,100) = 0$$
$$E(110,110) = I\{S + \frac{(P-S)}{2}\cos^2\theta\} \quad E(1\bar{1}0,110) = -I\,Q\,\cos^2\theta$$
$$E(111,111) = I\,\frac{P+2S}{3} \qquad E(11\bar{2},111) = I\{\frac{S-P}{3\sqrt{2}}\cos2\theta + \frac{2Q}{\sqrt{6}}\sin2\theta\}$$

Notice the effect of the Q term is to add a term in quadrature with the normal photon drag as can be evidenced in $E(11\bar{2},111)$. In general this will result in a field of the form $A\cos(2\theta+\phi)$ where ϕ is a measure of the strength of Q.

3. EXPERIMENT

The 10.6μ, measurements were done with a q-switched CO_2 laser, operating at 200 pulses per second and capable of giving 100kW peak power in 150nsec half width pulses. The output power of the laser was continuously monitored using a Ge photon-drag monitor. Signals from the GaAs were amplified with a Hewlett Packard pulse amplifier type HP 462A and displayed on an oscilloscope. For the 1.06μm measurements, a q-switched Nd^{+3}:YAG laser was used giving 3kW peak power in 150nsec pulses at a 1 kHz repetition rate.

The GaAs material was Cd doped with carrier concentration $p=3\times10^{16}/cm^3$ and a mobility of 225 cm^2/V-sec. Electrodes were alloyed onto polished samples by heat treatment of In:3% Zn contacts. If the electrodes did not alloy correctly, then large photoelectric effects would appear if the light were directed onto the contact region. These voltages were always more evident at 1.06μm than at 10.6μm.

In Fig.2. shown combined are results for measurements at 10.6μm and 1.06μm on a sample whose axes are $\hat{q} = [110]$, $\hat{a} = [001]$, $\hat{b} = [1\bar{1}0]$. Combining the results of section 2.1 and 2.2 we see that the form of the field should be:

$$E^{NL}(1\bar{1}0,110) = -I(D\sin2\theta + Q\cos^2\theta)$$

Clearly, the result show no $\cos^2\theta$ dependence, and the term Q for GaAs is much smaller than D. In a similar way, other orientations give relationships between the various terms. The results of the measurements at 10.6μm and 1.06μm are included in the table below, with corrections for reflection and absorption included.

Fig.2. Plot of experimental optical rectification versus angle. $0\leqslant\theta\leqslant\pi$: 1.06μm; $\pi\leqslant\theta\leqslant 2\pi$: 10.6μm. The curve is a plot of $A\sin2\theta$ where : A = 13.6 for 1.06μm points; A = 9.4 for 10.6μm points. Units are $10^{-8}\,\frac{V/cm}{W/cm^2}$.

Table 1. Value of the tensor coefficients at 300°K for GaAs at 1.06μm and 10.6μm (units are 10^{-10}m/A)

| | $|D|$ | S | P | $|P-S|$ | $|Q|$ |
|---|---|---|---|---|---|
| 1.06 μm | 14 | -1.4 | -2.2 | 0.8 | <0.6 |
| 10.6 μm | 15 | – | – | 8.8 | <0.6 |

4. DISCUSSION

From the table above, it will be noticed that S and P could not be determined at 10.6μm. An experimental difficulty was encountered in obtaining uniform electric field orientation in longitudinal samples. Coupled together with high absorption due to intervalence band transitions, this inevitably led to mixtures of transverse plus longitudinal field being measured.

The results of the measurements were all consistent with the photo-electric effect being described by a tensor of third rank plus a tensor of fourth rank. In section 2.1 it was pointed out that the optical rectification coefficient, D, is proportional to the linear electro-optic coefficient r_{jki} = R. The value of R is 1.6×10^{-12}m/V [7]. Using Eq.2. and the value for $|D|$ at 1.06μm given in the above table, R is computed to be 0.5×10^{-12}m/V, which is in good agreement with the directly measured value.

In [3] an expression for the photon-drag field using a classical treatment is given. Writing E^{NL}=TI, this expression reduces to T=αρμ/c where α, ρ and μ are the absorption coefficient, resistivity and mobility respectively. Using the parameters for the GaAs under investigation, T=13×10^{-10}m/A. If we compare this value with any of the values for S,P or P-S above, we see that the results at 10.6μm are in better agreement than those at 1.06μm. This difference in the photon drag results might best be explained by reference to the valence band structure of GaAs. The 1μm transition will proceed from the split off to the heavy hole band and will occur at larger values of k than the light to heavy hole transition at 10μm. The photon drag current is dependent not only on the scattering times of the excited carriers, but also on the derivative of the distribution function $\frac{\partial f(k)}{\partial k}$. The distribution function will be tailing off for the 1μm transition and as a consequence the drag current will be suppressed.

Ge as a detector at 10.6μm has been used extensively and has been fabricated as a monitor in the form of an optical bridge for determining pulse lengths [8]. Typical sensitivity for a 1cm^2 cross-section, 1cm long bar is 100μV/kW. Using a sample of P-type GaAs with ρ= 1Ω-cm, 90μ thick oriented in a [110],[1$\bar{1}$0],[001] configuration, the peak sensitivity would similarly be 100μV/kW. Used in the form of an optical bridge, the sample resolving time would be approximately 1psec. Coupled together with some thin film

non-linear absorber, very high resolution should be attainable at low cost, resulting in a useful tool for use with high power mode-locked Nd^{+3} lasers.

5. CONCLUSION

It has been shown that in a III-V semiconductor possessing $\bar{4}3m$ symmetry, a photoelectric effect possessing third and fourth rank characters is expected. The terms in this tensor have been measured for p-type GaAs. The optical rectification effect (3rd rank tensor) is of sufficient magnitude to serve as a fast detector at 1.06μm. Clearly it will never replace existing photodiodes in sensitivity, but properly fabricated, they can be used in an optical bridge for measuring pulse length.

The authors wish to thank Professor A.F. Gibson for many useful discussions throughout the period during which this work was undertaken.

6. REFERENCES

[1] K.H. Hermann and R. Vogel, Proc. of the Eleventh Conf. on the Physics of Semiconductors, Warsaw (1972).

[2] M. Bass, P.A. Franken, J.F. Ward and G. Weinreich; Phys. Rev. Lett <u>9</u>, p. 446, (1962).

[3] A.F. Gibson, M.F. Kimmitt and A.C. Walker; Appl. Phys. Lett <u>17</u>, p. 75. (1970).

[4] A.M. Danishevskii, A.A. Kastal'skii, S.M. Ryvkin and I.D. Yaroshetskii; Sov. Phys. JETP <u>31</u>, p. 292, (1970).

[5] J.A. Armstrong, N. Bloembergen, J. Ducuing and P.S. Pershan; Phys. Rev. <u>127</u>, p. 1918, (1962).

[6] A.C. Walker and D.R. Tilley; J. Phys. C : Solid St. Phys. <u>4</u> p. L376, (1971).

[7] A. Yariv, C.A. Mead and J.V. Parker; IEEE J. QE <u>QE-2</u>, p. 243, (1966).

[8] A.F. Gibson, C.A. Rosito, C.A. Raffo and M.F. Kimmitt, J. Phys D : Appl. Phys. <u>5</u>, p. 1800 (1972).

AUGER RECOMBINATION IN GaSb[+]

G. BENZ and R. CONRADT

Physikalisches Institut der Universität Stuttgart,
Stuttgart, Federal Republic of Germany

The observation of radiative recombination of Auger
excited holes in the split-off valence band with
shallow bound conduction band electrons in GaSb at
4.2K and 77K is reported. A discussion of the
luminescence lineshape indicates that the hole
distribution is a thermal one with a temperature
of about 250K. The lifetime t_s of the holes in the
split-off valence band is estimated to be
10^{-10} sec > t_s > 10^{-11} sec.

1. INTRODUCTION

Because of its bandstructure, GaSb should be a good model for
Auger-recombination: the spin-orbit splitting of its valence
band Δ_o = 0.75 eV [1] is nearly equal to the energy gap
E_g = 0.812 eV [2]. Therefore Auger-transitions into the split-
off valence band are possible with small momentum transfer and
small activation energy and, for this reason, with a large tran-
sition probability [3]. We have succeeded in getting indication
of these Auger excited holes in the split-off band by radiative
recombination with shallow bound conduction band electrons.

2. EXPERIMENT

GaSb crystals are immersed in liquid He and are excited by a
Q-switched Nd-YAG laser. A minority carrier density of about
10^{17} cm^{-3} is estimated. The crystals used in this experiment are
a solution grown sample [4], W15, p = 2.5x10^{16} cm^{-3}, and a sample,
WK32, grown by liquid phase epitaxy [5] on a substrate with high
conductivity. The luminescence is dispersed by a 0.85 m grating
double spectrometer and is detected with an RCA C31034A photo-
multiplier using a "digital boxcar integration" method described
elsewhere [6]. In addition, GaSb diodes are investigated. They
were fabricated by alloying Pb/Se in p-type GaSb [7] and mounted

[+]Supported by the Deutsche Forschungsgemeinschaft.

on a copper finger cooled to 77K. With current pulses a
carrier density of about 5×10^{16} cm^{-3} was generated. All spectra
are corrected for the spectral response of the system, for re-
absorption in the sample and for the transfer from the wave-
length scale to the energy scale.

3. RESULTS AND DISCUSSION

Fig. 1 shows the spectra of a solution grown sample at 4.2K.
A broad band (halfwidth 50 meV) is detected at 1,575 eV.

Fig. 1

High-energy lumin-
escence from GaSb.
The arrows indicate
the energy data for
$E_g+\Delta_o$ and $2E_g$.

Towards lower energies a weak continuum appears in the lumin-
escence spectra. For orientation the data of $E_g+\Delta_o$ and $2E_g$ are
drawn in Fig. 1. From the energetic position and the lineshape
of the new band at 1,575 eV, nonlinear frequency doubling of the
near bandedge emission as a possible explanation of the radia-
tion can be excluded. From additional measurements on GaSb
diodes at 77K which show a similar band at 1.56 eV and a tail
towards lower energies (see Fig. 2), two-step excitation can
also be eliminated. We suggest that the new emission band at
1,575 eV in GaSb is caused by the recombination of shallow bound
conduction band electrons with holes in the split-off valence
band which are generated by Auger-recombination. The continuum

in the luminescence spectra below this band is attributed to the
radiative recombination of Auger holes relaxing to the top of
the valence band.

Fig. 2
High-energy
luminescence
from a GaSb-
diode.

To confirm our interpretation, we calculated a theoretical
lineshape of the new band at $E_g+\Delta_o$ based on the following
assumptions:

1.) The carrier distribution in the split-off valence band is
 a thermal one, characterized by $F(E_2) \propto \sqrt{E} \exp(-E_2/kT)$.
2.) In the recombination, the k-selection rule is not valid.
 This is assumed because of the high carrier density and is
 supported by the measurements of Göbel in GaAs [8].
3.) The holes recombine with shallow bound conduction band
 electrons.

The lineshape of the luminescence band at $E_g+\Delta_o$ is then given
by folding the hole distribution with the experimental lineshape
of the E_g-luminescence, $I(E_1)$, which already takes into account
the binding energy of donors or excitonic states in the energy
balance. This yields the convolution integral:

$$J(E) \simeq \int_o^\infty \int_o^\infty I(E_1)F(E_2) \; \delta(E-E_1-E_2-\Delta_o)dE_1dE_2 \qquad (1)$$

In the calculation only two parameters must be fitted: the hole

temperature T and the spin-orbit-splitting Δ_o.

Fig. 3 shows the spectra of two different samples. The insert depicts the respective E_g-luminescence used in the linefit. Only the right part of the spectra (solid lines) which is caused by excitonic decay and by donor states [9] is considered. The broken line (sample WK32) is attributed to a band acceptor transition and is therefore not considered in the calculation.

Fig. 3

Luminescence spectra in the $E_g+\Delta_o$ and E_g region of GaSb.

From the appearance of shallow bound excitons a strong sample heating can be excluded. The band at $E_g+\Delta_o$ is measured at equal excitation as the E_g-spectra. The crosses represent the experimental values with corresponding error. The solid lines show the respective theoretical lineshapes calculated from equation (1). We get the fit for Δ_o(4.2K) = 0.752 eV for both samples, a hole temperature of 250K for sample W15 and 230K for sample WK32. The energetic position of the luminescence maxima of the two samples show a small difference of 5 meV. This is described in a fairly good manner by the theoretical linefit using equation (1), because the actual lineshape at E_g is used, which shows the same shift. Our value for Δ_o is in good agreement with magnetoreflectivity data of Reine et al [1], who determined

Δ_o (30K) = (0.749$\pm$0.002) eV. The good fit confirms our as-
sumption that the carrier distribution in the split-off band is
a thermal one with temperatures up to 250K, indicating that the
carriers are in equilibrium among themselves but not with the
lattice. For sample WK32 a slightly smaller hole temperature is
obtained. This is ascribed to the larger lifetime of the Auger-
holes in the split-off band in this purer sample and by this
means to a better thermalization of the carriers. From the
thermal carrier distribution with temperatures much higher than
the lattice temperature a lifetime t_s for the holes in the split-
off band can be established [10]: t_s must be shorter than the
radiative lifetime but longer than the relaxation time for
optical phonon scattering. This gives a hole lifetime t_s of
about 10^{-10} sec > t_s > 10^{-11} sec.

4. ACKNOWLEDGEMENTS

The authors wish to thank W. Jakowetz and C. Woelk for providing
the GaSb samples and K. Betzler for many discussions.

5. REFERENCES

[1] Reine, M.; Aggarwal, R.L.; Lax, B.: Solid State Commun. <u>8</u>,
 35 (1970).

[2] Baldereschi, A.; Lipari, N.O.: Phys. Rev. B <u>3</u>, 439 (1971).

[3] Beattie, A.R.; Landsberg, P.T.: Proc. Roy. Soc. A <u>249</u>, 16
 (1959).
 Takeshima, M.: J. Appl. Phys. <u>43</u>, 4114 (1972).

[4] Wolff, G.A.; Keck, P.H.; Broder, J.P.: Phys. Rev. <u>94</u>, 753
 (1954).

[5] Woelk, C.; Benz, K.W.: J. Crystal Growth, to be published.

[6] Betzler, K.; Weller, T.; Conradt, R.: Phys. Rev. B <u>6</u>, 1394
 (1972).

[7] Calawa, A.R.: J. Appl. Phys. <u>34</u>, 1660 (1963).

[8] Göbel, E.: Appl. Phys. Letters <u>24</u>, 492 (1974).

[9] Jakowetz, W.; Rühle, W.; Breuninger, K.; Pilkuhn, M.H.:
 Phys. Stat. Sol. (a) <u>12</u>, 169 (1972).
 Benoît à la Guillaume, C.; Lavallard, P.: Phys. Rev. B <u>5</u>,
 4900 (1972).

[10] Ulbrich, R.: Phys. Rev. B <u>8</u>, 5719 (1973).

SINGLY STIMULATED TWO-PHOTON EMISSION IN $Ga_{1-x}Al_xAs$

R. EYMARD

Centre National d'Etudes des Télécommunications

92220, Bagneux, France

The first observation of singly stimulated two-photon emission in a $Ga_{1-x}Al_xAs$ light emitting diode is reported. The diode is dc biased, and a Q-switched CO laser is used to stimulate emission in the lower energy side of the luminescence spectrum. It is shown that the amplitude, time dependence, and spectral position of the observed phenomenon can only be accounted for by two-photon emission.

INTRODUCTION

Two-photon emission processes have been observed in various gases, several years ago for spontaneous emission [1], and more recently for singly stimulated emission [2]. We have observed an increase of the luminescence of a $Ga_{1-x}Al_xAs$ (x ∿ 0.2) p-n junction in the low energy tail of the spectrum, when it was illuminated by a CO laser. The measurement was performed at an energy $h\omega$ lower than the main luminescence peak by 200 to 250 meV (the energy of the photon of the CO laser). The magnitude of the observed phenomenon, its time dependence, and the absence of any effect when using a CO_2 laser (120 meV) in place of the CO laser lead us to conclude that these results constitute the first experimental evidence of singly stimulated two-photon emission in a semiconductor, as an exact reverse process of the well-known [3] two-photon absorption.

EXPERIMENTAL METHODS AND RESULTS

The $Ga_{1-x}Al_xAs$ diode (from La Radiotechnique-Compelec) was operated at liquid N_2 temperature. It was dc biased by a 10 mA stabilized current. The peak of the spectrum is at 1.81 eV (6850 Å) and the line width is about 40 meV. The external quantum efficiency is about 2% . The laser light was gently focused on the diode, to avoid damaging of its surface.

Two lasers have been used : a Q-switched, liquid N_2 cooled CO laser, with an output peak power of 1 kW distributed over

several lines from 5.2 µ to 6 µ, and a Q-switched CO_2 laser,
with a peak power of about 2 kW at 10.6 µ. The pulse length was
0.2 µs, and the repetition rate was 200 Hz for both lasers. The
emitted light was focused on the cathode of an extended S 20
photomultiplier. Two filters were inserted in the beam : a
RG 780 Schott glass filter, and an interference filter, centered
at 7830 Å with a bandwidth of about 160 Å . The output voltage
of the photomultiplier was amplified, and detected with a sam-
pling and integrating unit. The gate width was 5 ns and the
integration time constant was about 1 mn . It was possible to
change the position of the gate with respect to the laser pulse,
and the output voltage of the detecting unit was recorded. The
time constant of the photomultiplier and its preamplifier was
about 150 ns.

Fig. 1 shows typical results obtained on the recorder chart when
the gate was opened simultaneously with the peak of the CO laser
pulse. It can be seen that the luminescence increase (c,e)
appears only when the sample is illuminated by the laser (c,e,f)
and, of course, when the current passes through the diode (b,c,
d,e). The observed noise arises from the continuous light back-
ground, due to the heavy doping of the junction, which gives
rise to low energy density-of-states tails (b and d compared to
a and f). The magnitude of the ratio between the observed increa-
se and the main luminescence peak is about 2×10^{-6} .
 Fig. 2 shows the variation of the signal when moving the gate
position with respect to the laser pulse (solid line). It is

clearly seen that, taking into account the time constant of the detector (150 ns), the time dependence of the signal is in agreement with that of the laser pulse.

An experiment has been performed with the CO_2 laser instead of the CO laser, the remaining parts of the experimental setup being unmodified. No variation of the luminescence has been detected.

DISCUSSION

Many mechanisms can be invoked to explain an increase of luminescence when illuminating the sample with an IR light :
i) Those involving impurity levels cannot explain our results because, due to the heavy doping of the material, the energy levels distribution is essentially continuous, and replacing the CO laser by the more powerful CO_2 one would have had a larger effect on the population of the involved levels.
ii) Heating of the diode would shift the luminescence spectrum towards lower energies and give an increase of luminescence at the observed wavelength. But the characteristic time constant of such an effect, which would depend on the thermal capacity and conductivity of the diode and its holder, would be of the order of tens of microseconds. As shown in Fig. 2, this effect is not noticeable in our experiment. Furthermore, a heating process would have been greater with the CO_2 laser, more powerful than the CO one.
iii) The nonlinear mixing process has been studied by many authors [4]. The order of magnitude of the relative efficiency ρ of this process can be computed, using the model of Armstrong et al. [5] for three waves mixing. This ratio is

$$\rho = \frac{128 \ \pi^3}{10^{-7}} \frac{\omega \ \omega_0 \ d^2}{c^3 n \ n_0 n_{IR}} \frac{P}{A} \frac{4 \sin^2 \left(\frac{1}{2} \left| \Delta k \right| L_c \right)}{\left| \Delta k \right|^2}$$

$\omega_0 = 2\pi/\lambda_0$ and $\omega = 2\pi/\lambda$ are the circular frequencies of the spontaneous light and of the replica ; n_0 , n and n_{IR} are the refractive indices at ω_0 , ω and the laser frequency ; P/A is the power density of the laser light incident on the sample ; L_c is the interaction length ; d is an effective nonlinear coef-

ficient, including the angular averaging on the directions of k and of the polarization vectors ; Δk is the phase mismatch : $\Delta k = k_0 - k - k_{IR}$. With the known values for GaAs [6], $n_0 = 3.6$, $n = 3.5$, $n_{IR} = 3.3$, one finds :

$$|\Delta k| = 2\pi \left| \frac{3.6}{\lambda_0} - \frac{3.5}{\lambda} - 3.3 \left(\frac{1}{\lambda_0} - \frac{1}{\lambda} \right) \right| \simeq 1.1\times10^4 \text{ cm}^{-1} .$$

Taking [7] $d = d_{14} = 3\times10^{-7}$ esu, $P/A = 10^5$ W/cm^2, one obtains $\rho < 10^{-7}$.

The inequality arises from taking $\sin^2 \left(\frac{1}{2} |\Delta k| L_c \right) = 1$ and $d = d_{14}$, i.e. neglecting the angular dependence of the effective nonlinear coefficient upon the directions of the beams in the crystal. It appears that the mixing process cannot explain the observed efficiency, which is at least 20 times greater.

iv) Therefore, we think that our data can be only explained by singly stimulated two-photon emission. Note that the spectral position and time dependence of the signal are in agreement with the proposed explanation. A crude estimate of the efficiency can be made in the following way. An order of magnitude of the two-photon absorption coefficient is 0.1 cm/MW [8] when the involved wavelengths are about the same as in our experiment, for a total energy of the transition 0.1 eV greater than the energy gap. The ordinary absorption coefficient is 10^4 cm^{-1} for the same energy [9]. For an incident power density about 0.1 MW/cm^2, the relative efficiency of the two-photon process is nearly 10^{-6}. This value is in good agreement with the observed efficiency and supports our interpretation of the new effects reported here.

Furthermore, Bepko [10] has calculated the efficiency of the mixing process compared with the efficiency of the two-photon process, and his results indicate that the mixing process is at least 2 orders of magnitude smaller.

CONCLUSION

We have observed a luminescence enhancement in the emission spectrum of a $Ga_{1-x}Al_xAs$ diode, when illuminated by a powerful IR laser, which cannot be explained either by a thermal or impurity effect or by a nonlinear mixing phenomenon. The amplitude of the enhancement is of the same order of magnitude than the

stimulated two-photon emission process, when computed using the experimental available data. We therefore conclude that the observed enhancement of the luminescence is due to a singly stimulated two-photon emission process. This interpretation is consistent with the observed time dependence and spectral position of the emitted light.

From the statistical nature of the noise in our results, it appears that any further improvement of the experiment, or practical application of the effect depends upon the availability of a better material, with high luminescence efficiency, and a well defined long wavelength cut-off in the emission spectrum. Such a material has not been available to us up to now. Therefore, other experiments such as polarization dependence measurements remain to be done, to obtain a definitive identification of the observed effect.

The author is grateful to Dr M. Bernard for having first suggested this experiment, to E. Batifol and F. Jalabert for experimental assistance, to Dr A.M. Jean-Louis, Dr J. Jerphagnon, Dr M. Voos, J. L. Oudar and A. Leclert for stimulating discussions.

REFERENCES

[1] Lipeles M., Novick R., Tolk N., Phys. Rev. Letters, 15 690 (1965)
[2] Shaul Yatsiv, Rokni M., Barak S., Phys. Rev. Letters, 20 1282 (1968)
[3] Worlock J.M., in Laser Handbook, F.T. Arecchi and E.O. Schulz-Dubois (North Holland Publ.Co., Amsterdam, 1972) 1323
[4] Warner J., Optoelectronics, 3, 37 (1971)
[5] Armstrong J.A., Bloembergen N., Ducuing J., Pershan P.S., Phys. Rev., 127, 1918 (1962)
[6] Seraphin B.O., Bennett H.E., in Semiconductors and Semimetals, R.K. Willardson and A.C. Beer (Academic Press, New-York, 1967), 3, 499
[7] Mc Fee J.M., Boyd G.D., Schmid P.H., Appl. Phys. Letters, 17, 57 (1970)
[8] Lee C.C., Fan H.Y., Phys. Rev., 9, 3502 (1974)
[9] Cardona M., in Semiconductors and Semimetals, R.K. Willardson and A.C. Beer (Academic Press, New York, 1967),3, 134
[10] Bepko S.J., to be published

PHOTOEMISSION

CONTENTS

VALENCE AND CONDUCTION BAND·STRUCTURE OF Ge
USING THEORETICAL AND EXPERIMENTAL PHOTOEMISSION SPECTRA FROM 6.5 TO 23 eV[+]

W. D. GROBMAN, D. E. EASTMAN, J. L. FREEOUF, AND J. SHAW

IBM Thomas J. Watson Research Center

Yorktown Heights, New York, U.S.A.

Experimental spectra for Ge obtained using synchrotron
radiation ($h\nu \leq 23$ eV) are compared with predictions of
a direct-transition theory which includes anisotropic
transmission through the (111) cleaved surface. The
direct transition-determined structure plot for Ge (peak
binding energies as a function of photon energy), as
well as other spectral features, are fit well up to
$h\nu \gtrsim 23$ eV using a d-nonlocal pseudopotential model. We
demonstrate that this model provides a proper descrip-
tion of conduction band topology far (~ 1 Ryd) above the
gap. Our ability to obtain these results shows that
features predicted by an anisotropic, direct optical-
transition model can determine one-electron energy bands
over a wide energy range even in the presence of short
escape depths and surface-associated emission.

1. INTRODUCTION

The present work first briefly reviews semiconductor one-electron
energy band positions using Ge as an example, obtained using optical
measurements and photoemission valence band overviews. We then illustrate
the additional valence and conduction band information available in low
energy spectra (below the "X-ray limit"[++]) by presenting an analysis of
such spectra based on a direct transition model which includes anisotropy
due to the fact that in a cleaved sample emission through a preferred
crystal face is obtained. Several new band positions high (~ 10 eV above
the gap) in the conduction bands are obtained and compared with the pre-
dictions of several band models. We also present new information concern-
ing the validity of a direct transition model for describing the data over
a wide photon energy range and for short inelastic scattering mean free

[+]Supported in part by AFOSR Contract No. F44620-70-C-0089.

[++]The "X-ray limit" is a photon energy region below which photoemission
spectra show peaks which are determined by the initial and final state
bands jointly, in the presence of direct ($\vec{k}$-conserving) transitions.
Below this limit final state band structure is important and manifests
itself via changing binding energy for dominant peaks as the photon energy
$h\nu$ varies. As $h\nu$ increases above this limit, the spectra gradually
are dominated by one-electron density of states features (weighted by
transition probabilities). See J. Freeouf, M. Erbudak, and D. E. East-
man, Solid State Commun. 13, 771 (1973).

paths.

Historically, measurements of $\varepsilon_2(\omega)$, the frequency-dependent imaginary part of the dielectric constant, have provided the most widely implemented technique for determining semiconductor energy band positions near the gap [1,2].

Additional band structure information which is present in frequency-dependent photoemission energy distributions (PED's) has been obtained only for a few band critical points at low energies [3,4]. This work did not use $\vec{k}$-space integration techniques coupled with full Brillouin zone energy band calculations to determine all the band structure information intrinsic to the low energy ($h\nu \lesssim 11.6$ eV) data. However, it has provided much useful information concerning certain symmetry line energy band eigenvalues.

More recently, photoemission overviews of the valence [5-7] and conduction bands [8] obtained at high photon energies have provided absolute band energies with respect to E_v, the valence band edge.

Some large-scale calculations of semiconductors and transition metals [9-11] have attempted to relate calculated energy bands to structure in PED's. The present paper presents new information on the relation of experiment to theory for a semiconductor (Ge) over a wide (~ 20 eV) photon energy range by A) giving for the first time experimental PED's for Ge obtained continuously up to $h\nu = 23$ eV using synchrotron radiation and B) providing a realistic large-scale model calculation which reproduces the direct-transition-induced structure seen in the experimental PED's.

2. THEORETICAL BAND MODELS FOR Ge

In this paper we use two model band calculations for Ge, for which the pseudopotential parameters were obtained by fitting to $\varepsilon_2(\omega)$ and to experimental photoemission valence band overviews [5]. These bands will then be used in a model calculation of $S(E_i, h\nu)$, the emission intensity as a function of initial electron energy E_i and photon energy $h\nu$.

Above the experimentally determined "X-ray limit" (about $h\nu \gtrsim 25$ eV for Ge) the valence band photoemission state density is obtained from $N_v(E_i) = S(E_i, h\nu)$. Main spectral features of the conduction band state density $N_c(E)$ can be obtained by measuring the partial yield $Y(E^*, h\nu)$ at fixed kinetic energy E^* and by varying $h\nu$ [8].

The experimental spectra N_v and N_c are shown in Figs. 1(a) and 1(b). From them, several band positions with respect to the valence band edge E_v have been determined [5], as given in Table I.

Theoretical results for these transitions and energy band widths and positions with respect to E_v for three separate calculations are also presented in Table I. The first of these, utilizing the pseudopotential of Cohen and Bergstresser [13] (CB model), was based only on optical transition data, e.g., on $\varepsilon_2(\omega)$. When photoemission-determined absolute bandwidths were obtained, it was realized that nonlocality had to be introduced into the pseudopotential to fit the new data.

Fig. 1: Experimental valence and conduction overviews N_v and N_c are compared with the corresponding d-nonlocal state densities (broadened by ~ 0.4 eV in 1(b)). A surface state peak and possible core exciton [8] are also shown.

We consider two such pseudopotentials. The first one is due to Phillips and Pandey [14] and adds an $\ell = 2$ nonlocal term to the pseudopotential. The second calculation adds to the potential a term linear in the energy [15]. Such nonlocality is explicit in the derivation of the pseudopotential equation [2]. We will refer to these as the "d-nonlocal" and the "energy-nonlocal" pseudopotential calculations. Valence and conduction band state densities for the d-nonlocal calculation are compared with the experimentally determined N_v and N_c in Fig. 1.

3. ANISOTROPIC DIRECT-TRANSITION MODEL FOR $S(E_i, h\nu)$

In order for theoretical spectra for $S(E_i, h\nu)$ to show the same elements of structure as experiment does, we have found it necessary to

	$\Sigma_{1\,min}$	L_1	$L_{2'}$	Γ_1	X_4	$X_{v \to c}$	$L_{v \to c}$	$L_{v \to c}$
Experiment	4.5[a]	7.7[c]	10.6[a]	12.6[a]	3.2[b]	17.0[b]	16.0[b]	20.5[b]
Theory [14]	4.7	7.5	10.5	12.7	3.4	15.0	15.0	19.0
Theory [15]	4.3	7.6	11.0	13.2	3.1	16.0	18.0	20.0
Theory [13]	3.8	6.9	9.9	12.0	2.57	14.2	15.4	--

Table I: Comparison of experiment with the d-nonlocal, energy-nonlocal and local theory band positions for Ge. The first five columns are symmetry point band distances below E_v in eV. The last three refer to the optical transitions shown in Fig. 5. "a" are from Ref. [5], while "b" refers to the present paper. The value for L_1 marked "c" is the correct value and through a typographical error was wrong in Ref. [5].

use an anisotropic model which yields only emission features corresponding to direct ($\vec{k}$-conserving) optical transitions for which the final state wave function describes electrons which can escape through a (111) cleaved surface.

In the independent particle formulation, the angle-resolved photoemission current $J(\hat{R}, E_i, h\nu)$ in a direction $\hat{R}$ away from the sample is given by[+] a "Fermi Golden Rule" formula [16]:

$$J(\hat{R}, E_i, h\nu) \propto \sum_I |\int \Phi_{-R} \mathcal{O} \psi_I \, d^3r|^2 \delta(E_I - E_i) \qquad (1)$$

where the sum is over all initial (I) occupied states (Bloch states modified by the proper surface boundary condition <u>and</u> surface states) of energy E_i, and Φ_{-R} is a low energy electron diffraction (LEED) wave function[++] corresponding to a plane wave <u>incident</u> on the solid in the direction $-\hat{R}$, of energy $E_i + h\nu$. $\mathcal{O} = \frac{1}{2} (\vec{A} \cdot \vec{p} + \vec{p} \cdot \vec{A})$ is the optical excitation operator. In the case where the escape depth (i.e., inelastic scattering mean free path $\lambda_{ee}(E, \vec{k})$) is $\gtrsim$ several layers, the bulk initial state contribution in equation (1) can be approximately written

$$S(E_i, h\nu) \propto \int d^3k \, T_{n'}(\vec{k}) \, D_{n'}(\vec{k}) \, [|P_{n,n'}(\vec{k})|^2$$

$$\times \, \delta(E_{n'}(\vec{k}) - E_n(\vec{k}) - h\nu) \, \delta(E_i - E_n(\vec{k}))]. \qquad (2)$$

In obtaining equation (2), we have assumed the collection of electrons emitted in all directions $\hat{R}$, $E_{n'}$ is a final, unoccupied Bloch state eigenvalue and $E_n(\vec{k})$ an occupied, initial state one. $P_{nn'}(\vec{k})$ is the optical dipole transition amplitude.

Equation (2) is in the form of a model where three steps, optical excitation, transport to the surface, and surface transmission are represented by the three terms (from right to left) in the integrand. However, we have not specialized (2) to the isotropic form used in the original "three-step model" of Berglund and Spicer [17].

We represent $D_{n'}(\vec{k})$, the effective escape depth [18] by

[+] In this paper we have omitted terms related only to the absolute quantum yield.

[++] The components of the LEED wave function Φ_{-R} of principal interest to us consist of a small number of (forward scattered) propagating Bloch-like waves inside the solid which match to the above described plane wave and which are exponentially attenuated by inelastic scattering. In contrast, the (small amplitude) reflected plane waves are the objects of interest in the LEED experiment.

$$D_{n'}(\vec{k}) \propto \hat{s} \cdot \vec{\nabla}_{\vec{k}} E_{n'}(\vec{k}) \, \tau(E_{n'}), \tag{3}$$

where $\hat{s}$ is a surface normal and τ is the inelastic scattering lifetime.

To calculate the surface transmission probability $T_{n'}(\vec{k})$ for the final state, we use a simplified LEED matching condition. We decompose the final pseudowavefunction into plane waves $|\vec{k} + \vec{G}>$:

$$\psi_{n'}(\vec{k}) = \sum_{\vec{G}} C_{\vec{G}}^{n'}(\vec{k}) \, |\vec{k} + \vec{G}> , \tag{4}$$

where $\vec{G}$ is a reciprocal lattice vector, and permit each $|\vec{k} + \vec{G}>$ to transmit according to the specular "nonreflecting barrier" condition:

$$T_{n'}(\vec{k}) = \sum_{\vec{G}} |C_{\vec{G}}^{n'}|^2 \, t_{\vec{k}+\vec{G}}; \quad t_{\vec{k}+\vec{G}} = \begin{cases} 1 \text{ for } E_{n'}(\vec{k}) - \dfrac{\hbar^2(\vec{k} + \vec{G})_{\|}^2}{2m} > V_0 \\ 0 \text{ otherwise,} \end{cases} \tag{5}$$

where V_0 is the surface barrier height.

We used a modification of a program due to Janak <u>et al</u>. [18] to do the $\vec{k}$-space integration and obtained theoretical PED's for Ge^+. For the d-non-local potential, Fig. 2 compares the experimental PED's and the theoretical ones obtained for both our anisotropic model and the corresponding isotropic model for $h\nu = 10$ and 12 eV. It is readily apparent that the overall spectral shapes of the anisotropic calculation are seen in the experiment while the isotropic model gives too many features, and with the wrong amplitudes.

4. NEW CONDUCTION AND VALENCE BAND FEATURES FOR Ge

In this section we illustrate the determination of band structure information from experimental PED's. Those for Ge are shown[++] in Fig. 3(b), where $S(E_i, h\nu)$ is plotted for a uniformly spaced series of 21 values of $h\nu$ in the 6.5 to 23 eV energy range. Direct transition features (as opposed to density of states features) give as a signature peaks whose

[+]In this calculation momentum matrix elements, wave function expansion coefficients, and energies and their gradients were calculated at over 1500 $\vec{k}$ values in the 1/48 of the reduced Brillouin zone (BZ). A symmetry transformation was used to permit the $\vec{k}$-space integral in (2) to remain in the reduced 1/48 BZ—i.e., to compensate for the breaking of the crystal symmetry by introduction of a surface [19]. A secondary electron distribution was also added to the spectra.

[++]Data for Ge were obtained as described in Ref. [5] and in a forthcoming more detailed paper [19].

Fig. 2: For the d-nonlocal energy bands, we compare experimental PED's with those predicted by the anisotropic and usual isotropic [18] theoretical models.

binding energy changes with hν. Such changing values of E_i as hν varies is illustrated graphically for several well-defined peaks and shoulders (dashed lines), and shows the preservation of direct-transition-induced structure even at high photon energies. A similar plot of $S(E_i, h\nu)$ using the anisotropic model described in Sec. 3 and the d-nonlocal energy bands shows the same evolution of peak positions as hν varies[+].

Fig. 3: $S(E_i, h\nu)$ for theory and experiment.

[+]The biggest discrepancy between theory and experiment (see Figs. 2 and 3) lies in surface-associated and intrinsic surface state emission which is seen in experiment but is not part of our bulk theoretical model. Such emission resembles the density of bulk and surface one-electron states and is most prominent near E_v for hν $\gtrsim$ 12 eV, gradually appearing stronger further below E_v as hν increases. That is, the one-electron state density contribution to equation (1) appears strongest where the inelastic scattering length λ_{ee} is smallest, and can be understood as follows. $\Phi_{-\vec{k}}$ in equation (1) couples to propagating final state Bloch waves, which for λ_{ee} large, decay slowly enough to contain several nodes, giving the $\vec{k}$-conserving direct-transition features. However, the damping means that such

Another comparison of theory and experiment is given by plotting the locus of $(E_i, h\nu)$ positions of major peaks. Such a "structure plot" [2,3] is shown in Fig. 4 for both theory and experiment. $S(E_i, h\nu)$ peaks correspond to <u>two</u>-dimensional critical points [20] in $\vec{k}$-space, and their $(E_i, h\nu)$ loci are what E. Kane [20] calls "$E_i - h\nu$ images" of photoemission critical points. Fig. 4 shows that experimentally determined energy band positions and topology are close to those of the theory.

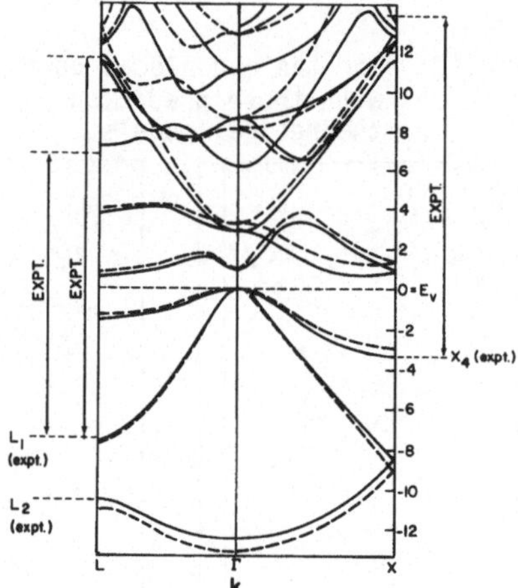

Fig. 4: Major peak positions in $S(E_i, h\nu)$ for experiment (open circles) and the d-nonlocal anisotropic theory (solid circles) are shown in the $E_i, h\nu$ plane.

Fig. 5: Energy bands along symmetry lines for the d-nonlocal (solid) and energy-nonlocal (dashed) energy bands. Experimental band positions described in the text are indicated.

$\vec{k}$-conservation is approximate, and there is some coupling to Φ_{-R} in (1) from <u>all</u> Ψ_I. Since the Ψ_I are distributed in energy according to the bulk plus surface spectral state densities, as λ_{ee} decreases the non-$\vec{k}$-conserving portion of (1) becomes greater, and additional one-electron state density emission is seen relative to the weaker (but still distinguishable) $\vec{k}$-conserving features. Also, the theoretical spectra are sharper, in part because we have not lifetime-broadened the final state bands in the calculation. As discussed in Sec. 5, such broadening would not affect the sharpness of $S(E_i, h\nu)$ in the E_i coordinate, but it would in the $h\nu$ coordinate.

We now briefly study one of the major experimental features, a peak whose E_i - hν image is labeled "B" in Fig. 4, and is seen from the theory to be composed of two E_i - hν images which converge at $E_i \sim$ - 3.4 eV at h$\nu \sim$ 16 eV. At this point, two other images also converge. Kane has shown that such a convergence of several E_i - hν images corresponds to emission at a symmetry point in $\vec{k}$-space. Indeed, both by looking at the energy bands for this calculation and from a computer analysis of the PED $\vec{k}$-space contributions, we identify this special point as corresponding to transitions from X_4 to a high-lying set of conduction bands at X (see Fig. 5). We then consider the experimental line "B" in Fig. 4, which turns at E_i = - 3.2 eV, h$\nu \sim$ 17 eV, which gives us E_i = - 3.2 eV and $\sim$ 14 eV above E_v as the position of a pair of bands at X. Returning to Fig. 5, we see that both the d-nonlocal and energy-nonlocal theories agree well with these experimental band positions.

Now consider emission from L_1 (the valence band emission peak at $E_i \sim$ - 7.4 eV in Fig. 3). The intensity of this peak is plotted in Fig. 6

as a function of hν for both experiment and the two models considered. The d-nonlocal model predicts behavior similar to that seen in the experiment, while the energy nonlocal calculation does not. The reason for

Fig. 6: Emission intensity vs. hν of L_1 peak (above secondary electron background) normalized to total intensity of peak corresponding to line "B" in Fig. 4 for experiment (X), d-nonlocal model (o), and energy-nonlocal model (□).

this behavior is shown in Fig. 5, where we indicate the energy band positions at L determined from the experimental thresholds. The lower threshold, corresponding to a final state band about 8 eV above E_v, is correctly predicted in the d-nonlocal model of Phillips and Pandey [14], while the topology of these high-lying conduction bands is qualitatively wrong for the energy-nonlocal calculation.

5. CONCLUSIONS

Space does not permit a more complete description of the analysis, which will be given in a longer paper [19]. The most important message is that short escape depths and final state lifetime broadening do not destroy direct transition features in high energy spectra. Rather, as discussed in Sec. 4, the short escape depth results in relatively featureless surface-associated emission (resembling the bulk plus surface state density) in high energy experimental spectra as shown in Fig. 3, upon which the bulk direct transition structure can be clearly identified in spite of its small amplitude[+].

Finally, while $\varepsilon_2(\omega)$ shows no sharp features at large $h\nu$ (since it convolves initial and final state lifetime broadening), we see sharp features in the E_i direction in the experimental spectra, limited only by the initial state <u>hole</u> lifetime broadening alone.

For this reason, as briefly illustrated here, ultraviolet photoemission spectroscopy can be used to determine the position of high-lying conduction bands which produce no structure in $\varepsilon_2(\omega)$. It adds an extra dimension to the optical determination of semiconductor energy bands over a wide energy range. Indeed, the one-electron picture of optical transitions appears to be valid over one Rydberg above the gap in semiconductors, a fact which was not obvious until these experimental and theoretical techniques we have briefly described were developed[++].

6. REFERENCES

[1] Brust, D.: Phys. Rev. <u>134</u>, A1337 (1964).

[2] Cohen, Marvin L.; Heine, V.: in <u>Solid State Physics</u>, H. Ehrenreich,
 F. Seitz, and D. Turnbull, editors (Academic Press, N.Y., 1970),
 Vol. 24, p. 38.

[3]. Spicer, W. E.; Eden, R. C.: Proc. of the IX Int. Conf. on the Physics
 of Semicond. (Navka, Moscow, 1968), p. 65.

[4] Spicer, W. E.: J. de Physique <u>34</u>, C6-19 (1973).

[5] Eastman, D. E.; Grobman, W. D.; Freeouf, J. L. ; Erbudak, M.: Phys.
 Rev. B<u>9</u>, 3473 (1974).

[+]The persistence of $h\nu$- (and crystal-face-) dependent structure even for $h\nu \sim 20 - 25$ eV requires that care be taken in assigning particular features in a single spectrum (obtained, for example, at 21.2 eV using a He I discharge) to intrinsic surface or bulk one-electron state density features.

[++]Direct transition-associated structure could be even more prominent in angle-resolved photoemission studies (depending on surface specularity) and may further increase the power of the techniques described here.

[6] Ley, L.; et al.: Phys. Rev. B9, 600 (1974).

[7] Shevchik, N. J.; et al.: Phys. Stat. Sol. (b) 59, 87 (1973).

[8] Eastman, D. E.; Freeouf, J. L.; Erbudak, M.: Proc. of Internat. Tetrahed. Bonded Semicond. Conf. (Yorktown Hts., N.Y., 1974).

[9] Similar attempts at low photon energies have been made by others, for example, Brust, and Saravia and Brust. See references in [5].

[10] Smith, N. V.: CRC Critical Rev. Solid State Sci. 2, 45 (1971).

[11] Williams, A. R.; et al.: private communication; and J. F. Janak, et al.: Phys. Rev. B6, 4367 (1972).

[12] Gudat, W.; Kunz, C.: Phys. Rev. Lett. 29. 169 (1972).

[13] Cohen, M. L.; Bergstresser, T. K.: Phys. Rev. 141, 789 (1966).

[14] Phillips, J. C.; Pandey, K. C.: Phys. Rev. Lett. 30, 787 (1973).

[15] Shaw, J.: private communication. The Schrödinger equation in this calculation is $(T + V + \alpha E)\psi = E\psi$ where T, V, and ψ are the kinetic energy operator, pseudopotential, and pseudowave function. The parameters used were $V(3)$, $V(8)$, $V(11)$ equal to $- 0.263$, 0.0442, and 0.026 Ryd. and $\alpha = 0.094$.

[16] Feibelman, P. J.; Eastman, D. E.: to be published.

[17] Berglund, C. N.; Spicer, W. E.: Phys. Rev. 136, A1030 and A1044 (1964).

[18] Janak, J.; et al.: Electronic Density of States, L. H. Bennett, editor (U.S. GPO, Wash., D.C., 1971). Special publication 323, p. 181.

[19] Grobman, W. D.; et al.: to be published.

[20] Kane, E. O.: Phys. Rev. 175, 1039 (1968).

X-RAY EMISSION FROM SEMICONDUCTORS
AND THE STRUCTURE OF THEIR VALENCE BANDS

K. UNGER

Sektion Physik, Karl-Marx-Universität Leipzig
701-DDR Leipzig, Arbeitsgemeinschaft "AIII-BV-Halbleiter"

Essential features of X-ray emission spectroscopy
of semiconductor valence bands are discussed on the
basis of exact computer calculations. In particular,
the influence of the symmetry of the states involved
and of the difference of the chemical components
within binary semiconducting compounds on these
spectra is analyzed and it is shown to what extent
effective nuclear charges, ionicity parameters, and
critical points of the four valence bands of diamond,
zincblende, wurtzite and rocksalt-type crystals can
be extracted from X-ray emission spectra.

1. INTRODUCTION

In contrast to the now widely used experimental method of
photo-electron spectroscopy (XPS, UPS), X-ray emission spectro-
scopy is a relatively old experimental method which has recently
attracted refreshed interest because reliable band structure
calculations of the total valence band became available for a
wide class of semiconductors.

X-ray emission spectroscopy gives insight into the structure of
the valence band, it reaches almost the same resolution power as
XPS (some 0.1 eV), but it does not reflect the state density of
the valence electrons in a simple way. The shape of the spectra
is greatly influenced by the transition probabilities for the
electrons in the valence band to empty states in the inner K or
L shell. (M spectra are in general of low intensity and re-
stricted to heavy elements only.)

The influence of the transition probability, making the evalu-
ation of the spectra more difficult, yields also some advantage
for the X-ray emission spectroscopy over XPS and UPS: We have
K and L emission spectra for each of the chemical components of
a semiconductor at our disposal. Different chemical species can
be used to 'scan' the valence band states providing different

spectra of the same semiconductor. This, we shall call hereafter the 'component difference' in the spectra of a compound semiconductor. As will be shown, this difference gives insight into the effective nuclear charges, ionicities, and related properties.

Already from the classical investigations of Wiech [1] on X-ray emission of silicon and silicon compounds, later those of Drahokoupil [2] on the spectra of the isoelectronic series Ge, GaAs, ZnSe, CuBr, two statements seemed to be clear:
(I) X-ray emission shows maxima in its spectra according to the energetic position of maxima in the state density and
(II) K emission reflects the p-like part and L emission the s-like part of the valence band state density.

To show the progress achieved in the understanding of the X-ray emission spectra, we want to concentrate our talk on three topics mainly: (I) the calculation of emission line shapes based on a pseudopotential method. There we consider especially the 'component difference' and examine whether this difference is due either to different core functions or to a different behaviour of the valence states at the positions of the different chemical components in the lattice. (II) We look for simple models which enable the understanding of the physical origin of the computer results obtained and which, after suitable generalization, would permit (III) to extract from the spectra information about the valence band structure.

Thus, we try to come from the _interpretation_ of the spectra using known band structures of the semiconductors to some rough _construction_ of the valence band structure of not so well investigated semiconductors from these experimental data.

2. OPW AND PSEUDOPOTENTIAL CALCULATIONS OF X-RAY EMISSION SPECTRA

2.1 ELEMENTAL SEMICONDUCTORS

Early in 1970, Klima [3] reported on OPW calculations of the spectra in silicon and germanium. He took into account energy-dependent Lorentzian broadening of the ideal spectra (due to finite resolution of the spectrometers, to the finite lifetime of the inner hole in the core states and due to Auger broadening). The agreement obtained with the experimental spectra was good.

2.2 BINARY SEMICONDUCTING COMPOUNDS

According to the remarks made in the introduction, the generalization on binary compounds seemed to be essential. BN has been calculated (with limited success) by Aljoschin and Smirnow [4] using the OPW method, a series of AIII-BV compounds has been investigated by Topol, Hess, Leonhardt, Unger [5...8] by means of an EPM-k.p method. The wave functions have been determined by applying the empirical pseudopotential method to the calculation of eigenvalues and eigenstates at point Γ and by extending them to other points $k \neq o$ by the k.p method. This procedure is advantageous for summing up the contributions of all k points within the irreducible segment of the first Brillouin zone (IBZ) to a given transition. The dependence of the intensity of the emission band on the photon energy $I(E)$ may be represented by

$$\tag{1}$$

where $P_{nc}(k)$ is the squared absolute value of the impulse matrix element (in dipole approximation) for the electron transition from the valence bands (n = 1...4 corresponding to increasing energy) to the core level (index c) of the corresponding atom. In order to get the true crystal wave function the pseudowave functions were completed by additional terms at each k-point to ascertain the orthogonality to the core states [9]. (Commonly the inverse procedure is used to define the pseudowave functions.) For the atomic core wave functions we adopted SCF wave functions

of Clementi [10] and have then got the matrix elements. It is essential to stress that in this manner it was possible to change core functions (neutral atomic functions or free ion functions for different chemical components) without changing the EPM band structure. The agreement with experiment was - after suitable broadening of the ideal theoretical spectra - very good. Two different spectra have been obtained for the two chemical components of the semiconducting compounds. Changing the core functions (without altering the band structure) we could answer the question posed in the introduction: The component difference stems not from different core functions (their influence on the spectra was found to be very small), but from the different behaviour of the valence wave functions at the positions of the atoms in the lattice!

For K-emission which reflects the p-states and should thus have high intensity near the top of the valence band (symmetry Γ_{15}), low intensity at the bottom of the valence band symmetry Γ_1), we found transition probabilities increasing with rising energy for the B component. However, these probabilities do not increase monotonically with rising energy for the A component. For L-emission the situation is analogous; the s-states (near Γ_1) have the highest transition probability, but for the A component there is also no monotonic variation of the transition probability with energy. The reason is quite simple. Due to the difference of the chemical constituents an inversion center lacks and the states X_1/X_3 split up yielding the antisymmetric gap E_{AS} within the valence band. For X_1/X_3 the factors $\exp|\pm\,i\vec{k}\vec{r}|$ of the wave function split into sine or cosine functions in the (100) direction yielding nocks or nodes at the positions of the A or B components in the zincblende lattice.

This splitting was hitherto known from many valley scattering [11], optical spectra [12] and occurs also at the L-point in the rocksalt structure [13,14].

Taking into account the s- or p-like character of the states at the edges of the antisymmetric gap (X_1, X_3) with respect to the positions of the A or B atom, the influence of the transition probability and especially the component difference of the X-ray

emission spectra can be explained thoroughly. For the symmetry
of the states besides the functions at the edges of the total
valence band (at Γ) also those at the edges of the antisymmetric
gap (at X or L) are important. Between them there is a smooth
energy dependence (cf. below) and no explicit $\vec{k}$-dependence of
the transition probability.

3. MODELS FOR THE CHARACTERIZATION OF THE VALENCE BAND

For a more detailed analysis of the structure of the valence
band and its reflection in X-ray emission spectra, two extreme
models will be considered, namely the dielectric theory of
crystals according to Phillips and Van Vechten [15,16] (which
stresses interactions of nearest neighbours within an approach
of an almost-free electron gas) and on the other hand tight-
binding approaches, where overlap integrals are taken as fitting
parameters. In view of the finite resolution of details in the
spectra, only relatively simple models seem to be adequate.

3.1 THE ANTISYMMETRIC GAP AND THE DIELECTRIC THEORY OF CRYSTALS

The basic parameters for the characterization of a crystal
within the dielectric theory, the homopolar part E_h and the
heteropolar part C of the Penn gap $E_g = (E_h^2 + C^2)^{1/2}$ can be
extracted from the spectra quite simply, using the following
relations for the antisymmetric gap E_{AS} and the width of the
total valence band E_{VB}

$$E_{AS} = E_{VB} \cdot \frac{2}{\pi} \arcsin \frac{C}{E_g} \, , \qquad (2)$$

$$E_{VB} = E_F - \frac{1}{2} E_g + \frac{2}{3} \frac{E_g^2}{E_F} \, . \qquad (3)$$

(E_F is the Fermi energy.) These relations are of surprisingly
wide validity [17] and cover also such elemental compounds
(Se, Te), which contain two kinds of bonds [18]. According to
(2), the Phillips ionicity parameter $f_i = (C/E_g)^{1/2}$ can be
determined directly from the spectra (cf. also [19]) and serves
as a good tool to test the antisymmetric form factors in the

EPM band structure calculation. (Care must be taken not to mix
the emission from the valence band with that from d levels
lying e.g. for AII-BVI compounds frequently within the anti-
symmetric gap.) The component difference in the X-ray emission
spectra is clearly a monotonously increasing function of the
antisymmetric gap or the antisymmetric EPM form factors: This
difference vanishes for diamond-like semiconductors and is
largest for e.g. rocksalt-type crystals of high ionicity.
Further, the component difference also tells us the sign of the
antisymmetric form factors: For example, because in cubic SiC
the lowest valence band appears clearly in the K spectrum of Si
and not in the K spectrum of C [20], we can conclude that the
C atoms are more attractive to the electrons than the Si atoms
in these crystals.

3.2 TIGHT-BINDING SCHEMES

The approaches due to Leman and Friedel [21] or Weaire and
Thorpe [22] and due to Stocker [23] are useful as interpolation
schemes to analyze the position of characteristic points within
the Brillouin zone for the four valence bands. For increasing
energy the sequence of important points in the Brillouin zone
for the zincblende structure is:

$$V_1 : \Gamma_1, L_1^{(1)}, X_1 \; ; \quad V_2 : X_3, L_1^{(2)}, \Gamma_{15} \; ; \quad V_3/V_4 : X_5, L_3, \Gamma_{15}$$

Γ and X represent the band edges, whereas L yields in most cases
the state density maxima. With increasing ionicity X_1 is shifted
towards Γ_1 more than $L_1^{(1)}$; X_3 is shifted towards Γ_{15} more than
$L_1^{(2)}$; thus X and L points come nearer, more sharp lines in the
spectra occur showing an asymmetric position of the maximum (L)
between the edge energies (X and Γ). According to Stocker [23]
taking into account only interactions between nearest bonds, the
following simple relation can be derived for the two lowest
valence bands [25]:

$$\sigma = \frac{L_1^{(1)} - \Gamma_1}{X_1 - \Gamma_1} = \frac{\Gamma_{15} - L_1^{(2)}}{\Gamma_{15} - X_3} = \left(1 - \frac{E_{AS}}{E_{VB}}\right)^{-1} \left[1 - \tfrac{1}{2}\sqrt{1 + 3\frac{E_{AS}^2}{E_{VB}^2}}\,\right] \quad (4)$$

This asymmetry parameter σ is bound to 0.5 ... 0.75, it can be
determined from the area ratio of the state density on both
sides of the line maximum [24] and serves as a test for the
correct position of the point Γ, which is otherwise often
difficult to locate (it does not correspond to the inflection
point in the spectrum as X does).

The same analysis is applicable to wurtzite-type crystals also
as far as only nearest-bond or nearest neighbour interactions
are concerned. (2) and (3) allows then to determine the two
overlap integral parameters. For rocksalt-type crystals, the
energetic position of L and X must be interchanged, the edges
of the antisymmetric gap being built from L points (cf. above).
In this way, three important critical points for each valence
band can be extracted from the spectra, whereby the spectra of
those components should be taken which represent the correspond-
ing valence band most clearly.

Fig. 1: p-like part of the state density.
———— covalent semiconductor, - - - - A component,
...... B component.

We indicate now in which manner the transition probability
i.e. the s- and p-symmetry parts of the state density in the
valence band F_s and F_p depend on their energy, using the LCAO
scheme of Weaire and Thorpe. For diamond-like semiconductors
$F_p(E) = 1 - F_s(E)$ is represented in Fig. 1 with full lines in
dependence on $v = V_1/V_2$. This parameter is related to the gap
energy E_g and the total valence band width E_{VB} according to
$E_g/E_{VB} = 1/(2v) - 1$ and therefore has usually the values
0.30 ... 0.45. $v > 0.5$ corresponds to a 'negative' energy gap;
in this case, the top of the valence band is performed from
s-symmetry states Γ_1. F_p then becomes zero at both edges of the
valence band. Such a situation should occur for e.g. PbSe, HgTe;
spectra of this type were already observed in SiO_2 (cf. [1]).

From Fig. 1 it is obvious that for high E_g/E_{VB} ($v \to 0$) a simple
linear interpolation between Γ_1 and Γ_{15} is valid for the F_p
energy dependence. The smaller $E_g = \Gamma_{1c} - \Gamma_{15v}$ the stronger the
interaction with the Γ_{1c} states, the steeper the curve near the
gap: Also according to the k.p theory,

$$\frac{dF_p(E)}{dE}\bigg|_{\Gamma_{15}} = \frac{1}{\Gamma_{15} - \Gamma_{1v}} + \frac{1}{\Gamma_{1c} - \Gamma_{15}} = \frac{1}{E_{VB}} + \frac{1}{E_g} \qquad (5)$$

is valid. For $v = \frac{1}{2}$ ($E_g = 0$) there is a linear dependence of
$F_p(E)$ between $\Gamma_{1v}(F_p = 0)$ and $\Gamma_{15c}(F_p = 1)$ according to the fact
that in the model used the total width of the conduction band
equals that of the valence band.

From the behaviour of F_p and its change with energy it becomes
obvious what will happen if an antisymmetric gap occurs within
the valence band: For the K spectra which reflect the p-like
part of the state density, there is at the edge of the anti-
symmetric gap p-symmetry at the top of V_1, s-symmetry at the
bottom of V_1, if the A component is concerned (broken lines in
Fig. 1) and vice versa for the B component (dotted lines). The
smaller the antisymmetric gap, the steeper is the symmetry
change, the slope of F_p at the edge of this gap is given by the
amount of E_{AS} (cf. Fig. 1). Thus, Fig. 1 represents clearly the
component difference in the spectra of the two components consid-
ered.

Using the curves of Fig. 1, it is readily possible to correct
the influence of the transition probability on the X-ray emission
spectra and to construct the total state density of the valence
band on the basis of these spectra.

4. CONCLUDING REMARKS

In conclusion, we note that the simple models used to explain
and to generalize the exact computer results rest mainly on
nearest-neighbour interactions. Indeed, they seem to give major
contributions to the behaviour of the spectra. Taking into
account second-nearest neighbour interactions explicitly,
5 parameters are to be determined by the spectra [25], which is
not easy considering the finite resolution of the spectra.
Finally, a general remark must be made: The empirical band
structure calculations, where were the basis of our discussion,
are merely fitted to optical transition energies across the
band gap. It is not self-evident that those calculations can
also explain the characteristic features of the total valence
band being 10 eV or more below the gap. Our experience up to
now is that the valence band data taken from X-ray spectroscopy
can be very useful to fit more precisely the pseudopotential
form factors not destroying the overall agreement between the
calculated band structure and optical transition data or
effective mass values. Of course, in some cases a recourse to
a nonlocal pseudopotential e.g. for diamond (cf. the X-ray
emission spectra and the comparison with existing band structure
calculations in [27]) or to relativistic corrections is necessary.
Beside collective excitations which cause special satellites [28]
being in most cases distinguishable from the valence band spectra,
the one-electron approximation seems to be sufficient to explain
the X-ray emission from the valence band of a semiconductor.

Thanks for good cooperation during the course of this work are
due to Dr. E. Heß, Mr. R. Hermann, Dr. K. Kreher, Dr. G. Leonhardt,
Dr. H. Neumann and last but not least to Dr. I. Topol (Rostow/
Don).

[1] Wiech, G.: Z. Phys. 216, 472 (1968)

[2] Drahokoupil, J.: J. Phys. C5, 2259 (1972)

[3] Klima, J.: J. Phys. C3, 70 (1970)

[4] Aljoschin, W.G.; Smirnow, W.P.: X-ray spectra and electronic structure of matter. Kiew 1969, p. 314

[5] Leonhardt, G.; Topol, I.; Unger, K.; Meisel, A.: Ann. Phys. (Leipzig) 28, 245 (1972)

[6] Topol, I.; Hess, E.: phys. stat. sol. (b) 56, K9 (1973)

[7] Topol, I.; Leonhardt, G.; Unger, K.; Hess, E.: phys. stat. sol. (b) 61, 285 (1974)

[8] Topol, I.; Unger, K.; Leonhardt, G.; Hess, E.: phys. stat. sol. (b) 61, 485 (1974)

[9] Unger, K.: Wiss. Z. Karl-Marx-Univ., Math.-nat. Reihe 20, 13 (1971)

[10] Clementi, E.: IBM J. Res. Developm. Suppl. 1, 1 (1965)

[11] Birman, J.L.; Lax, M.; Loudon, R.: Phys. Rev. 145, 620 (1966)

[12] Morgan, T.N.: Phys. Rev. Letters 21, 819 (1968)

[13] Cornwell, J.F.: phys. stat. sol. (b) 43, 763 (1971)

[14] Ferreira, L.G.: Phys. Rev. 137, A 1601 (1965)

[15] Phillips, J.C.: Rev. Mod. Phys. 42, 317 (1970)

[16] Van Vechten, J.A.: Phys. Rev. 182, 891 (1969); 187, 1007 (1969)

[17] Unger, K.; Neumann, H.: phys. stat. sol. (b) 64 (1974)

[18] Levine, B.F.: J. Chem. Phys. 59, 1463 (1973)

[19] Grobman, W.D.; Eastman, D.E.; Cohen, M.L.: Phys. Letters A43, 49 (1973)

[20] Nemoshkalenkow, W.W. et al: Fiz. tverd. Tela 12, 59 (1970); 15, 3465 (1973); Doklady Akad. Nauk. 214, 543 (1974)

[21] Leman, G.; Friedel, J.: J. Appl. Phys. Suppl. 33, 281 (1962)

[22] Weaire, D.; Thorpe, M.F.: Phys. Rev. B4, 2508 + 3518 (1971)

[23] Stocker, D.: Proc. Roy. Soc. A270, 397 (1962)

[24] Hermann, R.; Unger, K.: phys. stat. sol. (a), to be published

[25] Kreher, K.; Unger, K.: phys. stat. sol. (b), to be published

[26] Spicer, W.E.: J. Phys. 34, Suppl. 11 - 12, C6 (1973), See especially Fig. 12

[27] Umeno, M.; Wiech, G.: phys. stat. sol. (b) 59, 145 (1973)

[28] Hedin, L.; Lundqvist, B.I.: Sol. State Phys. 23, 1 (1969).

COMPUTED VALENCE BAND DENSITY OF STATES FOR GeTe, SnTe, AND PbTe AND INTERPRETATION OF PHOTOELECTRIC SPECTRA

G. CIUCCI and G.F. NARDELLI

Istituto di Fisica del Politecnico di Milano and Gruppo Nazionale di Struttura della Materia del Consiglio Nazionale delle Ricerche, Milano, Italy

Results of pseudopotential calculations of densities of states for the last three valence bands of GeTe, SnTe and PbTe are reported and discussed in the light of recent data on photoelectric spectroscopy. It is shown that the electron distribution curves (EDC) for u.v. photoelectrons in GeTe and SnTe agrees substantially with our valence-band density of states and this fact suggests that the EDC are practically unaffected by conduction band effects at least for these two materials. Minor agreement is instead found for PbTe. Finally, a $\vec{k}$-space analysis of the main structures occurring in the density of states is performed and it is found that large $\vec{k}$-space regions contribute to these structures.

1. INTRODUCTION

Recently attention has been called to Photoelectric Properties of IV-VI Semiconductors. Lead Calcogenides were considered by previous Authors [1] who already proposed the interpretation of these photoelectric spectra in terms of band structure. In this paper the attention is focussed on the IV-VI Tellurides and the valence-band density of states is computed for GeTe and SnTe. Although already available in the literature [2] results for PbTe are reported too, for reason of completeness in our calculation. Apart from some improvements, such as the use of fast subroutines in the solution of the algebraic eigenvalue problem, the numerical technique we have employed is essentially the same as that of Ref. [3] , i.e. the Empirical Pseudopotential Method with readjusted Cohen Form Factors [4] and spin orbit coupling coefficients fitted on atomic data. Details on the computing procedure used for the evaluation of the density of states starting from band structure in the Brillouin Zone may be found in Ref. [3] . The present values for Form Factors and Spin-Orbit Coefficients (in Rydberg) are:

V_4^S	V_8^S	V_{12}^S	V_3^A	V_{11}^A
-0.245	-0.023	0.032	0.061	0.015
-0.232	-0.026	0.018	0.052	0.021

$$-0.242 \qquad -0.0347 \qquad 0.0172 \qquad 0.052 \qquad 0.021$$

$$\lambda^S = 1.03.10^{-3} \qquad 1.09.10^{-3} \qquad 1.3.10^{-3}$$

$$\lambda^A = 5.227.10^{-4} \qquad 2.774.10^{-4} \qquad 6.64.10^{-4}$$

respectively for GeTe, SnTe and PbTe. The above values enables us to fit
the corresponding energy gaps (at point L) E = 0.21, 0.33, 0.19 (in ev),
as available in the standard literature.

2. NUMERICAL RESULTS

The calculated band structures are shown in Fig.1. As usual, for GeTe, we
did not take into account the small distortion with respect to rock salt
structure (GeTe is f.c. rhombic). Note that the poor agreement we found
for the energy position of the two lowest bands with respect to X-ray and
high energy u.v. data [1,5,6] is probably due to neglecting the non local
part of our pseudopotential [1,3] . As regards band structure of these
materials we just remark that an interesting symmetry line should be the
Q-line as it lies in a relevant plane on the surface of the Jones Zone [7].
In Fig.2 the density of states for the three highest valence bands are
reported and compared with u.v. photoelectric experimental data [1,8,9] .
It is worthwhile to note that the low energy u.v. photoelectric distribu-
tion curves (EDC) include in principle effects coming from conduction band.
Then, the fairly good agreement we found with these curves seems to
indicate that such effects, even if present, are not too relevant for the
interpretation of the main features of the spectra. The discrepancy in the
high energy part of the curve for GeTe does not surprise; indeed the sharp
energy peak at high energies in our density of states comes from regions
in the neighbourhood of the Δ-line and of the point W of the highest
valence band which are related to symmetry operations which disappears
when the distortion of the unit cell is considered.
For PbTe the comparison is made with the high energy u.v. photoelectric
spectrum. Apart from a discrepancy in the relative intensities of the peaks,
a fairly good agreement still exists for what concerns the energy position
of the most relevant structures.

The last remark concerns the origin of the peaks in the density of states. Our computed performed $\vec{k}$-space analysis gives evidence that large regions of $\vec{k}$-space contribute to these peaks so that is quite difficult to find a correspondence with definite critical points.[+]

3. REFERENCES

1 Cardona, M.; Langer, D.W.; Shevchik, N.J.; Tejeda, J: Phys. Stat. Sol. (b) 58, 127 (1973).
2 Kohn, S.E.; Yu, P.Y.; Petroff, Y.; Shen, Y.R.; Tsang, Y.; Choen, M.L.: Phys. Rev. B8, 1477 (1973).
3 Ciucci, G.; Nardelli, G.F.; Antoci, S.: Phys. Stat. Sol (b) vol.62,n.2.
4 Tung, Y.W.; Cohen, M.L.: Phys. Rev. 180, 823 (1969).
5 Shevchik, N.J.; Tejeda, J.; Langer, D.W.; Cardona, M.: Phys. Rev. Lett. 30, 659 (1973).
6 McFeely, F.R.; Kowalczyk, S.; Ley, L.; Pollak, R.A.; Sherley, D.A.: Phys. Rev. B7, 5228 (1973).
7 Ciucci, G.; Nardelli, G.F.: on print in Solid State Comm. Onodera, Y.: Solid State Comm. 11, 1397 (1972).
8 Fisher, G.B.; Spicer, W.E.: J. Non-Cryst. Solids 8-10, 978 (1972).
9 Abbati, I.; Braicovich, L.; De Michelis, B.: on print in J. Phys. C.

[+] All the computations of the present paper were performed by a UNIVAC 1106 computer making use of programs written in FORTRAN V and ALGOL 60.

ENERGY (EV)
PbTe
ENERGY (EV)
SnTe
ENERGY (EV)
GeTe
Γ Σ K W X Δ Γ Λ L K A L Q W

FIG. 1: Band structure along the main symmetry lines. Only 5 valence and 3 conduction bands are reported. Energies are referred to the top of the highest valence band at point L.[+]

FIG. 2: Curves labelled by 1 are the computed valence band densities of states (in arbitrary units);
curves labelled by 2 are the same of curves 1 with a broadening of 0.3 ev;
curves labelled by 3 are the experimental data (references [1], [9] and [8] respectively).
Energies are referred to the Fermi level.[+]

[+] All the figures have been plotted by a CALCOMP making use of software written in FORTRAN V and ALGOL 60 on the UNIVAC 1106 Computer at Centro di Calcolo dell'Università degli Studi di Milano.

UV - PHOTOEMISSION ON THE (111) - FACES OF GaP AND GaAs

K. JACOBI AND W. RANKE

Fritz-Haber-Institut der Max-Planck-Gesellschaft

Berlin/West, Fed. Rep. Germany

He I - excited photoelectron spectra were measured on clean polar (111)- and ($\overline{1}\overline{1}\overline{1}$)-faces of GaP and GaAs. In general the spectra agree with XPS-and other UPS-data, but in the uppermost valence bands the spectra have stronger structures and differ on both polar faces.

1. INTRODUCTION

Recently the knowledge of the density of valence band states (DOVS) of the III-V-compound semiconductors has been advanced both by experimental [1,2] and theoretical [3] work. The agreement between the UPS- and XPS-results is remarkable, from which can be concluded that the density of final states is of minor importance. The use of the He I - line (21,2 eV) as an excitation source gives results which can be interpreted in terms of theoretical DOVS, and has on the other hand the advantage of the low line width of a few meV, high light intensity, resulting in a better resolution of the photoelectron spectra.

Our UPS-measurements are taken from the polar (111)-faces of GaP and GaAs. The polar faces of the III-V-compound semiconductors show different behaviour in many respects, for example threshold energy for photoelectron emission and work function [4]. It is shown that the bond between the Ga- and As- respectively P-atoms in the bulk has an ionic contribution. Therefore, the question is how the (111) Ga-, ($\overline{1}\overline{1}\overline{1}$) P- and ($\overline{1}\overline{1}\overline{1}$) As- faces are stabilized. It was the aim of this work to look for differences between the polar faces by UPS.

2. EXPERIMENTAL

The cleaving planes of GaAs and GaP are the (110) - planes. Therefore, the polar (111)-faces must be prepared by Ar^+ - bombardment and annealing under UHV-conditions. The cleanliness and stoichiometry of the surfaces were controlled by Auger electron spectroscopy (AES). The same cylindrical mirror analyser is used

both for AES and UPS.

On GaAs no Ga-accumulation was found after Ar^+ - bombardment, whereas on GaP the first 5... 10 $\overset{o}{A}$ of the surface showed a Ga - accumulation up to a factor of two. Careful examination showed that this accumulation could be removed by annealing at 770 K. In an additional measurement it was checked, that the final values for the Auger peak heights of Ga and P on the polar faces were the same as for a surface prepared by breaking in the UHV-chamber.

3. RESULTS AND DISCUSSION

In the following the photoelectron spectra N(E) are shown. N(E) is obtained from the measured dN(E)/dE by computer integration. In the case of GaP the spectra are corrected for constant transmission. The spectra are drawn so that the peaks at E_B = 7, 0 eV coincide. E_B = 0 eV corresponds to the top of the valence band. Fig. 1 shows three energy spectra for the $(\bar{1}\bar{1}\bar{1})P$ - face for three different states of preparation.

FIG. 1 :

Photoelectron energy spectra for three different states of preparation

After Ar^+-bombardment the spectrum is smeared out. The peak at 9 eV vanishes
after a moderate annealing (curve 2). We suppose that this peak is caused by "broken
bonds". This view is supported by the fact that this peak appears again after an
oxidation of about 3 monolayers, when "broken bonds" can be expected for crystallo-
graphic reasons. The general trend when going from curve 1 to 3 is that the spectra
become more structured and that their amplitude increases. The variations at the
low energy side of the peak of inelastic scattered electrons is caused by variation
of the work function which is correlated with the Ga-accumulation. The emission
from -2 eV$< E_B \leq -0,5$ eV is caused by the 23,1 eV-line which always is emitted
together with the He I - line.

FIG. 2 :

Photoelectron energy spectra from
the polar faces of GaP

FIG. 3 :

Photoelectron energy spectra from
the polar faces of GaAs

Fig. 2 and 3 give the photoelectron spectra from the clean stoichiometric polar faces
of GaP and GaAs. The spectra of GaP are taken from a sample on which both polar
faces were prepared so that they could be analysed without changing the sample.
The spectra of GaAs are taken from two separate samples from the same cut of the
ingot. Therefore, in the case of GaAs the relative intensities from the two faces
are not comparable. Table 1 gives the energies of the observed peaks.

TABLE 1

	band I	band II	band III
GaP Ga(111)	(0.8) 1.9		(11.1) (12.5) 15.6
	4.4	7.0	
" P ($\overline{111}$)	1.2 3.2		(12.0) (13.4) 15.6
GaAs Ga(111)	1.8		(11.0) (12.4) (15.0)
	2.8 3.9	6.9	
" As($\overline{111}$)	1.0		(11.8) (13.3) –

The values given are the energies of the centre between positive and negative peaks
in the measured dN/dE - curves from which peak positions are seen with better
accuracy. With respect to the lower bands the agreement between the two faces is
rather good, but there are differences in the upper valence band. These differences
are completely reproducible and cannot be diminished by a controlled adsorption
of about 0.2 monolayers of oxygen or Ga-accumulation of a factor of 1.5. The com-
parison of the spectra from GaP and GaAs shows that the spectra from the corres-
ponding polar faces are very similar.

4. REFERENCES

[1] Ley, L.; Pollak, R.A.; McFeely, F.R.; Kowalczyk, S.P.; Shirley,D.A.:
 Phys. Rev.B-9,600 (1974)
[2] Shevchik, N.J.; Tejeda, J.; Cardona, M.: to be published
[3] Chelikowski , J.; Chadi, D.J.; Cohen, M.L.: Phys. Rev.B 8,2786 (1973)
[4] Ranke, W.; Jacobi, K.: Solid State Comm. 13, 705 (1973)

ELECTRONIC DENSITIES OF STATES OF AMORPHOUS
AND TRIGONAL SE AND TE*

J. D. JOANNOPOULOS,[†] M. SCHLÜTER[‡] AND MARVIN L. COHEN
Department of Physics and Inorganic Materials Research Division,
Lawrence Berkeley Laboratory, University of California
Berkeley, California 94720

We have used the Empirical Pseudopotential
Method to obtain electronic densities of states
and charge distributions for the valence
electrons in trigonal Se and Te. We have
explored the relation between density of states
structure and charge density. Specifically,
structures in recent photoemission spectra are
identified with particular interchain and
intrachain bonding states. These results are
then used to interpret the changes observed
in the experimental photoemission spectra of
amorphous Se and Te.

We present here new calculations of the electronic densities of
states of trigonal Se and Te (using the empirical pseudopoten-
tial method (EPM))which for the first time agree quantitatively
with all the observed structure in recent experimental photo-
emission spectra. In Fig. 1 we show the results of our EPM
calculations for the density of states along with the recent
photoemission results of Shevchik et al. [1] for Se, and
Schlüter et al. [2] for Te. The calculated spectra were convol-
uted with an energy dependent broadening function in order to
facilitate comparisons with experiment. The lowest band repre-
sents essentially the atomic s-like states of Se and Te. In Se
the density of states of this band strongly resembles that of a
one dimensional chain whereas in Te, there is, in addition,
structure resembling the density of states of a simple cubic
lattice. The next band contains p-like bonding states which lie
between -6.0 eV and -2.2 eV for Se and for Te. Finally the top
most valence states are predominantly non-bonding p-like in
nature.

*Supported in part by the National Science Foundation under
Grant No. GH 35688.
†Present address: Department of Physics, MIT, Cambridge, Mass.
‡Swiss National Science Foundation postdoctoral fellow.

FIG. 1: Calculated broadened density of states and photo-emission spectra for trigonal Se (top) and Te (bottom).

FIG. 2: Bonding charge of trigonal Se for the (a) lower and (b) upper p-like bonding states.

In order to understand the origin of the characteristic two-peak structure found in the p-like bonding states of both Se and Te we have calculated the electronic charge distributions of states in each peak. We then go one step further and isolate the short wavelength Fourier components from the long wavelength Fourier components. This introduces a new method of defining bonding charges and a way to separate out the effects of metallicity. The cutoff or boundary wavelength λ_o between short and long wavelength components was found to lie naturally at $\lambda_o = d$ where d is the nearest neighbor separation in Se and Te. The results obtained for Se by retaining only Fourier components with $\lambda \leq \lambda_o$ are shown in Fig. 2 (a) and (b). The results for Te are similar and will not be presented here [2]. The charge in the lower energy peak of the p-like bonding states is well localized between atoms belonging to the same chain and thus corresponds to states representing <u>intra-chain bonding</u>. In the upper or higher energy peak of the p-like bonding states the charge is

displaced out of the bonds and is concentrated in the region
between neighboring chains. This charge arises in part because
of the mixing and hybridization of p_x, p_y, s and d states.
Moreover we find that the amount of charge concentrated in the
right half of Fig. 2b depends strongly on the interchain separa-
tion and can thus be identified as an <u>inter-chain bonding</u>
charge.

If we compare the photoemission spectra of trigonal Se [1] and
Te [2] with those of amorphous Se [1] and Te [2] (see Fig. 3) we
find two very interesting differences occurring in the s-like
and bonding p-like states res-
pectively. In Se the lower
energy peak of the p-like bond-
ing states has become weaker
whereas the higher energy peak
has become stronger in the amor-
phous phase. From our analysis
of the crystalline case we
suggest that this reversal cor-
responds to decrease of the
number of pure intra-chain bond-
ing states. Thus there are now
more electrons occupying states
which are partially localized
<u>outside</u> the chains. In the s-
like region for Se we find a
very unusual effect. The dip
seems to be bigger in the amor-
phous phase than in the trigonal

FIG. 3: Photoemission spectra
on trigonal and amorphous Se
(top) and Te (bottom) as
obtained from refs. [1] and [2]
respectively.

phase. This suggests some very interesting structural proper-
ties. For example it could not be caused by just a breaking of
the infinite chains. This would only tend to fill up the dip
unless the chains were of order two which seems rather unlikely.
A possible alternative is the formation of some type of rings.
In particular the dip would increase if the rings were of order
three, five, six or seven. Rings of order four, eight, or five
and seven together, would certainly tend to fill up the dip.
Furthermore, since the bond angles in the trigonal phase are

about 104° we would suspect that the most likely ring structures would be of type five-fold and six-fold, or six-fold and seven-fold. We therefore propose that amorphous Se contains a substantial number of atoms in ring-like configurations with the rest of the atoms being members of chains.

The photoemission results for amorphous and trigonal Te obtained by Shevchik et al. [3] using the same sputtering technique as in the Se case give similar results. However these results differ considerably from photoemission data on amorphous Te prepared by argon bombardment [2]. The latter results are shown in Fig. 3 (bottom). In the bonding p-like region we now find, in contrast to Se, a shift of strength to _lower_ energies. This suggests an _increase_ in the number of the pure intra-chain bonding electrons which would be consistent with an increase in the covalency of Te in the amorphous phase. The structural information obtained from the s-like states is somewhat more difficult to discern since there is now a filling up of the dip in the amorphous case. It is difficult to make a conclusive statement about the structure of this sample of Te without better experimental resolution. One could speculate however that argon bombardment would tend to leave the system with atoms existing mostly in broken chain configurations. On the other hand, sputtering at room temperature with the deposition of thin films may favor the formation of rings.

REFERENCES

[1] Shevchik, N. J.; Tejeda, J.; Cardona, M.; Langer, D. W.:
Sol. State Comm. 12, 1785 (1973)
[2] Schlüter, M.; Joannopoulos, J. D.; Cohen, M. L.; Ley, L.;
Kowalczyk, S.; Pollak, R.; Shirley, D. A.: to be published
[3] Shevchik, J. J.; Cardona, M.; Tejeda, J.: Phys. Rev. B8,
2833 (1973)

DENSITIES OF VALENCE STATES AND RAMAN SCATTERING OF AMORPHOUS $CuGaSe_2$

W. BRAUN AND J.S. LANNIN

Max-Planck-Institut für Festkörperforschung,
Stuttgart, Fed. Rep. Germany

Electron and phonon densities of states infor-
mation has been obtained in the amorphous I-III-VI$_2$
system $CuGaSe_2$ by photoelectron spectroscopy and
Raman scattering. The partial p and d densities
of valence states are presented. The shape of the
d contribution is determined to be similar to that
in the copper halides.

1. INTRODUCTION

Recently the study of densities of valence states by photo-
electron spectroscopy and densities of vibrational states by
Raman scattering has been emphasized in the amorphous [1,2] and
crystalline [1,3] tetrahedral binary semiconductors. In this
contribution we report the first photoelectron and Raman
measurements of amorphous I-III-VI$_2$ systems and present the
experimental results for a-$CuGaSe_2$. As the local tetrahedral
bonding primarily determines the general features of both the
electronic and vibrational densities of states, the Raman
scattering and photoelectron spectroscopy results presented
here also yield information about the corresponding crystal. In
particular the degree of mixing of p and d valence electrons is
expected to be similar in the crystalline and amorphous forms.

2. EXPERIMENT

The thin film samples were prepared by sputtering from the poly-
crystalline compound onto a substrate held at room temperature.
Raman scattering and x-ray analysis indicated their amorphous
nature and suggest tetrahedral coordination. Annealing in vacu-
um at 300°C for 3 hrs. results in small crystallites of ~50 Å
in size.

The measurements for determining the densities of valence states (DOVS) were performed with a Vacuum Generators ESCA III system equipped with a He discharge lamp and an AlKα x-ray source. The resolution was 0.3 eV and 1.2 eV for uv and x-ray excitation, respectively. Raman measurements employed a Spex third monochromator system with 6471 Å (Kr) and 4880 Å (Ar) laser excitation. The Raleigh tail was estimated using an Al film.

3. RAMAN RESULTS

The Raman spectra of polycrystalline and amorphous $CuGaSe_2$ are shown in Figure 1 arbitrarily normalized to the same maximum intensity. The form of latter spectrum in particular its width as well as the position of the main low and higher frequency peaks does in fact indicate that the film is amorphous. The crystalline spectrum is dominated by a single peak of Γ_1 symmetry which is related to the W_1 mode in the zinc blende structure, while the lowest and highest modes, by analogy with c-$CuGaS_2$ [4], correspond to $TA(X_5)$ and $LO(\Gamma_{15})$. The amorphous spectrum in contrast is reasonably expected to yield a weighed density of vibrational states of c-$CuGaSe_2$, but broadened due to structural disorder as well as possible compositional disorder if Cu and Ga atoms are random distributed about the Se atoms.

FIG. 1: Comparison of the measured Raman intensities of crystalline (c) and amorphous (a) $CuGaSe_2$. Dashed curve is the reduced Raman spectrum.

The reduced Raman spectrum shown by the dashed curve has its major peak at $\sim$245 cm^{-1}, which is $\sim$20 cm^{-1} above the maximum of the broadened density of states of c-ZnSe, the II-VI analogue [5]. This is reasonable since the highest optical mode is $\sim$23 cm^{-1} higher in c-$CuGaSe_2$ than in c-ZnSe [6]. The low frequency region in contrast has a broad peak with an inflection $\sim$80 cm^{-1}.

An estimate of the amorphous density of states using an average
from c-CuBr [7] and c-GaAs yields, however, a peak $\sim$60 cm^{-1}. The
weak relative intensity of the low frequency region, its peak
position and the recent density of states determination of Axe
et al. [8] in a-Ge when compared with the reduced Raman intensi-
ty all suggest the importance of frequency dependent coupling
constant effects at frequencies beyond the elastic continu m
regime [9] . A variation of the coupling constant C between ω
and ω^2 is in fact required to yield the estimated peak in the
density of states.

4. PHOTOEMISSION RESULTS

Figure 2 shows the experimental photoelectron distribution
curves (EDC's) obtained from a-CuGaSe$_2$ at hν = 21.2, 40.8 and
1486.6 eV. The zero of binding energy in Fig. 2 was taken to be
the top of the valence band (TOVB). The estimated contribution
of secondary electrons is indicated by the dashed lines. The
structure in the 40.8 and 1486.6 eV curves of Fig.2 reflects
the cross section weighed density of d states; at these ener-
gies the excitation probability for p valence electrons is
small compared to that of d electrons [10,11]. At 21.2 eV, p
and d excitation probabilities are of the same order of magni-
tude [10,11] and features of the p density of states (peaks at
3.5, 5 and 7 eV) appear. It is also
clear that an appreciable fraction
of the d density of states is concen-
trated in a nearly core-like peak,
centered at 2.4 eV below the TOVB.
The shape of the EDC's shows a
strong dependence on photon energy
over the whole binding energy range
This feature indicates that the p-d
hybridization is not uniform over
the valence band. The same general
features have also been observed in
a-CuGaS$_2$.
In order to calculate p and d con-

Fig.2: Photoelectron
distribution curves for
amorphous CuGaSe$_2$ films
at photon energies 21.2,
40.8 and 1486.6 eV

tributions to the densities of valence states (DOVS) the experimental curves (shown in Fig.2) should be corrected for the analyser transmission factor and the escape depth of the electrons. However, for the small energy range of the valence bands these corrections can be neglected as a first approximation. The valence s levels in this compound are more tightly bound ($\sim$13 eV) than the p and d levels and thus do not mix appreciably with the portions of the valence bands shown in Fig.2. Thus, these VB's are essentially composed of the p orbitals of the anion and the d orbitals of the Cu ion. With the assumption of small interference between p and d absorption matrix elements [12] the experimental EDC's after correction for secondary electrons (in a tight binding approximation) is obtained [13]:

$$N(E, h\nu) = A\left[\rho_p(E)\sigma_p(h\nu)/6 + \rho_d(E)\sigma_d(h\nu)/10\right], \quad (1)$$

where A is a normalization constant, $\rho_{p,d}$ is the partial p and d DOVS and σ_i/n_i is the photoionization cross section per electron of orbital quantum number i.

Fig.3 : Partial and total densities of valence states (DOVS) for amorphous CuGaSe$_2$ obtained from the data of Fig.2.

Inserting into Eq.(1) for σ_p the experimental partial cross section of Kr [11] , the rare gas nearest in the periodic table to Se, and for σ_d the cross section for Cu [10] we obtain the partial and total DOVS as shown in Fig.3. The p DOVS extends over a range of 7.5 eV, 1.5 eV wider than in ZnSe [3]. This broadening of the band seems due to strong p-d mixing. The partial d DOVS in contrast exhibits a very close resemblance to the d-type VB of CuBr [13].This is a consequence of the approximately similar tetrahedral environment of the Cu atoms. From the partial DOVS analysis the average d admixture in the upper part of the VB ($\stackrel{<}{\sim}$ 4 eV) is estimated to be 55% $\pm$ 5%. The d admixtures obtained from the analysis of the spin-orbit splitting in CuGaSe$_2$ was found to be 36% [14] . However, for comparison

with these data, a correction has to be made for the contribution of the core-like d states at 2.4 eV [12]. This yields a d mixture at the upper valence bands of 37%, in good agreement with the optical data.

5. ACKNOWLEDGEMENTS

We wish to thank Mr. A. Barynin for assistance in preparing the films, Dr. A. Räuber for supplying the polycrystalline material, Mr. M. Bettini and Prof. M. Cardona for helpful discussions.

6. REFERENCES

[1] Shevchik, N.J.; Tejeda, J.; Cardona, M.: Phys.Rev.B9, 2627 (1974)

[2] Lannin, J.S.: Proc. Int. Topical Conference on Tetrahedral Amorphous Semiconductors, 1973, Yorktown Heights, N.Y. (to be published)

[3] Shevchik, N.J.; Tejeda, J.; Langer, D.W.; Cardona, M.: phys.stat.sol. (b) 60, 345 (1973)

[4] Bettini,M.(private communication)

[5] Hennion,B.; Moussa,F; Pepy,G; Kunc,K.: Phys.Lett. 36A, 376 (1971)

[6] Irwin,J.C.;and La Combe,J.: Can.Journal Phys. 50, 3596 (1972)

[7] Prevot,B.; Carabatos,C.; Schwab,C; Hennion,B.; and Moussa, F.: Sol. State Commun. 13, 1725 (1973)

[8] Axe, J.D.; Keating, D.T.; Cargill III, G.S.; Alben, R.: Proc. Int. Topical Conference of Tetrahedral Amorphous Semiconductors, 1973, Yorktown Heights, N.Y. (to be published)

[9] Lannin, J.S.: Sol. State Commun. 12, 947 (1973)

[10] Hagemann, H.J.; Gudat,W.; Kunz,C. (to be published)

[11] Kemeny, P.C.; Poole, R.I.; Jenkin, J.G.; Liesegang,J.; Lecky,G.: Phys.Rev. A, (to be published)

[12] Braun, W.; Goldmann,A.; Cardona,M. (to be published)

[13] Goldmann,A.; Tejeda,J.; Shevchik,N.J.; Cardona, M. (to be published)

[14] Shay, J.L.; Tell, B.; Kasper, H.M.; Schiavone, L.M.: Phys.Rev. B5, 5003 (1972)

DETAILED STRUCTURE OF ELECTRONIC SURFACE STATE DENSITIES IN THE GAP OF SEMICONDUCTORS BY PHOTOEMISSION MEASUREMENTS

C. SEBENNE, D. BOLMONT, G. GUICHAR and M. BALKANSKI

Laboratoire de Physique des Solides, associé au Centre
National de la Recherche Scientifique - Université Paris VI
75230 PARIS Cedex 05 - FRANCE

The accurate measurements of the photoemission yield, down to
the 10^{-10} range, as a function of photon energy in the threshold
region, can give a precise information on the electronic surface
state density in the band gap of semiconductors. The method is
illustrated with results obtained on the (111) face of silicon, clean
or after interaction with oxygen or hydrogen, and on the (10$\bar{1}$0)
face of CdS clean, or after interaction with oxygen.

1. INTRODUCTION.

In any material, the absolute threshold of electron photoemission cor-
responds to the work function φ, the energy between the Fermi level and
the vacuum level. It is also true for a semiconductor where the Fermi le-
vel is usually situated in the band gap. Thus, at photon energies between
φ and the ionization energy ϕ (energy from the top of the valence band
at the surface to the vacuum level), the photoemitted electrons, if any,
must come from surface or bulk localized levels, when the surface space
charge region is adequate, that is when the band bending is such that there
is no mixing of photoelectrons coming from localized gap states and bulk
valence states. This occurs when the semiconductor either has a low
enough doping for the space charge region depth to be large compared to
the penetration depth of the photons, or is heavily doped n type, which im-
plies a depletion layer for electrons.

From these considerations, it is clear that an accurate measurement
of the current of photoemitted electrons in a photon energy range covering
the interval from φ to ϕ with resolution lower than .01 eV can give a de-
tailed information on the localized state density between the Fermi level
and the top of the valence band, surface states being privileged compared
to localized bulk states, due to the small penetration depth of the photons.

The reason why such a method has not been developed so far is proba-
bly that the experimental requirements are difficult to meet : it is neces-
sary to measure currents in the range of a few electrons per second and
to eliminate any parasitic effect down to that level.

The experimental set up has been described elsewhere [1]. In the pre-
sent paper, the method is illustrated by some results obtained on ultra-
high vacuum cleaved samples : silicon (111) face, first clean, then exposed
to low pressures of oxygen or hydrogen, and CdS ($10\bar{1}0$) face first clean,
then exposed to oxygen at atmospheric pressure.

2. SILICON.

The electronic surface states of the clean cleaved and oxidized (111)
face of silicon have been extensively studied in the last few years (see re-
ference 1 and references therein). The main results can be summarized
as follows : the clean surface is characterized by a density of states which
is very low at the Fermi level and increases toward the valence band, the
top of which is about .50 eV below the Fermi level, whatever type and do-
ping of the sample, reaches a maximum about .50 eV below the top of the
valence band and then goes to zero. The surface covered with a monolayer
of oxygen shows a nearly complete disappearance of that band, which is
replaced by other structures at energies deeper in the valence band.

How to discriminate between photoemitted electrons coming from
either the surface states or the bulk valence states in the yield measure-
ments in the threshold region has been discussed elsewhere [1]. On Fig. 1,
we show the shape of the surface state distributions obtained from our
measurements assuming constant matrix elements for the involved transi-
tions. Since bulk effects dominate at photon energies higher than ϕ, the
results are given only in the band gap, below the Fermi level. Curve 1 re-
calls the distribution obtained for the clean (111) face after cleavage at a
pressure of 5×10^{-11} torr. Curve 2 corresponds to the surface covered
with a monolayer of oxygen : it shows a decrease of about one order of ma-
gnitude compared to the clean surface. However, this cannot be considered

as a definite conclusion since the transition probability has no reason to be the same for the clean and oxidized surfaces. Curve 3 shows the effect of hydrogen once an equilibrium has been reached after succesive inputs at increased pressures. The total hydrogen exposure was 450 Langmuirs as controlled with an ionization gauge, the ionization factor being taken into account. Besides the structural change which appears by comparison to the clean surface, Ψ decreases by about .25 eV and Φ by about .35 eV.

3. CADMIUM SULFIDE.

Electronic surface properties of CdS have been much less studied than those of silicon, and the present understanding is mostly qualitative. Without attempting to give a complete analysis of the literature, we shall just recall some features of these properties.

Photoemission threshold measurements, together with photovoltage, have been reported by Swank [2] and show an ionization energy of 7.26 eV and a band bending of about .1 eV upward. This last value is in good agreement with other determinations, including our own, but the ionization energy appears to be too high by about 1 eV, compared to the value obtained by Petroff [3] and our present results. No density of surface states has been proposed, up to now, and the most complete results have been reported by Łagowski, Balestra and Gatos [4], using photovoltage spectra : for the cleavage planes, structures in the spectra are found at .05 eV, 1.1 eV and 1.6 eV below the conduction band, and are attributed to discrete surface levels.

Preliminary results of photoemission threshold measurements on the clean ultrahigh vacuum cleaved $(10\bar{1}0)$ face of n-type CdS with 10^{15} to 10^{16} carriers per cm^3 are summarized on Fig. 2, curve 1, which shows the first derivative of the yield curve, in arbitrary units, as a function of photon energy. The first point to be noted is the starting point of the curve at low energy : it gives the highest possible value of Ψ at about 3.95 eV. Considering the high sensitivity of the measurements, this value will be adopted as the Fermi level position. With a band bending of about .10 eV upward and a Fermi level position at about .20 eV below the conduction band

in the bulk, the electron affinity is about 3.65 eV and the ionization energy 6.1 eV since the energy gap is 2.45 eV.

Using these values, the main peaks which appear on Fig. 2, curve 1, have been located with respect to the conduction band. They are given on the following table, in eV :

E_{peak}	4.1	4.45	4.65	5.1	5.5
$E_{CB} - E_{peak}$	.45	.8	1.0	1.45	1.85

The comparison with Łagowski et al. [4] results is not straightforward for two reasons, and the discrepancy may not be as bad as it seems. First of all, the photovoltage measurements probably give information more on the location of sharp structures (either peaks or dips) in the density of states than just on the maxima.

Besides, curve 1 of fig. 2 does not necessarily give the shape of the surface states distribution : contrary to silicon, i) different bands are obtained and the assumption of constant matrix elements for the transitions no longer holds, ii) CdS is a material known to contain numerous localized bulk states in the gap,and it appears quite difficult to discriminate between surface and localized bulk states. Curves 2 and 3 on fig. 2 illustrate this last point ; they correspond to a doped crystal with about 10^{18} carriers per cm^3 after cleavage and after exposure to 1 atmosphere of oxygen respectively. A comparison with curve 1 is possible, taking into account the necessary shifts in energy due to different band bendings and bulk distances between Fermi level and conduction band. Besides the structures observed on curve 1, which are still present on curve 2 and reduced by the effect of oxygen, new structures appear which are less affected by oxygen and can be attributed to bulk localized levels.

4. REFERENCES.

(1) C. SEBENNE, D. BOLMONT, G. GUICHAR and M. BALKANSKI, Proceedings of the 2nd ICSS, Kyoto, to be published (1974).
(2) R.K. SWANK, Phys. Rev. 153, 844 (1967).
(3) Y. PETROFF, Thèse d'Etat, Université de Paris (1970).
(4) J. ŁAGOWSKI, C.L. BALESTA and H.C. GATOS, Surf. Scien. 29, 213 (1972), and references therein.

FIGURE 1 : Effective density of surface states, in arbitrary units, of the cleaved (111) face of silicon, 1) clean, 2) covered with a monolayer of oxygen, 3) after exposure to 450 Langmuirs of hydrogen.

FIGURE 2 : First derivative of the yield curve on the ($10\bar{1}0$) face of CdS, 1) clean, pure sample, 2) clean doped sample, 3) doped sample after exposure to 1 atmosphere of oxygen.

INTERNAL PHOTOEMISSION: THEORY AND EXPERIMENT
DANIEL F. BLOSSEY
Xerox Webster Research Center
Webster, New York, USA

A one-dimensional Onsager theory is developed to
explain the field and temperature dependence of
carrier photoinjection from metals into semi-
conductors or insulators.

1. INTRODUCTION

Internal photoemission has been used largely for determination of barrier
heights at metal-semiconductor interfaces [1]. For proper determination
of barrier heights, it is desirable to work at fields above the saturation
field for the emission current; but, in most insulators, the saturation
field is too high to be accessible or in the region where barrier lowering
effects become important. Below the saturation field, the emission cur-
rent is observed to be both field and temperature dependent [2]. If the
barrier determination is to be done at a field below the saturation field,
it is important to understand which injected hot carriers are actually
collected and which are returned to the metal emitter. It is the purpose
of this paper to calculate the collection efficiency for injected hot
carriers and thereby develop a model to explain the field and temperature
dependence of internal photoemission currents in metal-insulator systems.

2. MODEL

The general features of the injection process are shown schematically in
Figure 1a for the case of electron injection. After the hot carrier is
injected into the solid, it will give up energy to the lattice and become
thermalized at some distance from the metal surface. The thermalized car-
rier can then drift under the influence of diffusion, its own image field,
and the applied field with the ultimate fate of either returning to the
metal or being collected. The problem thus divides naturally into two
parts a) a description of the thermalization process and b) collection of
the thermalized carriers. The thermalization of hot carriers we will desc-
ribe in a phenomenological fashion incorporating the microscopic mobility
whereas the motion after thermalization we will treat exactly. We will see
that the microscopic mobility determines the thermalization length in the
solid and that the collection efficiency is related to the microscopic
mobility through the thermalization length. The description of the thermal

carrier motion is simply a 1-D analog to a 3-D problem solved by Onsager [3] in which the thermalized carrier pair starts with an initial separation (thermalization length) and has the ultimate fate of either recombination or dissociation.

3. COLLECTION OF THERMAL CARRIERS

To describe the motion of thermal carriers we will start with the continuity equation which, for steady state currents, is given by

$$G(x) + \frac{d}{dx}(nv) = 0$$

where $G(x)$ defines the rate and position at which the hot injected carriers are thermalized and acts as a source term for the thermal carriers, n is the thermal carrier density and v is their velocity. The velocity of the thermal carriers is governed by diffusion, the image force, and the applied field, and is given by

$$v = -\frac{D}{n}\frac{dn}{dx} - \mu\frac{d\phi}{dx}, \quad e\phi = -\frac{e^2}{4\epsilon x} - eEx,$$

FIG. 1: (a) Schematic of carrier injection, thermalization, and collection. (b) Escape probability for thermal carriers vs. electric field at constant temperature for a series of initial positions.

where D = $\mu kT/e$ is the diffusion coefficient, μ is the carrier mobility. The first term in $e\phi$ is the image force potential and the second is the applied field potential. Combining the above equations we have that

$$\frac{e}{\mu kT} \, G(x) = \frac{d}{dx} \left\{ \exp\left(-\frac{e\phi}{kT}\right) \frac{d}{dx} \left[n \, \exp\left(\frac{e\phi}{kT}\right)\right]\right\}.$$

If there are sinks at both the origin (emitter) and infinity (collector) and the potential $e\phi$ becomes large negative at these points, by integrating the above expression twice, changing variables, and calculating the current J = nev at infinity, it may be shown that the current will have the form

$$J = \int_0^\infty dx e G(x) P(x), \quad P(x) = \frac{\int_0^x dx \, \exp\left(e\phi/kT\right)}{\int_0^\infty dx \, \exp\left(e\phi/kT\right)},$$

where the function $P(x)$ takes the role of an escape probability for a thermal carrier with initial position x. The escape probability P is shown as a function of dimensionless field at constant temperature in Figure 1b for a series of initial dimensionless starting positions. A typical value of E_T, at room temperature would be 10^4-10^5V/cm for most solids, a typical value for x_T would be 10-100Å.

4. THERMALIZATION OF HOT CARRIERS

To calculate thermalization lengths let us start with the equation of motion for an injected hot carrier

$$m \, \frac{dv}{dt} = mv \, \frac{dv}{dx} = -\frac{ev}{\mu} - e \, \frac{d\phi}{dx}$$

where we take the velocity of the particle to be an explicit function of x and the energy loss scattering process is included phenomenologically as a mobility dependent drag term. We have assumed in the above equation that the main forces in the deacceleration of the hot carriers are the potential and drag terms and have neglected diffusion. The condition for thermalization is taken to be when the carrier's kinetic energy is 1/2 kT. The above equation is non-linear and has to be solved numerically.

The treatment of the scattering of hot carriers is a very difficult problem. Rather than writing expressions in terms of electron-phonon coupling constants, final density of states, etc., we have chosen to parameterize the energy loss process as a mobility dependent drag term, this mobility being a parameter which is representative of the scattering processes in the

band. The above equation should be valid if the carrier mean path is small compared to the extent of the potential, x_E. Since it is only at small fields (large x_E) that field dependencies are important, we argue that the region of validity is self-defining in that it is valid for $E < E_{sat}$ and it does not enter into the problem for $E > E_{sat}$.

5. INTERNAL PHOTOEMISSION YIELD

To calculate the injection current J we must map the energy distribution of injected hot carriers into the source term G(x). But first we must calculate the kinetic energy distribution for the kinetic energy of hot carriers normal to the interface, E_n. For internal photoemission, the primary step is the creation of hot carriers in the metal. These hot carriers will have an equal probability of traveling in all directions whereby only a certain fraction of the carriers will have enough kinetic energy normal to the surface to get in. The standard assumption for internal photoemission calculations is that the transition matrix element in the metal does not change rapidly with photon energy over the region of interest. This means that excitations between 0 and $h\nu$ above the Fermi level are equally probable. If this approximation is valid, it is easy to show that the differential yield (carriers injected/photon/normal kinetic energy) is given by

$$\frac{dY}{dE_n} = \frac{1}{2h\nu}\left[\left(\frac{h\nu+E_F}{E_n}\right)^{1/2} - 1\right] \, , \; E_F < E_n < h\nu + E_F.$$

If the injection barrier (relative to E_F) is E_0 and scattering is neglected, integration of the above equation be-

FIG. 2: Comparison of theory and experiment of Mort, Schmidlin, Lakatos [2].
(a) Electron photoinjection from Cu into CdS.
(b) Hole photoinjection from Au into amorphous Se.

tween $E_F + E_0$ and $E_F + h\nu$ gives the usual field independent result for the photoemission yield,

$$Y = 1 - \frac{1}{2}\left(\frac{h\nu - E_0}{h\nu + E_F}\right) - \left[1 - \left(\frac{h\nu - E_0}{h\nu + E_F}\right)\right]^{1/2} \simeq \frac{1}{8}\left(\frac{h\nu - E_0}{h\nu + E_F}\right)^2$$

The above equation has been used as the basis for barrier determination, in that, by plotting the square root of the photoemission current versus photon energy near threshold, the threshold is determined by linear extrapolation of the curve to zero. To calculate the yield with scattering, we calculate the thermalization length $x = x(E_n)$ as a function of the normal kinetic energy, E_n. The yield $Y = J/eF$, where F is the photon flux, is then defined as

$$Y = \int_{E_F}^{h\nu + E_F} dE_n (dY/dE_n)\, P[x(E_n)]$$

where we have described the source term as $G(x) = (dY/dE_n)(dE_n/dx)$. This equation must be solved numerically. Comparison between theory and experiment [2] is shown for electron injection in Cu:CdS system in Figure 2a and for hole injection in Au:Se system in Figure 2b. The microscopic mobilities of 300 cm^2/volt-sec for CdS and 20 cm^2/volt-sec were chosen for best fits and agree quite well with crystalline values. The collected charge is in arbitrary units. This calculation is described in more detail in [4].

6. REFERENCES

[1] Williams, R.: Semicond. and Semimetals 8, 97 (1970).
[2] Mort, J.; Schmidlin, F.W.; Lakatos, A.I.: J. Appl. Phys. 42, 5761 (1971).
[3] Onsager, L.: Phys. Rev. 54, 554 (1938).
[4] Blossey, D.F.: Phys. Rev. B, June 15, 1974.

PHOTOEMISSION AND OPTICAL PROPERTIES OF GaSe[*]

P. Thiry, R. Pincheaux, D. Dagneaux, and Y. Petroff

University of Paris VI and L.U.R.E.[+] (Orsay)

We present a parallel study of the photoemission and the
optical properties of GaSe up to 30 eV using the synchro-
tron radiation of Orsay (L.U.R.E.). The reflectivity R
and dR were measured at various temperatures. In $dR/d\lambda$
region of the d band transitions (20-27 eV), 10 structures
are observed, giving an overview of the density of states
of the conduction band. Measurements with different ang-
les of incidence (between 20° and 60°) show a very large
anisotropy. From the photoemission we have deduced the
density of states of the valence up to 10 eV.

I. INTRODUCTION

The layer compound GaSe shows a strong structural anisotropy; it is composed
of thin tightly bound layers, stacked on top of each other. The layers con-
sist of four close-packed monoatomic sheets. The bonding between layers is of
the van der Waals type, and covalent with a small ionic contribution in the
layer.

A large number of experiments [1] have been reported in the last few years,
but very few above 10 eV [2]-[5]. The first band structure calculations per-
formed, [6],[7] were using the tight binding approximation. Recently,
Bourdon [8] and Schluter [9] have used a pseudopotential approach. In contrast
to the previous calculations, they included a interlayer interaction.

2. EXPERIMENTAL RESULTS AND DISCUSSION

For the reflectivity measurements, we used the radiation of the electron syn-
chrotron A.C.O. at Orsay, operated at 536 MeV with a current varying between
40 and 100 mA. The monochromator, a Seya Namioka type (MacPherson), with a
1200 lines/mm platinum grating had a resolution of 0.5 Å. The detector above
10 eV was an E.M.I. electron multiplier. The information, taken by a digital
voltmeter was sent directly to a small computer, giving $dR/d\lambda$ and $d^2R/d\lambda^2$.
The only noise detectable was in fact due to the vibrations of the line. For
the photoemission measurements, the photo-electrons were analyzed with a
spherical collector (resolution 0.1 eV). Below 11 eV, we used a hydrogen

[*] Research sponsored by the D.R.M.E.

[+] Laboratoire Commun au C.N.R.S. et à L.U.P.S.

lamp, above, the synchrotron radiation.

2.1 VALANCE BANDS

Figure 1 represents the energy distribution curve for GaSe when excited by a radiation of 24 eV; the contribution of the secondary electrons has been subtracted. The structure is dominated by three peaks at -2.8, -4.2, and -7.1 eV; shoulders appear at -0.8, -1.5, and -6.2 eV. In Bourdon's [8] and Schluter's [9] calculations, there is a gap around -3 eV; this shows that the first group of valence bands (Γ_4^-, Γ_6^+, Γ_5^-, Γ_5^+, Γ_6^-, Γ_1^+ at Γ) are too narrow (see Fig. 2). Also, the peak at -7.1 eV is in disagreement with the gap between 5 and 10 eV in Ref. [8].

Williams et al [4], have deduced from their photoemission experiments, a valence band of 8.5 eV.

FIG. 1. Photoemission energy distribution of GaSe excited by a radiation $\hbar\omega$ = 24 eV, after subtraction of the secondary electron emission.

FIG. 2. Schluter's band structure calculation of β-GaSe along a few symmetry axes.

It can be seen in FIG. 1 that the width is larger than 10 eV and can be esti-
mated [5] on the order of 15 eV.

2.2 CONDUCTION BANDS

To obtain information on the conduction bands, we have studied the transitions
between the d states of the Ga and the conduction bands. We have measured the
reflectivity R and calculated $d^2R/d\lambda^2$ between 20 and 27 eV. The results, for
an incidence $\theta = 20°$, are represented in FIG. 3. We have chosen to represent
$d^2R/d\lambda^2$ and not $dR/d\lambda$ because the shoulders of R are more easily identified

FIG. 3. Reflectivity R and $d^2R/d\lambda^2$ of
GaSe at T = 77°K between 20
and 27 eV, measured with an
incidence angle $\theta = 20°$.

FIG. 4. Second derivative of the pho-
toemission energy distribution
curves for various photon en-
ergy between 7.7 and 9.7 eV.
The zero of energy for the
electron is taken at the top
of the valence band.

by this way (down directed peaks in $dR^2/d\lambda^2$ correspond to peaks or shoulders
in R). The first structure is observed at 20.7 eV; if we subtract the value

of the band gap, this gives 18.6 eV for the energy between the top of the valence band and the level 5/2 of the d bands. The next structure is located at 21.15 eV; the energy between these two structures is 0.45 eV. This is very close to the spin orbit splitting of the core level of Ga (0.53 eV)[10]. This doublet is again observed at 21.6 and 23.05 eV. We tentatively [11] assign it to the spin-orbit of the d band of Ga. This gives two conduction levels at 0.9 and 2.35 eV above the bottom (in very good agreement with Schluter's calculation).

We have represented in FIG. 4 the second derivative of the photoemission distribution curves for various photon energy between 7.7 and 9.9 eV. The structure is dominated by three peaks (1, 2, 3) at 6.2, 6.8, and 7.6 above the top of the valence band. Although the first can be due to the scattering, these are all located in a region where $d^2R/d\lambda^2$ also show structures. The smaller structures 4, 5, and 6 follow a variation $\Delta E = \Delta\hbar\omega$. They correspond at relatively flat valence bands at -0.7, -1.5, and -2.7 eV.

In conclusion, these experiments will be used in a new band structure calculation, and we hope to be able to extract a correct density of states of the conduction band.

References

[1] Balzarotti, A; Piacentini, M.: Sol. Stat. Comm. 10, 421 (72) & Refs. in.

[2] Mamey, R.; Cazaux, J; Levegue, G; Raisin, C: Phys. Stat. Sol. (b) 62, 201 (1974).

[3] Perrin, J; Cazaux, J; Soukiassian, P: J. de Physique (to be published).

[4] Williams, R. H; Williams, G. P; Norris, C.; Howells, M. R.; Munro, I. H.: Journal of Physics C. Sol. State (to be published).

[5] Thomas, J. M.; Adams, I.; Williams, R. H.; Barber, M.: J. Chem. Soc. Faraday Trans II 68, 755 (1973).

[6] Bassani, F.; Pastori-Parravicini, G.: Il Nuovo Cimento 50B, 95 (1967).

[7] Kamimura, H.; Nakao, N.: J. Phys. Soc. Jap. 24, 1313 (1968).

[8] Bourdon, A.: Thesis, Paris (1972) unpublished.

[9] Schluter, M.: Il Nuovo Cimento 13, 313 (1973).

[10] Herman, F.; Skillman, S.: Atomic Structure Calculations (Prentice-Hall, Englewood Cliffs, N.J., 1963).

[11] Other similar experiments on GaAs, GaSb, PbS, PbSe, PbTe, where the matrix elements between the d bands and the different conduction bands can can be calculated, showed us that an assignment of the different peaks

in the Brillouin zone is possible only by including them. This will be
published elsewhere.

ULTRAVIOLET PHOTOEMISSION STUDIES OF THE BAND STRUCTURES
OF TRANSITION METAL DICHALCOGENIDES

F.R. SHEPHERD, P.M. WILLIAMS, D.A. YOUNG & C.B. SCRUBY

Department of Chemical Engineering & Chemical Technology
Imperial College, London, U.K.

Ultraviolet photoemission studies of layer transition
metal dichalcogenides show the progressive development
of 'd' bands from groups IV_a to VI_a. Charge density
waves in the V_a materials are shown to produce anoma-
lously broad 'd' bands. Nb doping of $MoSe_2$ produces
a shift in the photoemission threshold which is cor-
related with the free carrier screening of the
excitons in optical absorption spectra.

1. INTRODUCTION

Recent determinations of the valence band densities of states (d.o.s.) of
layer transition metal dichalcogenides by photoemission {1,2,3} have pro-
vided the first direct experimental test of band structure models for these
compounds {4,5}. For the metallic members of group V_a recent diffraction
studies {6,7} have revealed periodic lattice distortions in the form of
charge density waves resulting from a Kohn anomaly. In this paper we com-
pare the overall d.o.s. for the group IV_a, V_a and VI_a materials, and the
'd' band d.o.s. for both regular and distorted V_a compounds. The latter
results are discussed in the light of electron diffraction observations of
lattice distortions. Preliminary results are also presented on the doping
of $MoSe_2$ with Nb using both photoemission and optical absorption.

2. EXPERIMENTAL

Photoelectron spectra were recorded at 295 K using He excitation (40.8 and
21.2 eV) and a hemispherical analyser system (Vacuum Generators Ltd., ESCA
3). Electron diffraction patterns were recorded between 150 K and 500 K
in either JEOL JEM7 or JEM 100 B electron microscope. Optical absorption
spectra were recorded for Nb doped $MoSe_2$ at 295 and 77 K from the upper-
most cleavage face of the same crystal used in the photoemission experi-
ments. Absorption spectra were normalised for lamp and monochromator res-
ponse, but the crystal thickness, d, was not measured so that αd vs. E
is plotted in all spectra.

3. RESULTS & DISCUSSION

3.1 COMPARISON OF d.o.s. IN GROUPS IV_a, V_a and VI_a

Photoelectron spectra recorded using
He II photons (40.8 eV) are shown in
Figure 1 for MoS_2 (VI_a), $NbSe_2$ (V_a,
trigonal prism), $1T$-TaS_2 (V_a, octahed-
ral), and HfS_2 (IV_a). Immediately be-
low E_f (fixed by reference to the
sharp Ni 'd' band under identical con-
ditions) we see a gap prior to the
photoemission threshold for both HfS_2
and MoS_2, consistent with semiconduc-
ting or insulating behaviour and a
band gap of 1 to 2 eV. The thresh-
hold coincides with E_f for $NbSe_2$ and
$1T$-TaS_2, indicating metallic proper-

ties. The growth of a 'd' band immediately below E_f is clearly seen and
may be interpreted in terms of simple band filling. For example, the ad-
dition of the extra electron from $NbSe_2$ to MoS_2 completely fills the lowest
'd' band, split from the 'd' manifold by a hybridisation gap {4}. The 'd'
band in MoS_2 (FWHM 1.2 eV) is about twice as broad as that in $NbSe_2$, con-
sistent with this simple picture, although we note that the $1T$-TaS_2 'd'
band is again 1.5 eV wide as discussed below (§3.2). The deeper lying
states are based primarily on chalcogen 'p' electrons. Thus the 'p'
states in $1T$-TaS_2 closely resemble those in HfS_2 as expected on a rigid
band approach as both have octahedral coordination. On changing to tri-
gonal prism coordination, band calculations {4,5} predict a re-ordering
of p states with p_x/p_y bonding and antibonding combinations now above
corresponding p_z states. A distinct change in the 'p' d.o.s. is observed
in Figure 1 whilst the similarity between MoS_2 and $NbSe_2$ is again expected
on a rigid band model.

3.2 CHARGE DENSITY WAVES AND THE DENSITY OF STATES IN TaS_2 POLYTYPES

Figure 2 shows He II spectra for $1T$, $4H_b$ and $2HTaS_2$, the $4H_b$ polytype ex-
hibiting trigonal prism and octahedral coordination in alternate lamellae.
The 'p' bands in $1T$ strongly resemble HfS_2 while the 2H d.o.s. is very

similar to that in $NbSe_2$. The weighting of 'p' states in $4H_b$ is then consistent with an average over the 1T and 2H d.o.s. Considerable narrowing of the 'd' band is seen between 1T and 2H (see also Fig. 3) while that in $4H_b$ is split into two components resolved at HeI (21.2 eV, Fig. 3); the overall 'd' bandwidth in 1T and $4H_b$ is similar. These observations cannot necessarily be interpreted as reflecting the greater unsplit bandwidth of the lower

t_{2g} subset in the octahedral field, however, as both 1T and $4H_b$ are distorted by the presence of charge density waves {6,7}. Figure 4 shows (hk.0) electron diffraction parterns for 1T and $4H_b$ at the temperatures indicated. At higher temperatures in both cases extra reflections $\underline{q}$ are observed parallel to a^* but incommensurate with the lattice. These have been interpreted in terms of a Kohn anomaly for vectors $\underline{q}$ which just span parallel "nested" pieces of Fermi surface, producing a singularity in the static electronic susceptibility and hence a charge density wave coupled (in this case) to a periodic lattice distortion. At low temperatures $\underline{q}$ becomes commensurate, rotating to form a perfect $\sqrt{13} \times \sqrt{13}$ superlattice. In both cases, therefore, the broad 'd' band in photoemission may result from effects due to the CDW. The more "regular" V_a's ($NbSe_2$, $TaSe_2$, VSe_2) show narrower 'd' bands at 295 K but reveal similar diffraction effects at low temperatures; temperature-dependent photoemission experiments will be necessary to resolve this point.

Fig.4a 1T-TaS$_2$ 150 K

fig.4b 1T-TaS$_2$ 400 K

Fig. 3

fig.4c $4H_b TaS_2$ 295 K

fig.4d $4H_b TaS_2$ 400 K

3.3 PHOTOEMISSION & OPTICAL ABSORPTION STUDIES OF Nb DOPED $MoSe_2$

Figure 5 compares optical absorption spectra at 77 K with photoemission
results (at 295 K) for $Nb_{0.05}Mo_{0.95}Se_2$. The injection of holes into the
top of the valence band results in free carrier screening of the excitons,
as is readily observed in 5a and 5b. The corresponding photoemission
spectra show the occupied bands of the $MoSe_2$ to be almost totally unaf-
fected by the doping, however, and the most noticeable effect is that the
Fermi level is "pinned" at a different point within the energy gap follow-
ing the introduction of Nb. Previous results {1} for non-degenerate semi-
conductors indicate that band bending at the surface proceeds until con-
duction band edge and E_f coincide, the threshold then being equal to the
minimum gap. In the present case, however, the main Mo d-d gap is unal-
tered and we suggest that E_f is "pinned" within a new valence band d.o.s.,
based on Nb 'd' states, rather than at an otherwise arbitrary point with-
in the original gap. This is consistent with the contention of Wilson
and Yoffe {8} for Nb doped WSe_2 that the free carriers screening the ex-
citons are <u>not</u> introduced in a band occupied by either the electron or
the hole of the exciton. At present doping levels (5%) this low d.o.s.
is not observable and further measurements at higher doping levels are
planned.

fig 5

4. <u>REFERENCES</u>

{1} Williams, P.M.; Shepherd, F.R.: J.Phys.C $\underline{6}$, L36 (1973); and to
 be published.

{2} Murray, R.B.; Williams, R.H.: J.Phys.C ($\underline{1974}$) to be published.

{3} Wertheim, G.K.; diSalvo, F.J.; Buchanan, D.N.E.: Sol.St.Comm. $\underline{13}$,
 1225 (1973).

{4} Mattheiss, L.F.: Phys.Rev. B $\underline{8}$, 3719 (1973).

{5} Bromley, R.A.; Murray, R.B.; Yoffe, A.D.: J.Phys.C $\underline{5}$, 759 (1972).

{6} Williams, P.M.; Parry, G.S.; Scruby, C.B.: Phil.Mag. $\underline{29}$, 695
 (1974); and to be published.

{7} Wilson, J.A.; diSalvo, F.J.; Mahajan, S.: Phys.Rev.Letts. $\underline{32}$,
 882 (1974); and to be published.

{8} Wilson, J.A.; Yoffe, A.D.: Adv.Phys. $\underline{18}$, 193 (1969).

CONTRIBUTION OF PHONON-ASSISTED TRANSITIONS TO THE PHOTOEMISSION SPECTRA OF Si AND Ge.

L.D. LAUDE

Surface Physics Group, Astronomy Division,
European Space Research Organization,
Noordwijk, The Netherlands.

From the temperature dependence of the photoemission
spectra of (111) Si and Ge, evidence is obtained for phonon-
assisted (indirect) transitions to contribute these spectra.
A simple model is proposed which correctly describes the
observed effects and helps assigning such transitions to
critical points in the Brillouin Zone.

1. INTRODUCTION

Among the effects which may contribute the structure of the photo-
emission spectra of semiconductors, one can distinguish between those
which have an essentially optical character and those which are associated
with inelastic scattering processes. The former give a contribution to
the photoemission current which may be calculated in principle within
the band structure framework in terms of an "optical" energy distribution.
The latter are responsible for all observed departures from this
distribution, in a manner which depends on the characteristics of the
scattering processes involved. In this paper, an attempt is made at
investigating the contribution of electron-phonon scattering to the meas-
ured photoelectron energy distribution curves (EDC's) of vacuum-cleaved
(111) Si and Ge. This scattering process being the only one which may
depend on temperature, the study has been carried out at temperatures in
the range 100 to 300 $^{\circ}$K. Definite and qualitatively similar temperature
effects were observed on both Si and Ge. A simple theoretical approach
is proposed which allows to correctly describe these effects and to
associate them with bulk phonon-assisted (indirect) transitions. A
comprehensive report of this work is given elsewhere.

2. EXPERIMENTAL RESULTS

The samples were p-type, 10 to 50 Ωcm Ge and Si crystals which have
been cleaved along (111) in 7 x 10^{-11} torr vacuum at 300°K. The photo-
emission set-up used in this work has been described elsewhere[1] .

Higher derivative techniques were used to improve visibility of structure and the evolution of the spectra has been continuously traced with temperature.

The photoelectric threshold of freshly cleaved (111) Ge was measured to be E_T = 4.8eV. The EDC obtained at $h\nu$ = 7.72 eV and 300°K from such a sample is shown in fig. 1, together with its second-derivative. Upon cooling the sample to 100°K, this latter curve shows distinct changes (too small to show up on the EDC, [2]). The most noticeable is the weakening of the small peak at $\mathcal{E}$= 5.2eV, which recovers its full strength upon warming up to 300°K again.

The case of (111) Si is more complex, since an evolution of the photo-electric threshold (probably due to surface relaxation and dangling bond saturation) is observed, from E_T = 5.15 eV minutes after cleavage to E_T = 4.7 eV several hours later, in a 7 x 10^{-11} torr vacuum. The EDC and its second-derivative evolve then as shown in fig. 2 (top) which shows two EDC second-derivatives measured at hv = 7.72 eV and 300°K, respectively 20 min. (dashed line) and 20 hrs. (full line) after cleaving (111) Si. The latter is remarkably close to the corresponding spectrum obtained from (111) Ge right after cleavage, fig. 1. Whilst no important temperature dependence is observed when cooling the sample a few minutes after cleavage (dashed line), the relaxed spectrum (full line) and its EDC show an important and reversible evolution upon cooling to 100°K as shown in fig. 2. This evolution mostly effects the low-energy peak which has been gained upon ageing the sample, i.e. exactly the same piece of structure as observed in freshly cleaved Ge spectra. These observations should strongly suggest that the temperature effect reported here is essentially a bulk effect. In fact, since the sticking coefficient of residual gas molecules on the

FIG. 1: EDC and its second-derivative measured from (111) Ge at $h\nu$ = 7.72eV: T = 300 and 100°K. Energy $\mathcal{E}$ is the photoelectron kinetic energy plus E_T.

FIG. 2: EDC second-derivatives measured from (111) Si at $h\nu$ = 7.72 eV: 20 min after cleavage at T=300°K (dashed line), and 20 hrs, after cleavage at T=300, 200, 150 and 100°K (solid lines).

surfaces may be expected to increase upon cooling, the surface scattering probability would increase too, resulting into a strong and irreversible enhancement of the low-energy region of the spectra, at variance with the observed evolution.

3. INTERPRETATION AND CONCLUSIONS

Many mechanisms involving phonons could operate simultaneously. However, thermalization is only effective at energies below 2 eV [3]; changes in the photoelectron transport properties [4] should affect the EDC's whatever the value of E_T is, and it has been shown that the main effect on Si is observed when E_T = 4.7 eV and not at E_T = 5.15 eV; variations in band gaps should introduce marginal shifts, within instrumental accuracy. Therefore, a probable interpretation would deal with the existence of phonon-assisted transitions contributing these spectra around $\mathcal{E}$ = 5.2 eV.

A proper calculation of the indirect transitions probability in the whole Brillouin Zone (BZ) is impossible to perform. However, a simple way to test their eventuality is to subdivide BZ into two sub-zones, for instance : one limited to the high symetry part of BZ (small volume around the Λ axis and points Γ and L, quoted as $[\Lambda]$), the other consisting in the rest of BZ (quoted as $[NO\,\Lambda]$). Knowing the partial density-of-states (DOS) of both sub-zones [5], one calculates now a combined valence-conduction DOS of the form $\sum_{v,c} n_v (\mathcal{E} - h\nu) \times n_c (\mathcal{E})$, where no condition is imposed on $\vec{k}$, and n_v and n_c are alternatively the valence and conduction DOS within and outside the Λ region. One ends up with a set of two distributions represented in fig. 3 for both materials: one for $[NO\,\Lambda]_v \rightarrow [\Lambda]_c$ transitions (shaded) and the other for $[\Lambda]_v \rightarrow [NO\,\Lambda]_c$

transitions. A peak in these distributions at energy $\mathcal{E}$ should indicate approximately a high probability of phonon-assisted transitions contributing the EDC's around $\mathcal{E}$. Although such transitions could also occur within each sub-zone, the shaded distributions reproduce well the region of the EDC's which is temperature dependent in both materials, indicating that the observed fluctuation of the peak intensity at $\mathcal{E}$ = 5.2 eV could be attributed to indirect transitions initiating <u>outside</u> the [Λ] region and terminating <u>within</u> this region.

On the other hand, direct examination of the band structures of Si and Ge, [5] [6] [7], reveals that (i) Γ_2, is degenerate with conduction states about $K_{1,c}$ and $W_{2,c}$ at $\mathcal{E}$ = 4.8 to 5.0 eV in Si, and (ii) only $K_{1,c}$ and $W_{2,c}$ remain degenerate at that energy in Ge. These peculiarities involve a much higher probability (small energy denominator and higher density of conduction states about $\mathcal{E}$, n_c ($\mathcal{E}$)) of indirect transitions around $\mathcal{E}$ = 5.0 eV in Si than in Ge : namely from $K_{2,v}$ towards Γ_2, and $W_{2,c}$ via $K_{1,c}$ in Si, and from K_{2v} towards $W_{2,c}$ only via $K_{1,c}$ in Ge. It endorses the conclusions of the above model (fig. 3) and would seem to be sufficient to explain: (i) the non-temperature dependence of the Si spectra just after cleavage (when E_T = 5.15 eV, i.e. just above Γ_2,) and (ii) the relatively stronger effect observed in Si (when E_T = 4.7 eV) compared to Ge whilst, in the Debye approximation, one would expect an opposite behaviour (Debye temperature: Θ_D = 366°K in Ge and 658°K in Si). Further support to this interpretation is found in the phonon DOS of Ge and Si, which are the highest at all modes in the K-W-X region of BZ [8].

It has been shown in this work that phonon-assisted transitions may contribute the photoemission spectra

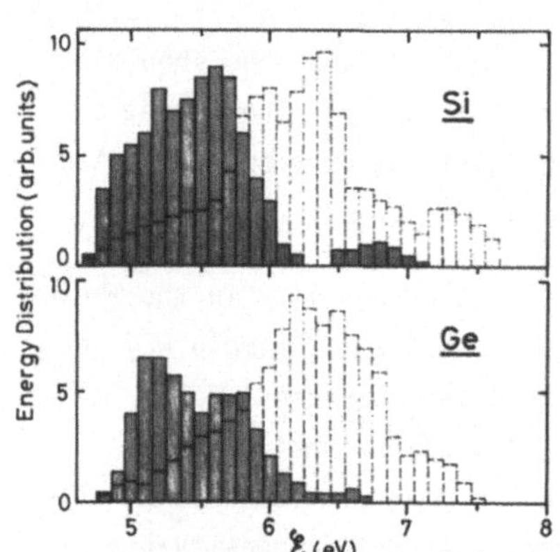

FIG. 3: Approximate probability distribution histograms for [NOΛ]$_v$→[Λ]$_c$ (shaded area) and [Λ]$_v$ → [NOΛ]$_c$ (dotted line) indirect transitions in Si(E_T = 4.7eV) and Ge (E_T = 4.8eV), at hν = 7.72eV.

of Si and Ge. Similar behaviour is expected in other semiconductors, which could be studied using the procedure described here.

The expert assistance of M. R. Barnes is gratefully acknowledged.

4. REFERENCES

[1] Laude, L.D.; Kramer, B.; Maschke, K. : Phys. Rev. B $\underline{8}$, 5794 (1973).

[2] Donovan, T.M.: Tech. Rep. No. 5221-2, Stanford Electronics Lab.,
 Stanford University (1970, unpublished).

[3] James, L.W.; Moll, J.L.: Phys. Rev. $\underline{183}$, 740 (1969).

[4] Kane, E.O. : Phys. Rev. $\underline{147}$, 335 (1966).

[5] Kramer, B. : Phys. Status Solidi (b) $\underline{47}$, 501 (1971).

[6] Saravia, L.R.; Casamayou, L. : J.Phys. Chem. Solids $\underline{32}$, 1075 (1971).
 ibid, $\underline{32}$, 1541 (1971).

[7] Chelikowsky, J.; Chadi, D.J.; Cohen, M.L.: Phys. Rev. B $\underline{8}$, 2786 (1973).

[8] Nilsson, G.; Nelin, G. : Phys. Rev. B $\underline{3}$, 364 (1971); ibid. $\underline{5}$
 3151 (1972).

CONCLUDING REMARKS

M. CARDONA

Max-Planck-Institut für Festkörperforschung
Stuttgart, Fed. Rep. Germany

When I was asked to give the concluding remarks at this con-
ference I felt confronted by a dilemma which must also have
confronted my illustrious predecessors: I was told I had to
submit the manuscript before the beginning of the conference.
The proceedings of the last conference held in Germany took
over two years to appear (in fact it is the only conference
whose proceedings had not yet appeared at the time of the
beginning of the next one!) and this time the organizing com-
mittee was not going to take the risk. So I decided to go back
and examine how my predecessors solved the problem and I was
struck by the diversity of possible approaches which very of-
ten reflected the personality of the men in charge. In
leafing through the proceedings of the previous conferences I
became fascinated by the historical development of our trade
and I decided to concentrate on this aspect: the highlights of
the past 11 conferences from the perspective of the twelfth
one and through the biased eye of my personal scientific inte-
rests. I have attended all conferences but one since Rochester
and I consider my scientific career strongly coupled to the
history of the semiconductors conferences. However, twenty four
years have gone by since the first conference was held at Exe-
ter and the torch has been passed to a new generation. Most of
the participants at this conference do not know what the first
ones were all about, let alone where they were held or how to
get their proceedings if needed. I decided to accompany my re-

port by a list of all conferences
and the proper references to their
proceedings plus a figure with the
number of papers presented at each
conference and the number of parti-
cipants (in brackets) if that in-
formation was available. The tre-
mendous growth in our trade from
1950 to 1960 can be connected
with the discovery of the transis-
tor (1948), the awarding of the
Nobel prize for that discovery
(1956), and the Sputnik effect
(1957). The size of the conference
has remained remarkably stable af-
ter 1960.

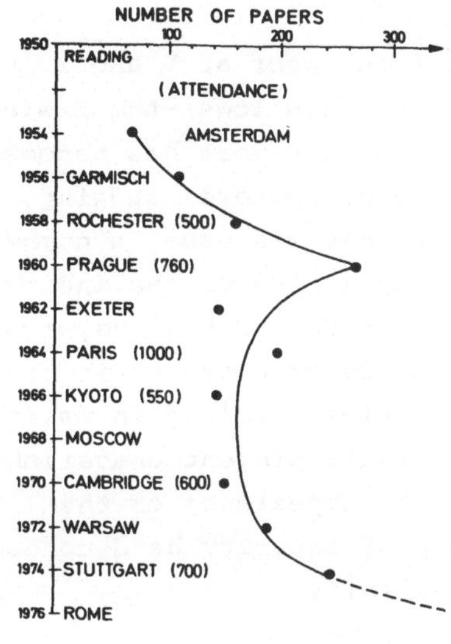

Up to the time of writing this manuscript I was not able to ob-
tain a copy of the proceedings of the first conference (I had
one as a graduate student which I used quite profusely). I shall
therefore start my historical remarks with the second conference,
held in Amsterdam in 1954. I should point out that, with the ex-
ception of Garmisch, the proceedings of all conferences appeared
very promptly,even at a time when photo-offset was not used.
They were always quoted profusely in the next conference.

The big thing at the Amsterdam Conference was cyclotron reso-
nance in Ge and Si, reported independently by the Berkeley and
the MIT groups. I consider this to be the dawn of modern semi-
conductor physics. It unraveled the complicated topological na-
ture of the band extrema in these materials, so different from
free electron behavior and also unaccountable for, at that time,
in terms of chemical bond pictures. Herman reported the first
OPW calculation for Ge. The obtained valence bands were in good
agreement with cyclotron resonance but the conduction band mi-
nimum was at the zone center (Γ). He examined the convergence
of his procedure and found to be good for the conduction band

at Γ but poor at L and X. He thus inferred that better convergence would lower the L minimum of Ge below Γ, thus laying the grounds for what has become known later as diddling with adjustable parameters, fudging, and also semiempirical techniques. Once this was done, a number of things fell into place, such as the switching of the indirect edges of Ge-Si alloys observed in 1954 at RCA. Being the proud owner of a collection of these alloys prepared at RCA in the early days, I am amazed at the difficulties involved in making single crystals of these materials and their present unavailability. Amsterdam was also the time of the appearance of the III-V compounds (Siemens), the discovery of impurity band conduction (Fritzsche) and the Burstein-Moss shift.

The third conference, held at Garmisch-Partenkirchen, Federal Republic of Germany, emphasized work on phosphors, a line of work of considerable tradition in Germany. This kind of research, of a very empirical nature, was quickly displaced thereafter, having been left out of the quantitative developments in semiconductor theory. Nevertheless the progress started with cyclotron resonance in Amsterdam continued at Garmisch. The Naval Research Lab. group reported infrared cyclotron resonance of electrons in InSb ($m^* = 0.013$, a value still believed nowadays). They also reported about preliminary measurements of the absorption edge of InSb in a magnetic field and the observation of Landau oscillations. The Siemens group (Folberth and Pfister) reported about semiconducting properties of ternary compounds of chalcopyrite structure, analogue to the III-V compounds ($ZnSiP_2$, $CdGaAs_2$...). This work, which lay dormant for a while, has become one of the main objects of semiconductors research in recent years (papers 15D06, 15D07, 15D08, 19D04 at this conference). Merten reported on model lattice dynamical calculations for InSb and ZnS thus initiating our theoretical understanding of the phonon spectra of these materials. There were also a few papers trying to correlate the various physical parameters becoming available, such as energy gap, effective masses, to thermodynamic properties and properties of the chemical bond. This

line of work was discredited by the giant developments in band theory at the time of the next conference. Instrumental in this process was the impossibility of accounting for the varied nature of conduction band extrema in terms of chemical bonds. The chemical bond aspect of semiconductors has received recently new impetus thanks to the work of Phillips. (papers 02, 03, 05, and 10, session 15B).

To me, the fourth conference, held in Rochester, New York, in 1968 represents the heyday of semiconductors physics. I was fortunate to attend the conference as a graduate student and to meet many prominent people in the field, including its three recent Nobel laureates (we now have four!) and the late soviet scientists Yoffe and Gross. I cannot resist telling the story of my sitting at the banquet table with Prof. Gross, trying to strike up a conversation, when a lady with a dress cut very low in her back went by. The décolleté was covered by a few horizontal straps, so I told Prof. Gross: "Look, Landau levels!" He added emphatically: "Yes, _very discrete_ Landau levels".

Landau levels were, of course, one of the big things at the conference, as the groups at MIT and at the Naval Research Lab. amply demonstrated the power of magnetooptical techniques. In fact, the supremacy of spectroscopy over transport methods became apparent at that time. The theoretical work of Elliott on excitons coincided with the work by the Royal Radar Establishment group on indirect and idrect excitons in Ge and Si. The power of high pressures for exploring electronic properties was established by W.Paul. Neutron diffraction was established as a tool for mapping out the dispersion curves of phonons by Brockhouse and the difficulties involved in correctly calculating the TA branches of germanium were pointed out by Herman. Ewald reported the growth of single crystals of gray tin, then believed to be a semiconductor. Spin-orbit splittings, usually neglected by theorists, were becoming experimentally accessible thanks to the work of Braunstein.

Feher positioned with his Endor technique the lowest minimum of
the conduction band of silicon. Surface physics was experiencing
a renaissance due to the development of techniques to prepare
clean surfaces. Considerable progress was being made in the theo-
retical field, let me mention Phillips' pseudopotential method
and Kane's k.p method.

As pointed out by Bardeen in his concluding remarks, the fifth
conference, held in Prague, was not one of those rare ones in
which great progress is made but rather the progress made in
the previous ones was consolidated. Attendants were overwhelmed
by the large number of papers presented and by the beauties of
Prague. An elaborate system of simultaneous translators exis-
ted to translate from and to any major European language plus
Czech. It was fun to play with the switches and go from one
language to another as the talks were being given, but I do
not regret the disappearance of such practices, as English has
become the scientific lingua franca.

Esaki tunneling had been discovered since the last conference,
and the possibilities of tunneling spectroscopy, including pho-
non induced tunneling, were presented by Hall. He suggested that
the zero bias anomaly was due to polaron effects. While the ex-
perimental fact has received plentiful confirmation, the early
interpretation remains controversial. The same Hall has pre-
sented in Stuttgart (18B08) one of the most exciting recent de-
velopent: the preparation of Ge with less than $10^{10} cm^{-3}$ active
impurities. Observations of negative mass cyclotron resonance
of holes in germanium attracted considerable attention; they
were never confirmed. One of the results which exerted most in-
fluence in later work was the observation and interpretation
(based on a suggestion by Phillips) by the Prague group of the
reflection spectra of Ge, Si, and III-V compounds in the funda-
mental region. Spectroscopists pried themselves loose from the
absorption edge and a new range of experimental possibilities
appeared. I was doing similar work at the time and was grateful
to be able to meet Prof. Tauc. In spite of his many responsabi-
lities as conference organizer he took time to discuss his work

with me, a very junior scientist at the time. Through the vaga-
ries of history he was to inherit, years later, my laboratory at
Brown University. Other interesting developments were the Franz-
Keldysh effect, the helicon wave propagation proposed by Agrain,
and the detailed work on the theory of uniaxial stress effects
on band structures reported by Picus and Bir. I recently disco-
vered, reading an article by Herring in the conference procee-
dings, that he pointed out at that time the near cancellation
of second order and iterated first order electron-phonon inter-
action. This idea has profoundly affected the recent work of my
group on resonant second order Raman scattering (paper 16B07).
We heard about it in the plenary talk by Ngai (17P2).

In his opening address at the sixth conference in Exeter (Eng-
land) Professor Mott presented his new ideas on the metal-insu-
lator transition (Mott Transition) recently observed. Such a
transition had been recently observed by Austin in V_2O_3 under
pressure. Mott discussed how such transition may take place as
the concentration of electrically active impurities is increased
in a semiconductor. His talk is full of chalenging ideas inclu-
ding a discussion of gaps and localized gap states in amorphous
solids. These ideas are even now by no means a closed subject
and occupied a number of talks at our meeting. Gibson said in
Exeter as a comment to a talk by Gershenzon that the emission
spectrum of GaP was on its way to becoming a goldmine for the
spectroscopist. This prediction had been confirmed by the time
of the next conference. In Exeter, reflection spectroscopy,
first discussed in Prague, appeared as a well established tech-
nique for studying electronic states above the band gap. The
II-VI compounds were receiving more interest. It became clear
that their lowest conduction band minimum was at k = o although
one had to wait for Paris before the semimetallic character of
HgSe and HgTe was discovered. The possibilities of laser spec-
troscopy were recognized in Exeter but no experimental work had
yet been presented.

The seventh conference was held in Paris in 1964. This conference showed a great detailed understanding of energy bands. Calculations were now possible with inclusion of spin-orbit interaction and other relativistic effects and Brillouin zone integrations yielded theoretical optical and photoemission spectra (Brust). Paul presented the picture of gray tin as a semimetal with inverted $\Gamma_{25'}-\Gamma_{2'}$ bands, thus opening the way for a similar interpretation in HgSe and HgTe. Bloembergen et al presented measurements of the non linear susceptibility X_{123} in a number of III-V compounds (Nd-YAG and ruby lasers). Gurevich and Firsov presented the theory of the magneto-phonon effect which became a powerful experimental tool to investigate effective masses and electron-phonon interaction (papers 15C08 and 15C09 at this conference). Impurity levels tied to conduction minima above the lowest were treated theoretically by Peterson. There has been considerable recent interest in this subject which remains somewhat elusive. Hopfield reviewed the very detailed work of excitons bound to impurities, the goldmine predicted by Gibson at Exeter. Frova and Handler presented electric field modulation data for the absorption edge of Ge obtained with p-n junctions, while Seraphin presented the first electroreflectance data for the direct edge of Ge. This was the beginning of the fruitful field of modulation spectroscopy which should flourish in Kyoto.

There was after the meeting a symposium on recombination radiation triggered by the discovery of the semiconductor laser.

The setting for the 8th conference was one of the most spectacular settings that could be imagined, the city of Kyoto. The field of optical properties, was experimenting a resurgence as a result of the advent of modulation techniques. Very detailed electroabsorption spectra were being obtained by Handler using p-n juntions while the techniques of magneto-piezo and magneto-electroreflectance were being used at the National Magnet Lab. My own group was concentrating on the electrolytic technique of electroreflectance. I felt rather reassured when, during a

visit to Nara, I put a coin in a fortune telling machine and got a bilingual slip of paper saying: "Your future lies in the use of water". The analysis of optical properties above the gap was helped by work reported by Toyozawa et al on the coexistence of exciton and band structure effects. The IBM group demonstrated the two-dimensional properties and the quantization of carriers bound to surface barriers in Si. We heard in Stuttgart several papers on this topic, including the beautiful far infrared work of Koch's group. Larsen and Johnson demonstrated the phonon anomalies in cyclotron resonance. We heard more about this subject in Stuttgart. There were in Kyoto lots of papers on acoustoelectric effect and on the coexistence of semiconduction and collective effects,such as magnetism and superconductivity. We had in Stuttgart an interesting session on magnetic semiconductors. As for superconducting semiconductors I am amazed at the way they have disappeared from the literature. Can nature in its multiplicity only provide us with three semiconducting-superconducting materials? We could profit from some "stamp collecting" in this field. I am also amazed at the small number of papers on spectroscopy using lasers at Kyoto. Obviously the high cost of commercial lasers and dwindling financial support was keeping their use low. The Gun effect made in Kyoto its official entrance into the semiconductors conferences. Phillips reviewed the results of the pseudopotential method and Herman warned experimentalists not to take these results uncritically. Well, they did and pseudopotentials dominated ever since the band structures of semiconductors. It is only recently that experimentalists, thanks to photoemission, have discovered the shortcomings of local pseudopotentials. They were quickly made non-local and the problem was solved. Pseudopotentials are here to stay although they indeed have to be taken critically (see review talk by Ngai).

By the time of the Moscow conference, laser spectroscopy, in particular light scattering had come of age. There were several papers on Raman scattering including predictions of the spin-flip laser in the talks of Lax and Wolff. We heard in Stuttgart

Brueck's talk on the subject. Brillouin scattering (Zucker et
al) had become a powerful technique to study acoustoelectric
phenomena. These phenomena occupied considerable space in Mos-
cow. The phenomenon of optical pumping was introduced by Lampel
in Moscow and reviewed by the same author in Stuttgart. Modula-
tion techniques in reflection and absorption spectroscopy had
become well established and in their tremendous development were
showing some signs of age. The details of the line widths and
field dependence were being studied and also the techniques were
being used profusely to achieve high resolution in magnetoopti-
cal work (see e.g., Pidgeon and Groves' work on inversion asym-
metry effects in InSb). Groves at al also measured the "negative"
gap of gray tin with magnetooptical methods while Koda et al in-
troduced the subject of stress-exchange splitting of excitons
which gave new impetus to work on excitons in II-VI compounds.
Baldereschi and Bassani presented the theory of magnetooptical
effects tied to extrema with non-positive-definite masses. Such
effects were observed later by Sari for InSb and remain to date
rather obscure; we heard Tuesday a contribution by the CNRS -
Max Planck group. I was quite amused by a paper presented by
Hall (ultrapure-Hall) in Moscow on photomechanical and electro-
mechanical effects (or the lack of them). He claimed that they
had been able to attribute all of their observations of those
effects to _psychological factors_! There were spirited objections
from the audience and, I believe, this subject is not yet closed.
In fact, I think it would be nice to have in Rome a session (pa-
nel discussion) on unsolved, elusive or misterious questions in-
volving semiconductors and to try to reach some consensus on
these problems. There were in Moscow many papers on injection
lasers and, on the more basic side, the KKR-muffin tin method es-
tablished itself as an effective band structure technique for
semiconductors, which easily took spin-orbit effects into ac-
count.Spicer's talk presented clearly the power of photoemission
for the study of energy bands. This technique has received great
impetus since then due largely to the commercial availability
of "ESCA" equipment and to the use of synchrotron radiation. A
full session and an invited talk were devoted to this subject in

Stuttgart, the most important recent development being probably
the ability to relate photoemission data to the structure of
clean surfaces and their theoretically calculated surface bands
(paper 17AO1). There was a review by the late Mrs. Goryunova on
ternary chalcopyrites, a field which was at the time experien-
cing great growth,and there was also a review by Rogachev on the
subject of exciton condensation which has occupied so much of
this conference. It is interesting to read Keldysh' concluding
remarks. He felt at that time that Ge, Si, and the III-V com-
pounds were in their way out. I counted in our conference 50 pa-
pers related to Ge and Si! As new experimental techniques, new
phenomena and new theoretical methods become available, these
rather well understood materials are ideal to test them. If one
wants to do experiments on electron-hole drops one takes Ge and
Si (possible even better Ge-Si alloys) which can be obtained
with great purity and have the multivalley structure favorable
to drop formation.If one wants to study electron-phonon interac-
tion one takes Ge and Si whose electrons and phonons are well
understood. We cannot guess what is yet to come out of Hall's
ultrapure Ge.

I shall be very brief about the next two conferences since many
of you attended them. The tenth was held in Cambridge, Mass., an
area of the US which had been very productive in the field of
semiconductors. Phillips' and Van Vechten's ideas on the chemical
bond were presented. Raman spectroscopy was becoming a mature
technique although tunable dye lasers were not yet beeing used.
Spin-flip laser action, predicted in Moscow, had been demonstra-
ted and two-photon spectroscopy was becoming a useful tool. The
use of CO_2 lasers opened the way to study the photon drag effect.
Synchrotron radiation was being used as a source for far UV
spectroscopy. I am actually disappointed at the small amount of
work with the use of this technique, reported in Stuttgart. The
closing of the CEA and NINA must be partly responsible for this.
The interest on amorphous materials was probably reaching its
peak in Cambridge while a large amount of attention was being
devoted to transition metal oxides and their Mott transitions.

One paper by Pankove et al on GaN reminded me of how little we know about the large band gap III-V compounds 18 years after Garmisch in spite of their possible application as light sources.

Laser spectroscopy was heavily represented in 1972 in Warsaw. Resonance Raman scattering appeared as a technique to investigate electron-phonon interaction. Raman scattering was being used to study the phonon spectra of amorphous materials. There were actually two long sessions devoted to amorphous materials in spite of topical conferences in 1971 and 1973. Considerable theoretical progress was being made in the field of lattice vibrations and electronic properties of these materials. The possible existence and effects of five-fold rings in tetrahedral semiconductors was introduced in Warsaw. This remains a rather controversial subject. Photoemission with synchrotron radiation (UV) and X-rays appeared as a novel technique to give information on density of valence states and core levels of semiconductors. Exciton condensation, drop formation and optical pumping were among the most popular subjects. A curiosity of the scientific literature, the Debye-Waller screening of pseudopotencial coefficients, had become established as a method to calculate temperature dependences of band structures. This technique first appeared in 1964 in a Ph.D. thesis by S.C.Yu (Harvard). Its practitioners have a copy of a handwritten manuscript by Yu and Brooks which has yet to appear in print. The limits of validity of the method, which works in many cases, are controversial.

It has become traditional for people giving introductory or concluding remarks to make predictions about the future of semiconductors physics. In the early sixties the predictions were that in the seventies all fundamental problems would have been solved and we would be mostly occupied in finding technological applications to semiconductors. I think that actually we are witnessing in the mid 70's a saturation in the technological interest in semiconductors brought about, in part, by the fact

that one material, silicon, seems to be able to solve most pro-
bles in the field. Fundamental studies remain, however, very
lively, mainly because of new experimental techniques adopted
from other domains of physics. Laser spectroscopy, synchrotron
radiation, and far UV and X-ray photoemission (ESCA)plus renewed
interest in amorphous materials have kept this and the past two
conferences of the decade very much alive. I feel progress in
the future will come from expanded experimental possibilities
such as the development of new tunable sources of radiation. Re-
cent advances in ultrahigh presure technology and the availabi-
lity of very high cw or pulsed magnetic fields is also bound to
produce new results. Developments in the field of material - ul-
trapure Ge on the one hand, the renewed interest <u>in organic</u> crys-
tals, in particular one dimensional structures, on the other -
are also bound to attract interest.

I have derived great pleasure in going through the proceedings
of the past conferences, in particular in reading the discus-
sions which were printed in most of them. They are the best of
the proceedings! It is with regret that I notice they are not
going to be printed this time; they have fallen victims of cost
escalation and paper crisis. I hope the custom will be restored
in Rome.

Let me <u>finally</u> conclude by saying that I hope to see you all in
Rome, barring new energy crises and increases in the air fares.

CONFERENCE PROCEEDINGS

I. Semiconducting Materials, ed. by K.H.Henish (Butterworths Scientific Publications, London, 1951)

II. Physica $\underline{20}$ (1954)

III. Halbleiter und Phosphore, ed. by M.Schön and H.Welker (F. Vieweg und Sohn, Braunschweig, 1958)

IV. J. Phys. Chem. Solids $\underline{8}$ (1958)

V. Proceedings of the International Conference on Semiconductor Physics (Publishing House of the Czechoslovak Academy of Sciences, Prague 1961)

VI. Report on the International Conference on the Physics of Semiconductors, ed. by A.C.Stickland (The Institute of Physics and The Physical Society, London, 1962)

VII. Physique des Semiconducteurs, ed. by M.Hulin (Dunod, Paris, 1964)

VIII. J. Phys. Soc. Japan, $\underline{21\ \text{Suppl.}}$ (1966)

IX. International Conference on the Physics of Semiconductors, Moscow, ed. by S.M.Ryvkin (Publishing House "Nauka", Leningrad, 1968)

X. Proceedings of the Tenth International Conference on the Physics of Semiconductors, ed. by S.P.Keller, J.C.Hensel, and F.Stern (US Atomic Energy Commission, 1970)

XI. International Conference on the Physics of Semiconductors Proceedings (PWN-Polish Scientific Publishers, Warsaw, 1972)

AUTHOR INDEX